T0235596

Lecture Notes in Computer Science 12540

More information about this subseries at http://www.springer.com/series/7412

Adrien Bartoli · Andrea Fusiello (Eds.)

Computer Vision – ECCV 2020 Workshops

Glasgow, UK, August 23–28, 2020
Proceedings, Part VI

Editors
Adrien Bartoli
University of Clermont Auvergne
Clermont Ferrand, France

Andrea Fusiello
Università degli Studi di Udine
Udine, Italy

ISSN 0302-9743   ISSN 1611-3349   (electronic)
Lecture Notes in Computer Science
ISBN 978-3-030-65413-9   ISBN 978-3-030-65414-6   (eBook)
https://doi.org/10.1007/978-3-030-65414-6

LNCS Sublibrary: SL6 – Image Processing, Computer Vision, Pattern Recognition, and Graphics

This Springer imprint is published by the registered company Springer Nature Switzerland AG
The registered company address is: Gewerbestrasse 11, 6330 Cham, Switzerland

Foreword

Hosting the 2020 European Conference on Computer Vision was certainly an exciting journey. From the 2016 plan to hold it at the Edinburgh International Conference Centre (hosting 1,800 delegates) to the 2018 plan to hold it at Glasgow's Scottish Exhibition Centre (up to 6,000 delegates), we finally ended with moving online because of the COVID-19 outbreak. While possibly having fewer delegates than expected because of the online format, ECCV 2020 still had over 3,100 registered participants.

Although online, the conference delivered most of the activities expected at a face-to-face conference: peer-reviewed papers, industrial exhibitors, demonstrations, and messaging between delegates. As well as the main technical sessions, the conference included a strong program of satellite events, including 16 tutorials and 44 workshops.

On the other hand, the online conference format enabled new conference features. Every paper had an associated teaser video and a longer full presentation video. Along with the papers and slides from the videos, all these materials were available the week before the conference. This allowed delegates to become familiar with the paper content and be ready for the live interaction with the authors during the conference week. The 'live' event consisted of brief presentations by the 'oral' and 'spotlight' authors and industrial sponsors. Question and Answer sessions for all papers were timed to occur twice so delegates from around the world had convenient access to the authors.

As with the 2018 ECCV, authors' draft versions of the papers appeared online with open access, now on both the Computer Vision Foundation (CVF) and the European Computer Vision Association (ECVA) websites. An archival publication arrangement was put in place with the cooperation of Springer. SpringerLink hosts the final version of the papers with further improvements, such as activating reference links and supplementary materials. These two approaches benefit all potential readers: a version available freely for all researchers, and an authoritative and citable version with additional benefits for SpringerLink subscribers. We thank Alfred Hofmann and Aliaksandr Birukou from Springer for helping to negotiate this agreement, which we expect will continue for future versions of ECCV.

August 2020

Vittorio Ferrari
Bob Fisher
Cordelia Schmid
Emanuele Trucco

Preface

Welcome to the workshops proceedings of the 16th European Conference on Computer Vision (ECCV 2020), the first edition held online. We are delighted that the main ECCV 2020 was accompanied by 45 workshops, scheduled on August 23, 2020, and August 28, 2020.

We received 101 valid workshop proposals on diverse computer vision topics and had space for 32 full-day slots, so we had to decline many valuable proposals (the workshops were supposed to be either full-day or half-day long, but the distinction faded away when the full ECCV conference went online). We endeavored to balance among topics, established series, and newcomers. Not all the workshops published their proceedings, or had proceedings at all. These volumes collect the edited papers from 28 out of 45 workshops.

We sincerely thank the ECCV general chairs for trusting us with the responsibility for the workshops, the workshop organizers for their involvement in this event of primary importance in our field, and the workshop presenters and authors.

August 2020

Adrien Bartoli
Andrea Fusiello

Organization

General Chairs

Vittorio Ferrari	Google Research, Switzerland
Bob Fisher	The University of Edinburgh, UK
Cordelia Schmid	Google and Inria, France
Emanuele Trucco	The University of Dundee, UK

Program Chairs

Andrea Vedaldi	University of Oxford, UK
Horst Bischof	Graz University of Technology, Austria
Thomas Brox	University of Freiburg, Germany
Jan-Michael Frahm	The University of North Carolina at Chapel Hill, USA

Industrial Liaison Chairs

Jim Ashe	The University of Edinburgh, UK
Helmut Grabner	Zurich University of Applied Sciences, Switzerland
Diane Larlus	NAVER LABS Europe, France
Cristian Novotny	The University of Edinburgh, UK

Local Arrangement Chairs

Yvan Petillot	Heriot-Watt University, UK
Paul Siebert	The University of Glasgow, UK

Academic Demonstration Chair

Thomas Mensink	Google Research and University of Amsterdam, The Netherlands

Poster Chair

Stephen Mckenna	The University of Dundee, UK

Technology Chair

Gerardo Aragon Camarasa	The University of Glasgow, UK

Tutorial Chairs

Carlo Colombo   University of Florence, Italy
Sotirios Tsaftaris   The University of Edinburgh, UK

Publication Chairs

Albert Ali Salah   Utrecht University, The Netherlands
Hamdi Dibeklioglu   Bilkent University, Turkey
Metehan Doyran   Utrecht University, The Netherlands
Henry Howard-Jenkins   University of Oxford, UK
Victor Adrian Prisacariu   University of Oxford, UK
Siyu Tang   ETH Zurich, Switzerland
Gul Varol   University of Oxford, UK

Website Chair

Giovanni Maria Farinella   University of Catania, Italy

Workshops Chairs

Adrien Bartoli   University Clermont Auvergne, France
Andrea Fusiello   University of Udine, Italy

Workshops Organizers

W01 - Adversarial Robustness in the Real World

Adam Kortylewski   Johns Hopkins University, USA
Cihang Xie   Johns Hopkins University, USA
Song Bai   University of Oxford, UK
Zhaowei Cai   UC San Diego, USA
Yingwei Li   Johns Hopkins University, USA
Andrei Barbu   MIT, USA
Wieland Brendel   University of Tübingen, Germany
Nuno Vasconcelos   UC San Diego, USA
Andrea Vedaldi   University of Oxford, UK
Philip H. S. Torr   University of Oxford, UK
Rama Chellappa   University of Maryland, USA
Alan Yuille   Johns Hopkins University, USA

W02 - BioImage Computation

Jan Funke   HHMI Janelia Research Campus, Germany
Dagmar Kainmueller   BIH and MDC Berlin, Germany
Florian Jug   CSBD and MPI-CBG, Germany
Anna Kreshuk   EMBL Heidelberg, Germany

Peter Bajcsy	NIST, USA
Martin Weigert	EPFL, Switzerland
Patrick Bouthemy	Inria, France
Erik Meijering	University New South Wales, Australia

W03 - Egocentric Perception, Interaction and Computing

Michael Wray	University of Bristol, UK
Dima Damen	University of Bristol, UK
Hazel Doughty	University of Bristol, UK
Walterio Mayol-Cuevas	University of Bristol, UK
David Crandall	Indiana University, USA
Kristen Grauman	UT Austin, USA
Giovanni Maria Farinella	University of Catania, Italy
Antonino Furnari	University of Catania, Italy

W04 - Embodied Vision, Actions and Language

Yonatan Bisk	Carnegie Mellon University, USA
Jesse Thomason	University of Washington, USA
Mohit Shridhar	University of Washington, USA
Chris Paxton	NVIDIA, USA
Peter Anderson	Georgia Tech, USA
Roozbeh Mottaghi	Allen Institute for AI, USA
Eric Kolve	Allen Institute for AI, USA

W05 - Eye Gaze in VR, AR, and in the Wild

Hyung Jin Chang	University of Birmingham, UK
Seonwook Park	ETH Zurich, Switzerland
Xucong Zhang	ETH Zurich, Switzerland
Otmar Hilliges	ETH Zurich, Switzerland
Aleš Leonardis	University of Birmingham, UK
Robert Cavin	Facebook Reality Labs, USA
Cristina Palmero	University of Barcelona, Spain
Jixu Chen	Facebook, USA
Alexander Fix	Facebook Reality Labs, USA
Elias Guestrin	Facebook Reality Labs, USA
Oleg Komogortsev	Texas State University, USA
Kapil Krishnakumar	Facebook, USA
Abhishek Sharma	Facebook Reality Labs, USA
Yiru Shen	Facebook Reality Labs, USA
Tarek Hefny	Facebook Reality Labs, USA
Karsten Behrendt	Facebook, USA
Sachin S. Talathi	Facebook Reality Labs, USA

W06 - Holistic Scene Structures for 3D Vision

Zihan Zhou	Penn State University, USA
Yasutaka Furukawa	Simon Fraser University, Canada
Yi Ma	UC Berkeley, USA
Shenghua Gao	ShanghaiTech University, China
Chen Liu	Facebook Reality Labs, USA
Yichao Zhou	UC Berkeley, USA
Linjie Luo	Bytedance Inc., China
Jia Zheng	ShanghaiTech University, China
Junfei Zhang	Kujiale.com, China
Rui Tang	Kujiale.com, China

W07 - Joint COCO and LVIS Recognition Challenge

Alexander Kirillov	Facebook AI Research, USA
Tsung-Yi Lin	Google Research, USA
Yin Cui	Google Research, USA
Matteo Ruggero Ronchi	California Institute of Technology, USA
Agrim Gupta	Stanford University, USA
Ross Girshick	Facebook AI Research, USA
Piotr Dollar	Facebook AI Research, USA

W08 - Object Tracking and Its Many Guises

Achal D. Dave	Carnegie Mellon University, USA
Tarasha Khurana	Carnegie Mellon University, USA
Jonathon Luiten	RWTH Aachen University, Germany
Aljosa Osep	Technical University of Munich, Germany
Pavel Tokmakov	Carnegie Mellon University, USA

W09 - Perception for Autonomous Driving

Li Erran Li	Alexa AI, Amazon, USA
Adrien Gaidon	Toyota Research Institute, USA
Wei-Lun Chao	The Ohio State University, USA
Peter Ondruska	Lyft, UK
Rowan McAllister	UC Berkeley, USA
Larry Jackel	North-C Technologies, USA
Jose M. Alvarez	NVIDIA, USA

W10 - TASK-CV Workshop and VisDA Challenge

Tatiana Tommasi	Politecnico di Torino, Italy
Antonio M. Lopez	CVC and UAB, Spain
David Vazquez	Element AI, Canada
Gabriela Csurka	NAVER LABS Europe, France
Kate Saenko	Boston University, USA
Liang Zheng	The Australian National University, Australia

Xingchao Peng   Boston University, USA
Weijian Deng   The Australian National University, Australia

W11 - Bodily Expressed Emotion Understanding

James Z. Wang   Penn State University, USA
Reginald B. Adams, Jr.   Penn State University, USA
Yelin Kim   Amazon Lab126, USA

W12 - Commands 4 Autonomous Vehicles

Thierry Deruyttere   KU Leuven, Belgium
Simon Vandenhende   KU Leuven, Belgium
Luc Van Gool   KU Leuven, Belgium, and ETH Zurich, Switzerland
Matthew Blaschko   KU Leuven, Belgium
Tinne Tuytelaars   KU Leuven, Belgium
Marie-Francine Moens   KU Leuven, Belgium
Yu Liu   KU Leuven, Belgium
Dusan Grujicic   KU Leuven, Belgium

W13 - Computer VISion for ART Analysis

Alessio Del Bue   Istituto Italiano di Tecnologia, Italy
Sebastiano Vascon   Ca' Foscari University and European Centre for Living Technology, Italy
Peter Bell   Friedrich-Alexander University Erlangen-Nürnberg, Germany
Leonardo L. Impett   EPFL, Switzerland
Stuart James   Istituto Italiano di Tecnologia, Italy

W14 - International Challenge on Compositional and Multimodal Perception

Alec Hodgkinson   Panasonic Corporation, Japan
Yusuke Urakami   Panasonic Corporation, Japan
Kazuki Kozuka   Panasonic Corporation, Japan
Ranjay Krishna   Stanford University, USA
Olga Russakovsky   Princeton University, USA
Juan Carlos Niebles   Stanford University, USA
Jingwei Ji   Stanford University, USA
Li Fei-Fei   Stanford University, USA

W15 - Sign Language Recognition, Translation and Production

Necati Cihan Camgoz   University of Surrey, UK
Richard Bowden   University of Surrey, UK
Andrew Zisserman   University of Oxford, UK
Gul Varol   University of Oxford, UK
Samuel Albanie   University of Oxford, UK

| Kearsy Cormier | University College London, UK |
| Neil Fox | University College London, UK |

W16 - Visual Inductive Priors for Data-Efficient Deep Learning

Jan van Gemert	Delft University of Technology, The Netherlands
Robert-Jan Bruintjes	Delft University of Technology, The Netherlands
Attila Lengyel	Delft University of Technology, The Netherlands
Osman Semih Kayhan	Delft University of Technology, The Netherlands
Marcos Baptista-Ríos	Alcalá University, Spain
Anton van den Hengel	The University of Adelaide, Australia

W17 - Women in Computer Vision

Hilde Kuehne	IBM, USA
Amaia Salvador	Amazon, USA
Ananya Gupta	The University of Manchester, UK
Yana Hasson	Inria, France
Anna Kukleva	Max Planck Institute, Germany
Elizabeth Vargas	Heriot-Watt University, UK
Xin Wang	UC Berkeley, USA
Irene Amerini	Sapienza University of Rome, Italy

W18 - 3D Poses in the Wild Challenge

Gerard Pons-Moll	Max Planck Institute for Informatics, Germany
Angjoo Kanazawa	UC Berkeley, USA
Michael Black	Max Planck Institute for Intelligent Systems, Germany
Aymen Mir	Max Planck Institute for Informatics, Germany

W19 - 4D Vision

Anelia Angelova	Google, USA
Vincent Casser	Waymo, USA
Jürgen Sturm	X, USA
Noah Snavely	Google, USA
Rahul Sukthankar	Google, USA

W20 - Map-Based Localization for Autonomous Driving

Patrick Wenzel	Technical University of Munich, Germany
Niclas Zeller	Artisense, Germany
Nan Yang	Technical University of Munich, Germany
Rui Wang	Technical University of Munich, Germany
Daniel Cremers	Technical University of Munich, Germany

W21 - Multimodal Video Analysis Workshop and Moments in Time Challenge

Dhiraj Joshi	IBM Research AI, USA
Rameswar Panda	IBM Research, USA
Kandan Ramakrishnan	IBM, USA
Rogerio Feris	IBM Research AI, MIT-IBM Watson AI Lab, USA
Rami Ben-Ari	IBM-Research, USA
Danny Gutfreund	IBM, USA
Mathew Monfort	MIT, USA
Hang Zhao	MIT, USA
David Harwath	MIT, USA
Aude Oliva	MIT, USA
Zhicheng Yan	Facebook AI, USA

W22 - Recovering 6D Object Pose

Tomas Hodan	Czech Technical University in Prague, Czech Republic
Martin Sundermeyer	German Aerospace Center, Germany
Rigas Kouskouridas	Scape Technologies, UK
Tae-Kyun Kim	Imperial College London, UK
Jiri Matas	Czech Technical University in Prague, Czech Republic
Carsten Rother	Heidelberg University, Germany
Vincent Lepetit	ENPC ParisTech, France
Ales Leonardis	University of Birmingham, UK
Krzysztof Walas	Poznan University of Technology, Poland
Carsten Steger	Technical University of Munich and MVTec Software GmbH, Germany
Eric Brachmann	Heidelberg University, Germany
Bertram Drost	MVTec Software GmbH, Germany
Juil Sock	Imperial College London, UK

W23 - SHApe Recovery from Partial Textured 3D Scans

Djamila Aouada	University of Luxembourg, Luxembourg
Kseniya Cherenkova	Artec3D and University of Luxembourg, Luxembourg
Alexandre Saint	University of Luxembourg, Luxembourg
David Fofi	University Bourgogne Franche-Comté, France
Gleb Gusev	Artec3D, Luxembourg
Bjorn Ottersten	University of Luxembourg, Luxembourg

W24 - Advances in Image Manipulation Workshop and Challenges

Radu Timofte	ETH Zurich, Switzerland
Andrey Ignatov	ETH Zurich, Switzerland
Kai Zhang	ETH Zurich, Switzerland
Dario Fuoli	ETH Zurich, Switzerland
Martin Danelljan	ETH Zurich, Switzerland
Zhiwu Huang	ETH Zurich, Switzerland

Hannan Lu	Harbin Institute of Technology, China
Wangmeng Zuo	Harbin Institute of Technology, China
Shuhang Gu	The University of Sydney, Australia
Ming-Hsuan Yang	UC Merced and Google, USA
Majed El Helou	EPFL, Switzerland
Ruofan Zhou	EPFL, Switzerland
Sabine Süsstrunk	EPFL, Switzerland
Sanghyun Son	Seoul National University, South Korea
Jaerin Lee	Seoul National University, South Korea
Seungjun Nah	Seoul National University, South Korea
Kyoung Mu Lee	Seoul National University, South Korea
Eli Shechtman	Adobe, USA
Evangelos Ntavelis	ETH Zurich and CSEM, Switzerland
Andres Romero	ETH Zurich, Switzerland
Yawei Li	ETH Zurich, Switzerland
Siavash Bigdeli	CSEM, Switzerland
Pengxu Wei	Sun Yat-sen University, China
Liang Lin	Sun Yat-sen University, China
Ming-Yu Liu	NVIDIA, USA
Roey Mechrez	BeyondMinds and Technion, Israel
Luc Van Gool	KU Leuven, Belgium, and ETH Zurich, Switzerland

W25 - Assistive Computer Vision and Robotics

Marco Leo	National Research Council of Italy, Italy
Giovanni Maria Farinella	University of Catania, Italy
Antonino Furnari	University of Catania, Italy
Gerard Medioni	University of Southern California, USA
Trivedi Mohan	UC San Diego, USA

W26 - Computer Vision for UAVs Workshop and Challenge

Dawei Du	Kitware Inc., USA
Heng Fan	Stony Brook University, USA
Toon Goedemé	KU Leuven, Belgium
Qinghua Hu	Tianjin University, China
Haibin Ling	Stony Brook University, USA
Davide Scaramuzza	University of Zurich, Switzerland
Mubarak Shah	University of Central Florida, USA
Tinne Tuytelaars	KU Leuven, Belgium
Kristof Van Beeck	KU Leuven, Belgium
Longyin Wen	JD Digits, USA
Pengfei Zhu	Tianjin University, China

W27 - Embedded Vision

| Tse-Wei Chen | Canon Inc., Japan |
| Nabil Belbachir | NORCE Norwegian Research Centre AS, Norway |

Stephan Weiss   University of Klagenfurt, Austria
Marius Leordeanu   Politehnica University of Bucharest, Romania

W28 - Learning 3D Representations for Shape and Appearance

Leonidas Guibas   Stanford University, USA
Or Litany   Stanford University, USA
Tanner Schmidt   Facebook Reality Labs, USA
Vincent Sitzmann   Stanford University, USA
Srinath Sridhar   Stanford University, USA
Shubham Tulsiani   Facebook AI Research, USA
Gordon Wetzstein   Stanford University, USA

W29 - Real-World Computer Vision from inputs with Limited Quality and Tiny Object Detection Challenge

Yuqian Zhou   University of Illinois, USA
Zhenjun Han   University of the Chinese Academy of Sciences, China
Yifan Jiang   The University of Texas at Austin, USA
Yunchao Wei   University of Technology Sydney, Australia
Jian Zhao   Institute of North Electronic Equipment, Singapore
Zhangyang Wang   The University of Texas at Austin, USA
Qixiang Ye   University of the Chinese Academy of Sciences, China
Jiaying Liu   Peking University, China
Xuehui Yu   University of the Chinese Academy of Sciences, China
Ding Liu   Bytedance, China
Jie Chen   Peking University, China
Humphrey Shi   University of Oregon, USA

W30 - Robust Vision Challenge 2020

Oliver Zendel   Austrian Institute of Technology, Austria
Hassan Abu Alhaija   Interdisciplinary Center for Scientific Computing
                          Heidelberg, Germany
Rodrigo Benenson   Google Research, Switzerland
Marius Cordts   Daimler AG, Germany
Angela Dai   Technical University of Munich, Germany
Andreas Geiger   Max Planck Institute for Intelligent Systems
                          and University of Tübingen, Germany
Niklas Hanselmann   Daimler AG, Germany
Nicolas Jourdan   Daimler AG, Germany
Vladlen Koltun   Intel Labs, USA
Peter Kontschieder   Mapillary Research, Austria
Yubin Kuang   Mapillary AB, Sweden
Alina Kuznetsova   Google Research, Switzerland
Tsung-Yi Lin   Google Brain, USA
Claudio Michaelis   University of Tübingen, Germany
Gerhard Neuhold   Mapillary Research, Austria

Matthias Niessner	Technical University of Munich, Germany
Marc Pollefeys	ETH Zurich and Microsoft, Switzerland
Francesc X. Puig Fernandez	MIT, USA
Rene Ranftl	Intel Labs, USA
Stephan R. Richter	Intel Labs, USA
Carsten Rother	Heidelberg University, Germany
Torsten Sattler	Chalmers University of Technology, Sweden and Czech Technical University in Prague, Czech Republic
Daniel Scharstein	Middlebury College, USA
Hendrik Schilling	rabbitAI, Germany
Nick Schneider	Daimler AG, Germany
Jonas Uhrig	Daimler AG, Germany
Jonas Wulff	Max Planck Institute for Intelligent Systems, Germany
Bolei Zhou	The Chinese University of Hong Kong, China

W31 - The Bright and Dark Sides of Computer Vision: Challenges and Opportunities for Privacy and Security

Mario Fritz	CISPA Helmholtz Center for Information Security, Germany
Apu Kapadia	Indiana University, USA
Jan-Michael Frahm	The University of North Carolina at Chapel Hill, USA
David Crandall	Indiana University, USA
Vitaly Shmatikov	Cornell University, USA

W32 - The Visual Object Tracking Challenge

Matej Kristan	University of Ljubljana, Slovenia
Jiri Matas	Czech Technical University in Prague, Czech Republic
Ales Leonardis	University of Birmingham, UK
Michael Felsberg	Linköping University, Sweden
Roman Pflugfelder	Austrian Institute of Technology, Austria
Joni-Kristian Kamarainen	Tampere University, Finland
Martin Danelljan	ETH Zurich, Switzerland

W33 - Video Turing Test: Toward Human-Level Video Story Understanding

Yu-Jung Heo	Seoul National University, South Korea
Seongho Choi	Seoul National University, South Korea
Kyoung-Woon On	Seoul National University, South Korea
Minsu Lee	Seoul National University, South Korea
Vicente Ordonez	University of Virginia, USA
Leonid Sigal	University of British Columbia, Canada
Chang D. Yoo	KAIST, South Korea
Gunhee Kim	Seoul National University, South Korea
Marcello Pelillo	University of Venice, Italy
Byoung-Tak Zhang	Seoul National University, South Korea

W34 - "Deep Internal Learning": Training with no prior examples

Michal Irani	Weizmann Institute of Science, Israel
Tomer Michaeli	Technion, Israel
Tali Dekel	Google, Israel
Assaf Shocher	Weizmann Institute of Science, Israel
Tamar Rott Shaham	Technion, Israel

W35 - Benchmarking Trajectory Forecasting Models

Alexandre Alahi	EPFL, Switzerland
Lamberto Ballan	University of Padova, Italy
Luigi Palmieri	Bosch, Germany
Andrey Rudenko	Örebro University, Sweden
Pasquale Coscia	University of Padova, Italy

W36 - Beyond mAP: Reassessing the Evaluation of Object Detection

David Hall	Queensland University of Technology, Australia
Niko Suenderhauf	Queensland University of Technology, Australia
Feras Dayoub	Queensland University of Technology, Australia
Gustavo Carneiro	The University of Adelaide, Australia
Chunhua Shen	The University of Adelaide, Australia

W37 - Imbalance Problems in Computer Vision

Sinan Kalkan	Middle East Technical University, Turkey
Emre Akbas	Middle East Technical University, Turkey
Nuno Vasconcelos	UC San Diego, USA
Kemal Oksuz	Middle East Technical University, Turkey
Baris Can Cam	Middle East Technical University, Turkey

W38 - Long-Term Visual Localization under Changing Conditions

Torsten Sattler	Chalmers University of Technology, Sweden, and Czech Technical University in Prague, Czech Republic
Vassileios Balntas	Facebook Reality Labs, USA
Fredrik Kahl	Chalmers University of Technology, Sweden
Krystian Mikolajczyk	Imperial College London, UK
Tomas Pajdla	Czech Technical University in Prague, Czech Republic
Marc Pollefeys	ETH Zurich and Microsoft, Switzerland
Josef Sivic	Inria, France, and Czech Technical University in Prague, Czech Republic
Akihiko Torii	Tokyo Institute of Technology, Japan
Lars Hammarstrand	Chalmers University of Technology, Sweden
Huub Heijnen	Facebook, UK
Maddern Will	Nuro, USA
Johannes L. Schönberger	Microsoft, Switzerland

Pablo Speciale   ETH Zurich, Switzerland
Carl Toft   Chalmers University of Technology, Sweden

W39 - Sensing, Understanding, and Synthesizing Humans

Ziwei Liu   The Chinese University of Hong Kong, China
Sifei Liu   NVIDIA, USA
Xiaolong Wang   UC San Diego, USA
Hang Zhou   The Chinese University of Hong Kong, China
Wayne Wu   SenseTime, China
Chen Change Loy   Nanyang Technological University, Singapore

W40 - Computer Vision Problems in Plant Phenotyping

Hanno Scharr   Forschungszentrum Jülich, Germany
Tony Pridmore   University of Nottingham, UK
Sotirios Tsaftaris   The University of Edinburgh, UK

W41 - Fair Face Recognition and Analysis

Sergio Escalera   CVC and University of Barcelona, Spain
Rama Chellappa   University of Maryland, USA
Eduard Vazquez   Anyvision, UK
Neil Robertson   Queen's University Belfast, UK
Pau Buch-Cardona   CVC, Spain
Tomas Sixta   Anyvision, UK
Julio C. S. Jacques Junior   Universitat Oberta de Catalunya and CVC, Spain

W42 - GigaVision: When Gigapixel Videography Meets Computer Vision

Lu Fang   Tsinghua University, China
Shengjin Wang   Tsinghua University, China
David J. Brady   Duke University, USA
Feng Yang   Google Research, USA

W43 - Instance-Level Recognition

Andre Araujo   Google, USA
Bingyi Cao   Google, USA
Ondrej Chum   Czech Technical University in Prague, Czech Republic
Bohyung Han   Seoul National University, South Korea
Torsten Sattler   Chalmers University of Technology, Sweden
                            and Czech Technical University in Prague,
                            Czech Republic
Jack Sim   Google, USA
Giorgos Tolias   Czech Technical University in Prague, Czech Republic
Tobias Weyand   Google, USA

Xu Zhang   Columbia University, USA
Cam Askew   Google, USA
Guangxing Han   Columbia University, USA

W44 - Perception Through Structured Generative Models

Adam W. Harley   Carnegie Mellon University, USA
Katerina Fragkiadaki   Carnegie Mellon University, USA
Shubham Tulsiani   Facebook AI Research, USA

W45 - Self Supervised Learning – What is Next?

Christian Rupprecht   University of Oxford, UK
Yuki M. Asano   University of Oxford, UK
Armand Joulin   Facebook AI Research, USA
Andrea Vedaldi   University of Oxford, UK

Contents – Part VI

W41 - Fair Face Recognition and Analysis

W44 - Perception Through Structured Generative Models

W36 - Beyond mAP: Reassessing the Evaluation of Object Detection

W36 - Beyond mAP: Reassessing the Evaluation of Object Detection

The BMREOD workshop was conducted to address the future of how the computer vision community compares and analyses object detectors, and promote the new field of probabilistic object detection (PrOD). It had been identified that, in recent years, through a focus on evaluation using challenges, the object detection community had been able to quickly identify the effectiveness of different methods by standardized testing and examining performance metrics. However, it was seen as important to reassess these standardized procedures to ensure they provided enough insight into the specific aspects of object detection that can be analyzed, and that performance adequately aligned with how they might be used in practical applications. A key concept that drove this aspect of the workshop was that quantitative results should be easily reconciled with a detector's performance in applied tasks. This workshop gave the computer vision community a platform to assess the current evaluation standards, find any weaknesses therein and propose new directions forward to improve object detection evaluation. The workshop also hosted the 3rd PrOD challenge, aimed at promoting object detection which provides both spatial and semantic uncertainty estimates accurately. Participants that achieved high scores were provided with the opportunity to supply a paper to the workshop and receive monetary prizes. The final workshop comprised of 12 presentations, including keynote presentations, workshop paper presentations, and PrOD challenge presentations. Three keynote presentations were provided by invited speakers Assoc. Prof. Emre Akbas, Assoc. Prof. Walter Scheirer, and Dr Larry Zitnick, providing their thoughts on the future of object detection evaluation research. Two presentations were given by workshop organizer Dr David Hall, one adding to the general object detection evaluation discussion, and one summarizing the PrOD challenge. Including all workshop papers and PrOD challenge papers, nine papers were submitted to the workshop. After review by the workshop organizers, seven were accepted for the workshop, comprizing of four workshop papers and three PrOD challenge papers. We would like to thank all our fellow researchers for their contributions to the BMREOD workshop, and both the ECCV general and workshop chairs for providing the material and support for this workshop. This workshop was supported by the Australian Research Council Centre of Excellence for Robotic Vision (project number CE140100016), and the QUT Centre for Robotics, Australia.

August 2020

David Hall
Niko Sünderhauf
Feras Dayoub
Gustavo Carneiro
Chunhua Shen

Assessing Box Merging Strategies and Uncertainty Estimation Methods in Multimodel Object Detection

Felippe Schmoeller Roza[1]([✉]) [iD], Maximilian Henne[1], Karsten Roscher[1], and Stephan Günnemann[2]

[1] Fraunhofer IKS, Munich, Germany
`felippe.schmoeller.da.roza@iks.fraunhofer.de`
[2] Technical University of Munich, Munich, Germany

Abstract. This paper examines the impact of different box merging strategies for sampling-based uncertainty estimation methods in object detection. Also, a comparison between the almost exclusively used softmax confidence scores and the predicted variances on the quality of the final predictions estimates is presented. The results suggest that estimated variances are a stronger predictor for the detection quality. However, variance-based merging strategies do not improve significantly over the confidence-based alternative for the given setup. In contrast, we show that different methods to estimate the uncertainty of the predictions have a significant influence on the quality of the ensembling outcome. Since mAP does not reward uncertainty estimates, such improvements were only noticeable on the resulting PDQ scores.

Keywords: Uncertainty estimation · Deep ensembles · Object detection

1 Introduction

Uncertainty estimation has got increasing attention in the research community during the last years as an utterly important feature for Deep Neural Networks (DNNs) embedded in safety-critical applications, such as autonomous driving, robotics, and medical image classification.

Bayesian neural networks allow to inherently express uncertainty by learning a distribution over the weights, with the caveat of being extremely expensive in terms of computation and therefore an unfeasible option for most practical problems. Several methods were proposed to allow uncertainty estimation on computer vision tasks like image classification. Sampling-based methods, such as Monte Carlo Dropout and Deep Ensembles ([4,8]) approximate Bayesian models by combining multiple predictions for the same input. Other works focus on training a model to predict the uncertainties for out-of-distribution detection or to calibrate the usually overconfident softmax confidence scores (respectively [2,6]).

© Springer Nature Switzerland AG 2020
A. Bartoli and A. Fusiello (Eds.): ECCV 2020 Workshops, LNCS 12540, pp. 3–10, 2020.
https://doi.org/10.1007/978-3-030-65414-6_1

Adapting such methods to object detection architectures brings the challenge to a new complexity level because not only uncertainty regarding the label assignment has to be estimated but also concerning the spatial uncertainty of the bounding box coordinates. [9] and [3] presented object detectors based on the dropout approximation. Uncertainty regarding both label and box position is mostly based on the softmax-output confidences of networks which are often sampled from several forward passes through different model parameterizations ([4,8]).

In [1], the authors propose a method that aggregates the ensemble detections by means of the intersection over union (IoU) between the predicted objects. Then, the merging method consists in discarding the clusters with the number of elements below a certain threshold. Three thresholds were tested, 1, $m/2$, and m, which is the number of models in the ensemble. In the end, non-maximum suppression is used to get the final box predictions. [7] recently introduced a loss function that was named KL-loss and integrate the usual bounding box regression loss with a variance estimation, increasing the performance of object detection models.

Even though existing works show how the quality of the uncertainty estimations can be improved by different methods, there is, to the best of our knowledge, no work that compares network softmax confidence scores and variance estimates as effective features to help on adjusting the final bounding box predictions.

In this paper, we compare variance and confidence as potential sources of information to get reliable uncertainty estimations and compare different box merging strategies using these predictions to obtain a final box prediction out of multiple models. Also, two different uncertainty estimation methods are compared. The results are evaluated using mAP, the dominant evaluation metric for object detection models, and PDQ, which encompasses uncertainty on both localization and classification tasks.

2   Methods

In this section, different methods used to combine the detections in the ensemble to either get better detection estimates or improve the uncertainty estimation are presented. Methods to combine the bounding boxes will be classified as merging strategies while the uncertainty estimation methods use different detections to assess more meaningful estimates. If we consider the bounding box coordinates as Gaussian distributions over the pixels, the different merging strategies allow shifting the mean coordinate values whereas the variance estimates represent the standard deviation values.

$\mathcal{D} = [D_1, D_2, ..., D_n]$, is a detection vector that contains the detections provided by n detectors trained independently. Each detection $D_i = [B_i, c_i, s_i]$, with $i \in \{1...n\}$, consists of the bounding box coordinates B_i, the class label c_i, and the respective softmax confidence score s_i. To ensure that the detections match the same object, they must have an IoU above a given threshold t_{IoU}

and only one detection per model is taken in an ensemble $\mathcal{D}$. The vector of bounding boxes will be represented as $\mathcal{B} = [B_1, B_2, ..., B_n]$, the class label vector $\mathcal{C} = [c_1, c_2, ..., c_n]$, and confidence score vector $\mathcal{S} = [s_1, s_2, ..., s_n]$. The final merged detection is represented as $\hat{D} = [\hat{B}, \hat{c}, \hat{s}]$. The detections can be extended by adding uncertainty estimates, $\hat{D}^\sigma = [\hat{B}, \hat{\sigma}, \hat{c}, \hat{s}]$.

Since this paper focuses on improving the bounding box estimation, only the spatial uncertainty will be considered. For the same reason, different methods for improving the merging of the label predictions and the final scores will not be discussed. In this paper, the resulting class $\hat{c}$ is the mode of all predicted labels in the ensemble, and the resulting score $\hat{s}$ is the mean score.

2.1 Spatial Uncertainty Estimation

The spatial uncertainty estimation consists in finding the variances for each coordinate of the bounding boxes, $\sigma = [\sigma_{x1}, \sigma_{y1}, \sigma_{x2}, \sigma_{y2}]$. Two different approaches were considered.

Ensemble Variances: Considering that the ensemble provides boxes that, most likely, have differences in the coordinates, variance estimates can be obtained by calculating the covariance matrices from the bounding boxes coordinates:

$$\sigma_b^2 = \frac{1}{n-1} \sum_{k=1}^{n} (b_k - \bar{b})^2,$$

where b represents a vector with all values for a single box coordinate in the ensemble $\mathcal{B}$, i.e., $b \in \{x_0, y_0, x_1, y_1\}$, where $x_0 = [x_{01}, x_{02}, ..., x_{0n}]$ and so on; and $\bar{b}$ is the mean value of the vector b.

KL-var: Another approach consists of changing the object detector architecture to produce localization variance estimates. In this paper, all variance estimation models utilize the KL-loss in combination with variance voting, as introduced and explained in [7]. With this method, the bounding box predictions are treated as Gaussian distributions whereas the ground truths are represented as a Dirac delta. The regression loss is then the KL divergence of prediction and ground truth distributions, allowing the model to learn the bounding box regression and uncertainty at once. These models will be referred to as KL-var models.

For the KL-var ensembles , the resulting variance is the mean of the variances present in the ensemble, $\hat{\sigma}_{KL} = [\bar{\sigma}_{x0}, \bar{\sigma}_{y0}, \bar{\sigma}_{x1}, \bar{\sigma}_{y1}]$.

2.2 Box Merging

The merging strategy for the final box consists in a method to combine the ensemble detections to get an improved resulting bounding box $\hat{B}$. It can be obtained by the weighted sum of the ensemble elements, i.e., $\hat{B} = W \cdot \mathcal{B}^\intercal$, where W is a vector of weights: $W = [w_1, w_2, ..., w_n]$.

Max: Taking always the box with the highest confidence score is a simple algorithm, although highly dependent on a good calibration of the confidences concerning the bounding box spatial accuracy.

Mean: The resulting bounding box can also be obtained as the mean of the coordinates from the ensemble detections. : $W = \left[\frac{1}{n}, \frac{1}{n}, ..., \frac{1}{n}\right]$.

WBF: Another method called weighted boxes fusion (WBF) was presented in [11]. It uses the normalized confidence scores as weights:

$$W = \frac{S}{\sum\limits_{i=1}^{n} s_i}.$$

Var-WBF: To further investigate the comparison between confidence scores and variance estimates as a good indicator of the detected bounding box correctness, WBF was adapted by replacing the normalized confidences with the normalized inverted variances as weights (with the variance represented as the mean from the 4 coordinate variances).

Variance Voting: Var-voting was introduced in [7] to adjust the box coordinates during non-maximum suppression (NMS) based on the variance estimates. Since the task of merging the outputs of an ensemble of models $\mathcal{D}^\sigma$ is so closely related to what is done in NMS algorithms, in both the goal is to select and/or combine several bounding boxes to find a better box representative, var-voting was adapted as one of the merging strategies.

3   Results

The results were obtained using Efficientdet-D0 and YoloV3 frameworks ([10,12]) trained on the Kitti dataset [5]. Vanilla and KL-var ensembles were trained for both frameworks. Each ensemble consists of 7 models trained independently. The dataset was randomly split as follows: 80% of the images were used for training and 20% for testing. The IoU threshold was $t_{IoU} = 0.5$ for all the experiments. The model was not fine-tuned in any specific way as the primary interest in this paper is to examine the difference of merging strategies and uncertainty estimation methods and not to improve the state-of-the-art performance. Also, objects detected by a single model were discarded, because such cases will not benefit from any of the considered methods.

3.1   Confidence Versus Variances

The correlation between the quality of each predicted bounding box and the confidence and variance estimates was investigated. The comparison was done using the KL-var ensemble of Efficientdet-D0 models, that output both confidence scores and coordinate variances for each prediction. The quality measure was hereby defined as the IoU of the prediction and the corresponding ground-truth. Good estimates, with a high IoU, should present a high confidence/low variance, and a low IoU when the opposite is true.

The results, shown in Fig. 1, demonstrate that the variances are a better indicator of the detection spatial correctness, with a more pronounced correlation (negative since high variance translates as a high spatial uncertainty). The confidences show a weaker, albeit still present, correlation. It is important to notice that poor estimates are much more recurrent on the confidences, as shown in the scatter plots Fig. 1(a) and (b). Although in both cases a more dense mass of points is close to the top-right and top-left corners respectively and therefore representing good estimates, the confidences are almost uniformly distributed for the remaining of the points. On the other hand, the variances proved more accurate, being more concentrated around the first-order polynomial fitted to the samples.

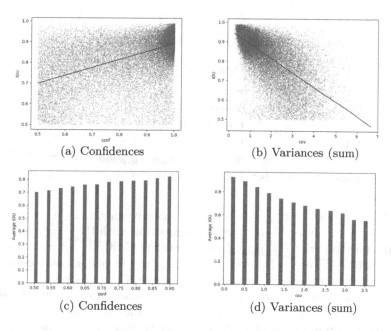

(a) Confidences

(b) Variances (sum)

(c) Confidences

(d) Variances (sum)

Fig. 1. Correlation between IOU and confidences/variances. Remark: the variance values are obtained by the sum of the σ vector elements.

3.2 Merging Strategies and Uncertainty Estimation

Tables 1 and 2 show the results obtained from the testing set regarding mAP, average IoU and PDQ. All merging strategies presented in the Sect. 2.2 were compared using vanilla ensembles without an uncertainty estimation, with uncertainty obtained through the ensemble variance method and the KL-var ensembles, which provides out-of-the-box uncertainty estimates. mAP and Avg. IoU do not take the uncertainty estimates into account and are therefore presented only once.

Table 1. EfficientDet-D0 ensemble results.

	No uncertainty estimation			Ens. var.	KL-var
Method	mAP	Avg. IoU	PDQ	PDQ	PDQ
Max	0.3206	0.8217	0.1253	0.2859	0.3489
Mean	0.3371	0.8356	0.1480	0.3007	0.3518
WBF	0.3373	0.8360	0.1482	0.3015	0.3520
Var-WBF*	0.3381	0.8366	0.1484	0.3074	**0.3521**
Var-voting*	**0.3390**	**0.8372**	**0.1488**	**0.3113**	0.3519

Table 2. YoloV3 ensemble results.

	No uncertainty estimation			Ens. var.	KL-var
Method	mAP	Avg. IoU	PDQ	PDQ	PDQ
Max	0.3001	0.8064	0.0725	0.3143	0.3388
Mean	0.3279	**0.8271**	0.1203	0.3391	**0.3679**
WBF	0.3280	0.8268	0.1263	0.3378	0.3673
Var-WBF*	0.3284	0.8268	0.1269	**0.3399**	0.3677
Var-voting*	**0.3285**	0.8253	**0.1274**	0.3384	0.3678

Var-WBF and Var-voting depend on the ensemble having individual variance for each box in the ensemble and are not compatible with the vanilla ensembles. All results for these methods were therefore obtained with the KL-var ensembles, but the variances were only used for merging the boxes and not used as an uncertainty estimate, except for the results on the KL-var column.

The results show that the different merging strategies perform almost the same with a slight improvement between taking the maximum confidence box and all other merging strategies, showing that trusting only in the confidences is not a good method. However, despite the variances presenting a superior estimate than the confidence scores, the Var-WBF did not perform better than the regular WBF method. None of the methods performed significantly better than taking the mean, which is the simplest of the merging strategies. One of the reasons is that most of the detections have high confidence or low variance, since non-maximum suppression and var voting suppress the low scored ones. After normalization, the weight distributions for both methods become close to the mean weights.

Whilst the influence of the merging strategies appears negligible, a more pronounced difference can be observed when comparing the different uncertainty estimation methods presented in Sect. 2. A big step in the PDQ values is already expected when providing meaningful uncertainty estimates since the variances are treated as zeroes when no uncertainty estimate is available, as if the model is completely certain about its bounding box predictions. With the ensemble variances as an uncertainty estimate, PDQ values already drastically improve. Even though this simple method does not require any modification of the base object detectors it more than doubles the detection quality if uncertainty is taken into account. However, if the models themselves predict their individual uncertainties, results improve even further for all cases and both model architectures.

4   Conclusions and Future Work

In this paper different strategies to merge bounding boxes from different models in an ensemble-based object detection setting were investigated. Furthermore, two approaches to estimate the uncertainty of the final box were presented and evaluated.

The results demonstrate that predicted variances are better correlated with the resulting IOUs than the almost uniquely used confidence values for object detectors and should be interpreted as a better spatial correctness estimate. However, when comparing merging strategies based on the variance and the confidence scores, the improvement was not as pronounced as expected and all the averaging methods showed similar results. In this regard, simply taking the mean values can be considered a good baseline, with competitive performance on all metrics and simple implementation.

We also observed significantly better results when employing variance predictions for bounding box uncertainty estimation, which is only rewarded by the PDQ metric. Out of the two proposed approaches, utilizing predicted variances from the individual detectors led to a better performance than relying on the variance of the detected boxes by the different models.

With uncertainty estimation playing an important role towards the goal of embedding AI in safety-critical systems, PDQ stands out as a more suited evaluation metric in opposition to mAP, commonly used and accepted as the object detection metric. The results show that there is still great room for improvement when working with the uncertainty estimations, with an expressive increase for all metrics in both presented methods.

As future work, we would like to apply the same methods on different datasets, such as COCO and Pascal VOC, that have different classes and different aspect ratios, which may play an important factor in the merging results. Methods to obtain uncertainty estimates on the classification task can also be integrated to improve the PDQ scores. Finally, we would like to test if combining the individual box variances with the ensemble variance will result in a more refined uncertainty estimation.

Acknowledgments. This work was funded by the Bavarian Ministry for Economic Affairs, Regional Development and Energy as part of a project to support the thematic development of the Institute for Cognitive Systems.

References

1. Casado-García, A., Heras, J.: Ensemble methods for object detection (2019). https://github.com/ancasag/ensembleObjectDetection
2. DeVries, T., Taylor, G.W.: Learning confidence for out-of-distribution detection in neural networks (2018). arXiv preprint arXiv:1802.04865
3. Feng, D., Rosenbaum, L., Dietmayer, K.: Towards safe autonomous driving: Capture uncertainty in the deep neural network for LiDAR 3d vehicle detection. In: 2018 21st International Conference on Intelligent Transportation Systems (ITSC), pp. 3266–3273. IEEE (2018)
4. Gal, Y., Ghahramani, Z.: Dropout as a Bayesian approximation: Representing model uncertainty in deep learning. In: International Conference on Machine Learning, pp. 1050–1059 (2016)
5. Geiger, A., Lenz, P., Stiller, C., Urtasun, R.: Vision meets robotics: the KITTI dataset. Int. J. Robot. Res. **32**(11), 1231–1237 (2013)
6. Guo, C., Pleiss, G., Sun, Y., Weinberger, K.Q.: On calibration of modern neural networks (2017). arXiv preprint: arXiv:1706.04599
7. He, Y., Zhu, C., Wang, J., Savvides, M., Zhang, X.: Bounding box regression with uncertainty for accurate object detection. In: Proceedings of the IEEE Conference on Computer Vision and Pattern Recognition, pp. 2888–2897 (2019)
8. Lakshminarayanan, B., Pritzel, A., Blundell, C.: Simple and scalable predictive uncertainty estimation using deep ensembles. In: Advances in Neural Information Processing Systems, pp. 6402–6413 (2017)
9. Miller, D., Nicholson, L., Dayoub, F., Sünderhauf, N.: Dropout sampling for robust object detection in open-set conditions. In: 2018 IEEE International Conference on Robotics and Automation (ICRA), pp. 1–7. IEEE (2018)
10. Redmon, J., Farhadi, A.: Yolov3: An incremental improvement (2018). arXiv preprint: arXiv:1804.02767
11. Solovyev, R., Wang, W.: Weighted boxes fusion: ensembling boxes for object detection models (2019). arXiv preprint: arXiv:1910.13302
12. Tan, M., Pang, R., Le, Q.V.: Efficientdet: Scalable and efficient object detection. In: Proceedings of the IEEE/CVF Conference on Computer Vision and Pattern Recognition, pp. 10781–10790 (2020)

Implementing Planning KL-Divergence

Jonah Philion[✉], Amlan Kar, and Sanja Fidler

NVIDIA, Vector Institute, University of Toronto, Toronto, Canada
jphilion@nvidia.com

Abstract. Variants of accuracy and precision are the gold-standard by which the computer vision community measures progress of perception algorithms. One reason for the ubiquity of these metrics is that they are largely task-agnostic; we in general seek to detect zero false negatives or positives. The downside of these metrics is that, at worst, they penalize all incorrect detections equally without conditioning on the task or scene, and at best, heuristics need to be chosen to ensure that different mistakes count differently. In this paper, we revisit "Planning KL-Divergence", a principled metric for 3D object detection specifically for the task of self-driving. The core idea behind PKL is to isolate the task of object detection and measure the impact the produced detections would induce on the downstream task of driving. We summarize functionality provided by our python package planning-centric-metrics that implements PKL. nuScenes is in the process of incorporating PKL into their detection leaderboard and we hope that the convenience of our implementation encourages other leaderboards to follow suit.

1 Introduction

In the past, raw accuracy and precision sufficed as canonical evaluation metrics for measuring progress in computer vision. Today, researchers should additionally try to evaluate their models along other dimensions such as robustness [9], speed [7], and fairness [11], to name a few. In real robotics systems such as self-driving, it is critical that perception algorithms be ranked according to their ability to enable the downstream task of driving. An object detector that achieves higher accuracy and precision on a dataset is not guaranteed to lead to safer driving. For example, failing to detect a parked car far away in the distance, spanning perhaps only a few pixels in an image or a single LIDAR point, is considered equally bad as failing to detect a car slamming the breaks just in front of the ego-car. Ideally, our perception-evaluation metrics would more accurately translate to the real downstream driving performance.

One way to evaluate performance is by evaluating the complete driving system either by having it drive in the real world or in simulation. Collecting real data is surely cumbersome and time consuming; as self-driving systems continue to improve, one needs to collect statistics over a very large pool of driven miles in order to get an accurate measurement. Even so, the scenarios the autonomous driving car finds itself in vary each time, and typically it is the very sparse edge cases that lead to failures. Repeatability in the real world is thus a major issue that may lead to noisy estimates. An alternative of course is to build a perfect driving simulator in which we could sample realistic and

© Springer Nature Switzerland AG 2020
A. Bartoli and A. Fusiello (Eds.): ECCV 2020 Workshops, LNCS 12540, pp. 11–18, 2020.
https://doi.org/10.1007/978-3-030-65414-6_2

challenging scenes and measure how different detectors affect collision rates, driving smoothness, time to destination, and other high level metrics that self-driving systems are designed to optimize for as a whole. Although progress has been made in this direction [1,4–6,10], these simulators currently can only provide biased estimates of real-world performance.

Planning KL-Divergence [8] is a metric that provides task-specific feedback on perception quality while being more computationally lightweight than a full simulator. The main idea of PKL is to use neural motion-planners to induce a metric on detection quality. We summarize the functionality of our open-source implementation of PKL as part of the `planning-centric-metrics` library in this paper.

2 Planning KL-Divergence

In this section, we revisit the definition of the PKL metric [8]. More details about PKL including empirical comparison of PKL to NDS [3] and human studies that show PKL aligns with human notions of safety in driving can be found in "Learning to Evaluate" [8]. While we present PKL here as an object detection metric, the approach can be extended to other common perception outputs such as lane detection, drivable area segmentation, or human-pose estimation.

2.1 "Planning KL-Divergence (PKL)"

Let $s_1, ..., s_t \in S$ be a sequence of raw sensor observations, $o_1^*, ..., o_t^* \in O$ be the corresponding sequence of ground truth object detections, and $x_1, ..., x_t$ be the corresponding sequence of poses of the ego vehicle. Let $A : S \rightarrow O$ be an object detector that predicts o_t conditioned on s_t. We define the PKL at time t as

$$\text{PKL}(A) = \sum_{0 < \Delta \leq T} D_{KL}(p_\theta(x_{t+\Delta}|o_{\leq t}^*) \| p_\theta(x_{t+\Delta}|A(s_{\leq t}))) \tag{1}$$

where $p_\theta(x_t|o_{\leq t})$ models the distribution of ground truth trajectories in the dataset D,

$$\theta = \arg\min_{\theta'} \sum_{x_t \in D} - \log p_{\theta'}(x_t|o_{\leq t}^*). \tag{2}$$

Intuitively, the PKL is a way to measure how similar a set of detections in a scene are from the ground truth detections. It does so by measuring how differently the ego car would plan if it only saw the predicted objects versus seeing the actual objects in the scene.

We model the marginal likelihoods of future positions with a similar approach to other end-to-end planning architectures [2,12]. Under default settings from the `planning-centric-metrics` library, we discretize the grid -17.0 m behind the ego to 60.0 m in front of the ego and ± 38.5 on either side into cells of size 0.3 m by 0.3 m. The first channel is the union of "road" and "lane" layers, the second channel is the "road divider" layer, the third channel is the "lane divider" layer, the fourth channel is the ego vehicle, and the fifth channel is all detections at the current timestep t. We predict the future position of the ego car at times $\{t + 0.25k \mid 1 \leq k \leq 16$. All coordinates are transformed to the frame of the ego car from time t (Fig. 1) .

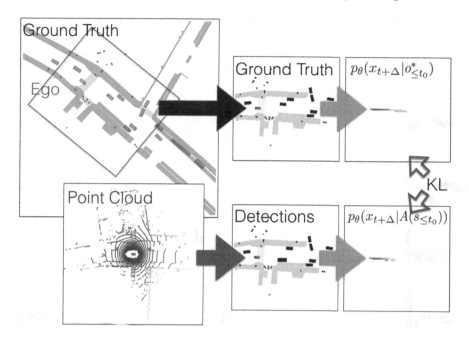

Fig. 1. PKL. We model $p_\theta(x_t|o_{\leq t})$ in the local frame of each vehicle with a CNN (green). $o_{\leq t}$ includes all map data and detected objects prior to the current time t. For a detector A (red), our metric is defined by $\text{PKL}(A) = D_{KL}(p_\theta(x_t|o^*_{\leq t}) \parallel p_\theta(x_t|A(s_{\leq t})))$ where s_t includes sensor modalities that the object detector A requires and o^* includes ground truth detections. If the detector is perfect, the PKL is 0. See Sect. 2 for details (Color figure online).

3   Experiments

In this section, we demonstrate certain capabilities of our implementation of PKL in the `planning-centric-metrics` library. The library exposes a single function called `calculate_pkl` which we use to perform all experiments shown below. Code for all examples below is included in the accompanying repository.

3.1   Neural Planner

We visualize the input channels and output channels of the "neural motion planner" in Fig. 2. The planner is a fully convolutional network that takes as input a rasterized representation of drivable area, road dividers, lane dividers, detections, and ego vehicle location. The output of the planner is a tensor of size $T \times H \times W$ representing the distribution over the location of the ego vehicle for T timesteps into the future.

A subset of predictions of the default planner used by our library on are shown in Fig. 3. The model is capable of expressing the multi-modal nature of future prediction. Note that each trajectory corresponds to T future timestamps so trajectories that appear shorter in length imply that the planner predicts the ego car will travel at a slow velocity, usually because the the car is at an intersection or waiting behind another car.

| Drivable | Road Divider | Lane Divider | Detections | Ego | **Trajectory** |

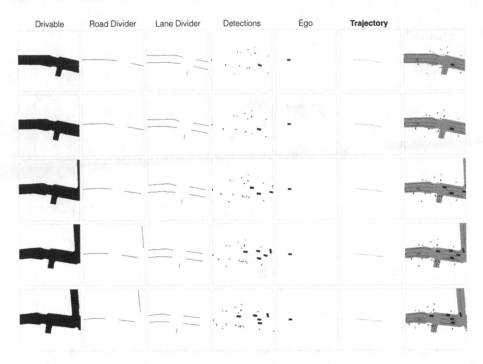

Fig. 2. Inputs and Outputs. Our planner takes as input drivable area (first column), road dividers (second column), lane dividers (third column), detections (fourth column), and ego vehicle location (fifth column). The planner is trained to predict a distribution over the location of the ego vehicle at T future timesteps (sixth column). All input channels plotted together are shown in the rightmost column.

3.2   Using PKL to Evaluate MEGVII [13] Detections

PKL is a per-timestamp metric. To extract a "PKL" across an entire dataset, one needs to define how to accumulate PKL values across multiple timestamps.

We show in Fig. 4 how the mean and median PKL behaves across the nuScenes [3] validation set for state-of-the-art MEGVII [13] detections. The distribution of PKL values is long-tailed. In many cases, the timestamps that correspond to high PKL are timestamps in which the detector truly made a dangerous mistake as shown in Fig. 5. Occasionally, the high PKL is the result of poor planning due to the distribution shift between the ground-truth detections that the model was trained on and the detections output by the detector. Neural planners trained on more data or industrial hard-coded planners would rectify this issue.

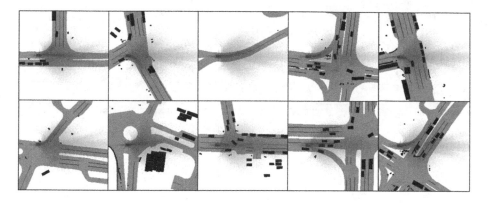

Fig. 3. Neural Planner. We visualize the distribution output by the planner. We plot heatmaps for all timesteps in purple with transparency determined by the output probability. The planner generally predicts that the ego vehicle will stay in its lane while stopping for cars in front of it.

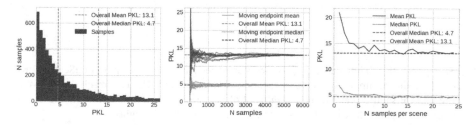

Fig. 4. Evaluating MEGVII. [13] **with PKL** On the left, we visualize the distribution of PKL values for MEGVII [13] detections on the validation split of nuScenes. In the center, we visualize the mean and median PKL as more samples from the validation set are used for 10 random orderings of the validation set. On the right, we visualize the mean and median PKL as the number of samples per scene varies.

3.3 Detection "Importance" from PKL

Since PKL uses a neural planner, PKL measurements can be somewhat uninterpretable in the sense that it's hard to know what mistakes are maximally penalized by PKL. We probe PKL by testing how it PKL changes under different false positives and false negatives. In Fig. 6, we color each object according to how much the PKL increases if the object was missed by a detector. Objects that are near the ego vehicle or about to pass in front of the ego vehicle are more "important" to detect according to PKL.

In Fig. 7, we color each spatial position according to the amount that PKL increases if a false positive is detected at that position. PKL is especially sensitive to false positives located in intersections. In agreement with intuition, the worst false positives appear directly in front of the ego vehicle.

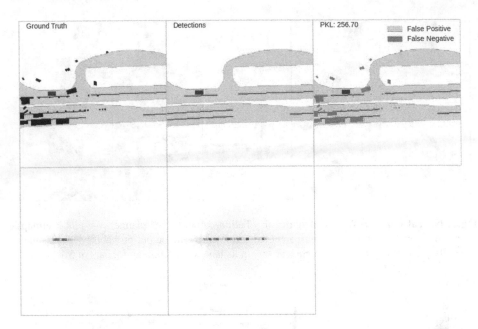

Fig. 5. Highest PKL Timestamp. If no objects are detected on the mini split of nuScenes, PKL identifies the timestamp above as the timestamp with maximum PKL. This result is intuitive in that the ego car should brake if the car in front of the ego is detected (bottom left) but the car would drive straight if the car in front of the ego was missed (bottom right).

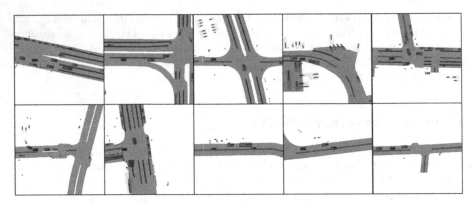

Fig. 6. False Negatives. We color each ground-truth object according to the PKL if the object was not detected. Objects that result in a high PKL are generally in the lane of the ego vehicle or turning into the ego vehicle's likely future path.

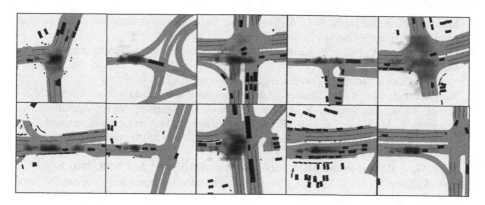

Fig. 7. False Positives. We color each position across a grid of positions according to the PKL if an object was falsely detected at that position. Positions that result in a high PKL are generally directly in front of the ego vehicle or in the middle of an intersection.

4   Conclusion

Planning KL-Divergence is a metric that aligns how the computer vision community evaluates self-driving perception with downstream driving performance. It does so by learning from recorded expert trajectories how different percepts affect driving, thereby lossening the requirement for researchers to design heuristics themselves that they intuitively suspect are important for driving. Our python library planning-centric-metrics is an attempt to make PKL as easy as possible to calculate in the hopes that datasets and organizations will consider incorporating PKL or a variant of PKL in their perception leaderboards. We are in the process of incorporating PKL into the nuScenes detection leaderboard and we hope the convenience of our library will encourage other popular leaderboards to do the same.

References

1. Nvidia drive constellation. https://www.nvidia.com/en-us/self-driving-cars/drive-constellation/. Accessed 14 Oct 2019
2. Bansal, M., Krizhevsky, A., Ogale, A.S.: Chauffeurnet: Learning to drive by imitating the best and synthesizing the worst. CoRR abs/1812.03079 (2018). http://arxiv.org/abs/1812.03079
3. Caesar, H., et al.: nuscenes: A multimodal dataset for autonomous driving. CoRR abs/1903.11027 (2019). http://arxiv.org/abs/1903.11027
4. Dosovitskiy, A., Ros, G., Codevilla, F., Lopez, A., Koltun, V.: CARLA: An open urban driving simulator. In: Proceedings of the 1st Annual Conference on Robot Learning, pp. 1–16 (2017)
5. Kar, A., et al.: Meta-Sim: learning to generate synthetic datasets. In: ICCV (2019). http://arxiv.org/abs/1904.11621
6. Manivasagam, S., et al.: LiDARsim: Realistic LiDAR simulation by leveraging the real world (2020)

7. Mao, H., Yang, X., Dally, W.J.: A delay metric for video object detection: What average precision fails to tell (2019)
8. Philion, J., Kar, A., Fidler, S.: Learning to evaluate perception models using planner-centric metrics (2020)
9. Szegedy, C., et al.: Intriguing properties of neural networks (2013)
10. Yang, Z., et al.: SurfelGAN: Synthesizing realistic sensor data for autonomous driving (2020)
11. Zemel, R., Wu, Y., Swersky, K., Pitassi, T., Dwork, C.: Learning fair representations. In: Dasgupta, S., McAllester, D. (eds.) ICML, Proceedings of Machine Learning Research, Vol. 28, pp. 325–333. PMLR, Atlanta, Georgia, USA (17–19 Jun 2013). http://proceedings.mlr.press/v28/zemel13.html
12. Zeng, W., et al.: End-to-end interpretable neural motion planner. In: CVPR (June 2019)
13. Zhu, B., Jiang, Z., Zhou, X., Li, Z., Yu, G.: Class-balanced grouping and sampling for point cloud 3d object detection (2019)

ODIN: An Object Detection and Instance Segmentation Diagnosis Framework

Rocio Nahime Torres$^{(\boxtimes)}$, Piero Fraternali, and Jesus Romero

Politecnico di Milano, Piazza Leonardo da Vinci, 32, Milano, Italy
{rocionahime.torres,piero.fraternali}@polimi.it,
jesusmaria.romero@mail.polimi.it

Abstract. Object detection and instance segmentation are major tasks in Computer Vision and have substantially progressed after the introduction of Deep Convolutional Neural Network (DCNN). Analyzing the performance of DCNNs is an open research issue, addressed with attention techniques that inspect the response of inner network layers to input stimuli. A complementary approach relies on the black-box diagnosis of errors, which exploits ad hoc metadata on the input data set and factors the performance into indicators sensible or impacted by specific facets of the input (e.g., object size, presence of occlusions, image acquisition conditions, etc.). In this paper we present an open source error diagnosis framework for object detection and instance segmentation that helps model developers to add meta-annotations to their data sets, to compute performance metrics split by meta-annotation values, and to visualize diagnosis reports. The framework accepts the popular PASCAL VOC and MS COCO input formats, is agnostic to the training platform, and can be extended with application- and domain-specific meta-annotations and metrics with almost no coding.

Keywords: Object detection · Instance segmentation · Metrics · Evaluation

1 Introduction

Object detection is a key problem in Computer Vision, defined as the task of locating objects of a given class in images [12]. The output of object detection is the specification of the image region where the object appears, normally encoded as the coordinates of a rectangular bounding box. A finer-grain object detection task is instance segmentation, in which the output is an annotation of the image that associates each pixel to the object instance it belongs to [8]. Object detection and image segmentation have innumerable applications, including medical image analysis, surveillance, environment monitoring, autonomous driving, and more [1]. Performances of object detection and instance segmentation methods have rapidly increased, thank to Deep Learning methods and tools. Progress is not only ascribed to improved model architecture and training methods but also

© Springer Nature Switzerland AG 2020
A. Bartoli and A. Fusiello (Eds.): ECCV 2020 Workshops, LNCS 12540, pp. 19–31, 2020.
https://doi.org/10.1007/978-3-030-65414-6_3

to the availability of data sets and open challenges. Data sets widely accepted as standard for benchmarking performance include MS COCO [10], OpenImages [9], and PASCAL VOC [3]. Most benchmarks rely on standardized metrics to enable the comparison of competing methods. The most widely used ones stem from the indicators used for classification: precision, recall, accuracy and their derivatives. True positives are determined by Intersection over Union (IoU), a metric that quantifies the "goodness" of a detection [13]. The availability of recognized performance evaluation standards has played a key role in the progress of the field, enabling researches to compare their results on a fair ground. However, the metrics used for assessing the end-to-end performance of a method and for comparing methods with a black-box approach may not be the most adequate ones for understanding the inner behavior of an architecture and for optimizing it for a certain data set or task. The analysis of a DCNN can be pursued in two different and complementary ways. On one side, model interpretation techniques aim at "opening the box" to assess the relationship between the input, the inner layers and the output. A notable example is the use of attention models that capture the essential region of the input that have most impact on the inference [18]. On the other hand, it is possible to investigate the performance of a model by enriching the annotations of the images with metadata (or *meta-annotations*, i.e., annotations of the annotations) that do not contribute to model training but can be exploited for understanding performance. Such performance-driven meta-annotations enable the computation of task- and data set-specific metrics that may help the diagnosis. As an example of meta-annotations used to fine tune standard metrics, the MS COCO data set differentiates the AP metrics based on the size of the detected object (small, medium or large), permitting researchers to focus their improvement on the sub-classes of objects where they expect the most gain. The work in [7] is a pioneering effort to systematically integrate meta-annotations in the evaluation of detectors. The authors design a tool suite that allows developers, provided a data set with certain object characteristics, exploit such metadata to evaluate the influence on detection of such dimensions as occlusion, object size, aspect ratio, visibility of parts, viewpoint, localization error, confusion with other objects and with background.

In this paper, we follow the line of [7] and implement a novel tool suite for supporting the diagnosis of errors in object detection and instance segmentation components. The contribution of our work can be summarized as follows:

– We realize an integrated framework for error diagnosis in object detection and instance segmentation components. The framework applies to image data sets in the popular MS PASCAL VOC and COCO formats. It accepts ground truth annotations in the form of bounding boxes and segmentation masks. Its meta-annotation user interface (Fig. 1) supports the enrichment of ground truth annotations with custom meta-annotations representing features of interest. It computes and visualizes custom metrics that highlight the influence and impact of object characteristics on detection or segmentation performance (Figs. 3, 4, 5 and 6).

– We give the framework a plug&play architecture, whereby model developers can easily add their own meta-annotations and custom metrics with no coding besides the actual implementation of the metrics function.
– We showcase the use of the proposed tool suite in the diagnosis of segmentation performances in the RacePlane dataset [16]. The data set contains several meta-annotations such as sunlight intensity, time of the day, or weather conditions.
– We release the code of the framework publicly[1]. The tool suite is developed in Python, integrates directly with the MS COCO data set format and with the PASCAL VOC format via parsing, and is agnostic to the model training platforms.

2 Related Work

Object detection is the task of recognizing an object of a given class and localize it in the image. A detector distinguishes the object from the background and outputs its position as a bounding box [12]. Instance segmentation has the same goal as object detection, but outputs a pixel-level segmentation mask of each object instance instead of a bounding box per object [8]. Early methods started by computing areas of the image most likely to contain an object (Regions of Interest, ROIs), extracted descriptors for each ROI using features such as SIFT [11] or HOG [2] and then fed such descriptors to a classifier to predict a class label for each ROI. The use of region proposals based on fixed-size sliding windows and of hand-crafted descriptors were the most relevant limitations of such first-generation approaches [6]. The breakthrough of CNNs in the image classification task boosted a novel generation of object detectors, in which both features extraction and region proposals identification were formulated as learning tasks [5]. A further progress occurred with the elimination of the distinct steps for ROI extraction and classification, given the unified prediction of bounding boxes and class probabilities directly from images [15]. The improvement in architectures and models has been fostered by the availability of benchmarking data sets and challenges, thanks to the common practice of publishing standardized methods to measure performance. The popular Pascal VOC 2012 [3] data set for object detection features 20 categories, 11.500+ annotated images and 10.000+ images without annotations. MS COCO [10] contains 80 categories with 123.000+ annotated images and 40.670 images without annotations. Object detection performance is measured with standard metrics such as mean Average Precision (mAP), which relies on the Intersection Over Union (IoU) measure to determine True and False Positives, based on the overlap between a detected and a ground truth object. For example, PASCAL VOC 2012 employs a IoU threshold of 0.5 and MS COCO uses both 0.5 and 0.75 values. Other commonly used metrics are True Positive Rate (TPR) and False Positives Per Image (FPPI). A survey of the metrics commonly employed for object detection and a discussion of the problems associated with their implementation is performed in [13],

---

[1] https://github.com/rnt-pmi/odin.

which also provides a standard implementation easily adapted to the different image annotation formats found in commonly used object detection data sets. The above mentioned indicators afford a quantitative assessment of the overall performance but do not provide insight on how object characteristics affect performance. A first step in the direction of supporting deeper insight by means of finer grain metrics is found in MS COCO, in which mAP is differentiated based on object size into mAP_{small}, mAP_{medium} and mAP_{big}. A more general approach is presented in [7], which specifically focuses on the design of metrics supported by a software framework for object detection error diagnosis. The authors designed an open-source framework for evaluating objects proposals in greater detail, thanks to metadata that expose characteristics that may affect performance, such as occlusion, object size, aspect ratio, visibility of parts, viewpoint, localization error, confusion with other objects and with background. The framework is implemented in MatLab and is equipped with a meta-annotated data set to put the diagnosis approach to work.

In this paper, we proceed along the line of [7]. We extend the set of implemented metrics and visualizations to add support for instance segmentation, modernize the code adapting it the most popular data formats, and give the framework hooks for the easy integration of novel meta-annotations and metrics.

3   An Object Detection and Instance Segmentation Error Diagnosis Framework

The ODIN framework supports the development of object detection and instance segmentation models by enabling designers to add application-specific meta-annotations to the input, evaluate standard metrics on inputs grouped by meta-annotation values, assess custom metrics that exploit meta-annotations, and visualize the results.

3.1   Input and Meta-Annotations

The ODIN framework accepts as input an image data set (PASCAL VOC and MS COCO format) and a list of objects predictions. Objects predictions can be expressed as bounding boxes for object detection tasks or as pixel masks for instance segmentation tasks. Besides the standard annotations consisting of the object class and location, developers can provide meta-annotations that denote task- or domain-specific characteristics of the whole image or of a ROI. Meta-annotations can convey additional information about:

- ROI size (extra-small, small, medium, large, extra-large) or aspect ratio (extra-tall, tall, medium, wide, extra-wide). If not provided, values can be calculated by the framework based on the areas and shapes of the annotations.
- Object visibility conditions: for example, occlusion level (none, low, medium, high), truncation (no, yes), viewpoint (bottom, front, top, side, rear).

– Object parts, which may denote specific parts of the object in view.
– Image context: e.g., image technical parameters or acquisition conditions such as weather or lighting.

The addition of meta-annotation is supported by a Jupyter Notebook that given a set of images and a set of valid meta-annotation values allows the developer to iterate on the images and select the appropriate value, which is saved in the format used for evaluation. Figure 1 shows the interface of the meta-annotation editor.

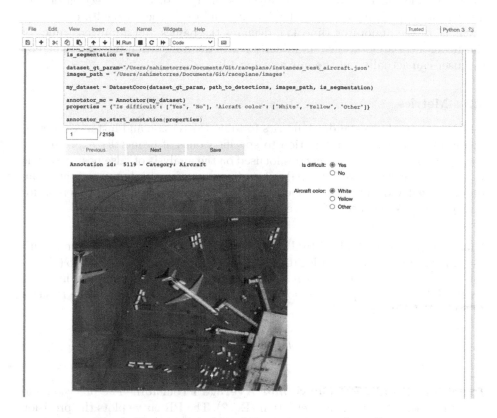

Fig. 1. The user interface of the meta-annotation editor

The following example shows the code needed to create a meta-annotation session for two custom properties (Is difficult and Airplane color).

```
1  from emd_with_classes.classes import Dataset, Annotator
2
3  path_to_detections = "../detections"
4  dataset_gt_path="../coco.json'
5  images_path = "../images'
6
```

```
7  is_segmentation = True
8
9  my_dataset = Dataset(dataset_gt_path,
10                       path_to_detections, images_path, is_segmentation)
11 annotator_mc = Annotator(my_dataset)
12 properties = {"Is difficult": ["Yes", "No"],
            "Aicraft color": ["White", "Yellow", "Other"]}
13 annotator_mc.start_annotation()
```

The code imports the necessary modules (line 1), declares the physical paths to inputs and outputs (lines 3–5), specifies the type of task (detection or segmentation, line 7), instantiates the data set and task object (line 9), creates an instance of the annotator (line 11), declares the names and admitted values of the novel meta-annotations, and finally starts the annotation interface whereby the user can actually tag the images (line 13).

3.2   Metrics

ODIN supports both standard metrics for object detection and instance segmentation assessment, their restriction to specific values of properties expressed by the meta-annotations, and metrics focused on the analysis of false positives. The values of all metrics are reported using diagrams of multiple types, which can be visualized and saved. Developers can run all metrics, or a subset thereof, for a single class or for a set of classes.

Intersection over Union IoU is the intersection between the ground truth B_{gt} and the predicted object location B_{pr} divided by the union of the two (Eq. 1). IoU is a base measure from which all the other metrics are computed. A threshold on the IoU value of a predicted location is used to consider it as a True Positive (TP) or False Positive (FP).

$$IoU = \frac{B_{GT} \cap B_{pr}}{B_{GT} \cup B_{pr}} \tag{1}$$

Precision, Recall, PR Curve and Average Precision. The precision and recall metrics have the usual definition (Eq. 2). The PR curve plots the precision vs recall for all the values of the IoU threshold and can be computed for all the classes, per class, or on a subset of the classes [14].

$$Precision = \frac{TP}{TP + FP}, Recall = \frac{TP}{TP + FN} \tag{2}$$

Average Precision (AP) summarizes the PR curve as a single value, i.e., the precision averaged for all recall values from 0 to 1, equivalent to the Area Under the PR Curve. A common approximation (Interpolated AP) used in the PASCAL VOC data set [4] is computed as the sum of the maximum precision values for recall greater than the current sampling value, weighted by the recall delta (Eqs. 3 and 4).

$$\sum_{r=0}^{1}(r_{n+1} - r_n)\, p_{interpol}(r) \tag{3}$$

with

$$p_{interpol}(r) = \max_{\tilde{r}:\tilde{r}\geq r_{n+1}}\, p(\tilde{r}) \tag{4}$$

where $p(\tilde{r})$ is precision at recall $\tilde{r}$.

An alternative definition of AP, normalized AP, is defined to cope with class unbalance [7]. Normalized AP uses a definition of precision in which the cardinality of the class is replaced by the average cardinality of all the classes in the data set. Normalized AP is the default metrics used in ODIN for multi-class diagnosis.

AP per Property. For a given set of classes (from 1 to all) the AP can be computed only for object proposals having a certain property value.

Property Sensitivity and Impact. For each property value, the worst and best performing image subsets can be computed, with the maximum and minimum AP achieved. The difference between the maximum and minimum AP highlights the sensitivity of AP w.r.t. the property, while the maximum w.r.t. the overall AP provides insight on the impact of the property onto the AP.

False Positive Impact. The per-class analysis of wrong object detection is supported, including:

– Confusion with background or with unlabeled objects (IoU lower than 0.1)
– Poor localization (L) for predictions with the correct class with an IoU lower than the threshold (default is 0.5); the indicator also counts duplicated predictions for the same ground truth object.
– Confusion with similar objects (S), when a similarly annotation is provided.
– Confusion with other objects (O).

For each identified issue the percentage of predictions that fall into that category is reported and also the absolute AP improvement by removing all false positives of each type.

3.3   Data Set Visualizator

A visualization interface, realized as a Jupyter Notebook, enables the inspection of the data set and of the results of a diagnosis session. The visualization can be executed at multiple levels:

– All the images.
– All the images of a certain class.
– All the images with a certain meta-annotation value.
– All the images of a given class with a meta-annotation value.

3.4   Extending the Framework

ODIN has been designed with a "plug&play" architecture, which facilitates the addition of metrics. Adding a new metrics requires extending the provided Analyzer class and implementing the wrapper method that calls the computation of the metrics.

```
class MyCustomAnalyzer(Analyzer):
    def _evaluation_metric(self, gt, proposals, match):
        # Returns the desired metric value and the std_devivation
        # To be used by the different plots

        # Parameters:
        #    gt: contains the a list of the GT objects
        #    proposals: contains a list of proposals
        #    both contain: {image, category, segemetation/bbox}
        #    match: association between gt objects and proposals

        # Returns:
        #    metric_value: the calculated value in the set
        #    std_deviation: of the metric (None: does not apply)

        #TODO: call metrics computation code...
        metric_value = #...
        std_deviation = # ...
        return metric_value, std_deviation
```

The following example shows the code needed to create an instance of the novel metrics class and execute the analysis.

```
1   from emd_with_classes.classes import Dataset
2   import MyCustomAnnotator
3
4   path_to_detections = "../detections"
5   dataset_gt_path="../coco.json'
6   images_path = "../images'
7   res_path = "../results"
8   is_segmentation = True
9
10  my_dataset = Dataset(dataset_gt_path, path_to_detections,
            images_path, is_segmentation)
11
12  my_analyzer =
13  MyCustomAnalyzer(my_dataset, results_path=res_path)
14  my_analyzer.analyze_property( "Is difficult")
```

The code imports the necessary modules (lines 1–2), declares the physical paths to inputs and outputs of the detector (lines 4–7), specifies the type of task (line 8), creates an instance of the data set (line 10), creates an instance of the analyzer (line 12) and finally starts the analysis for some of the meta-annotations previously added to the data set (line 14).

In practice, the addition of new meta-annotations and metrics to ODIN requires only a simple wrapper around the implementation of the metrics.

4   Use Case

In this section we illustrate the functionality of ODIN by applying it to an instance segmentation benchmark: the RarePlanes [16] data set of satellite images for aircraft detection. RarePlanes comprises real and synthetic images associated with object detection and instance segmentation annotations provided in MS COCO format. It contains 50,000 images and $\approx 630,000$ (14,707 real and 629,551 synthetic) annotations. It also features multiple meta-annotations: aircraft length, wing span, wing shape, wing position, Federal Aviation Administration (FAA) wing span class, propulsion, number of engines, number of vertical stabilizers, canards, aircraft type and weather acquisition conditions (Fig. 2). We performed the output analysis of the task of aircraft instance segmentation for small, medium and large civil transportation vehicles. To this end, we replicated the instance segmentation experiments of [16] using Detectron2 [17] with the configuration files associated with the data set[2]. The predictions obtained by the Detectron2.COCOEvaluator component were processed by ODIN.

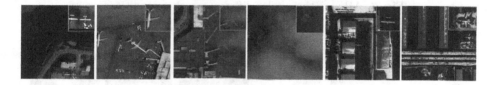

Fig. 2. Example of RarePlanes images with different weather conditions: Clear (the first two), Cloud or Haze (the middle ones), Snow (the last two).

Figure 2 shows a few examples of images acquired in different weather conditions, a meta-annotation that affects the visibility of the planes. Figure 3 reports the AP in different weather conditions for the different classes of civil airplanes. The diagram highlights that the conditions with more difficulties are snow for the small objects and cloud for mid size ones. Since in the approach of [16] data can be augmented by generating images synthetically, this finding can be taken into account when creating new examples.

Figure 4 shows how the model responds to the aspect ratio of the masks. The different shapes and sizes affect each category in a specific way. For example, the small civil aircraft class has better performances when objects are extra-wide, while the other two categories exhibit a different pattern. From a visual inspection of the examples of each class with extra wide aspect ratio (Fig. 4) we observed that while an elongated geometry is prevalent for the small civil airplanes, for medium and large ones it seems correlated to the mask orientation combined with truncation. Based on such observation, one can design measures to improve the detection of this type of images, by applying suitable data augmentation techniques.

---

[2] Models and files published at: https://github.com/aireveries/RarePlanes/tree/mast er/models.

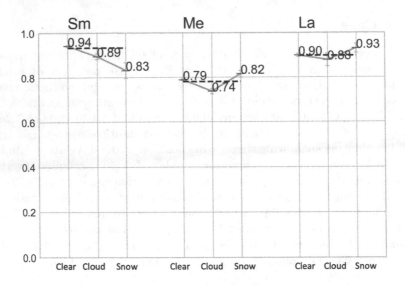

Fig. 3. AP variation under different weather conditions

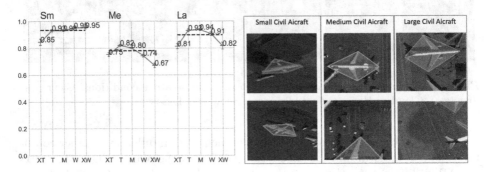

Fig. 4. AP variation under different aspect ratios for each category and examples of small medium and large airplanes with extra wide aspect ratio

Figure 5 (left) shows the relative impact of different objects characteristics. Propulsion, wing type and number of engines display much larger influence on the overall performance. The inspection of the propulsion property shows that it has a high imbalance across the aircraft classes (Fig. 5 right). Figure 5 also shows that the impact of truncation is lower than that of the aspect ratio, which suggest to prioritize data augmentation by giving precedence to image rotation over the creation of artificial truncation.

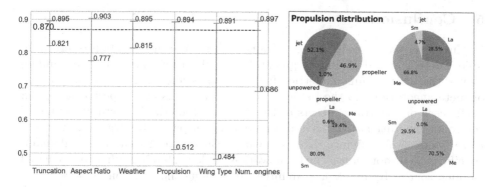

Fig. 5. Relative impact of different properties and distribution of the propulsion values (propeller, jet, unpowered) across all images and per class (small, mid, large)

Figure 6 reports the false positive distribution across classes and the distribution of the different error types explained in Sect. 3.2 for each class. In all cases, localization is responsible of a minority of errors, with the minimum incidence for large airplanes. This is an indicator of the good localization capabilities of the model and thus shifts the focus of improvement to the classification step. The small civil aircraft objects are the ones with more confusion with background or unlabeled objects, while the large civil aircraft ones present less errors of such types. To some extent, the real dimensions of the objects impact on how likely it is to confuse them with other unlabeled objects.

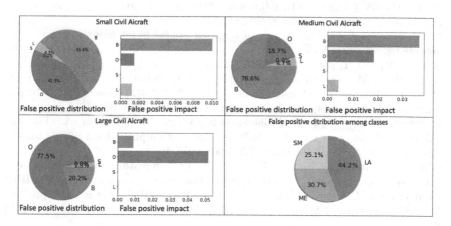

Fig. 6. Analysis of False Positives. L (light blue): location error, B: (orange) confusion with background or unlabeled objects B, O (purple) confusion with other classes (Colour figure online).

5   Conclusions

In this paper we have described a framework for the investigation of errors in object detection and instance segmentation applications. The framework implements the most common metrics for such tasks and enables the association of meta-annotations to the input, so that the analysis of the performance can be focused on arbitrary subsets of the input characterized by critical values of the meta-annotations. We have illustrated the output of the framework on the RarePlanes satellite image data set, which features both generic and domain-specific meta-annotations. The framework is implemented in Python and released as open source. Its plug&play architecture permits the addition of novel meta-annotations and custom metrics with minimal coding effort. Our future work will concentrate on adding support to classification tasks and on extending the library of metrics implementations with further classes for specific applications, e.g., human pose detection. In addition, we will add support for the automatic extraction of specific types of meta-annotations, such as the geographical coordinates, date and time of image acquisition, and the orientation of the instance segmentation mask. We also plan to integrate the analysis of attention, by computing the position and extent of the CAM w.r.t. to the object bounding box or segmentation mask to support the optimization of weakly supervised models.

References

1. Abiodun, O.I., Jantan, A., Omolara, A.E., Dada, K.V., Mohamed, N.A., Arshad, H.: State-of-the-art in artificial neural network applications: A survey. Heliyon 4(11), e00938 (2018). 10.1016/j.heliyon.2018.e00938, http://www.sciencedirect.com/science/article/pii/S2405844018332067
2. Dalal, N., Triggs, B.: Histograms of oriented gradients for human detection. In: 2005 IEEE computer society conference on computer vision and pattern recognition (CVPR'05). vol. 1, pp. 886–893. IEEE (2005)
3. Everingham, M., Van Gool, L., Williams, C.K.I., Winn, J., Zisserman, A.: The PASCAL Visual Object Classes Challenge 2012 (VOC2012) Results (2012). http://www.pascal-network.org/challenges/VOC/voc2012/workshop/index.html
4. Everingham, M., Gool, L., Williams, C.K., Winn, J., Zisserman, A.: The Pascal visual object classes (VOC) challenge. Int. J. Comput. Vision 88(2), 303–338 (2010). https://doi.org/10.1007/s11263-009-0275-4
5. Girshick, R., Donahue, J., Darrell, T., Malik, J.: Rich feature hierarchies for accurate object detection and semantic segmentation. In: Proceedings of the 2014 IEEE Conference on Computer Vision and Pattern Recognition, CVPR 2014, pp. 580–587. IEEE Computer Society, USA (2014). https://doi.org/10.1109/CVPR.2014.81
6. Harzallah, H., Jurie, F., Schmid, C.: Combining efficient object localization and image classification. In: 2009 IEEE 12th International Conference On Computer Vision, pp. 237–244. IEEE (2009)
7. Hoiem, D., Chodpathumwan, Y., Dai, Q.: Diagnosing error in object detectors. In: Fitzgibbon, A., Lazebnik, S., Perona, P., Sato, Y., Schmid, C. (eds.) ECCV 2012. LNCS, vol. 7574, pp. 340–353. Springer, Heidelberg (2012). https://doi.org/10.1007/978-3-642-33712-3_25

8. Kirillov, A., He, K., Girshick, R., Rother, C., Dollar, P.: Panoptic segmentation. In: Proceedings of the IEEE/CVF Conference on Computer Vision and Pattern Recognition (CVPR) (June 2019)
9. Kuznetsova, A., et al.: The open images dataset V4. Int. J. Comput. Vis. **128**(7), 1956–1981 (2020). https://doi.org/10.1007/s11263-020-01316-z
10. Lin, T.Y., et al.: Microsoft COCO: common objects in context. In: Fleet, D., Pajdla, T., Schiele, B., Tuytelaars, T. (eds.) Computer Vision - ECCV 2014, pp. 740–755. Springer International Publishing, Cham (2014)
11. Lindeberg, T.: Scale invariant feature transform (2012)
12. Liu, L., et al.: Deep learning for generic object detection: A survey. Int. J. Comput. Vis. **128**(2), 261–318 (2020). https://doi.org/10.1007/s11263-019-01247-4
13. Padilla, R., Netto, S., da Silva, E.: A survey on performance metrics for object-detection algorithms. In: International Conference on Systems, Signals and Image Processing (IWSSIP) (07 2020). https://doi.org/10.1109/IWSSIP48289.2020
14. Raghavan, V., Bollmann, P., Jung, G.S.: A critical investigation of recall and precision as measures of retrieval system performance. ACM Trans. Inform. Syst. (TOIS) **7**(3), 205–229 (1989)
15. Redmon, J., Divvala, S.K., Girshick, R.B., Farhadi, A.: You only look once: Unified, real-time object detection. In: 2016 IEEE Conference on Computer Vision and Pattern Recognition, CVPR 2016, Las Vegas, NV, USA, June 27–30, 2016. pp. 779–788. IEEE Computer Society (2016). https://doi.org/10.1109/CVPR.2016.91
16. Shermeyer, J., Hossler, T., Van Etten, A., Hogan, D., Lewis, R., Kim, D.: Rareplanes: Synthetic data takes flight (2020). arXiv preprint: arXiv:2006.02963
17. Wu, Y., Kirillov, A., Massa, F., Lo, W.Y., Girshick, R.: Detectron2 (2019). https://github.com/facebookresearch/detectron2
18. Zhang, J., Bargal, S.A., Lin, Z., Brandt, J., Shen, X., Sclaroff, S.: Top-down neural attention by excitation backprop. Int. J. Comput. Vision **126**(10), 1084–1102 (2018). https://doi.org/10.1007/s11263-017-1059-x

Shift Equivariance in Object Detection

Marco Manfredi$^{(\boxtimes)}$ and Yu Wang

TomTom, Amsterdam, The Netherlands
{marco.manfredi,yu.wang}@tomtom.com

Abstract. Robustness to small image translations is a highly desirable property for object detectors. However, recent works have shown that CNN-based classifiers are not shift invariant. It is unclear to what extent this could impact object detection, mainly because of the architectural differences between the two and the dimensionality of the prediction space of modern detectors.

To assess shift equivariance of object detection models end-to-end, in this paper we propose an evaluation metric, built upon a greedy search of the lower and upper bounds of the mean average precision on a shifted image set. Our new metric shows that modern object detection architectures, no matter if one-stage or two-stage, anchor-based or anchor-free, are sensitive to even one pixel shift to the input images.

Furthermore, we investigate several possible solutions to this problem, both taken from the literature and newly proposed, quantifying the effectiveness of each one with the suggested metric. Our results indicate that none of these methods can provide full shift equivariance.

Measuring and analyzing the extent of shift variance of different models and the contributions of possible factors, is a first step towards being able to devise methods that mitigate or even leverage such variabilities.

Keywords: Convolutional Neural Networks · Object detection · Network robustness · Shift equivariance

1 Introduction

Convolutional Neural Networks (CNNs) have achieved impressive results on many computer vision tasks, like classification [26], detection [5] and segmentation [6]. However, in safety critical applications, like autonomous driving or medical imaging, CNNs need to be not just accurate, but also reliable and robust to image perturbations.

The evaluation of the robustness of CNNs has been a very active research field, with several works addressing the impact of image noise [12], image transformations (translation [3], rotation [8], scale [22]), object inter-dependencies [1, 21] and adversarial attacks [27].

Focus of this paper is the robustness of modern object detectors to (small) image translations. Modern CNN architectures are not shift invariant by design.

M. Manfredi and Y. Wang—Equal contribution.

© Springer Nature Switzerland AG 2020
A. Bartoli and A. Fusiello (Eds.): ECCV 2020 Workshops, LNCS 12540, pp. 32–45, 2020.
https://doi.org/10.1007/978-3-030-65414-6_4

Commonly used down-sampling approaches, like max pooling, are one cause for shift variance [28], since these operations are against the classic sampling theorem.

Small transformations can cause a significant drop in classification accuracy for classification models [8]. Furthermore, with the huge receptive field, rather than learning to be shift invariant, modern CNNs filters can derive and exploit absolute spatial location all over the image from zero-padding [14,16].

Unlike classification, in object detection we need to measure *equivariance* and not *invariance*, since we expect the output of a detector to reflect the translation applied to its input. Moreover, each prediction cannot be evaluated independently from all others, given the non trivial assignment between predictions and ground truth instances and the impact of aggregation strategies like non maximum suppression. To the best of our knowledge, no previous work

(a)   (b)

Fig. 1. Qualitative results of shift variance for RetinaNet on the COCO validation set, with confidence threshold set to 0.4 for visualization purposes. The predictions in column a) and b) are computed on images that differ only for a small shift of at maximum one pixel. In the second row we see how both classification and localization of the instances are deeply affected. The third row shows an example of a class swap, where the same instance is predicted as an elephant or a cow.

has proposed a quantitative evaluation of the robustness of detectors to image translations.

We address this gap by designing an evaluation pipeline to quantify shift equivariance of object detection models in an end to end manner.

Previous works also focused on identifying the root causes of the lack of shift invariance, and tried to address them [13,23,28]. However, the impact that these factors might have on object detectors has not been investigated yet. We believe this has non-trivial implications, since detection networks extend classification backbones with additional modules like regression/classification heads (both anchor-based and anchor-free), Region Proposals Networks or ROI Pooling/Align modules. We therefore conduct experiments to dissect translation equivariance on object detection networks: anti-aliasing the intermediate features, densifying the output resolution, using dataset which is free of photographer's bias and test-time augmentation.

Our contributions are:

1. We design an evaluation pipeline and a metric, AP variations (ΔAP), specifically targeted to shift equivariance in object detectors;
2. We test several modern object detectors with the proposed metric and we report a severe lack of robustness to translations as small as one pixel, on two different datasets;
3. We measure the effectiveness of different techniques to enhance robustness, and show that they provide only a partial answer to the problem.

2   Related Work

Network Robustness. Convolutional neural networks are reported to have brittle robustness to perturbations. Adversarial attacks [10,15], which are intentionally designed noises to fool the network to make wrong predictions, are used to investigate network failures. Recent work has found that models are also brittle to less extreme perturbations that are not adversarially constructed, such as image quality distortions, i.e. blur, noise, contrast and compression [7]. CNNs are reported to not be robust to scale variations [22] and highly sensitive to pose perturbations [2]. Networks fail to even small rotations and translations [3,8,28]. Imperceptible variations and natural transformations across consecutive video frames would induce instability for networks as well [3,11].

There are also works on the robustness of object detectors. Image corruptions such as pixel noise, blur, varying weather conditions lead to significant performance drops in object detection models [4,19]. Object detectors are not robust against scale variations [22]. Replacing image sub-regions with another sub-image is shown to have non-local impact on object detection, which may occasionally make other objects undetected or misclassified [21].

Network Robustness Evaluation. The robustness is often evaluated for classification networks in terms of classification label swap or mean absolute score changes. For example attack success rates (changes of top-1 prediction) across several distortion sizes are used to measure robustness with a broader threat model and diverse differentiable attacks [15]. Azulay et al. quantify both top-1 and mean average score changes of the classifier to small translations or re-scaling of the input images [3]. [12] quantifies the classifiers mean and relative error rate on corruptions with different severity level to show the networks robustness against a variety of common corruptions. To measure robustness against geometric transformations, Manitest defines the invariance as the minimal geodesic distance between the identity transformation and a transformation in Lie group τ which is sufficient to change the predicted label of the classifier [9]. In our paper, we focus on evaluating shift equivariance of object detectors, considering both classification and localization performance.

Approaches to Improve Shift Invariance. One approach to improve the robustness of the model is to apply data-augmentation such as random cropping to make the network learn to be invariant to shifts [13]. However, [3] has shown that data augmentation teaches the network to be invariant to translations but only for images that are visually similar to typical images seen during training, i.e. images that obey the photographer's bias. Another way to mitigate shift variance is to apply anti-aliasing operations by blurring the representations before downsampling [28], as often done in signal processing. Although [28] observed increased accuracy in ImageNet classification, [3] has shown that blurring would degrade model performance, especially in datasets that obey the photographer's bias. GaussNet CNN architecture, which represents convolutional kernels with an orthogonal Gauss-Hermite basis whose basis coefficients are learned in convolution layers without a sub-sampling layer, leads to fully translation invariant representations that keeps the number of parameters in kernels across layers constant [23]. In our paper, we also investigate several solutions to alleviate translation variance, focusing the analysis on object detection models.

3   Measuring Translation Equivariance for Object Detectors

A good measure of translation equivariance should capture how the output of a detector changes when a shift is applied to the input image. The ideal detector would output the same predictions, but shifted.

In order to measure translation equivariance we need to define: an experimental setting to generate shifted images and a metric to measure the amount of change in the detector output. For the experimental setting, we follow other works on shift invariance for classification networks [3], please refer to our experimental section for further details.

The shift invariance metrics for classification measure the changes in per-image class scores, for example reporting the gap between the highest and lowest class scores predicted by the network on shifted images. In object detection, it is possible to compute these metrics at the granularity of single instances, evaluating the probability score variations of each ground truth object. However, object detection is a more complex task than classification, and the metrics to measure robustness to translation should be tailored specifically for it. We believe a good metric for translation equivariance in object detection should: i) consider all predicted boxes, also the ones on background regions (i.e. potential false positives) and ii) include localization accuracy in its formulation. In general, this robustness metric should be a clear indicator of the performance variations we might get when perturbing the input with small translations.

This motivates the introduction of AP variations as our proposed indicator of translation equivariance in object detectors.

3.1   AP Variations

Average precision is the most common way of evaluating object detectors [18]. A natural formulation for a shift equivariance metric is thus the variations of average precision over a (shifted) dataset. This metric captures the performance gap between the worst and best case.

Given a set $\mathbf{X} = (X_1, X_2, \cdots, X_N)$ of N images and a maximum shift range M (e.g. 1 pixel), we compute a shifted set, where each image X_i is present multiple times, slightly shifted horizontally and vertically, $\mathbf{X}_i^\Delta = \{X_i^\delta \mid \delta \in \{(0,0),(0,1),\cdots,(M,M)\}\}$.

The AP variation over the shifted dataset is defined as the difference between the highest and lowest achievable APs: $\Delta AP = AP_{best} - AP_{worst}$. As an example, AP_{best} is the maximum AP achievable by selecting the best shift δ_i^* for each image. In other words, for each image the shift that contributes to the highest overall AP is selected. In practice, we use AP_{50} as the metric, that is, using 0.5 as the IoU threshold to define a positive match between a prediction and a ground truth object.

Greedy Approximation of AP Variations. Average precision is computed on the entire dataset by sorting the predictions of all images by confidence score, and evaluating the ranking of true positives and false positives. This means that we can't find the best (and worst) shift for each image independently from all other images. In theory, finding the best AP would involve computing the AP of all combinations of shifts across images. This is computationally infeasible, since the number of combinations would be $(M+1)^{2N}$. For the validation set of COCO, and a maximum shift of 1 pixel, that would be 2^{10000} total AP computations.

We propose an approximation to this optimal solution using a greedy algorithm that iteratively finds the best shift δ_i^* for each sample X_i while keeping fixed the solutions of all other samples $X_{j,j\neq i}$. Algorithm 1 details the procedure to compute AP_{best}, by switching $\arg\max$ with $\arg\min$ we can compute AP_{worst}.

The number of AP computations becomes now tractable and is equal to $K \times (M+1)^2 \times N$. In practice, setting the number of iterations $K = 1$ is enough for the algorithm to converge to a stable solution, since in our experiments further iterations contributed for less than 1% to the total AP difference.

Input:
$\mathbf{X}$: the set of images
$\Delta = (\delta_0, \delta_1, \cdots)$: set of shifts per samples
K: number of iterations
e: evaluation function
f: detector inference function
Initialize $\delta_i^* = \delta_0$ (for each sample i, shift (0,0) is selected)
for k *in* K **do**
$\quad$**for** i *in* N **do**
$\quad\quad AP = [\,]$
$\quad\quad$**for** *each* δ *in* Δ **do**
$\quad\quad\quad AP_\delta = e(f(X_i^\delta \cup \tilde{\mathbf{X}})), \ \tilde{\mathbf{X}} = \{X_j^{\delta_j^*} \mid j \neq i, j \in \{1, \cdots, N\}\}$
$\quad\quad\quad AP.\text{append}(AP_\delta)$
$\quad\quad$**end**
$\quad\quad$#update the best performing shift for sample X_i
$\quad\quad \delta_i^* = \underset{\delta \in \Delta}{\arg\max} \ AP$
$\quad$**end**
end
$AP_{best} = e(f(\mathbf{X}^*)), \ \mathbf{X}^* = \{X_i^{\delta_i^*} \mid i \in \{1, \cdots, N\}\}$

Algorithm 1: Greedy algorithm to approximate the best Average Precision by iteratively selecting the best shift for each sample.

4 Experiments

We evaluate shift equivariance for three common object detectors: RetinaNet [17] (one-stage, anchor-based), CenterNet [20] (one-stage, anchor-free) and Faster-RCNN [29] (two-stage) on the COCO object detection dataset [18]. For RetinaNet and FasterRCNN, we use the pretrained models with ResNet-50 backbone available in the detectron2 framework [24], for CenterNet we use the pretrained model with ResNet-101 with deformable convolutions from the official repository[1].

In order to evaluate shift equivariance, we create a shifted validation set, where each image appears multiple times, shifted horizontally and vertically. We follow the experimental setting proposed in [3], where each image is embedded in a black background image at different locations. This setting prevents to crop out any context from the image, ensuring the only difference is a small translation, see Fig. 2. This setting is particularly appropriate for object detection, since objects are located anywhere in the image, and cropping even a small portion of it might imply removing relevant object parts.

Given a maximum shift, we compute predictions for all shifted images and the best and worst overall APs using the proposed method. ΔAP is obtained as the simple difference of the two. In Table 1, the results for the entire COCO validation set are reported for a maximum shift of one pixel.

---

[1] github.com/xingyizhou/CenterNet.

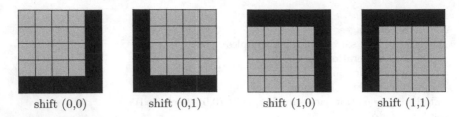

shift (0,0)   shift (0,1)   shift (1,0)   shift (1,1)

Fig. 2. Creating the dataset for shift equivariance evaluation. An example 4×4 image (in green) is embedded into a black image at different locations (maximum shift $= 1$) to create the 4 translated versions (Color figure online).

All three methods show a surprising lack of shift equivariance, with CenterNet being the least robust. We show some qualitative examples in Fig. 1. CenterNet is the only method of the three that does not use NMS to aggregate overlapping predictions. Instead, it uses a max-pooling layer to extract local peaks in the keypoint heatmap. This is a simpler implementation of NMS that might be less effective to retain the best boxes.

The ΔAPs are comparable to the performance gaps between these methods, meaning that a one pixel shift is enough to change what we might consider state-of-the-art performance. CenterNet is the only method trained with extensive data augmentations, including random cropping. Its poor performance in terms of shift equivariance confirms the conclusions from [3]: data augmentation is not enough to improve robustness.

Table 1. AP variations on the COCO validation set (the lower, the better), with a maximum shift of one pixel.

Method	AP	$^{worst}/_{best}$ AP	ΔAP	AP_{50}	$^{worst}/_{best}$ AP_{50}	ΔAP_{50}
RetinaNet	36.5	35.3/37.5	2.2	56.7	53.9/59.0	5.1
FasterRCNN	37.6	36.5/39.4	2.9	59.0	55.7/62.1	6.4
CenterNet	34.6	32.9/36.3	3.4	53.0	49.2/57.3	8.1

One other important result is that all methods are able to achieve better performance than the baseline by carefully selecting the best shifts for each validation image. This implies that shift variance could be leveraged to boost performance, as we show in a simple experiment in Sect. 5.4.

4.1   Increasing Shift Range

We showed how a simple one-pixel shift can drastically affect the performance of several state-of-the-art detectors. But how do the detectors behave for increasingly high shifts? In this section we report the results of shift equivariance of

RetinaNet with increasing maximum shift, from 0 to 15. Since the number of images grows quadratically with the number of shifts (256 images for a maximum shift of 15), to save computation we perform the analysis on a randomly sampled validation set of 100 images. Results are reported in Table 2, in Fig. 3 we illustrate the AP difference with respect to the un-shifted baseline. The AP variation grows sub-linearly with the maximum shift, reaching an impressive 21.5 ΔAP_{50} for a maximum shift of 15. The worst and best APs are symmetric with respect to the baseline AP, meaning that the baseline performance can be regarded as an *average case* among the shifts. It can be noted that ΔAP_{50} grows monotonically with the maximum shift, but ΔAP does not. This is a result of our evaluation metric, that uses AP_{50} to find best and worst shifts.

Table 2. AP variations of RetinaNet on 100 images from the COCO validation set. Results are reported for increasingly high shifts in pixels.

Max shift	$^{worst}/_{best}\ AP$	ΔAP	$^{worst}/_{best}\ AP_{50}$	ΔAP_{50}
0 - baseline	43.8	-	64.2	-
1	41.8/43.4	1.6	60.7/66.9	6.2
3	41.2/46.4	5.2	58.0/69.5	11.5
7	39.6/46.0	6.4	55.4/72.1	16.7
15	38.7/47.0	8.3	53.6/75.1	21.5

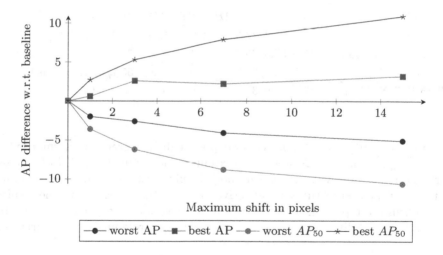

Fig. 3. Difference between baseline results and best/worst APs computed by our method with increasing maximum shift. Results are reported for RetinaNet on 100 random images from the COCO validation set.

5   Dissecting Translation Equivariance

So far, we showed how modern object detectors suffer from a lack of shift equivariance. A small shift in the input image can have huge effects on the output predictions. In the literature, many attempts have been made to discover the root causes of the lack of shift invariance and to propose solutions to counter it. To the best of our knowledge, no previous attempts have been made to evaluate how these can impact the performance of object detectors.

In the following, we dissect shift equivariance performance under the light of several factors that can potentially affect it.

5.1   Anti-aliasing in Downsampling Layers

We retrain RetinaNet and FasterRCNN using Blur Pool [28] in their backbone network, ResNet-50. Blur Pool affects all downsampling layers, by limiting aliasing effects introduced by max pooling and strided convolutions. In practice we use a 3×3 blur filter, as recommended in the original paper. We report the results of ΔAP on the COCO validation set in Table 3, compared to baseline methods.

Table 3. Effect of anti-aliased features on AP variations on the COCO validation set, with a maximum shift of one pixel.

Method	AP	$worst/best$ AP	ΔAP	AP_{50}	$worst/best$ AP_{50}	ΔAP_{50}
RetinaNet	36.5	35.3/37.5	2.2	56.7	53.9/59.0	5.1
RetinaNet+blurpool	35.2	34.3/35.7	1.4	55.1	51.4/58.6	3.4
FasterRCNN	37.6	36.5/39.4	2.9	59.0	55.7/62.1	6.4
FasterRCNN+blurpool	37.8	36.6/38.6	2.0	58.7	56.3/60.9	4.6

On the un-shifted validation set, blur pool slightly degrades performance for RetinaNet, while keeping FasterRCNN results basically unchanged. Robustness to translation is improved for both methods in all metrics of about 30%, suggesting that anti-aliased features are effective also for object detection. The results also confirm the reports from [3], that anti-aliasing is only a partial solution to the problem, and different factors might be playing a role in shift equivariance.

5.2   Densifying the Output Space

One possible reason to the lack of shift equivariance for object detection is the mismatch between input and output resolution. For example, the finest output resolution Retinanet can reach is eight times smaller compared to that of the input image. As for one-layer-output CenterNet, the output resolution is four times smaller. The mismatch results in discontinuities in the ground truth and anchor matching: moving even one pixel implies an anchor from being assigned from positive to negative and vice versa for a neighboring anchor. To investigate to what extent this can contribute to the shift variance, we densify the output space of CenterNet by adding two extra upsampling layers with transposed convolutions to match the output resolution to the input images. CenterNet is selected because it uses a single layer to make predictions instead of multiple layers as RetinaNet. Densifying the output space leads to two problems: i) the imbalance between foreground and background in the output space becomes even more severe given that increasing output resolution leads to 16 times more negative predictions and ii) the memory footprint due to a full-resolution output poses a problem in training efficiency. The only hyperparameter changes we apply to compensate for the higher resolution are: i) decreased $loss_{size}$ weight to 0.02 and ii) increased max pooling kernel size to 11 for NMS. Dense CenterNet achieves an AP of 33.4 (baseline is 34.6). However, we couldn't see many benefits for shift equivariance: the ΔAP_{50} for the dense CenterNet is 7.6 (baseline 8.1) and the AP difference is 3.3 (baseline 3.4).

5.3   Removing Photographer's Bias

Photographer's bias is present in datasets composed of images taken by humans [3]. Humans tend to put objects at the center of the frame, and usually take pictures by standing vertically, almost perpendicular to the ground. Absolute location in the image frame becomes then a useful feature to recognize object categories (e.g. the sky is up, a sofa is on the floor, ...). As [16] pointed out, it is possible for a CNN to exploit such biases, and to introduce shift variance as a mean to capture absolute object location.

Fig. 4. Example crops from the DOTA object detection dataset. Instance annotation is available for rotated as for horizontal bounding boxes.

We therefore aim to measure the importance of the photographer's bias for shift equivariance in detection. To this end, we extended our quantitative analysis to the DOTA v1.0 object detection dataset [25]. DOTA is the largest object detection dataset in aerial images, it contains 2806 large size images with $188,282$ object instances belonging to 15 categories, some sample crops are shown in Fig. 4. Aerial images do not suffer from the photographer's bias, since objects are not arranged in any particular order in the image frame. The dataset defines two tasks: detection with oriented bounding boxes and detection with horizontal bounding boxes. We compute results for the latter, to keep the methods comparable to previous experiments.

Table 4. Shift equivariance evaluation on DOTA validation set, with a maximum shift of one pixel.

Method	AP	$^{worst}/_{best}\ AP$	ΔAP	AP_{50}	$^{worst}/_{best}\ AP_{50}$	ΔAP_{50}
RetinaNet	34.1	33.5/35.4	1.9	58.0	55.7/60.5	4.8
FasterRCNN	36.7	36.0/38.7	2.7	60.3	57.4/63.2	5.8
CenterNet	34.6	33.5/35.7	2.2	57.3	54.4/60.2	5.8

Following the experimental setting of [25], images from the training set are split into patches of size 1024×1024 with overlap of 512 pixels. This is necessary in order to fit the images to a CNN, since the original images have dimensions of 4000×4000 pixels or more. In the original experimental setting, validation images are also split into smaller crops, and the predictions from each crop are aggregated to compute the final performance. We retrained all detectors following the suggested setting and we obtained performance comparable with the results reported in [25].

We evaluate shift equivariance on the validation crops, that can fit in memory for both the baseline and the shifted variants. Shift equivariance results on the validation set, for a maximum shift of one pixel, are reported in Table 4.

Shift equivariance results are comparable to the ones of Table 1, however, all methods show an improvement compared to the COCO setting. CenterNet is the most different, with ΔAP going from 3.4 to 2.2 and ΔAP_{50} from 8.1 to 5.8. Although an increase in robustness compared to the COCO dataset is measured, photographer's bias does not seem to be a major factor in translation equivariance.

5.4   Test-Time Augmentation

Test-time augmentation is commonly used to improve performance by computing predictions on several variations of the same image and then aggregating the predictions before evaluation. One example is multi-scale testing [22], where each validation image is resized to different dimensions, predictions are computed for each scale and then mapped to a common reference frame, where non maximum suppression (NMS) is performed to aggregate them.

In the following, we perform test-time augmentation on the translated images. The predictions of each shift are mapped back to the original image coordinates, and aggregated using NMS. As for all other experiments, we use a maximum shift of one pixel, leading to 4 translated versions of each image.

Table 5. Detection performance on COCO validation set. For each method we report baseline results on the un-shifted images, together with test-time augmentation results and best results obtained by the proposed method with maximum shift of one pixel.

Method	Baseline		w/ tta		Best APs	
	AP	AP_{50}	AP	AP_{50}	AP	AP_{50}
RetinaNet	36.5	56.7	36.7	56.9	37.5	59.0
FasterRCNN	37.6	59.0	38.4	59.5	39.4	62.1
CenterNet	34.6	53.0	35.6	55.2	36.3	57.3

We want to measure if test-time augmentation is capable of retaining the predictions from the best performing shifts, leading to better results than the baseline. In this experiment we don't evaluate the shift equivariance, since all shifts are considered at the same time during aggregation. Results are reported in Table 5.

Test-time augmentation improves results for all methods, bringing them closer to the best achievable APs on the shifted validation set. For CenterNet, we report an improvement of 1.0 AP and 2.2 AP_{50} over the baseline. Test-time augmentation is a very simple way of exploiting shift variance, and it comes with a considerable computational overhead (4 shifts per image, leading to 4× inference time). We believe there is room for better ways of leveraging the lack of shift equivariance, that can turn out to be an exploitable property, instead of an undesirable side effect of modern CNNs.

6 Conclusions

We showed that object detection models exhibit considerable variation with even small translations applied to input images, which has been observed and reported in image classification tasks [3,8,28]. In order to quantitatively evaluate shift equivariance of object detection models end-to-end, we proposed a greedy algorithm to search the lower and upper bounds of mean average precision on a test set. We found consistent large performance gaps on multiple modern object detection architectures for even a single-pixel shift. Apart from providing the evaluation pipeline, we investigated and demonstrated ways to mitigate this problem. With blur-pool, we improved the Δ_{AP} by around 30% on both one-stage and two-stage detectors. By densifying the output space, we could improve shift equivariance by a small margin on ΔAP_{50}. In addition, we observed improved robustness on the DOTA aerial imaging dataset, that doesn't suffer from the photographer's bias. Furthermore, we showed that test-time augmentation could leverage the variance introduced by input translations and improve the network performance at the cost of significantly higher inference time.

However, none of the aforementioned solutions could remove shift variance completely. Further investigations on architecture improvements, loss formulations or data augmentation techniques would shed more light on this topic.

References

1. Agarwal, V., Shetty, R., Fritz, M.: Towards causal VQA: revealing and reducing spurious correlations by invariant and covariant semantic editing. In: Proceedings of the IEEE/CVF Conference on Computer Vision and Pattern Recognition, pp. 9690–9698 (2020)
2. Alcorn, M.A., et al.: Strike (with) a pose: neural networks are easily fooled by strange poses of familiar objects. In: Proceedings of the IEEE/CVF Conference on Computer Vision and Pattern Recognition (CVPR) (2019)
3. Azulay, A., Weiss, Y.: Why do deep convolutional networks generalize so poorly to small image transformations? J. Mach. Learn. Res. **20**(184), 1–25 (2019)
4. von Bernuth, A., Volk, G., Bringmann, O.: Simulating photo-realistic snow and fog on existing images for enhanced CNN training and evaluation. In: 2019 IEEE Intelligent Transportation Systems Conference (ITSC), pp. 41–46. IEEE (2019)
5. Cai, Z., Vasconcelos, N.: Cascade R-CNN: delving into high quality object detection. In: Proceedings of the IEEE Conference on Computer Vision and Pattern Recognition, pp. 6154–6162 (2018)
6. Chen, L.C., Papandreou, G., Kokkinos, I., Murphy, K., Yuille, A.L.: Deeplab: semantic image segmentation with deep convolutional nets, atrous convolution, and fully connected CRFs. IEEE Trans. Pattern Anal. Mach. Intell. **40**(4), 834–848 (2017)
7. Dodge, S., Karam, L.: Understanding how image quality affects deep neural networks. In: 2016 Eighth International Conference on Quality of Multimedia Experience (QoMEX), pp. 1–6. IEEE (2016)
8. Engstrom, L., Tsipras, D., Schmidt, L., Madry, A.: A rotation and a translation suffice: fooling CNNs with simple transformations (2017). arXiv preprint arXiv:1712.02779
9. Fawzi, A., Frossard, P.: Manitest: are classifiers really invariant? In: British Machine Vision Conference (BMVC), pp. 106.1–106.13 (2015)
10. Goodfellow, I.J., Shlens, J., Szegedy, C.: Explaining and harnessing adversarial examples (2014). arXiv preprint arXiv:1412.6572
11. Gu, K., Yang, B., Ngiam, J., Le, Q., Shlens, J.: Using videos to evaluate image model robustness (2019). arXiv preprint arXiv:1904.10076
12. Hendrycks, D., Dietterich, T.G.: Benchmarking neural network robustness to common corruptions and surface variations (2018). arXiv preprint arXiv:1807.01697
13. Hendrycks, D., Mu, N., Cubuk, E.D., Zoph, B., Gilmer, J., Lakshminarayanan, B.: Augmix: a simple data processing method to improve robustness and uncertainty. In: International Conference on Learning Representations (2019)
14. Islam, M.A., Jia, S., Bruce, N.D.: How much position information do convolutional neural networks encode? In: International Conference on Learning Representations (2019)
15. Kang, D., Sun, Y., Hendrycks, D., Brown, T., Steinhardt, J.: Testing robustness against unforeseen adversaries (2019). arXiv preprint arXiv:1908.08016
16. Kayhan, O.S., Gemert, J.C.v.: On translation invariance in CNNs: convolutional layers can exploit absolute spatial location. In: Proceedings of the IEEE/CVF Conference on Computer Vision and Pattern Recognition, pp. 14274–14285 (2020)
17. Lin, T.Y., Goyal, P., Girshick, R., He, K., Dollár, P.: Focal loss for dense object detection. In: Proceedings of the IEEE International Conference on Computer Vision, pp. 2980–2988 (2017)

18. Lin, T.-Y., et al.: Microsoft COCO: common objects in context. In: Fleet, D., Pajdla, T., Schiele, B., Tuytelaars, T. (eds.) ECCV 2014. LNCS, vol. 8693, pp. 740–755. Springer, Cham (2014). https://doi.org/10.1007/978-3-319-10602-1_48
19. Michaelis, C., et al.: Benchmarking robustness in object detection: autonomous driving when winter is coming (2019). arXiv preprint arXiv:1907.07484
20. Ren, S., He, K., Girshick, R., Sun, J.: Faster R-CNN: towards real-time object detection with region proposal networks. In: Advances in Neural Information Processing Systems, pp. 91–99 (2015)
21. Rosenfeld, A., Zemel, R.S., Tsotsos, J.K.: The elephant in the room (2018). arXiv preprint arXiv:1808.03305
22. Singh, B., Davis, L.S.: An analysis of scale invariance in object detection SNIP. In: Proceedings of the IEEE Conference on Computer Vision and Pattern Recognition, pp. 3578–3587 (2018)
23. Sundaramoorthi, G., Wang, T.E.: Translation insensitive CNNs (2019). arXiv preprint arXiv:1911.11238
24. Wu, Y., Kirillov, A., Massa, F., Lo, W.Y., Girshick, R.: Detectron2 (2019). https://github.com/facebookresearch/detectron2
25. Xia, G.S., et al.: Dota: a large-scale dataset for object detection in aerial images. In: Proceedings of the IEEE Conference on Computer Vision and Pattern Recognition, pp. 3974–3983 (2018)
26. Xie, S., Girshick, R., Dollár, P., Tu, Z., He, K.: Aggregated residual transformations for deep neural networks. In: Proceedings of the IEEE Conference on Computer Vision and Pattern Recognition, pp. 1492–1500 (2017)
27. Zhang, H., Wang, J.: Towards adversarially robust object detection. In: Proceedings of the IEEE International Conference on Computer Vision, pp. 421–430 (2019)
28. Zhang, R.: Making convolutional networks shift-invariant again. In: International Conference on Machine Learning, pp. 7324–7334 (2019)
29. Zhou, X., Wang, D., Krähenbühl, P.: Objects as points (2019). arXiv preprint arXiv:1904.07850

Probabilistic Object Detection with an Ensemble of Experts

Dimity Miller[✉]

Australian Centre for Robotic Vision, Queensland University of Technology,
Brisbane, Australia
dimity.miller@hdr.qut.edu.au

Abstract. Probabilistic object detection requires detectors to localise
and classify objects in an image, while also providing accurate spatial and
semantic uncertainty. In this work, we present an 'Ensemble of Experts'
as a method to solve this challenging problem. This technique utilises
a ranked ensembling association process and leverages the individual
strengths of each expert detector to create a final set of detections with
a meaningful spatial and semantic uncertainty. Our approach placed first
place in the 3rd Probabilistic Object Detection Challenge with a PDQ
of 22.848.

1 Introduction

CNN-based object detectors have shown impressive performance at *localising*
and *classifying* objects within an image [1–3]. As a result, they can be a useful
perception tool aboard robots and autonomous systems.

Standard CNN-based object detectors currently do not express uncertainty
in their localisation, and often report unreliable or miscalibrated uncertainty in
their classification [4,5]. Without the ability to express a reliable uncertainty,
questions have been raised about the robustness and safety of object detectors
for use aboard autonomous systems or robots [6,7].

Probabilistic object detection was recently introduced as an extension of the
classic object detection task [8], where a detector must also provide a reliable
spatial and *semantic* uncertainty with each detection. To assess performance on
this task, the Probability-based Detection Quality (PDQ) metric was introduced
[8]. Unlike previous object detection metrics, such as mean Average Precision
(mAP) [9], PDQ formally evaluates and rewards meaningful spatial and semantic
uncertainties, as well as general object detection performance [8].

A number of previous approaches to probabilistic object detection have
utilised either a Monte Carlo (MC) Dropout technique [10] or Deep Ensembles technique [11] to extract both spatial and semantic uncertainty from an
object detector [4,12–16]. Other works have used generative models to estimate
spatial uncertainty [17] and various calibration techniques to calibrate semantic
uncertainty [5] for the application of autonomous driving.

© Springer Nature Switzerland AG 2020
A. Bartoli and A. Fusiello (Eds.): ECCV 2020 Workshops, LNCS 12540, pp. 46–55, 2020.
https://doi.org/10.1007/978-3-030-65414-6_5

In this paper, we describe the approach behind our submission to the 3rd Probabilistic Object Detection challenge. We utilise an 'Ensemble of Experts' technique, where we create an ensemble of high-performing object detectors [1–3] that provide detections with spatial and label uncertainty for the task of probabilistic object detection. We also utilise an adaptive gamma correction technique to preprocess the data for more robust performance in a range of lighting conditions.

2   Our Approach

In this section, we describe the approach behind our submission to the 3rd Probabilistic Object Detection Challenge. We took two main approaches, namely:

1. Adaptively gamma correcting images to provide a more consistent brightness across the challenge scenes.
2. Ensembling a number of 'expert' object detectors to utilise the best qualities of each and provide a meaningful spatial uncertainty.

2.1   Adaptive Gamma Correction of Data

The challenge dataset consists of a number of indoor scenes viewed from various camera heights and lighting conditions. Images with low levels of lighting posed a particular challenge for our object detectors, with fewer objects detected in total and lower label confidence for the objects that were correctly detected. To overcome this, we utilised a gamma correction technique, which had been successfully used by previous challenge submissions to brighten dark images in the dataset [14]. Gamma correction is a nonlinear function that adjusts the luminance values within an image:

$$V_{out} = AV_{in}^{\gamma}. \tag{1}$$

While a previous PrOD challenge submission [14] employed a hard threshold to detect dark images and a fixed γ value for correction, we instead implement an adaptive γ value. For each image in the dataset, we calculate the mean of the *value* channel (μ_v) of the HSV image, using this as an approximation of the mean luminance of the image. We can then specify a desired mean luminance of the image (d_v)), and use this to calculate the image-specific γ value required to adjust the existing mean luminance:

$$\gamma = \log_{\mu_v}(d_v) \tag{2}$$

Ideally, d_v should be equivalent to the mean of the value channel of the training dataset for any given detector. For this work, we simply used $d_v = 0.5$ and a gamma correction constant $A = 1$. We also limited the range of our adaptive γ to between 0.4 and 1, as a value below 0.4 tended to create grainy image quality and values above 1 would undesirably *decrease* the luminance of an image. Figure 1 shows a number of example images with our adaptive gamma correction applied.

$\gamma = 0.40$ $\qquad$ $\gamma = 0.40$ $\qquad$ $\gamma = 0.49$ $\qquad$ $\gamma = 0.74$ $\qquad$ $\gamma = 1.00$

Fig. 1. Examples of validation images before (top row) and after (bottom row) our adaptive gamma correction is applied.

2.2 Ensemble of Experts

Our approach ensembles a set of existing high-performing 'expert' object detectors, namely single-stage detector EfficientDet D7 [2] and multi-stage detectors DetectoRS [3] and CascadeRCNN [1]. Each of these detectors have their own strengths in object detection (as explored in Sect. 3.1), and by ensembling their detections, we hope to leverage their best qualities and extract meaningful spatial and label uncertainty [11]. This is a similar approach to a previous challenge submission [15], however, as explained below, our approach differs in that we use a ranked ensembling process and extract spatial uncertainty from clusters of associated detections.

For any given test image, each object detector provides a set of detections $D = [d_1, \ldots, d_n]$. For the following explanation, we represent each individual detection d_i as a dictionary, with a field that contains the bounding box coordinates $d_i.bbox = [x1, y1, x2, y2]$, a field with the label probabilities of the known classes $d_i.probs = [p_0, \ldots, p_{29}]$ and a field with the spatial uncertainty for the bounding box coordinates $d_i.covars$.

A key difficulty in any sampling-based uncertainty technique for object detection, including ensembles, is correctly associating and merging detections that represent the same object [18]. Previous work that addressed this difficulty found that a Basic Sequential Algorithmic Scheme (BSAS) clustering technique, with detection similarity measured by the intersection over union (IoU) between bounding boxes and a 'same label' semantic constraint, to be the optimal approach [18]. We follow this approach, with some adaptations, as described below.

While previous work [12,18] utilised only one variety of object detector, our ensemble consists of object detectors with different architectures and training procedures. As a result, the detectors can have significantly contrasting responses to an input image and achieve varied performance on the challenge. For this reason, we found the 'same label' semantic constraint recommended in [18] to be ineffective. In addition, we do not consider each detector equal during the

---

Algorithm 1: Ranked Expert Ensembling and Merging Function

Result: Final detections, D_{final}

$[D_1, D_2, \ldots, D_N]$, detections from N Expert Detectors, ordered by best performance;

$D_{final} = []$;

for $i \leftarrow 1$ *to* N **do**
    $D_i, [D_{i+1}, \ldots, D_N] = Merge(D_i, [D_{i+1}, \ldots, D_N])$;
    D_{final}.append(D_i);
end

Function Merge(A, $[B_1, \ldots, B_{end}]$)**:**
    for $a \in A$ **do**
        $a_{merge} = [a]$;
        for $B_i \in [B_1, \ldots, B_{end}]$ **do**
            for $b \in B_i$ **do**
                if $IoU(a.bbox, b.bbox) > \theta$ **then**
                    a_{merge}.append(b);
                    B_i.remove(b);
            end
        end
        $a.bbox =$ Mean($a_{merge}.bbox$);
        $a.probs =$ Mean($a_{merge}.probs$);
        $a.covars =$ Var($a_{merge}.bbox$)
    end
    return A, $[B_1, \ldots, B_{end}]$;

---

merging process, and introduce a ranked ensembling procedure to prioritise the detections of higher performing detectors.

Our ranked ensembling procedure and detection merging process are described in Algorithm 1. In brief, we initially rank our detectors from best to worst performing on the validation dataset for the challenge. Starting with the best performing detector, we update the detections with any correctly associated detections from all lower-performing detectors. Detections are correctly associated if they have bounding box IoU greater than a threshold θ. The original detection is updated by averaging with associated bounding boxes and label probabilities, and a spatial uncertainty is extracted by taking the variance of bounding box coordinates and forming a diagonal covariance matrix. This procedure is then repeated for the next best performing detector, and all its lower-performing detectors, where only detections that have not been previously associated and merged are considered. The ensembling process ends when the worst performing detector is reached, and its remaining unassociated detections are simply discarded.

2.3   Implementation Details

We use DetectoRS [3] with a ResNeXt-101-32x4d [19] backbone and Cascade RCNN [1] with a ResNeXt-101-64x4d [19] FPN [20] backbone. For each detector

Table 1. Challenge test dataset results for competing methods and our submission. Submissions with a * indicate submissions for the current 3rd iteration of the challenge, where details of the method have not yet been published. Best performance is shown in bold.

Method	PDQ	pPDQ	Spatial	Label	FPQ	TP	FP	FN
MCD MR-CNN [13]	14.927	0.472	0.455	0.673	0.349	115041	103690	181326
Sub.4*	15.959	0.455	0.480	0.561	**0.639**	138325	271642	158042
NSE [15]	20.019	**0.655**	**0.497**	**1.000**	0.000	118678	91735	177689
INHA-CVIP*	20.261	0.547	0.468	0.832	0.050	145579	101318	150788
Ours*	**22.848**	0.507	0.482	0.638	0.444	**151535**	**72342**	**144832**

we used pre-trained weights for COCO, but only considered the label probabilities for the 30 known classes present in the challenge. EfficientDet D7 [2] uses a sigmoid activation to provide the final label probabilities, rather than the standard softmax function, and as a result, does not produce a distribution over the labels as required by the challenge. To overcome this, we normalise the EfficientDet D7 label probabilities to sum to 1.

When computing the spatial uncertainty for each bounding box corner, we clamped the variance to be between 5 and 100. This was particularly important for providing a minimum spatial uncertainty for any detections that were not associated, and thus unable to compute a spatial uncertainty. A minimum spatial uncertainty (as opposed to *zero* spatial uncertainty was shown to be beneficial in previous competition submissions [14, 21].

An IoU threshold θ of 0.7 was used to associate detections, as recommended by previous studies on the merging process for sampling-based uncertainty techniques for object detection[18]. In practice, we found the worst-performing expert Cascade RCNN's label probabilities [1] to decrement the performance of our Expert Ensemble, and therefore did not include its label distribution when merging associated detections.

3 Results and Discussion

Our submission on the test dataset achieved first place in the 3rd Probabilistic Object Detection challenge, hosted at ECCV2020, with a PDQ of 22.848. The breakdown of this score, along with the scores of competitor submissions, are shown in Table 1. As shown, our submission had the best number of true positive, false positive and false negative counts, as well as the second best spatial quality and false positive quality.

3.1 Ablation Study

In this section, we quantify the individual performance and 'expertise' of each object detector, as well as how these qualities combine to affect our final 'Expert

Table 2. Results of each object detector alone and as an ensemble when tested on the gamma corrected validation data of the challenge. Best performance among the individual detectors is indicated in bold. Individual detectors are evaluated with a fixed covariance with diagonal elements set to 15 pixels.

Method	PDQ	pPDQ	Spatial	Label	FPQ	TP	FP	FN
EfficientDet D7	**23.32**	0.491	0.432	0.651	**0.541**	**46003**	14629	**44230**
DetectoRS	21.66	0.590	0.466	**0.848**	0.293	36107	11485	54126
CascadeRCNN	18.30	**0.601**	**0.497**	0.811	0.321	29329	**8989**	60905
Expert Ensemble	25.25	0.523	0.463	0.682	0.485	47741	16942	42492

Ensemble'. The validation dataset performance of each object detector alone and as an ensemble of experts is shown in Table 2. In order to give reasonable comparison results for the individual detectors, which had no method for obtaining spatial uncertainty, we used a fixed covariance with diagonal elements set to 15 pixels.

As shown in Table 2, each detector has its advantage in the PDQ metric. While Cascade RCNN [1] generates highly spatially accurate detections and few false positives, EfficientDet D7 [2] generates a much greater number of true detections and has less overconfident false positive detections. DetectoRS [3] appears to be a middle ground between these two detectors and obtains the greatest label quality. By ensembling these detectors, we are able to achieve a better PDQ than alone. Our final Expert Ensemble achieves the greatest true positive detection count and consistently substantial scores for each component of detection quality.

We now further examine the performance of each individual detector with respect to confidence calibration, object size and object class.

Confidence Calibration. Confidence calibration measures how well a model's confidence reflects the true probability it is correct. A perfectly calibrated network should produce a confidence score that exactly represents a true probability, e.g. detections with 80% confidence should be correct 80% of the time.

Reliability diagrams provide a visualisation of model calibration [22], where prediction confidences are sorted into interval bins and the overall accuracy of each bin is computed. We follow the approach in [23] and define the network's confidence as the maximum score in the label probability distribution. For object detection, we must additionally define what constitutes a 'positive' detection; while simpler classification tasks only require that the winning label matches the ground truth image label [23], object detection also must consider whether the detection is accurately localised to a ground truth object. We define 'positive' detections with respect to the PDQ metric [8]. This means that the winning label does not necessarily represent the ground truth object label, but that at least some probability has been placed on the ground truth label. The resulting reliability diagram for our detectors are shown in Fig. 2.

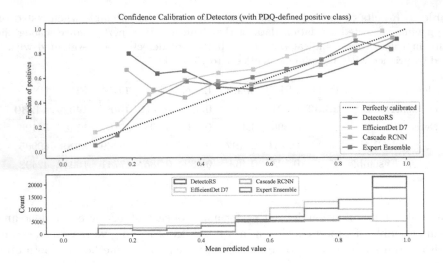

Fig. 2. Reliability diagram of the maximum label probability of detections. Detections were considered positive if they detected a ground truth object (as computed by the PDQ metric). Perfect calibration is represented by the dotted line, below this represents overconfidence and above represents underconfidence.

Alone, each of the expert object detectors have miscalibrated confidences. While EfficientDet D7 [2] is consistently underconfident, DetectoRS [3] and CascadeRCNN [1] present both underconfident detections at lower confidence levels and overconfident detections at higher confidence levels. When combining all detectors, our final Expert Ensemble achieves a more consistent calibration and improves upon the calibration of EfficientDet D7 (the best performing detector on the validation dataset).

Object Size. In Fig. 3, we compare our detector's overall detection rate (% of all instances correctly detected), spatial quality and label quality for detections of differently sized objects. We measure object size by their bounding box size in a given image.

As can be seen, each detector has its own performance strength; EfficientDet D7 [2] has the greatest detection rate, Cascade RCNN [1] has the greatest spatial quality and DetectoRS [3] has the greatest label quality. For each of their areas of strength, the detectors shows an ability to more consistently maintain performance across the object sizes, where the other detectors performance are more sensitive to object size. Interestingly, order of performance in each quality category is consistent, with the best detector for small objects also being the best for large objects. By combining the detectors, our expert ensemble is able to achieve the greatest detection rate across all object sizes ensemble and a spatial and label quality that is improved upon from EfficientDet D7 alone.

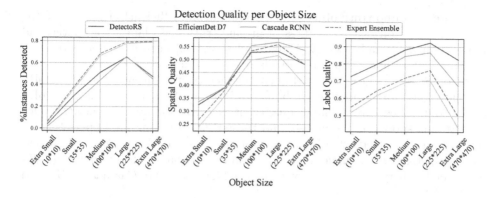

Fig. 3. Performance of each detector on various object sizes in terms of detection rate (%instances detected), detection spatial quality and detection label quality. The mean bounding box size for each object size category is also reported.

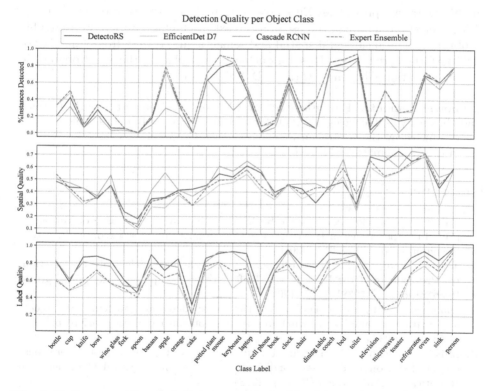

Fig. 4. Performance of each detector on various object classes in terms of detection rate (%instances detected), detection spatial quality and detection label quality.

Object Class. In Fig. 4, we compare our detector's overall detection rate (% of all instances correctly detected), spatial quality and label quality for detections of different ground truth object classes. Once again, EfficientDet D7 [2] has the greatest detection rate, regardless of ground truth class label. For spatial and label quality, Cascade RCNN [1] and DetectoRS [3] both exhibit greatly contrasting performances depending on class label. While earlier results showed that Cascade RCNN had the best overall spatial quality, it is surpassed by DetectoRS for the classes fork, spoon, cake, book, and toaster. Similarly, while DetectoRS achieved the best overall label quality, it is surpassed by CascadeRCNN for classes cup, spoon, apple, mouse, and toaster. Once again, our Expert Ensemble achieves a better detection rate than any detector alone across all classes. While our Expert Ensemble improves upon the spatial and label quality of EfficientDet D7, it typically does not reach the higher levels of Cascade RCNN and DetectoRS.

4   Conclusion

In this work, we present an Ensemble of Experts for the task of probabilistic object detection. Through the use of a ranked ensembling process, we were able to obtain a more meaningful spatial and semantic uncertainty from a number of high performing detectors. This work achieved first place in the 3rd Probabilistic Object Detection Challenge, with a PDQ of 22.848.

Acknowledgements. The authors acknowledge continued support from the Queensland University of Technology (QUT) through the Centre for Robotics. This research was conducted by the Australian Research Council Centre of Excellence for Robotic Vision (project number CE140100016).

References

1. Cai, Z., Vasconcelos, N.: Cascade R-CNN: delving into high quality object detection. In: Proceedings of the IEEE Conference on Computer Vision and Pattern Recognition, pp. 6154–6162 (2018)
2. Tan, M., Pang, R., Le, Q.V.: Efficientdet: scalable and efficient object detection. In: Proceedings of the IEEE/CVF Conference on Computer Vision and Pattern Recognition, pp. 10781–10790 (2020)
3. Qiao, S., Chen, L.C., Yuille, A.: Detectors: detecting objects with recursive feature pyramid and switchable atrous convolution (2020). arXiv preprint arXiv:2006.02334
4. Miller, D., Nicholson, L., Dayoub, F., Sünderhauf, N.: Dropout sampling for robust object detection in open-set conditions. In: IEEE International Conference on Robotics and Automation (ICRA) (2018)
5. Feng, D., Rosenbaum, L., Glaeser, C., Timm, F., Dietmayer, K.: Can we trust you? On calibration of a probabilistic object detector for autonomous driving (2019). arXiv preprint arXiv:1909.12358
6. Amodei, D., Olah, C., Steinhardt, J., Christiano, P., Schulman, J., Mané, D.: Concrete problems in AI safety (2016). arXiv preprint arXiv:1606.06565
7. Sünderhauf, N., et al.: The limits and potentials of deep learning for robotics. Int. J. Robot. Res. **37**(4–5), 405–420 (2018)

8. Hall, D., et al.: Probabilistic object detection: definition and evaluation. In: The IEEE Winter Conference on Applications of Computer Vision, pp. 1031–1040 (2020)
9. Lin, T.-Y., et al.: Microsoft COCO: common objects in context. In: Fleet, D., Pajdla, T., Schiele, B., Tuytelaars, T. (eds.) ECCV 2014. LNCS, vol. 8693, pp. 740–755. Springer, Cham (2014). https://doi.org/10.1007/978-3-319-10602-1_48
10. Gal, Y., Ghahramani, Z.: Dropout as a Bayesian approximation: representing model uncertainty in deep learning. In: International Conference on Machine Learning (ICML), pp. 1050–1059 (2016)
11. Lakshminarayanan, B., Pritzel, A., Blundell, C.: Simple and scalable predictive uncertainty estimation using deep ensembles. In: Advances in Neural Information Processing Systems, pp. 6393–6395 (2017)
12. Miller, D., Sünderhauf, N., Zhang, H., Hall, D., Dayoub, F.: Benchmarking sampling-based probabilistic object detectors. In: CVPR Workshops, Vol. 3 (2019)
13. Morrison, D., Milan, A., Antonakos, E.: Uncertainty-aware instance segmentation using dropout sampling (2019)
14. Wang, C.W., Cheng, C.A., Cheng, C.J., Hu, H.N., Chu, H.K., Sun, M.: AugPOD: augmentation-oriented probabilistic object detection. In: CVPR Workshop on the Robotic Vision Probabilistic Object Detection Challenge (2019)
15. Li, D., Xu, C., Liu, Y., Qin, Z.: TeamGL at ACRV robotic vision challenge 1: probabilistic object detection via staged non-suppression ensembling. In: CVPR Workshop on the Robotic Vision Probabilistic Object Detection Challenge (2019)
16. Harakeh, A., Smart, M., Waslander, S.L.: Bayesod: A Bayesian approach for uncertainty estimation in deep object detectors (2019). arXiv preprint arXiv:1903.03838
17. Wang, Z., Feng, D., Zhou, Y., Zhan, W., Rosenbaum, L., Timm, F., Dietmayer, K., Tomizuka, M.: Inferring spatial uncertainty in object detection (2020). arXiv preprint arXiv:2003.03644
18. Miller, D., Dayoub, F., Milford, M., Sünderhauf, N.: Evaluating merging strategies for sampling-based uncertainty techniques in object detection. In: 2019 International Conference on Robotics and Automation (ICRA), pp. 2348–2354. IEEE (2019)
19. Xie, S., Girshick, R., Dollár, P., Tu, Z., He, K.: Aggregated residual transformations for deep neural networks (2016). arXiv preprint arXiv:1611.05431
20. Lin, T.Y., Dollár, P., Girshick, R., He, K., Hariharan, B., Belongie, S.: Feature pyramid networks for object detection. In: Proceedings of the IEEE Conference on Computer Vision and Pattern Recognition, pp. 2117–2125 (2017)
21. Ammirato, P., Berg, A.C.: A mask-RCNN baseline for probabilistic object detection (2019). arXiv preprint arXiv:1908.03621
22. DeGroot, M.H., Fienberg, S.E.: The comparison and evaluation of forecasters. J. R. Stat. Soc. Ser. D (The Statistician) 32(1–2), 12–22 (1983)
23. Guo, C., Pleiss, G., Sun, Y., Weinberger, K.Q.: On calibration of modern neural networks (2017). arXiv preprint arXiv:1706.04599

EPrOD: Evolved Probabilistic Object Detector with Diverse Samples

Jaewoong Choi[1], Sungwook Lee[2], Seunghyun Lee[2],
and Byung Cheol Song[1,2(✉)]

[1] Department of Future Vehicle Engineering, Inha University,
Incheon 22212, South Korea
chlwodnd500@naver.com
[2] Department of Electrical and Computer Engineering, Inha University,
Incheon 22212, South Korea
lsw2646@gmail.com, lsh910703@gmail.com, bcsong@inha.ac.kr

Abstract. Even small errors in object detection algorithms can lead
to serious accidents in critical fields such as factories and autonomous
vehicles. Thus, a so-called probabilistic object detector (PrOD) has been
proposed. However, the PrOD still has problems of underestimating the
uncertainty of results and causing biased approximation due to limited
information. To solve the above-mentioned problems, this paper proposes
a new PrOD composed of four key techniques, i.e., *albedo extraction,
soft-DropBlock, stacked NMS, and inter-frame processing*. In terms of
Probability-based Detection Quality (PDQ) metric, the proposed object
detector achieved 22.46, which is 4.46 higher than a backbone method,
for the Australian Centre for Robotic Vision (ACRV) validation dataset
consisting of four videos.

Keywords: Probabilistic object detector · Albedo extraction · Monte
Carlo method · Non-maximum suppression

1 Introduction

With the recent rapid development of convolutional neural networks (CNNs),
CNN-based object detectors [5, 8, 11] have been applied to various fields such as
industry, military, and home. In conventional object detectors, the higher uncertainty, the greater risk of misjudgement because the results are overconfident.
In order to prevent this risk in advance, probabilistic object detectors (PrOD)
that take into account the uncertainty of results have appeared [2, 6, 9, 10]. PrOD
excludes results with high uncertainty, so it can prevent the risk of misjudgement beforehand. Particularly, by applying the Monte Carlo method to existing
object detectors [6, 9, 10], PrODs could consider uncertainty while ensuring high
performance. Here, the Monte Carlo method allows object detectors to obtain

---

J. Choi and S. Lee—Equal contribution.

© Springer Nature Switzerland AG 2020
A. Bartoli and A. Fusiello (Eds.): ECCV 2020 Workshops, LNCS 12540, pp. 56–66, 2020.
https://doi.org/10.1007/978-3-030-65414-6_6

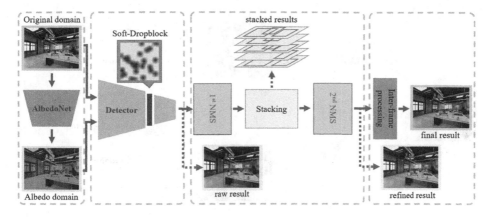

Fig. 1. The overall framework of the proposed EPrOD. It consists of four major blocks: Albedo extraction, the object detector with soft-DropBlock applied, the stacked NMS, and the inter-frame processing.

approximation and uncertainty simultaneously by using results from several samples. However, the Monte Carlo method-based PrOD still underestimates the uncertainty of results, and uses too limited information to calculate the approximation.

To overcome this shortcoming, we propose an evolved PrOD (EPrOD) which is composed of the following four key techniques: First, the albedo extraction produces albedo images through AlbedoNet, which allows EPrOD to detect objects in a multi-domain. Second, the soft-DropBlock removes the local information of feature maps, and takes diverse samples. The above two techniques make EPrOD prevent biased approximation. Third, the stacked non-maximum suppression (NMS) that is two-step post-processing with additional constraints mitigates uncertainty-underestimated results. Fourth, in order to reduce false positives in albedo images, the inter-frame processing removes the bounding boxes that are not common between the current frame and adjacent frames. Since the albedo extraction and inter-frame processing are highly dependent on each other, this paper combines them and calls as albedo processing. Note that the four techniques positively affect the proposed EProD without additional training. The experimental results demonstrate that the EProD noticeably improves the PDQ score [6].

To summarize, this paper has three contribution points as follows:

- The albedo processing overcomes the limitation of single-domain detection by providing multi-domain images that induce diverse detection results.
- The soft-DropBlock effectively removes the local information from feature maps to collect diverse samples.
- A novel NMS called 'stacked NMS', which imposes additional restrictions, suppresses uncertainty-underestimated results while leaving reliable results.

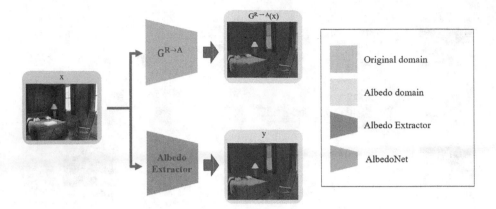

Fig. 2. The training procedure of AlbedoNet. To obtain a paired dataset, the albedo extractor uses [1] and $\mathbf{G}^{R\to A}$ uses U-Net [12]. A refers to the albedo domain, and R refers to the original image domain.

2   Method

This section describes the overall operation of EPrOD and explains four key techniques of EPrOD in detail. As a backbone of EPrOD, we adopt a well-known one-stage object detector YOLOv3 with Adaptively Spatial Feature Fusion (ASFF) [7]. Figure 1 describes the EPrOD framework. In the albedo extraction stage, AlbedoNet extracts the albedo from an input image. The extracted albedo image and the corresponding original image are input to the subsequent object detector. Next, an object detector with soft-DropBlock applied to feature maps takes some samples for the Monte Carlo method. The detected samples are stacked and refined by the following two stages: the stacked NMS and the inter-frame processing. The details of each stage is described as follows.

2.1   Albedo Extraction

Conventional PrOD [10], which calculates approximation only in a single domain image, carries the risk producing biased results caused by a light source. To overcome this problem, we propose to minimize the influence of the light source by additionally using an albedo image with shading information is removed from the original image. Bell et al. first proposed a method to extract albedo images [1]. However, Bell's method has a disadvantage that it takes a long time due to the procedure of repeating the pixel-wise operation. To shorten the albedo extraction time, we propose AlbedoNet that converts an input image to the albedo domain. AlbedoNet is designed based on U-Net [12], and it is trained with a paired dataset which is composed of MS COCO dataset and the albedo images extracted from MS COCO. The training procedure is shown in Fig. 2. The objective function for training AlbedoNet is shown in Eq. (1).

$$\mathcal{L}_1 = \mathbb{E}[\|G^{R\to A}(x) - y\|_1] \tag{1}$$

Fig. 3. A pair of original and albedo images. The first row shows originals and the second row indicates the corresponding albedos.

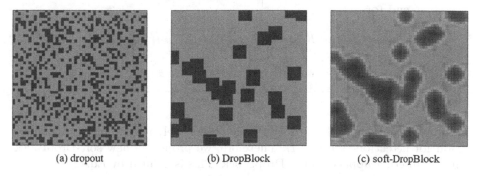

(a) dropout   (b) DropBlock   (c) soft-DropBlock

Fig. 4. Examples of each drop method. (a), (b), and (c) are all drop masks obtained using a drop rate of 0.3. From left to right, they are (a) dropout, (b) DropBlock, and (c) soft-DropBlock.

where $G^{R \to A}$ is AlbedoNet using U-Net [12]. x and y indicate the original and the albedo image, respectively. As shown in Fig. 3, the albedo extracted by AlbedoNet is no longer affected by the light source because shading is removed. As a result, AlbedoNet additionally produces potential objects in a multi-domain, hence the approximation can be derived from diverse detection results.

2.2 Soft-DropBlock

The Monte Carlo method-based PrOD applies dropout [13] to some feature maps and infers several times to obtain random samples [3,9,10]. However, feature maps with high inter-pixel correlation are hardly affected by dropout, which hinders sample diversity. We experimentally observed some examples where dropout did not effectively remove information from feature maps (see Table 3). On the other hand, DropBlock which can effectively remove the local information of feature maps through a block-wise approach was proposed [4]. However, the dis-

---

Algorithm 1. Procedure of stacked NMS

Data: Dataset D, Object detector F, IOU threshold τ_1, τ_2, Grouped boxes threshold β, number of inferences I

  for all frame $d \in D$ **do**

    $R, O \leftarrow \emptyset$ //final results and observations of each inference

    for all $i < I$ **do**

      $B_i = F(d)$

      $O = O \cup NMS(B_i, \tau_1)$

    end for

    $selected_{all} \leftarrow False$

    for all $o_l \in O$ **do**

      $G_l \leftarrow \emptyset$ //groups containing closely located elements

      if o_l has no group **then**

        for all arbitrary o_x without group **do**

          $G_l \leftarrow$ put o_x into G_l where $IoU(o_l, o_x) \geq \tau_2$

        end for

      end if

      put $mean(G_l)$ **into** R **where** $n(G_l) \geq \beta$

    end for

  end for

  return R

---

continuous boundary of the DropBlock may excessively remove important information for detection (see Fig. 4). Therefore, we propose a new soft-DropBlock with box filtering applied to the DropBlock. This is defined by Eq. (2).

$$\mathbf{M}_{SDB}(u,v) = \frac{1}{k^2} \sum_{i=u-\frac{k}{2}}^{u+\frac{k}{2}} \sum_{j=v-\frac{k}{2}}^{v+\frac{k}{2}} \mathbf{M}_{DB}(i,j) \qquad (2)$$

where $\mathbf{M}$ means drop mask, and u and v represent coordinates in the drop mask. SDB and DB indicate soft-DropBlock and DropBlock, respectively. k means the block size to drop. The appearance of each drop mask is seen in Fig. 4. We can observe that the soft-DropBlock produces smooth boundary that can reduce the loss of information for detection while removing the local information.

2.3 Stacked NMS

Conventional NMS, which refines detection results, extracts the final result under an assumption that all detection results are reliable. In general, inaccurate object candidates increase the risk of misjudgment. If object detection is performed with some information removed, the bounding box is generally loosened, i.e., inaccurate bounding boxes are produced. So, if conventional NMS is applied to these inaccurate boxes, the accuracy can be severely degraded because stacking and overlapping inaccurate bounding boxes makes refining complicated. To solve this problem, the stacked NMS refines bounding boxes through a two-step

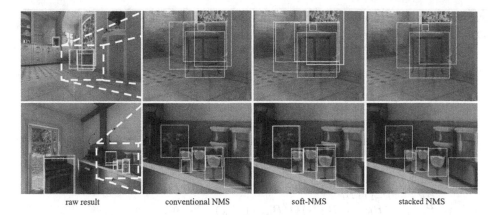

raw result   conventional NMS   soft-NMS   stacked NMS

Fig. 5. Column 1 is the stacked detection result. Column 2 is the result of the conventional NMS. Several boxes are assigned to an object. Column 3 is the result of soft-NMS. Note that the result is worse than the conventional NMS. Column 4 is the proposed stacked NMS. Here, each object is assigned a box.

Fig. 6. Detected objects from original and albedo images. We can find that more TPs are obtained from the albedo image.

post-processing, and then calculates the frequency for bounding boxes to decide whether to save. The pseudo-code of stacked NMS is described in Algorithm 1.

The proposed stacked NMS has two advantages. First, the entire complexity is alleviated by performing a preliminary refinement before inaccurate bounding boxes are overlapped. Second, bounding boxes with underestimated uncertainty can be further removed based on the detection frequency. The results of each NMS can be seen in Fig. 5.

2.4   Inter-frame Processing

We found that when albedo images are used with existing images, FPs as well as TPs increase (see Table 2). In addition, it was confirmed that most of the FPs detected in the albedo images were not related to adjacent frames. Therefore, to improve detection performance in multi-domain by reducing FPs detected in albedo images, we propose the inter-frame processing. The procedure of inter-frame processing is shown in Algorithm 2.

Figure 6 compares the results according to whether albedo processing is used or not. By applying albedo processing, additional TPs could be secured

---

Algorithm 2. Procedure of inter-frame processing

Data: Original dataset D_{ori}, albedo of dataset D_{albedo}
   object detector F, AlbedoNet A, number of frames K

   $D_{albedo} = A(D_{ori})$ //Translate D_{ori} to D_{albedo} using AlbedoNet A
   for all frame $k \in K$ **do**
      $B_k = F(D_{ori}^k, D_{albedo}^k)$ //Class list of detection results are stored in B_k
      if k is 1 **then**
         pass
      else if k is 2 **then**
         $R_1 = B_1 \cap B_2$ //Apply class list intersection to R
      else if k is K **then**
         $R_K = B_{K-1} \cap B_K$
         $R_{K-1} = (B_{K-2} \cap B_{K-1}) \cup (B_{K-1} \cap B_K)$
      else
         $R_{k-1} = (B_{k-2} \cap B_{k-1}) \cup (B_{k-1} \cap B_k)$
      end if
   end for
   return R

---

through multi-domain images. Unintended FPs generated by the side-effect can be removed by inter-frame processing, reducing the possibility of misjudgment.

3 Experiments

This section shows several experimental results to verify the performance of the proposed method. First, we analyze the effect of each component technique, and evaluate the performance of EPrOD to which all these techniques are applied. The quantitative evaluation metric is PDQ [6] score. For fair comparison, the hyper-parameters used in PrOD were fixed. In order to evaluate the performance without additional learning, weights pre-trained with the MS COCO dataset were applied to the object detector [7]. As the evaluation dataset, the ACRV validation dataset consisting of four videos was used.

To measure the PDQ score, the covariance matrix of two corner coordinates of the predicted box representing the uncertainty and the label score of the predicted class corresponding to the box are required. However, in comparing the proposed method based on multiple inferences and the baseline based on single inference, the above-mentioned two factors are greatly affected by the number of inferences. This makes it difficult to grasp the effectiveness of the proposed method. To mitigate these negative effects, we present a way that is independent of the number of inferences. First, the covariance matrix for quantifying the spatial uncertainty is defined using 20% of the height and width of the box as in Eq. (3). And, in the case of a label score containing the uncertainty of the predicted class, the probability of all the predicted classes per box was squared so that the class with the highest probability stood out. Both the proposed method and the baseline are subject to similar penalty due to the uncertainty of the predicted results. Eventually, the difference caused by the number of inferences can be resolved.

$$\Sigma = \begin{bmatrix} 0.2 \cdot width & 0 \\ 0 & 0.2 \cdot height \end{bmatrix} \tag{3}$$

3.1 Effect of Each Technique on the Overall Performance

This section analyzes the effect of each technique on the overall performance in terms of PDQ. Baseline is YOLOv3 with ASFF. Table 1 shows that the PDQ score improves when each technique is applied.

Table 1. The effect of each technique on the entire performance. Bold means the best score.

Albedo processing	Soft-DropBlock	Stacked NMS	PDQ
			17.909
✓			19.788
	✓		17.455
✓	✓		19.566
	✓	✓	21.885
✓	✓	✓	**22.466**

Fig. 7. Comparison of detection results before and after the application of the proposed method. The top row represents the baseline and the bottom row represents the proposed method. Note that the proposed method rarely generates FPs even though it detects more objects.

Looking at the third row, if only the soft-DropBlock is applied, the PDQ score is lower than before application. However, seeing the fifth row, when the soft-DropBlock is used with the stacked NMS, FPs whose uncertainty is underestimated are effectively removed, resulting in improved performance. A detailed analysis of this is shown in Sects. 3.3 and 3.4. When all the techniques are applied, we can observe that the PDQ score of 4.56 is finally improved (see the last row). Figure 7 shows the qualitative results. We can see that the proposed method detects more objects than the baseline with less misjudgements.

3.2 Albedo Processing

In this experiment, the effect of albedo processing is verified. The second row of Table 2 gives the result of using the albedo image obtained by AlbedoNet. We can find that the additional use of albedo images secures more TPs, and eventually improves the PDQ score. However, due to the domain difference between the original image and the albedo image, a large number of FPs occurred, so the score was not so much increased as expected. However, note that when albedo extraction and inter-frame processing are used together, the number of FPs are reduced significantly and the PDQ score is improved as in the third row.

Table 2. Quantitative evaluation of albedo processing. Bold means the best score.

Albedo extraction	Inter-frame processing	PDQ	Avg_pPDQ	TPs	FPs	FNs
		17.909	0.553	36,329	24,014	53,904
✓		19.478	0.494	52,044	44,208	38,189
✓	✓	**19.788**	0.509	49,600	38,712	40,633

3.3 Soft-DropBlock

In this experiment, the detection results for several drop masks are analyzed, and the effectiveness of the soft-DropBlock is verified through the following metric. The drop rate for each method was set to 0.3. The block sizes of dropout and DropBlock were set to 1 and 5, respectively. Drop rate and block size were determined experimentally. For qualitative performance comparison between drop masks, we propose the dropping quality score γ as shown in Eq. (4).

$$\gamma = \frac{\text{Before TPs}}{\text{After TPs}} \cdot \frac{\text{TPs reductions}}{\text{FPs reductions}} \tag{4}$$

After dropping, the better the TPs are maintained or the FPs are removed at a higher rate than the TPs, the lower the dropping quality score is. The experiment results according to the drop mask are shown in Table 3. The means and standard deviations were calculated through a total of five experiments. It can be seen that soft-DropBlock dominates in terms of dropping quality score. In other words, soft-DropBlock effectively removes feature map information than dropout, while preserving information necessary for detection better than DropBlock. In addition, standard deviation is additionally shown to evaluate the diversity of samples. Soft-DropBlock has a higher standard deviation than dropout. This indicates more diverse samples were used.

Table 3. Quantitative comparison of dropping strategies

Strategy	PDQ	Avg_pPDQ	TPs(std.)	FPs(std.)	FNs	γ
Default	17.909	0.553	36,329	24,014	53,904	-
Dropout	17.573	0.562	35,343(36.36)	23,650(86.37)	54,890	2.784
DropBlock	16.466	0.561	32,430(92.35)	20,910(144.93)	57,803	1.407
Soft-DropBlock	17.455	0.564	34,594(80.36)	22,287(105.90)	55,640	1.054

3.4 Stacked NMS

In this section, we examine the extent to which FPs are erased according to NMS. Here, as samples for the Monte Carlo method, the results obtained by testing 12 times are employed. The conventional NMS used the IoU threshold as 0.55, and the stacked NMS used 0.45 and 0.55 for different IoU thresholds, respectively. In order to eliminate the infrequent result, the threshold for the number of detection boxes used was experimentally set to 5. Table 4 shows the performance when different NMSs are applied to each drop mask. It was confirmed that the detection of FPs increased because conventional NMS accepted even boxes with high uncertainty as final results. On the other hand, the stacked NMS significantly eliminated FPs by removing the less frequent detection boxes from each sample, contributing to the increase of the PDQ score. Also, we can find that the γ and the standard deviation of each drop mask affect the PrOD result.

Table 4. Quantitative comparison of NMS strategies

Strategy	Drop mask	PDQ	Avg_pPDQ	TPs	FPs	FNs
Conventional NMS	Dropout	19.653	0.534	52,722	55,468	37,511
	DropBlock	19.565	0.553	51,664	53,018	38,569
	Soft-DropBlock	20.432	0.539	50,674	45,608	39,559
Stacked NMS	Dropout	21.870	0.547	46,185	26,635	44,048
	DropBlock	21.683	0.555	43,783	22,911	46,450
	Soft-DropBlock	21.885	0.547	46,193	26,459	44,040

4   Conclusion

PrOD is a crucial method to solve the problem that the object detector over-confidences the detection result. This paper proposes a new PrOD to solve this problem. The proposed method improves the performance of PrOD without additional learning, and the PDQ score is 4.56 higher than the previous method. It was experimentally confirmed that TPs were increased by the proposed method, and FPs were effectively suppressed. However, PrOD using the Monte Carlo method still has a problem that depends on randomness. In future research, mathematical and statistical analysis is required.

Acknowledgement. This work was supported by the Industrial Technology Innovation Program funded By the Ministry of Trade, industry & Energy (MI, Korea) [20006483, Development on Deep Learning based 4K30P Edge Computing enabled IP Camera System], and Industrial Technology Innovation Program through the Ministry of Trade, Industry, and Energy (MI, Korea) [Development of Human-Friendly Human-Robot Interaction Technologies Using Human Internal Emotional States] under Grant 10073154.

References

1. Bell, S., Bala, K., Snavely, N.: Intrinsic images in the wild. ACM Trans. Graph. (TOG) **33**(4), 1–12 (2014)
2. Choi, J., Chun, D., Kim, H., Lee, H.J.: Gaussian yolov3: an accurate and fast object detector using localization uncertainty for autonomous driving. In: Proceedings of the IEEE International Conference on Computer Vision, pp. 502–511 (2019)
3. Gal, Y., Ghahramani, Z.: Dropout as a Bayesian approximation: representing model uncertainty in deep learning. In: International Conference on Machine Learning, pp. 1050–1059 (2016)
4. Ghiasi, G., Lin, T.Y., Le, Q.V.: Dropblock: a regularization method for convolutional networks. In: Advances in Neural Information Processing Systems, pp. 10727–10737 (2018)
5. Girshick, R., Donahue, J., Darrell, T., Malik, J.: Rich feature hierarchies for accurate object detection and semantic segmentation. In: Proceedings of the IEEE Conference on Computer Vision and Pattern Recognition, pp. 580–587 (2014)
6. Hall, D., et al.: Probabilistic object detection: definition and evaluation. In: The IEEE Winter Conference on Applications of Computer Vision, pp. 1031–1040 (2020)
7. Liu, S., Huang, D., Wang, Y.: Learning spatial fusion for single-shot object detection (2019). arXiv preprint arXiv:1911.09516
8. Liu, W., et al.: SSD: single shot multibox detector. In: Leibe, B., Matas, J., Sebe, N., Welling, M. (eds.) ECCV 2016. LNCS, vol. 9905, pp. 21–37. Springer, Cham (2016). https://doi.org/10.1007/978-3-319-46448-0_2
9. Miller, D., Dayoub, F., Milford, M., Sünderhauf, N.: Evaluating merging strategies for sampling-based uncertainty techniques in object detection. In: 2019 International Conference on Robotics and Automation (ICRA), pp. 2348–2354. IEEE (2019)
10. Miller, D., Nicholson, L., Dayoub, F., Sünderhauf, N.: Dropout sampling for robust object detection in open-set conditions. In: 2018 IEEE International Conference on Robotics and Automation (ICRA), pp. 1–7. IEEE (2018)
11. Redmon, J., Divvala, S., Girshick, R., Farhadi, A.: You only look once: unified, real-time object detection. In: Proceedings of the IEEE Conference on Computer Vision and Pattern Recognition, pp. 779–788 (2016)
12. Ronneberger, O., Fischer, P., Brox, T.: U-Net: convolutional networks for biomedical image segmentation. In: Navab, N., Hornegger, J., Wells, W.M., Frangi, A.F. (eds.) MICCAI 2015. LNCS, vol. 9351, pp. 234–241. Springer, Cham (2015). https://doi.org/10.1007/978-3-319-24574-4_28
13. Srivastava, N., Hinton, G., Krizhevsky, A., Sutskever, I., Salakhutdinov, R.: Dropout: a simple way to prevent neural networks from overfitting. J. Mach. Learn. Res. **15**(1), 1929–1958 (2014)

Probabilistic Object Detection via Deep Ensembles

Zongyao Lyu, Nolan Gutierrez, Aditya Rajguru,
and William J. Beksi(✉)

University of Texas at Arlington, Arlington, TX, USA
{zongyao.lyu,nolan.gutierrez,aditya.rajguru}@mavs.uta.edu,
william.beksi@uta.edu

Abstract. Probabilistic object detection is the task of detecting objects in images and accurately quantifying the spatial and semantic uncertainties of the detection. Measuring uncertainty is important in robotic applications where actions based on erroneous, but high confidence visual detections, can lead to catastrophic consequences. We introduce an approach that employs deep ensembles for estimating predictive uncertainty. The proposed framework achieved 4th place in the ECCV 2020 ACRV Robotic Vision Challenge on Probabilistic Object Detection.

Keywords: Object detection · Uncertainty estimation · Deep ensembles

1 Introduction

Object detection is a fundamental task in computer vision and an essential capability for many real-world robotic applications. These applications allow robots to localize and recognize objects in the environment in which they operate. Models for object detection have achieved high mean average precision (mAP) on datasets such as ImageNet [17], the PASCAL Visual Object Classes [6,7], and the Microsoft Common Objects in Context (COCO) [13]. However, these models may fail in real-world scenarios. The mAP metric encourages detectors to output many detections for each image, yet it does not provide a consistent measure of confidence other than a higher score indicating that the system is more confident. Nonetheless, robotic vision systems must be able to cope with diverse operating conditions such as variations in illumination and occlusions in the surrounding environment.

The ACRV Robotic Vision Challenge on Probabilistic Object Detection (PrOD) [18] introduces a variation on traditional object detection tasks by requiring estimates of spatial and semantic uncertainty. This is done through a new evaluation metric called the probabilistic-based detection quality (PDQ) [9]. Compared to mAP, the PDQ jointly evaluates the quality of spatial and label uncertainty, foreground and background separation, and the number of true positive (correct), false positive (spurious), and false negative (missed) detections.

© Springer Nature Switzerland AG 2020
A. Bartoli and A. Fusiello (Eds.): ECCV 2020 Workshops, LNCS 12540, pp. 67–75, 2020.
https://doi.org/10.1007/978-3-030-65414-6_7

Additionally, the PrOD challenge provides video data consisting of a variety of scenes that has been generated from high-fidelity simulations.

In this work, we explore the use of deep ensembles to address the issue of uncertainty estimation. We apply this technique jointly on several detection models including Cascade R-CNN [2], RetinaNet [12], Grid R-CNN [14], and Hybrid Task Cascade (HTC) [3], to produce probabilistic bounding boxes. Our system attained a PDQ score of 15.959 in the ECCV 2020 PrOD challenge.

The remainder of this paper is organized as follows. We describe our PrOD approach in Sect. 2. In Sect. 3, we report our results on the PrOD challenge. We conclude in Sect. 4 with a discussion of potential future work.

2   Method Description

In this section, we describe the key components of our system for the PrOD challenge. We begin by describing the data preprocessing and augmentation steps in Sect. 2.1. Next, we introduce the detection models that we used for the challenge in Sect. 2.2. Then, we explain how our system utilizes deep ensembles (Sect. 2.3) for uncertainty estimation followed by a discussion of the heuristics we applied to improve the PDQ scores (Sect. 2.4). Finally, we report how the predictions are averaged over an ensemble of deep neural networks in Sect. 3. An outline of our system is shown in Fig. 1.

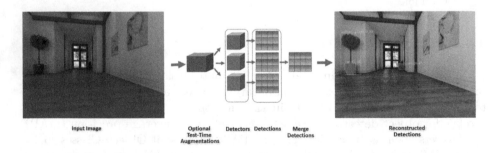

Input Image   Optional   Detectors Detections   Merge   Reconstructed
                Test-Time   Detections   Detections
                Augmentations

Fig. 1. An overview of our deep ensemble architecture for probabilistic object detection.

2.1   Data Preprocessing

The testing set of the PrOD challenge is a subset of the COCO dataset [13]. COCO consists of 80 distinct classes of objects and we use the dataset as our training data. We first filter out the 50 classes in COCO that are not included in the challenge. The test and validation sets of the challenge come from video sequences captured from simulation. The videos span multiple environments including daytime and nighttime, and miscellaneous camera heights. Object detection is problematic in these videos due to variations in brightness levels, especially from images captured at night.

To help alleviate this problem, we apply several data augmentation techniques including random gamma adjustments, changing the brightness and contrast values, and channel shuffle. We also applied augmentation techniques during testing to evaluate the effectiveness of adaptive gamma correction (AGC) [20]. This preprocessing step heuristically amends images that are either too dark or too bright. However, we found that this increases the number of false positives and decreases the average label quality resulting in lower PDQ scores. Examples of raw input images and preprocessed augmented images are shown in Fig. 2.

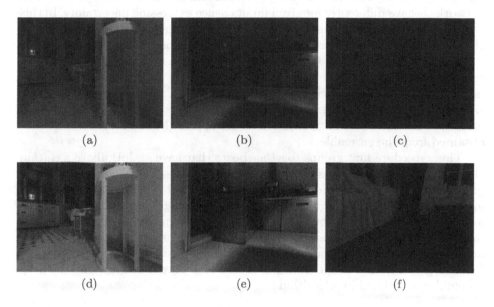

(a)   (b)   (c)

(d)   (e)   (f)

Fig. 2. Examples of raw input images (a–c) and their corresponding augmented images (d–f) at testing time.

2.2   Detection Models

For the first stage detection task, we make use of four detection models proposed in recent years that have proven to perform well at object detection tasks. The first model is Cascade R-CNN [2], a multistage detection architecture. It consists of a sequence of detectors trained with increasing intersection over union (IoU) thresholds. The detectors are trained stage by stage, leveraging the observation that the output of a detector is a good distribution for training the next higher-quality detector. The second model is RetinaNet [12], a fully convolutional one-stage detector that uses focal loss. Focal loss applies a modulating term to the cross entropy loss in order to direct learning on hard negative examples. The third model is Grid R-CNN [14] which replaces the traditional box offset regression strategy in object detection by a grid-guided mechanism for high-quality localization. The fourth model is Hybrid Task Cascade (HTC) [3], a new cascade architecture for instance segmentation and object detection.

2.3 Deep Ensembles

Traditionally, Bayesian neural networks (BNNs) [5,15] have provided a mathematically grounded foundation to measure model uncertainty. However, BNNs come with a prohibitive computational cost. Lakshminarayanan et al. [10] proposed a non-Bayesian solution as an alternative to BNNs by training an ensemble of several networks independently with random parameter initialization. The networks behave differently for given inputs hence expressing uncertainty. In this sense, dropout [19] can be interpreted as an ensemble model combination where the samples (predictions) are averaged over a collection of neural networks.

Our approach to averaging the predictions from an ensemble of object detectors consists of merging the bounding boxes using an approach similar to non-maximal suppression (NMS) [11]. This method was inspired by the dropout sampling proposed by Miller et al. [16] in which image detections are grouped into observations. To average predictions, we applied NMS to a list of all boxes obtained from the ensemble.

This procedure first groups together boxes that have an IoU above a certain threshold and each group is sorted by its confidence score. Next, each of these groups is assigned the label of the highest scoring label. Then, the boxes for each group are reduced to a single box by simple averaging to obtain a list of boxes and labels. Lastly, for each bounding box and label from the list of the ensemble's merged detections, the covariance matrices of both the bounding box's top left and bottom right corners are computed as if the list of bounding boxes and labels resulted from a single neural network. This process for merging the deep ensembles is outlined in Algorithm 1.

2.4 Heuristics for Improving PDQ Scores

In this challenge there are certain metrics which can greatly affect the PDQ score. For example, as Ammirato and Berg [1] have shown, higher PDQ scores may be obtained by reducing the number of false negatives produced after comparing the object detector's predictions with the ground truth. Additionally, last year's competitors found that setting the label scores of all detections to one can increase the average label quality of the detections and result in an inflated PDQ score. In the following paragraphs we describe the implementation details of each heuristic employed to improve our PDQ score.

Threshold Filtering: The most significant heuristic that can be used once the bounding boxes are computed is removing boxes with confidence scores below a threshold. This heuristic produces a wide range of PDQ scores that largely affects the number of true positives, false positives, and false negatives. We empirically found that a successful threshold for HTC was 0.35.

Algorithm 1. Merging of Deep Ensembles

```
1: procedure MERGE_DETECTIONS
2:     I = Input image
3:     Matches = Bounding box clusters
4:     D = d₁, d₂, . . . , dₖ                                    ▷ dᵢ: iᵗʰ detector
5:     IoUThreshold = [0, 1]
6:     BBoxes = empty_list()
7:     Labels = empty_list()
8:     for each detector in D do
9:         detection = detector(I)
10:        bboxes, labels = process_detection(detection)
11:        BBoxes = concatenate(BBoxes, bboxes)
12:        Labels = concatenate(Labels, labels)
13:    Matches = nms_match(BBoxes, IoUThreshold)
14:    boxes = empty_list()                          ▷ List of merged bounding boxes
15:    labels = empty_list()                   ▷ List of labels for merged bounding boxes
16:    for each match in Matches do
17:        newBBoxes = empty_list()
18:        for id in match do
19:            newBBoxes.append(BBoxes[id])
20:        bboxes.append(mean(newBBoxes))                         ▷ Reduce along axis 0
21:        labels.append(Labels[match[0]])         ▷ match[0] has highest probability
       return bboxes, labels
```

Bounding Box Reduction: As noted by Ammirato and Berg [1], the PDQ score is substantially reduced by labeling background pixels with high probabilities. To counter this effect, we first reduce the width and height of the bounding box by a certain percent while maintaining the center of the bounding box. Reducing the size of the bounding box results in boxes more tightly enclosing objects in the foreground and thus increases the PDQ score. We gained better results with a reduction ratio of 0.15 for each bounding box's height and width.

Improving False Positive Quality: If a detection is incorrect and the correct label is assigned a low probability, then we can expect the quality of the false positives to be reduced. We developed a heuristic which allowed us to obtain the highest false positive label quality in the challenge. To do this, we set the other scores in each bounding box's score vector to $(1 - S)/30$ where S is the confidence of the highest scoring label. This evenly distributes the labels across the rest of the classes in the score vector resulting in higher average false positive label qualities.

Covariance Matrix Calculation: Although we experimented with fixed covariances and traditional covariance calculations, we improved our results with a heuristic that simply sets the elements of the covariance matrix proportional to the bounding box's width and height [1]. For example, if the bounding box's width and height are 40 and 50 with a proportion of 0.25, then we set the x variance to 10 and the y variance to 12.5. This results in a covariance matrix of $[[10, 0], [0, 12.5]]$ for both the top left and bottom right corners. We empirically attained a higher PDQ score with a proportion of 0.2.

3   Results

In this section, we present our results from participating in the PrOD challenge. Our implementations are based on MMDetection [4], a popular open-source object detection toolbox. We experimented with covariance matrices dependent on the widths and heights of the bounding boxes. The effectiveness of this technique was demonstrated by Ammirato and Berg [1] and we use it in our final results. In our implementation, the top left and bottom right elements of the covariance matrix are equal to 0.2 times the box width and 0.2 times the box height, respectively.

The results of using HTC with various threshold parameters on the validation set are shown in Table 1. Table 2 shows the results of RetinaNet, Cascade R-CNN, and Grid R-CNN with the same parameter values on the validation set. Table 3 shows the efficacy of the HTC and Grid R-CNN ensembles before and after using AGC on the validation set. From the results we can see that AGC helps improve the PDQ score. Table 4 shows the results of our five submissions to the competition evaluation server and Table 5 shows the parameter values used for each submission. For both tables, NMS means that for each image the model's predicted bounding boxes are passed through NMS once more.

An unexpected discovery was that the behavior of deep ensembles turned out to be worse than that of a single detection model. After examining the differences between deep ensembles and HTC, we observed that the ensemble

Table 1. The results of HTC using different threshold values on the validation set (SQ = spatial quality, LQ = label quality, TP = true positives, FP = false positives, FN = false negatives).

Threshold	PDQ Score	Avg. pPDQ	Avg. FP	Avg. SQ	Avg. LQ	TPs	FPs	FNs
0.2	20.545	0.544	0.495	0.495	0.736	37367	21637	50722
0.35	20.629	0.560	0.373	0.497	0.767	35838	14611	52251
0.5	19.665	0.590	0.255	0.506	0.823	31610	9123	56479

Table 2. The results of different detection models using the same parameter values on the validation set (SQ = spatial quality, LQ = label quality, TP = true positives, FP = false positives, FN = false negatives).

Method	PDQ Score	Avg. pPDQ	Avg. FP	Avg. SQ	Avg. LQ	TPs	FPs	FNs
Cascade R-CNN	15.857	0.569	0.358	0.502	0.772	26325	9915	61764
Grid R-CNN	15.433	0.548	0.349	0.476	0.793	26912	11401	61177
RetinaNet	11.875	0.480	0.548	0.485	0.566	22475	5957	65614

Table 3. The results of ensembles before and after using AGC on the validation set (SQ = spatial quality, LQ = label quality, TP = true positives, FP = false positives, FN = false negatives).

Method	PDQ Score	Avg. pPDQ	Avg. FP	Avg. SQ	Avg. LQ	TPs	FPs	FNs
Ensembles	17.487	0.559	0.267	0.501	0.758	39182	50656	48907
Ensembles + AGC	18.602	0.565	0.379	0.499	0.781	32241	15691	55848

Table 4. The final results of our methods on the test set (SQ = spatial quality, LQ = label quality, TP = true positives, FP = false positives, FN = false negatives).

Method	PDQ Score	Avg. pPDQ	Avg. FP	Avg. SQ	Avg. LQ	TPs	FPs	FNs
HTC	15.959	0.455	0.638	0.480	0.561	138325	271642	158042
HTC	13.506	0.467	0.300	0.465	0.613	104022	90658	192345
HTC and Grid R-CNN	12.144	0.342	0.710	0.452	0.354	126582	208356	169785
HTC w/ NMS	11.946	0.311	0.774	0.422	0.335	140266	308088	156101
HTC w/ NMS	10.655	0.390	0.550	0.442	0.467	93999	107583	202638

Table 5. The parameter values used on the test set to produce our final results (CovP = covariance percent, BoxR = ratio to reduce bounding boxes, Thr = threshold for confidence filtering, IoUThr = IoU threshold for NMS and merging detections).

Method	CovP	BoxR	Thr	IoUThr
HTC	0.2	0.15	0.36	1
HTC	0.2	0.15	0.05	1
HTC and Grid R-CNN	0.2	0.15	0.35	0.5
HTC w/ NMS	0.2	0.15	0.05	0.5
HTC w/ NMS	0.2	0.15	0.05	0.4

has similar true positive, false negative, and false positive scores, but lower label qualities. We believe this discrepancy might arise from the way in which we average the scores in the ensemble and that a different reduction method could provide higher performance.

4   Conclusion

In this paper, we introduced an approach to measure uncertainty via deep ensembles for the PrOD challenge. Our framework utilized a combination of HTC and Grid R-CNN object detection models, and obtained 4th place in the challenge with a PDQ score of 15.959. For future work, we plan to combine Monte Carlo dropout [8] with deep ensembles for uncertainty estimation. In addition, we will investigate the effectiveness of AGC when enabled for both training and testing. We also intend to try other test-time augmentation techniques to improve the performance of the detection task.

References

1. Ammirato, P., Berg, A.C.: A mask-rcnn baseline for probabilistic object detection. arXiv preprint arXiv:1908.03621 (2019)
2. Cai, Z., Vasconcelos, N.: Cascade R-CNN: Delving into high quality object detection. In: Proceedings of the IEEE Conference on Computer Vision and Pattern Recognition, pp. 6154–6162 (2018)
3. Chen, K., et al.: Hybrid task cascade for instance segmentation. In: Proceedings of the IEEE conference on computer vision and pattern recognition, pp. 4974–4983 (2019)
4. Chen, K., et al.: Mmdetection: open mmlab detection toolbox and benchmark. arXiv preprint arXiv:1906.07155 (2019)
5. Denker, J.S., LeCun, Y.: Transforming neural-net output levels to probability distributions. In: Advances in Neural Information Processing Systems, pp. 853–859 (1991)
6. Everingham, M., Eslami, S.A., Van Gool, L., Williams, C.K., Winn, J., Zisserman, A.: The pascal visual object classes challenge: a retrospective. Int. J. Comput. Vis. 111(1), 98–136 (2015)
7. Everingham, M., Van Gool, L., Williams, C.K., Winn, J., Zisserman, A.: The pascal visual object classes (VOC) challenge. Int. J. Comput. Vis. 88(2), 303–338 (2010)
8. Gal, Y., Ghahramani, Z.: Dropout as a bayesian approximation: representing model uncertainty in deep learning. In: International Conference on Machine Learning, pp. 1050–1059 (2016)
9. Hall, D., et al.: Probabilistic object detection: definition and evaluation. In: The IEEE Winter Conference on Applications of Computer Vision, pp. 1031–1040 (2020)
10. Lakshminarayanan, B., Pritzel, A., Blundell, C.: Simple and scalable predictive uncertainty estimation using deep ensembles. In: Advances in Neural Information Processing Systems, pp. 6402–6413 (2017)
11. Li, D., Xu, C., Liu, Y., Qin, Z.: Teamgl at acrv robotic vision challenge 1: probabilistic object detection via staged non-suppression ensembling. In: CVPR Workshop on the Robotic Vision Probabilistic Object Detection Challenge (2019)
12. Lin, T.Y., Goyal, P., Girshick, R., He, K., Dollár, P.: Focal loss for dense object detection. In: Proceedings of the IEEE International Conference on Computer Vision, pp. 2980–2988 (2017)
13. Lin, T.-Y., et al.: Microsoft COCO: common objects in context. In: Fleet, D., Pajdla, T., Schiele, B., Tuytelaars, T. (eds.) ECCV 2014. LNCS, vol. 8693, pp. 740–755. Springer, Cham (2014). https://doi.org/10.1007/978-3-319-10602-1_48
14. Lu, X., Li, B., Yue, Y., Li, Q., Yan, J.: Grid R-CNN. In: Proceedings of the IEEE Conference on Computer Vision and Pattern Recognition, pp. 7363–7372 (2019)
15. MacKay, D.J.: A practical bayesian framework for backpropagation networks. Neural Comput. 4(3), 448–472 (1992)
16. Miller, D., Nicholson, L., Dayoub, F., Sünderhauf, N.: Dropout sampling for robust object detection in open-set conditions. In: 2018 IEEE International Conference on Robotics and Automation (ICRA), pp. 1–7. IEEE (2018)
17. Russakovsky, O., et al.: Imagenet large scale visual recognition challenge. Int. J. Comput. Vis. 115(3), 211–252 (2015)
18. Skinner, J., Hall, D., Zhang, H., Dayoub, F., Sünderhauf, N.: The probabilistic object detection challenge. arXiv preprint arXiv:1903.07840 (2019)

19. Srivastava, N., Hinton, G., Krizhevsky, A., Sutskever, I., Salakhutdinov, R.: Dropout: a simple way to prevent neural networks from overfitting. J. Mach. Learn. Res. **15**(1), 1929–1958 (2014)
20. Wang, C.W., Cheng, C.A., Cheng, C.J., Hu, H.N., Chu, H.K., Sun, M.: Augpod: augmentation-oriented probabilistic object detection. In: CVPR Workshop on the Robotic Vision Probabilistic Object Detection Challenge (2019)

W37 - Imbalance Problems in Computer Vision

W37 - Imbalance Problems in Computer Vision

The performance of learning-based methods is adversely affected by imbalance problems at various levels, including the input, intermediate or mid-level stages of the processing, and the objectives to be optimized in a multi-task setting. Currently, researchers tend to address these challenges in their particular contexts with problem-specific solutions and with limited awareness of the solutions proposed for similar challenges in other computer vision problems.

Imbalance problems pertain to almost all computer vision problems and therefore, the Imbalance Problems in Computer Vision (IPCV) workshop is highly relevant and interesting for a broad community. Since learning-based approaches dominate the current methodologies and systems, a workshop that targets general issues in such learning approaches is highly beneficial. In addition, interest in imbalance problems by the community has significantly increased in recent years.

Despite being the first edition, IPCV has received very encouraging attention from the community: There were 13 paper submissions, and thanks to the detailed reviews written by the Program Committee members, we were able to accept 7 papers for presentation and inclusion in this proceeding. Moreover, IPCV hosted six invited speakers, who also participated in a panel discussion on "balancing the imbalance problems":

Bernt Schiele, Max Planck Institute for Informatics, Saarland University, Germany
Boqing Gong, Google, USA
Kemal Oksuz, Middle East Technical University, Turkey
Ming-Hsuan Yang, UC Merced; Google, USA
Tengyu Ma, Stanford University, USA
Vittorio Ferrari, The University of Edinburgh; Google, UK

The Program Committee members were Brandon Leung, Baris Can Cam, Chih-Hui Ho, Emre Akbas, Jiacheng Cheng, Kemal Oksuz, Pei Wang, Sinan Kalkan, Pedro Morgado, Tz-Ying Wu, and Yi Li. We thank them for providing detailed reviews on submitted papers.

We hope that the workshop was beneficial to the community for developing better solutions for addressing imbalance problems.

August 2020

Emre Akbas
Baris Can Cam
Sinan Kalkan
Kemal Oksuz
Nuno Vasconcelos

A Machine Learning Approach to Assess Student Group Collaboration Using Individual Level Behavioral Cues

Anirudh Som[1]([✉]), Sujeong Kim[1], Bladimir Lopez-Prado[2], Svati Dhamija[1], Nonye Alozie[2], and Amir Tamrakar[1]

[1] Center for Vision Technologies, SRI International, Menlo Park, USA
anirudh.som@sri.com
[2] Center for Education Research and Innovation, SRI International, Menlo Park, USA

Abstract. K-12 classrooms consistently integrate collaboration as part of their learning experiences. However, owing to large classroom sizes, teachers do not have the time to properly assess each student and give them feedback. In this paper we propose using simple deep-learning-based machine learning models to automatically determine the overall collaboration quality of a group based on annotations of individual roles and individual level behavior of all the students in the group. We come across the following challenges when building these models: (1) Limited training data, (2) Severe class label imbalance. We address these challenges by using a controlled variant of Mixup data augmentation, a method for generating additional data samples by linearly combining different pairs of data samples and their corresponding class labels. Additionally, the label space for our problem exhibits an ordered structure. We take advantage of this fact and also explore using an ordinal-cross-entropy loss function and study its effects with and without Mixup.

Keywords: Education · Collaboration assessment · Limited training data · Class imbalance · Mixup data augmentation · Ordinal-cross-entropy loss.

1 Introduction

Collaboration is identified by both the Next Generation Science Standards [21] and the Common Core State Standards [7] as a required and necessary skill for students to successfully engage in the fields of Science, Technology, Engineering and Mathematics (STEM). Most teachers in K-12 classrooms instill collaborative skills in students by using instructional methods like project-based learning [17] or problem-based learning [8]. For a group of students performing a group-based collaborative task, a teacher monitors and assesses each student based

---

A. Som—The author is currently a student at Arizona State University and this work was done while he was an intern at SRI International.

A. Bartoli and A. Fusiello (Eds.): ECCV 2020 Workshops, LNCS 12540, pp. 79–94, 2020.
https://doi.org/10.1007/978-3-030-65414-6_8

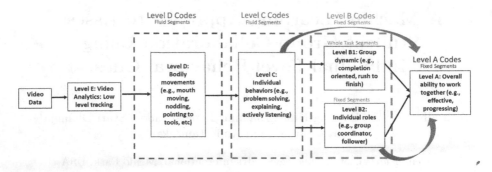

Fig. 1. The collaboration assessment conceptual model. In this paper we focus on building machine learning models that map features from Level B2 ⟶ Level A and Level C ⟶ Level A as indicated by the red arrows.

on various verbal and non-verbal behavioral cues. However, due to the wide range of behavioral cues, it can often be hard for the teacher to identify specific behaviors that contribute to or detract from the collaboration effort [18,20,22]. This task becomes even more difficult when several student groups need to be assessed simultaneously. To better assist teachers, in our previous work we proposed an automated collaboration assessment conceptual model that provides an assessment of the collaboration quality of student groups based on behavioral communication at individual and group levels [2,3]. The conceptual model illustrated in Fig. 1 represents a multi-level, multi-modal integrated behavior analysis tool. The input to this model consists of Video or Audio+Video data recordings of a student group performing a collaborative task. This was done to test if visual behaviors alone could be used to estimate collaboration skills and quality. Next, low level features like facial expressions, body-pose are extracted at Level E. Information like joint attention and engagement are encoded at Level D. Level C describes complex interactions and individual behaviors. Level B is divided into two categories: Level B1 describes the overall group dynamics for a given task; Level B2 describes the changing individual roles assumed by each student in the group. Finally, Level A describes the overall collaborative quality of the group based on the information from all previous levels. This paper focuses on building machine learning models that predict a group's collaboration quality from individual roles (Level B2) and individual behaviors (Level C) of the students, indicated by red arrows in Fig. 1.

Deep-learning algorithms have gained increasing attention in the Educational Data Mining (EDM) community. For instance, the first papers to use deep-learning for EDM were published in 2015, and the number of publications in this field keeps growing with each year [12]. Despite their growing popularity, deep-learning methods are difficult to work with under certain challenging scenarios. For example, deep-learning algorithms work best with access to large amounts of representative training data, *i.e.*, data containing sufficient variations of each class label pattern. They also assume that the label distribution of the training

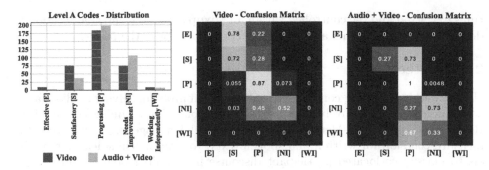

Fig. 2. (left) Distribution of Level A codes which also represent the target label distribution for our classification problem. (middle, right) Aggregate confusion matrix of Multi Layer Perceptron (MLP) classification models that have been subjected to class-balancing during the training process. Even with class-balancing the MLP models are unable to overcome the bias in the training data. Note, the confusion matrix is normalized along each row, with the number in each cell representing the percentage of data samples that are classified to each class.

data is approximately uniform. If either case is not satisfied then deep-learning methods tend to perform poorly at the desired task. Challenges arising due to limited and imbalanced training data is clearly depicted in Fig. 2. For our classification problem the label distribution appears to be similar to a bell-shaped normal distribution. As a result, for both Video and Audio+Video modality cases we have very few data samples for *Effective* and *Working Independently* codes, and the highest number of samples for *Progressing* code. Figure 2 also shows the aggregate confusion matrix over all test sets after training Multi Layer Perceptron (MLP) classification models with class-balancing (*i.e.*, assigning a weight to the training data sample that is inversely proportional to the number of training samples corresponding to that sample's class label). The input feature representations used were obtained from Level B2 and Level C. We observe that despite using class-balancing the predictions of the MLP model are biased towards the *Progressing* code.

Contributions: To address the above challenges, in this paper we explore using a controlled variant of Mixup data augmentation, a simple and common approach for generating additional data samples [24]. Additional data samples are obtained by linearly combining different pairs of data samples and their corresponding class labels. Also note that the label space for our classification problem exhibits an ordered relationship. In addition to Mixup, we also explore the value in using an ordinal-cross-entropy loss function instead of the commonly used categorical-cross-entropy loss function.

Outline of the Paper: Section 2 discusses related work. Section 3 provides necessary background on categorical-cross-entropy loss, ordinal-cross-entropy loss and Mixup data augmentation. Section 4 provides description of the dataset, features extracted and the controlled variant of Mixup data augmentation. Section 5 describes the experiments and results. Section 6 concludes the paper.

2   Related Work

Use of machine learning concepts for collaboration problem-solving analysis and assessment is still relatively new in the Educational Data Mining community. Reilly *et al.* used Coh-Metrix indices (a natural language processing tool to measure cohesion for written and spoken texts) to train machine learning models to classify co-located participant discourse in a multi-modal learning analytics study [19]. The multi-modal dataset consisted of eye-tracking, physiological and motion sensing data. They analyzed the collaboration quality between novice programmers that were instructed to program a robot to solve a series of mazes. However, they studied only two collaborative states thereby making it a binary classification problem. Huang *et al.* used an unsupervised machine learning approach to discover unproductive collaborative states for the same multi-modal dataset [14]. For input features they computed different measures for each modality. Using an unsupervised approach they were able to identify a three-state solution that showed high correlation with task performance, collaboration quality and learning gain. Kang *et al.* also used an unsupervised learning approach to study the collaborative problem-solving process of middle school students. They analyzed data collected using a computer-based learning environment of student groups playing a serious game [15]. They used *KmL*, an R package useful for applying k-means clustering on longitudinal data [10]. They too identified three different states using the proposed unsupervised method. In our paper we define five different group collaboration quality states in a supervised learning setup. The above studies discuss different ways to model positive collaboration between participants in a group. For Massive Open Online Courses (MOOCs), Alexandron *et al.* proposed a technique to detect cheating in the form of unauthorized collaboration using machine learning classifiers trained on data of another form of cheating (copying using multiple accounts) [1].

Guo and Barmaki used a deep-learning based object detection approach for analysis of pairs of students collaborating to locate and paint specific body muscles on each other [11]. They used a Mask R-CNN for detecting students in video data. This is the only paper that we found that used deep-learning for collaboration assessment. They claim that close proximity of group participants and longer time taken to complete a task are indicators of good collaboration. However, they quantify participant proximity by the percentage of overlap between the student masks obtained using the Mask R-CNN. The amount of overlap can change dramatically across different view points. Also, collaboration need not necessarily be exhibited by groups that take a longer time to complete a task. In this paper, the deep-learning models are based off of the systematically designed multi-level conceptual model shown in Fig. 1. The proposed approach utilizes features at lower levels of our conceptual model but we go well beyond these and also include higher level behavior analysis as well roles taken on by students to predict the overall group collaboration quality.

We propose using Mixup augmentation, an over-sampling approach together with an ordinal cross-entropy loss function to better handle limited and imbalanced training data. Over-sampling techniques have been commonly used to make the different label categories to be approximately equal. SMOTE is one of the oldest and most widely cited over-sampling methods proposed by Chawla et al. [5]. The controlled variant of Mixup that we propose is very similar to their approach. However, ordinal loss functions have not received as much attention since the label space of most current classification problems of interest do not exhibit an ordered structure or relationship. We refer interested readers to the following papers that talk about ordinal loss functions for deep ordinal classification [4,13]. In this paper we propose a simple variant of the regular cross-entropy loss that takes into account the relative distance of the predicted samples from their true class label location.

3   Preliminaries

3.1   Classification Loss Functions

Let us denote the input variables or covariates as $\mathbf{x}$, ground-truth label vector as $\mathbf{y}$ and the predicted probability distribution as $\mathbf{p}$. The cross-entropy loss a.k.a. the categorical-cross-entropy loss function is commonly used for training deep-learning models for multi-class classification. Given a training sample $(\mathbf{x}, \mathbf{y})$, the cross-entropy loss can be represented as $\mathrm{CE}_{\mathbf{x}}(\mathbf{p}, \mathbf{y}) = -\sum_{i=1}^{C} \mathbf{y}_i \log(\mathbf{p}_i)$.

Here, C represents the number of classes. For a classification problem with C label categories, a deep-learning model's softmax layer outputs a probability distribution vector $\mathbf{p}$ of length C. The i-th entry in $\mathbf{p}_i$ represents the predicted probability of the i-th class. The ground-truth label $\mathbf{y}$ is one-hot-encoded and represents a binary vector whose length is also equal to C. Note, $\sum_i \mathbf{y}_i = 1$ and $\sum_i \mathbf{p}_i = 1$. For an imbalanced dataset, the learnt weights of a deep-learning model will be greatly governed by the class having the most number of samples in the training set. Also, if the label space exhibits an ordinal structure, the cross-entropy loss focuses only on the predicted probability of the ground-truth class and ignores the relative distance between an incorrectly predicted data sample and its true class label. A simple variant of the cross-entropy loss that is useful for problems exhibiting an ordered label space is shown in Eq. 1.

$$\mathrm{OCE}_{\mathbf{x}}(\mathbf{p}, \mathbf{y}) = -(1 + w) \sum_{i=1}^{C} \mathbf{y}_i \log(\mathbf{p}_i), \quad w = |\mathrm{argmax}(\mathbf{y}) - \mathrm{argmax}(\mathbf{p})| \quad (1)$$

Here, $(1 + w)$ is an additional weight that is multiplied with the regular cross-entropy loss. Within w, argmax returns the index of the maximum valued element in the vector and $|.|$ denotes the absolute value. During the training process, $w = 0$ for training samples that are correctly classified, with the ordinal-cross-entropy loss being the same as the cross-entropy loss. However, the ordinal-cross-entropy loss will be higher than cross-entropy loss for misclassified samples and the increase in loss is proportional to how far the samples have been misclassified from their true label locations. We later go over the benefit of using the ordinal-cross-entropy loss function in Sect. 5.1.

3.2 Mixup Data Augmentation

Despite best data collection practices, bias exists in most training datasets resulting from time or resource constraints. These biases, and the resulting performance problems of machine learning models trained on this data, are directly correlated with the problem of class imbalance. Class imbalance refers to the unequal representation or number of occurrences of different class labels. If the training data is more representative of some classes than others, then the model's predictions would systematically be worse for the under-represented classes. Conversely, with too much data or an over-representation of certain classes can skew the decision toward a particular result. Mixup is a simple data augmentation technique that can be used for imbalanced datasets [24]. It is used for generating additional training samples and encourages the deep-learning model to behave linearly in-between the training samples. It extends the training distribution by incorporating the prior knowledge that linear interpolations of input variables $\mathbf{x}$ should lead to linear interpolations of the corresponding target labels $\mathbf{y}$. For example, given a random pair of training samples $(\mathbf{x}^1, \mathbf{y}^1)$, $(\mathbf{x}^2, \mathbf{y}^2)$, additional samples can be obtained by linearly combining the input covariate information and the corresponding class labels. This is illustrated in Eq. 2.

$$\tilde{\mathbf{x}} = \lambda \mathbf{x}^1 + (1 - \lambda)\mathbf{x}^2$$
$$\tilde{\mathbf{y}} = \lambda \mathbf{y}^1 + (1 - \lambda)\mathbf{y}^2 \tag{2}$$

Here, $(\tilde{\mathbf{x}}, \tilde{\mathbf{y}})$ is the new generated sample. $\lambda \in [0, 1]$ and is obtained using a Beta(α, α) distribution with $\alpha \in (0, \infty)$. Figure 3 shows different Beta(α, α) distributions for $\alpha = 0.1, 0.4, 0.7, 1.0$ respectively. If α approaches 0 the λs obtained have a higher probability of being 0 or 1. If α approaches 1 then the Beta distribution looks more like a uniform distribution. Based on the suggestions and findings in other papers [23,24], for our experiments we set $\alpha = 0.4$. Apart from improving the classification performance on various image classification benchmarks [24], Mixup also leads to better calibrated deep-learning models [23]. This

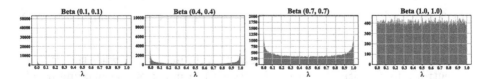

Fig. 3. Beta(α, α) distributions for $\alpha = 0.1, 0.4, 0.7, 1.0$ respectively. Each Beta distribution plot has a different y-axis range and represents a 500-bin histogram of 200000 randomly selected λs. Note, most λs for Beta(0.1,0.1) are at 0 and 1.

means that the predicted softmax scores of a model trained using Mixup are better indicators of the actual likelihood of a correct prediction than models trained in a regular fashion. In Sect. 5.2, we explore the benefit of using Mixup with and without ordinal-cross-entropy loss.

4   Dataset Description, Feature Extraction and Controlled Mixup Data Generation

4.1   Dataset Description

Audio and video data was collected from 15 student groups across five middle schools [2,3]. Each group was given an hour to perform 12 open-ended life science and physical science tasks that required the students to construct models of different science phenomena. This resulted in 15 hours of audio and video recordings. Out of the 15 groups, 13 groups had 4 students, 1 group had 3 students, and 1 group had 5 students. For Level A and Level B2, each video recording was coded by three human annotators using ELAN (an open-source annotation software) under two different modality conditions: 1) Video, 2) Audio+Video. For a given task performed by a group, each annotator first manually coded each level for the Video modality and later coded the same task for the Audio+Video modality. This was done to prevent any coding bias resulting due to the difference in modalities. A total of 117 tasks were coded by each annotator. Next, the majority vote (code) from the three coders was used to determine the ground-truth Level A code. For cases where a clear majority was not possible the median of the three codes was used as the ground-truth. We used the same code ordering depicted in Fig. 2. For example, if the three coders assigned *Effective, Satisfactory, Progressing* for a certain task then *Satisfactory* would be selected as the ground-truth label. Note, out of the 117 tasks within each modality we did not observe a majority Level A code for only 2 tasks. The distribution of the Level A target labels is shown in Fig. 2. For learning mappings from Level B2 $\longrightarrow$ Level A we had access to only 351 data samples (117 tasks $\times$ 3 coders) to train the machine learning models. The protocol used for generating training-test splits is described in Sect. 5. In the case of Level C, each video recording was coded by just one annotator. Because of this we only had access to 117 data samples (117 tasks coded) for training the machine learning models to learn mappings from Level C $\longrightarrow$ Level A. This makes it an even more challenging classification problem. Note, the distribution of the Level A labels for this classification setting is similar to the distribution shown in Fig. 2, with the difference being that each label class now has just one-third of the samples.

4.2   Level B2 and Level C Histogram Representation

For the entire length of each task, Level B2 was coded using fixed-length one minute segments and Level C was coded using variable-length segments. This is illustrated in Fig. 4. As shown in Table 1, Level B2 and Level C consist of 7 codes

Table 1. Coding rubric for Level B2 and Level C.

Level B2 Codes	Level C Codes		
Group guide/Coordinator [GG]	Talking	Recognizing/Inviting others contributions	Joking/Laughing
Contributor (Active) [C]	Reading	Setting group roles and responsibilities	Playing/Horsing around/Rough housing
Follower [F]	Writing	Comforting, encouraging others/Coralling	Excessive difference to authority/leader
Conflict Resolver [CR]	Using/Working with materials	Agreeing	Blocking information from being shared
Conflict Instigator/Disagreeable [CI]	Setting up the physical space	Off-task/Disinterested	Doing nothing/Withdrawing
Off-task/Disinterested [OT]	Actively listening/Paying attention	Disagreeing	Engaging with outside environment
Lone Solver [LS]	Explaining/Sharing ideas	Arguing	Waiting
	Problem solving/Negotiation	Seeking recognition/Boasting	

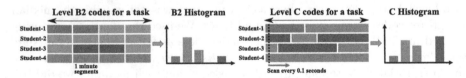

Fig. 4. Histogram feature generation for Level B2 and Level C. Different colors indicate different codes assigned to each segment. Level B2 codes are represented as fixed-length 1 min segments. Level C codes are represented as variable-length segments. A B2 histogram is generated for each task by compiling all the codes from all the students in the group. Similarly, level C histogram was generated by compiling all the codes observed after every 0.1 s over the duration of the task.

and 23 codes respectively. Our objective in this paper is to be able to determine the overall collaboration quality of a group by summarizing all of the individual student roles and behaviors for a given task. A simple but effective way to do this is by generating histogram representations of all the codes observed in each task. Figure 4 also provides a simple illustration of the histogram generation process. While it is straightforward to generate histograms for Level B2, in the case of Level C we compile all the codes observed after every 0.1 s for generating the histogram. Once the histogram is generated we normalize it by dividing by the total number of codes in the histogram. Normalizing the histogram in a way removes the temporal component of the task. For example, if group-1 took 10 min to solve a task and group-2 took 30 min to solve the same task, but both groups were assigned the same Level A code despite group-1 finishing the task sooner. The raw histogram representations of both these groups would look different due to the difference in number of segments coded. However, normalized histograms would make them more comparable.

4.3 Controlled Mixup

We described the simplicity and benefits of Mixup augmentation in Sect. 3.2. Here, we describe a controlled variant of Mixup and how it is used for our dataset. From Fig. 2, we know that our dataset has an imbalanced label distribution. Conventional Mixup selects a random pair of samples and interpolates them by a λ that is determined using a Beta distribution. However, this results in generating samples that have the same imbalanced class distribution. We want to be able to generate a fixed number of samples for a specific category. To do this we first limit the range of λ, i.e., $\lambda \in [\tau, 1]$, with τ being the threshold.

Next, to generate additional samples for a specific class, we pair that class with its adjacent or neighboring classes. For example, let us use the following denotation: (primary-class, [adjacent-class-1, adjacent-class-2]), where primary-class represents the class for which we want to create additional samples; adjacent-class-1 and adjacent-class-2 represent its neighbors. We create the following pairs: (*Effective*, [*Satisfactory,Progressing*]), (*Satisfactory*, [*Effective,Progressing*]), (*Progressing*, [*Satisfactory,Needs Improvement*]), (*Needs Improvement*, [*Progressing,Working Independently*]) and (*Working Independently*, [*Progressing,Needs Improvement*]). The final step consists of generating n samples for the primary-class using Mixup. We do this by randomly pairing samples from the primary-class with samples from the adjacent-classes. This process is repeated n times. Note that for Mixup augmentation, λ is always multiplied with the primary-class sample and $(1 - \lambda)$ is multiplied with the adjacent-class sample. For our experiments we explore the following values of τ: 0.55, 0.75 and 0.95. Setting $\tau > 0.5$ guarantees that the generated sample would always be dominated by the primary-class.

5 Experiments

Network Architecture: We used a 5-layer Multi Layer Perceptron (MLP) model whose design was based on the MLP model described in [9]. It contains the following layers: 1 input layer, 3 middle dense layers and 1 output dense layer. The normalized histogram representations discussed in Sect. 4.2 are passed as input to the input layer. Each dense middle layer has 500 units with ReLU activation. The output dense layer has a softmax activation and the number of units is equal to 5 (total number of classes in Level A). We also used dropout layers between each layer to avoid overfitting. The dropout-rate after the input layer and after each of the three middle layers was set to 0.1, 0.2, 0.2,

0.3 respectively. We try three different types of input data: B2 histograms, C histograms and concatenating B2 and C histograms (referred to as B2+C histograms). The number of trainable parameters for B2 histogram is 507505, C histogram is 515505, B2+C histogram is 519005. Our models were developed using Keras with TensorFlow backend [6]. We used the Adam optimizer [16] and trained all our models for 500 epochs. The batch-size was set to one-tenth of the number of training samples during any given training-test split. We saved the best model that gave us the lowest test-loss for each training-test split.

Training and Evaluation Protocol: We adopt a round-robin leave-one-group-out cross validation protocol. This means that for each training-test split we use data from $g - 1$ groups for training and the gth group is used as the test set. This process is repeated for all g groups. For our experiments $g = 14$ though we have histogram representations for each task performed by 15 student groups. This is because in the Audio+Video modality setting all samples corresponding to the *Effective* class were found only in one group. Similarly, for the Video modality all samples corresponding to *Working Independently* class were also found in just one group. Due to this reason we do not see any test samples for the *Effective* class in Audio+Video and the *Working Independently* class in Video in the confusion matrices shown earlier in Fig. 2. Note, for Level B2 $\longrightarrow$ Level A we have 351 data samples, and for Level C $\longrightarrow$ Level A we only have 117 data samples (discussed in Sect. 4.1).

5.1   Effect of Ordinal-Cross-Entropy Loss

The ordinal-cross-entropy loss shown in Eq. 1 takes into account the distance of the highest predicted probability from its one-hot encoded true label. This is what separates it from the regular cross-entropy loss which only focuses on the predicted probability corresponding to the ground-truth label. In this section we explore the following four variations: cross-entropy loss only, cross-entropy loss with class balancing, ordinal-cross-entropy loss only and ordinal-cross-entropy loss with class balancing. Here class balancing refers to weighting each data sample by a weight that is inversely proportional to the number of data samples corresponding to that sample's class label.

Figure 5 illustrates the average weighted F1-score classification performance for the four variations under different parameter settings. We only varied the patience and minimum-learning-rate (Min-LR) parameter as we found that these two affected the classification performance the most. These parameters were used to reduce the learning-rate by a factor of 0.5 if the loss did not change after a certain number of epochs indicated by the patience parameter. Compared to the two cross-entropy-loss variants we clearly see that the two ordinal-cross-entropy loss variants help significantly improve the F1-scores across all the parameter settings. We consistently see improvements across both modality conditions and for the different histogram inputs. Using class balancing we only see marginal improvements for both loss functions. Also, the F1-scores for Video modality is always lower than the corresponding settings in Audio+Video modality. This is expected as it shows that annotations obtained using Audio+Video recordings are more cleaner and better represent the student behaviors.

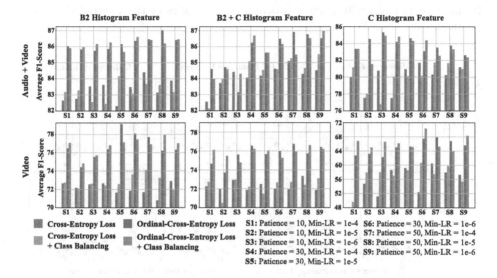

Fig. 5. Comparison of the average weighted F1-Score performance between cross-entropy loss and ordinal-cross-entropy loss, with and without class balancing and under different parameter settings **S1–S9**.

5.2 Effect of Controlled Mixup Data Augmentation

We use different parameter settings (listed in Fig. 5) for each type of histogram input. For B2 histogram we use parameter setting S5 for both modalities. For C histogram we use parameter setting S6 for Video and S3 for Audio+Video. For B2+C histogram we use parameter setting S7 for both Video and Audio+Video. For each modality and each type of histogram input we vary threshold τ in the range 0.55, 0.75, 0.95, and also vary n (the number of controlled Mixup samples generated per class) in the range of 200, 500, 1000. Note, $n = 200$ implies that the training dataset contains 1000 samples ($n \times$ number of classes). Note, this form of Mixup does not retain the original set of training samples. Figure 6 shows the effect of controlled Mixup with and without ordinal-cross-entropy loss. Across both modality settings, we observe that Mixup augmentation with ordinal-cross-entropy loss is better than Mixup with regular cross-entropy loss for all cases in B2 histogram and for most cases in C histogram and B2+C histogram. This implies that controlled Mixup and ordinal-cross-entropy loss complement each other in most cases. We also observed that having a larger n does not necessarily imply better performance. For Audio+Video modality we observe F1-scores to be similar irrespective of the value of n. However, in the Video modality case we observe that F1-score decreases as n increases. This could be attributed to the noisy nature of codes assigned by the annotators due to the lack of Audio modality. We also notice better performance using $\tau = 0.75$ or $\tau = 0.95$ for Audio+Video modality and $\tau = 0.55$ for Video modality for B2 and B2+C histogram. However, we see the opposite effect in the case of C histogram. In the next section we will discuss two different variants of the controlled Mixup augmentation.

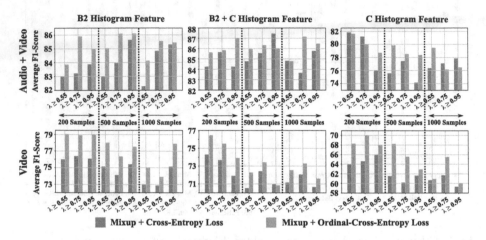

Fig. 6. Comparison of the average weighted F1-Score performance of using controlled Mixup augmentation, with and without ordinal-cross-entropy loss. Here, 200, 500, 1000 samples refer to the number of samples generated per class (n) using controlled Mixup.

5.3   Full Mixup Vs Limited Mixup

For the controlled Mixup experiment described in the previous section, the MLP models were trained using the n generated samples per class which do not retain the original set of training samples. Let us refer to this as Full-Mixup. In this section we explore training MLP models with the original set of training samples and only generate samples needed to reach n samples per class using controlled Mixup. For example, let us assume that the *Effective* class already has m training samples, then we only compute $n - m$ samples using controlled Mixup to reach the required n samples per class. This process makes sure that we always have the original set of training samples. Let us refer to this as Limited-Mixup. Figure 7 shows the average weighted F1-score comparing Full-Mixup and Limited-Mixup. We only show results for $n = 200$ using B2 histogram feature as we observed similar trends in the other cases as well. We see that Full-Mixup and Limited-Mixup have similar F1-scores. This implies that we can generate the n samples per class only using controlled Mixup protocol described in Sect. 3.2 without much noticeable difference in F1-score performance.

5.4 Additional Analysis and Discussion

In this section we discuss in more detail the behavior of different classification models seen in the previous sections. Under each modality, Table 2 shows the weighted precision, weighted recall and weighted F1-score results for the best MLP models under different experimental settings. Here, the best MLP models were decided based on the weighted F1-score since it provides a more summarized assessment by combining information seen in both precision and recall. Values in the table represent the longest bars observed in Figs. 5 and 6. Note, weighted recall is equal to accuracy. We also show results using an SVM classifier. For SVM we explored linear and different non-linear kernels with different C parameter settings and only showed the best result in Table 2.

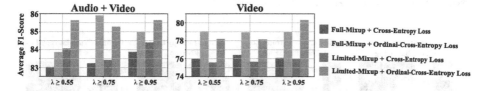

Fig. 7. Full-Mixup Vs Limited Mixup evaluation using different loss functions. Average weighted F1-score shown only for B2 histogram feature input and $n = 200$.

Table 2. Weighted precision, weighted recall and weighted F1-score Mean ± Std for the best MLP models under different experimental settings. The best models were selected based on the weighted F1-score. Bold values indicate the top two methods with the highest weighted precision under each modality condition.

Feature	Classifier	Video			Audio + Video		
		Weighted Precision	Weighted Recall	Weighted F1-Score	Weighted Precision	Weighted Recall	Weighted F1-Score
B2 Histogram	SVM	74.60±11.27	62.67±9.42	63.84±11.18	84.45±13.43	73.19±16.65	76.92±15.39
	MLP - Cross-Entropy Loss	76.90±12.91	73.95±11.02	72.89±13.22	83.72±16.50	86.42±10.44	84.40±13.85
	MLP - Cross-Entropy Loss + Class-Balancing	77.08±13.03	73.84±13.27	74.12±13.59	83.93±17.89	85.29±14.37	84.16±16.23
	MLP - Ordinal-Cross-Entropy Loss	81.51±13.44	79.09±13.62	79.11±13.96	86.96±14.56	88.78±10.36	87.03±13.16
	MLP - Ordinal-Cross-Entropy Loss + Class-Balancing	80.78±14.12	78.70±11.98	77.93±14.05	86.73±14.43	88.20±9.66	86.60±12.54
	MLP - Cross-Entropy Loss + Mixup	**81.61±12.81**	**73.56±10.31**	**76.40±11.00**	**88.51±12.32**	**83.58±14.14**	**85.64±13.23**
	MLP - Ordinal-Cross-Entropy Loss + Mixup	**83.30±10.06**	**76.57±9.42**	**79.06±9.66**	**89.59±10.15**	**84.93±13.20**	**86.09±12.94**
C Histogram	SVM	59.27±27.00	49.76±20.69	46.85±22.26	72.33±20.33	60.15±19.45	63.25±17.96
	MLP - Cross-Entropy Loss	63.24±20.78	65.73±16.34	60.46±17.57	81.15±16.90	84.16±11.67	81.70±14.41
	MLP - Cross-Entropy Loss + Class-Balancing	63.82±22.08	64.77±18.51	60.64±19.89	80.44±18.11	84.88±11.70	81.67±15.06
	MLP - Ordinal-Cross-Entropy Loss	68.16±27.13	72.59±17.88	67.88±23.01	86.05±14.11	86.90±11.43	85.33±13.07
	MLP - Ordinal-Cross-Entropy Loss + Class-Balancing	71.74±24.34	74.10±16.75	70.37±20.94	85.24±13.54	86.11±11.65	84.94±12.52
	MLP - Cross-Entropy Loss + Mixup	72.27±23.29	64.45±19.55	66.02±20.35	84.25±13.78	81.91±13.68	81.82±13.93
	MLP - Ordinal-Cross-Entropy Loss + Mixup	75.11±21.63	69.54±18.64	70.03±20.01	82.94±14.63	81.91±14.68	81.63±14.46
B2+C Histogram	SVM	72.49±15.35	61.89±13.21	64.95±14.15	82.32±16.53	73.32±15.27	76.65±15.65
	MLP - Cross-Entropy Loss	76.15±15.81	74.59±15.02	73.35±16.08	83.38±19.42	87.75±14.68	85.09±17.12
	MLP - Cross-Entropy Loss + Class-Balancing	75.75±17.23	73.81±16.50	73.11±17.17	84.71±16.57	88.68±11.04	85.52±15.01
	MLP - Ordinal-Cross-Entropy Loss	78.05±17.94	77.88±16.16	76.73±17.65	85.51±17.28	89.25±12.19	86.91±14.99
	MLP - Ordinal-Cross-Entropy Loss + Class-Balancing	78.10±17.70	77.33±17.02	76.61±17.96	86.90±15.83	88.82±11.50	86.99±14.15
	MLP - Cross-Entropy Loss + Mixup	77.99±17.42	72.86±14.32	74.29±16.08	**90.48±11.20**	**86.57±14.07**	**87.45±13.65**
	MLP - Ordinal-Cross-Entropy Loss + Mixup	77.92±16.66	75.82±15.27	76.45±16.02	**90.05±10.80**	**85.91±14.00**	**87.01±13.18**

For both modalities and for each type of histogram input, if we focus only on the weighted F1-scores we notice that there is little or no improvement as we go towards incorporating controlled Mixup and ordinal-cross-entropy loss. For this reason we also show the corresponding weighted precision and weighted recall values. We observe that the average weighted precision increases and the standard-deviation of the weighted precision decreases as we go towards the proposed approach. For an imbalanced classification problem the objective is to be able to predict more true positives. Thus a higher precision indicates more true positives as it does not consider any false negatives in its calculation. The bold values in Table 2 indicate the top two methods with highest weighted precision values in each modality. We find that the Cross-Entropy loss + Mixup and Ordinal-Cross-Entropy loss + Mixup methods show the highest weighted Precision using the B2 histogram input in the Video modality and the B2+C histogram input in the Audio+Video modality.

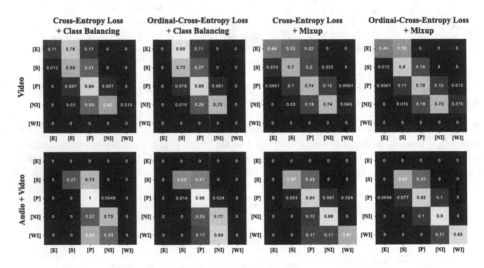

Fig. 8. Aggregate confusion matrix illustrations of the MLP classification model under different experimental conditions. The confusion matrix for each method corresponds to the best MLP model described in Table 2. The confusion matrices are normalized along each row. Note, the number in each cell represents the percentage of samples classified to each class.

The higher weighted precision is better illustrated using the confusion matrices shown in Fig. 8. Here, we show confusion matrices for Video modality using the B2 histogram features and for Audio+Video modality using the B2+C histogram, as these showed the best weighted precision values in Table 2. As seen earlier in Sect. 5.1, ordinal-cross-entropy loss did show significant improvements in terms of weighted F1-score. However, even with class balancing we notice that the best MLP model is still biased towards the class with the most training samples. If we look at the controlled Mixup variants with either cross-entropy loss or ordinal-cross-entropy loss we notice a better diagonal structure in the confusion matrix, indicating more number of true positives. Note, we do not see any test samples for the *Effective* class in Audio+Video and the *Working Independently* class in Video in the confusion matrices. Between Cross-Entropy loss + Mixup and Ordinal-Cross-Entropy loss + Mixup, we notice that ordinal-cross-entropy loss helps minimize the spread of test sample prediction only to the nearest neighboring classes.

6 Conclusion

In this paper we built simple machine learning models to determine the overall collaboration quality of a student group based on the summary of individual roles and individual behaviors exhibited by each student. We come across challenges like limited training data and severe class imbalance when building these models. To address these challenges we proposed using an ordinal-cross-entropy loss function together with a controlled variation of Mixup data augmentation. Ordinal-cross-entropy loss is different from the regular categorical cross-entropy loss as it takes into account how far the training samples have been classified from their true label locations. We proposed a controlled variant of Mixup allowing us to generate a desired number of data samples for each label category for our problem. Through various experiments we studied the behavior of different machine learning models under different experimental conditions and realized the benefit of using ordinal-cross-entropy loss with Mixup. For future work, we would like to explore building machine learning models that learn mappings across the other levels described in Fig. 1 and also explore the temporal nature of the annotation segments.

References

1. Alexandron, G., Ruipérez-Valiente, J.A., Pritchard, D.E.: Towards a general purpose anomaly detection method to identify cheaters in massive open online courses (2020)
2. Alozie, N., Dhamija, S., McBride, E., Tamrakar, A.: Automated collaboration assessment using behavioral analytics. In: International Society of the Learning Sciences (ISLS) (2020)
3. Alozie, N., McBride, E., Dhamija, S.: Collaboration conceptual model to inform the development of machine learning models using behavioral analytics. AERA Ann. Meet. (2020)
4. Beckham, C., Pal, C.: Unimodal probability distributions for deep ordinal classification (2017). arXiv preprint: arXiv:1705.05278
5. Chawla, N.V., Bowyer, K.W., Hall, L.O., Kegelmeyer, W.P.: Smote: synthetic minority over-sampling technique. J. Artif. Intell. Res. **16**, 321–357 (2002)
6. Chollet, F., et al.: Keras (2015). https://keras.io
7. Daggett, W.R., GendroO, D.S.: Common core state standards initiative. In: International Center (2010)

8. Davidson, N., Major, C.H.: Boundary crossings: Cooperative learning, collaborative learning, and problem-based learning. J. Excell. Coll. Teach. **25**, 7–55 (2014)

9. Ismail Fawaz, H., Forestier, G., Weber, J., Idoumghar, L., Muller, P.-A.: Deep learning for time series classification: a review. Data Min. Knowl. Discov. **33**(4), 917–963 (2019). https://doi.org/10.1007/s10618-019-00619-1

10. Genolini, C., Falissard, B.: KML: A package to cluster longitudinal data. Computer Methods Prog. Biomed. **104**(3), e112–e121 (2011)

11. Guo, Z., Barmaki, R.: Collaboration analysis using object detection. In: EDM (2019)

12. Hernández-Blanco, A., Herrera-Flores, B., Tomás, D., Navarro-Colorado, B.: A systematic review of deep learning approaches to educational data mining. Complexity **2019**, 1–23 (2019)

13. Hou, L., Yu, C.P., Samaras, D.: Squared earth mover's distance-based loss for training deep neural networks (2016). arXiv preprint: arXiv:1611.05916

14. Huang, K., Bryant, T., Schneider, B.: Identifying collaborative learning states using unsupervised machine learning on eye-tracking, physiological and motion sensor data. In: International Educational Data Mining Society (2019)

15. Kang, J., An, D., Yan, L., Liu, M.: Collaborative problem-solving process in a science serious game: exploring group action similarity trajectory. In: International Educational Data Mining Society (2019)

16. Kingma, D.P., Ba, J.: Adam: A method for stochastic optimization (2014). arXiv preprint: arXiv:1412.6980

17. Krajcik, J.S., Blumenfeld, P.C.: Project-based learning. na (2006)

18. Loughry, M.L., Ohland, M.W., DeWayne Moore, D.: Development of a theory-based assessment of team member effectiveness. Educ. Psychol. Measur. **67**(3), 505–524 (2007)

19. Reilly, J.M., Schneider, B.: Predicting the quality of collaborative problem solving through linguistic analysis of discourse. In: International Educational Data Mining Society (2019)

20. Smith-Jentsch, K.A., Cannon-Bowers, J.A., Tannenbaum, S.I., Salas, E.: Guided team self-correction: impacts on team mental models, processes, and effectiveness. Small Group Res. **39**(3), 303–327 (2008)

21. States, N.L.: Next Generation Science Standards: For States, by States. The National Academies Press, Washington, DC (2013)

22. Taggar, S., Brown, T.C.: Problem-solving team behaviors: development and validation of BOS and a hierarchical factor structure. Small Group Res. **32**(6), 698–726 (2001)

23. Thulasidasan, S., Chennupati, G., Bilmes, J.A., Bhattacharya, T., Michalak, S.: On mixup training: Improved calibration and predictive uncertainty for deep neural networks. In: Advances in Neural Information Processing Systems, pp. 13888–13899 (2019)

24. Zhang, H., Cisse, M., Dauphin, Y.N., Lopez-Paz, D.: mixup: Beyond empirical risk minimization (2017). arXiv preprint: arXiv:1710.09412

Remix: Rebalanced Mixup

Hsin-Ping Chou[1]([✉]), Shih-Chieh Chang[1], Jia-Yu Pan[2], Wei Wei[2],
and Da-Cheng Juan[2]

[1] Department of Computer Science, National Tsing-Hua University, Hsinchu, Taiwan
alan.durant.chou@gmail.com
[2] Google Research, Mountain View, CA, USA

Abstract. Deep image classifiers often perform poorly when training
data are heavily class-imbalanced. In this work, we propose a new
regularization technique, Remix, that relaxes Mixup's formulation and
enables the mixing factors of features and labels to be disentangled.
Specifically, when mixing two samples, while features are mixed in the
same fashion as Mixup, Remix assigns the label in favor of the minority class by providing a disproportionately higher weight to the minority class. By doing so, the classifier learns to push the decision boundaries towards the majority classes and balance the generalization error
between majority and minority classes. We have studied the state-of-the-
art regularization techniques such as Mixup, Manifold Mixup and Cut-
Mix under class-imbalanced regime, and shown that the proposed Remix
significantly outperforms these state-of-the-arts and several re-weighting
and re-sampling techniques, on the imbalanced datasets constructed by
CIFAR-10, CIFAR-100, and CINIC-10. We have also evaluated Remix on
a real-world large-scale imbalanced dataset, iNaturalist 2018. The experimental results confirmed that Remix provides consistent and significant
improvements over the previous methods.

Keywords: Imbalanced data · Mixup · Regularization · Image
classification

1 Introduction

Deep neural networks have made notable breakthroughs in many fields such as
computer vision [20,27,35], natural language processing [9,18,19] and reinforcement learning [28]. Aside from delicately-designed algorithms and architectures,
training data is one of the critical factors that affects the performance of neural
models. In general, training data needs to be carefully labeled and designed in a
way to achieve a balanced distribution among classes. However, a common problem in practice is that certain classes may have a significantly larger presence in

Electronic supplementary material The online version of this chapter (https://
doi.org/10.1007/978-3-030-65414-6_9) contains supplementary material, which is available to authorized users.

A. Bartoli and A. Fusiello (Eds.): ECCV 2020 Workshops, LNCS 12540, pp. 95–110, 2020.
https://doi.org/10.1007/978-3-030-65414-6_9

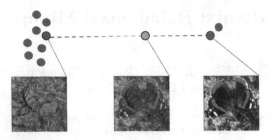

Fig. 1. Assume the label "butterfly" belongs to the minority class and the label "yellow plant" belongs to the majority class. All linear combinations of the two images are on the dashed line. For a mixed image that is 70% "butterfly" and 30% "yellow plant" like the middle one, Mixup will assign the label to be 70% "butterfly" and 30% "yellow plant". However, Remix would assign a label that is in favor of the minority class, e.g., 100% "butterfly". For more details on how exactly Remix assigns labels for the mix images, please refer to Sect. 3 and 4.

the training set than other classes, making the distribution skewed. Such a scenario is referred to as data imbalance. Data imbalance may bias neural networks toward the majority classes when making inferences.

Many previous works have been proposed to mitigate this issue for training neural network models. Most of the existing works can be split into two categories: re-weighting and re-sampling. Re-weighting focuses on tuning the cost (or loss) for different classes. Re-sampling focuses on reconstructing a balanced dataset by either over-sampling the minority classes or under-sampling the majority classes. Both re-weighting and re-sampling have some disadvantages when used for deep neural networks. Re-weighting tends to make optimization difficult under extreme imbalance. Furthermore, it has been shown that re-weighting is not effective when no regularization is applied [4]. Re-sampling is very useful in general especially for over-sampling techniques like SMOTE [2]. However, it is hard to integrate into modern deep neural networks where feature extraction and classification are performed in an end-to-end fashion while over-sampling is done subsequent to feature extraction. This issue is particularly difficult to overcome when training with large-scale datasets.

In order to come up with a solution that is convenient to incorporate for large-scale datasets, we focus on regularization techniques which normally introduce little extra costs. Despite the recent success in regularizations [10,32,36,37], these advanced techniques are often designed for balanced data and evaluated on commonly used datasets (e.g., CIFAR, ImageNet ILSVRC 2012) while real-world datasets tend to have a long-tailed distribution of labels [14,24]. As a result, our motivation is to make the commonly used regularization techniques such as Mixup [37], Manifold Mixup [32] and CutMix [36] perform better in the real-world imbalanced scenario.

The key idea of Mixup is to train a neural model with mixed samples virtually created via the convex combinations of pairs of features and labels. Specifically,

Mixup assumes that linear interpolations of feature vectors should come with linear interpolations of the associated labels using the same mixing factor λ. We observe that this assumption works poorly when the class distribution is imbalanced. In this work, we propose Remix that relaxes the constraint of using the same mixing factor and allows the mixing factors of the features and labels to be different when constructing the virtual mixed samples. Figure 1 illustrates the difference between Remix and Mixup. Note that the mixing factor of labels is selected in a way to provide a better trade-off between the majority and minority classes.

This work brings the following contributions. (a) We propose Remix, a computationally cheap regularization technique to improve the model generalization when training with imbalanced data. (b) The proposed Remix can be applied to all Mixup-based regularizations and can be easily used with existing solutions against data imbalance to achieve better performance. (c) We evaluate Remix extensively on various imbalanced settings and confirm that Remix is general and effective for different scenarios.

2   Related Works

Re-weighting. Re-weighting (cost-sensitive learning) focuses on tuning cost or loss to redefine the importance of each class or sample [3,6,15,34]. In particular, early works [4,11] study on how re-weighting affects the decision boundary in the binary classification case. The naive practice of dealing with an imbalanced dataset is weighted by the inverse class frequency or by the inverse square root of class frequency. Motivated by the observation that each sample might cover a small neighboring region rather than just a single point, Cui *et al.* [7] introduced the concept of "effective number" of a class, which takes the class overlapping into consideration for re-weighting. In general, re-weighting methods perform poorly when the classes are extremely imbalanced, where the performance of majority classes is significantly compromised. As an example, Cao *et al.* [5] show that re-weighting can perform even worse than vanilla training in the extreme setting.

Re-sampling. Re-sampling methods can be summarized into two categories: over-sampling the minority classes [16,25] and under-sampling the majority classes. Both of these methods have drawbacks. Over-sampling the minority classes may cause over-fitting to these samples, and under-sampling majority samples discards data and information, which is wasteful when the imbalance is extreme. For over-sampling, instead of sampling from the same group of data, augmentation techniques are applied to create synthetic samples. Classical methods include SMOTE [2] and ADASYN [12]. The key idea of such methods is to find the k nearest neighbors of a given sample and use the interpolation to create new samples.

Alternative Training Objectives. Novel objectives are also proposed to fight against class imbalance. For example, Focal Loss [21] identifies the class imbalance in object detection task, and the authors proposed to add a modulating term to cross entropy in order to focus on hard negative examples. Although Focal Loss brought significant improvements in object detection tasks, this method is known to be less effective for large-scale imbalanced image classification [5]. Another important work for designing an alternative objective for class imbalance is the Label-Distribution-Aware Margin Loss [5], which is motivated by the recent progress on various margin-based losses [22,23]. Cao *et al.* [5] derived a theoretical formulation to support their proposed method that encourages minority classes to have larger margins and encourage majority classes to have smaller margins.

Other Types. Two competitive state-of-the arts [17,38] focus on the representation learning and classifier learning of a CNN. Kang *et al.* [17] found that it is possible to achieve strong long-tailed recognition ability by adjusting only the classifier. Zhou *et al.* [38] proposed a Bilateral-Branch Network to take care both representation and classifier learning.

Mixup-Based Regularization. Mixup [37] is a regularization technique that proposed to train with interpolations of samples. Despite its simplicity, it works surprisingly well for improving generalization of deep neural networks. Mixup inspires several follow-up works like Manifold Mixup [32], RICAP [29] and Cut-Mix [36] that have shown significant improvement over Mixup. Mixup also shed lights upon other learning tasks such as semi-supervised learning [1,33], adversarial defense [31] and neural network calibration [30].

3 Preliminaries

3.1 Mixup

Mixup [37] was proposed as a regularization technique for improving the generalization of deep neural networks. The general idea of Mixup is to generate mixed sample $\tilde{x}^{MU}$ and $\tilde{y}$ by linearly combining an arbitrary sample pair $(x_i, y_i; x_j, y_j)$ in a dataset $\mathcal{D}$. In Eq. 1, this mixing process is done by using a mixing factor λ which is sampled from the beta distribution.

$$\tilde{x}^{MU} = \lambda x_i + (1 - \lambda)x_j \tag{1}$$

$$\tilde{y} = \lambda y_i + (1 - \lambda)y_j \tag{2}$$

3.2 Manifold Mixup

Instead of mixing samples in the feature space, Manifold Mixup [37] performs the linear combination in the embedding space. This is achieved by randomly performing the linear combination at an eligible layer k and conducting Mixup on $(g_k(x_i), y_i)$ and $(g_k(x_j), y_j)$ where $g_k(x_i)$ denotes a forward pass until layer k. As a result, the mixed representations which we denoted as $\tilde{x}^{MM}$ can be thought of "mixed samples" that is forwarded from layer k to the output layer. Conducting the interpolations in deeper hidden layers which captures higher-level information provides more training signals than Mixup and thus further improve the generalization.

$$\tilde{x}^{MM} = \lambda g_k(x_i) + (1 - \lambda)g_k(x_j) \tag{3}$$

$$\tilde{y} = \lambda y_i + (1 - \lambda)y_j \tag{4}$$

3.3 CutMix

Inspired by Mixup and Cutout [10], rather than mixing samples on the entire input feature space like Mixup does, CutMix [36] works by masking out a patch of it B when generating the synthesized samples where patch B is a masking box with width $r_w = W\sqrt{1-\lambda}$ and height $r_h = H\sqrt{1-\lambda}$ randomly sampled across the image. Here W and H are the original width and height of the image, respectively. The generated block makes sure that the proportion of the image being masked out is equal to $\frac{r_w r_h}{WH} = 1 - \lambda$. A image level mask M is then generated based on B with elements equal to 0 when it is blocked by B and 1 otherwise. CutMix is defined in a way similar to Mixup and Manifold Mixup in Eq. 6. Here $\odot$ is element-wise multiplication and we denote M to be generated by a random process that involves W, H, and λ using a mapping $f(\cdot)$

$$\tilde{x}^{CM} = \mathbf{M} \odot x_i + (\mathbf{1} - \mathbf{M}) \odot x_j \tag{5}$$

$$\tilde{y} = \lambda y_i + (1 - \lambda)y_j \tag{6}$$

$$\mathbf{M} \sim f(\cdot | \lambda, W, H) \tag{7}$$

4 Remix

We observe that both Mixup, Manifold Mixup, and CutMix use the same mixing factor λ for mixing samples in both feature space and label space. We argue that it does not make sense under the imbalanced data regime and propose to disentangle the mixing factors. After relaxing the mixing factors, we are able to assign a higher weight to the minority class so that we can create labels that are in favor to the minority class. Before we further introduce our method. We first show the formulation of Remix as below:

$$\tilde{x}^{RM} = \lambda_x x_i + (1 - \lambda_x)x_j \tag{8}$$

$$\tilde{y}^{RM} = \lambda_y y_i + (1 - \lambda_y)y_j \tag{9}$$

The above formulation is in a more general form compares to other Mixup-based methods. In fact, $\tilde{x}^{RM}$ can be generated based on Mixup, Manifold Mixup, and CutMix according to Eq. 1, Eq. 3 and Eq. 5 respectively. Here we use Mixup

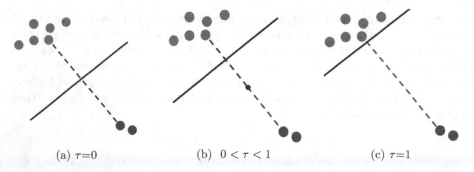

(a) $\tau=0$ (b) $0<\tau<1$ (c) $\tau=1$

Fig. 2. A simple illustration of how the hyper-parameter τ affects the boundary. Blue and red dots are majority and minority samples in the feature space. The dashed line represents all the possible mixed samples and the solid black line represents the decision boundary. When τ is set to 0, mixed samples are linearly mixed and labelled as the original Mixup, Manifold Mixup, and Cutmix algorithms. But when τ is set to a value larger than zero, then part of the mixed samples on the red dashed line will be labelled as the minority class. In the most extreme case for fighting against data imbalance, τ is set to 1 where all mixed samples are labelled as the minority class. (Color figure online)

for the above formulation as an example. Note that Eq. 9 relaxes the mixing factors which are otherwise tightly coupled in the original Mixup's formulation. Mixup, Manifold Mixup, and Cutmix are a special case when $\lambda_y = \lambda_x$. Again, λ_x is sampled from the beta distribution and we define the exact form of λ_y as in Eq. 10.

$$\lambda_y = \begin{cases} 0, & n_i/n_j \geq \kappa \text{ and } \lambda < \tau; \\ 1, & n_i/n_j \leq 1/\kappa \text{ and } 1-\lambda < \tau; \\ \lambda, & \text{otherwise} \end{cases} \tag{10}$$

Here n_i and n_j denote the number of samples in the corresponding classification class from sample i and sample j. For example, if $y_i = 1$ and $y_j = 10$, n_i and n_j would be the number of samples for class 1 and 10, which are the class that these two samples represent. κ and τ are two hyper-parameters in our method. To understand what Eq. 10 is about, we first define the κ-majority below.

Definition 1. κ-*Majority. A sample* (x_i, y_i), *is considered to be* κ-*majority than sample* (x_j, y_j), *if* $n_i/n_j \geq \kappa$ *where* n_i *and* n_j *represent the number of samples that belong to class* y_i *and class* y_j, *respectively.*

The general idea in Eq. 10 shows when exactly Remix assigns the synthesized labels in favor to the minority class. When x_i is κ-majority to x_j and the other condition is met, λ_y is set to 0 which makes the synthesized labels 100% contributed by the minority class y_j. Conversely, when x_j is κ-majority to x_i along with other conditions, λ_y to 1 which makes the synthesized labels 100% contributed by the minority class y_i.

The reason behind this choice of making the synthesized samples to be labeled as the minority class is to move the decision boundary towards the majority class. This is aligned with the consensus in the community of imbalanced classification problem. In [5], the authors gave a rather theoretical analysis illustrating that how exactly the decision boundary should be pushed towards the majority classes by using a margin loss. Because pushing the decision boundary towards too much may hurt the performance of the majority class. As a result, we don't want the synthesized labels to be always pointing to the minority class whenever mixing a majority class sample and a minority class sample. To achieve that we have introduced another condition in Eq. 10 controlled by parameter τ that is conditioned on λ. In both conditions, extreme cases will be rejected and λ_y will be set to λ when λ_x is smaller than τ. The geometric interpretation of this condition can be visualized in Fig. 2. Here we see that when τ is set to 0, our approach will degenerate to the base method, which can be Mixup, Manifold Mixup, or Cutmix depending on the choice of design. When τ is set to a value that is larger than 0, synthesized samples $\tilde{x}^{RM}$ that are close to the minority classes will be labelled as the minority class, thus benefiting the imbalanced classification problems. To summarize, τ controls the extent that the minority samples would dominate the label of the synthesized samples. When the conditions are not met, or in other words, when there is no trade-off to be made, we will just use the base method to generate the labels. This can be illustrated in the last condition in Eq. 10, when none of i or j can claim to be κ-majority over each other and in this case λ_y is set to λ_x.

Attentive readers may realize that using the same τ for various pairs of a majority class and a minority class implies that we want to enforce the same trade-off for those pairs. One may wonder why not introduce τ_{ij} for each pair of classes? This is because the trade-off for multi-class problem is intractable to find. Hence, instead of defining τ_{ij} for each pair of classes, we use a single τ for all pairs of classes. Despite its simplicity, using a single τ is sufficient to achieve a better trade-off.

Remix might look similar to SMOTE [2] and ADASYN [12] at first glance, but they are very different in two perspectives. First, the interpolation of Remix can be conducted with any two given samples while SMOTE and ADASYN rely on the knowledge of a sample's same-class neighbors before conducting the interpolation. Moreover, rather than only focusing on creating new data points, Remix also pays attention to labelling the mixed data which is not an issue to SMOTE since the interpolation is conducted between same-class data points. Secondly, Remix follows Mixup to train the classifier only on the mixed samples while SMOTE and ADASYN train the classifier on both original data and synthetic data.

To give a straightforward explanation of why Remix would benefit learning with imbalance datasets, consider a mixed example $\tilde{x}$ between a majority class and a minority class, the mixed sample includes features of both classes yet we mark it as the minority class more. This force the neural network model to learn that when there are features of a majority class and a minority class appearing in

---

Algorithm 1. Remix

Require: Dataset $\mathcal{D} = \{(x_i, y_i)\}_{i=1}^{n}$. A model with parameter θ
1: Initialize the model parameters θ randomly
2: **while** θ is not converged **do**
3:      $\{(x_i, y_i), (x_j, y_j)\}_{m=1}^{M} \leftarrow \text{SamplePairs}(\mathcal{D}, M)$
4:      $\lambda_x \sim Beta(\alpha, \alpha)$
5:      **for** $m = 1$ to M **do**
6:          $\tilde{x}^{RM} \leftarrow \text{RemixImage}(x_i, x_j, \lambda_x)$ according to Eq. 8
7:          $\lambda_y \leftarrow \text{LabelMixingFactor}(\lambda_x, n_i, n_k, \tau, \kappa)$ according to Eq. 10
8:          $\tilde{y}^{RM} \leftarrow \text{RemixLabel}(y_i, y_j, \lambda_y)$ according to Eq. 9
9:      **end for**
10:      $\mathcal{L}(\theta) \leftarrow \frac{1}{M} \sum_{(\tilde{x}, \tilde{y})} L((\tilde{x}, \tilde{y}); \theta)$
11:      $\theta \leftarrow \theta - \delta \nabla_\theta \mathcal{L}(\theta)$
12: **end while**

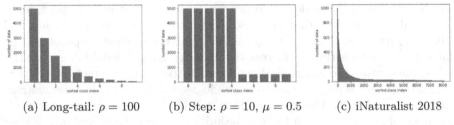

(a) Long-tail: $\rho = 100$ (b) Step: $\rho = 10$, $\mu = 0.5$ (c) iNaturalist 2018

Fig. 3. Histograms of three imbalanced class distributions. (a) and (b) are synthesized from long-tail and step imbalance, respectively. (c) represents the class distribution of iNaturalist 2018 dataset.

the sample, it should more likely to consider it as the minority class. This means that the classifier is being less strict to the minority class. Please see Sect. 5.5 for qualitative analysis. Note that Remix method is a relaxation technique, and thus may be integrated with other techniques. In the following experiments, besides showing the results of the pure Remix method, we also show that the Remix method can work together with the re-weighting or the re-sampling techniques. Algorithm 1 shows the pseudo-code of the proposed Remix method.

5   Experiments

5.1   Datasets

We compare the proposed Remix with state-of-the-art methods fighting against class imbalance on the following datasets: (a) artificially created imbalanced datasets using CIFAR-10, CIFAR-100 and CINIC-10 datasets, and (b) iNaturalist 2018, a real-world and large-scale imbalanced dataset.

Imbalanced CIFAR. The original CIFAR-10 and CIFAR-100 datasets both contain 50,000 training images and 10,000 validation images of size 32×32, with

10 and 100 classes, respectively. We follow [5] to construct class-imbalanced datasets from the CIFAR-10 and CIFAR-100 datasets with two common imbalance types "long-tailed imbalance" and "step imbalance". The validation set is kept balanced as original. Figure 3(a)(b) illustrates the two imbalance types. For a dataset with long-tailed imbalance, the class sizes (number of samples in the class) of the dataset follow an exponential decay. For a dataset with step imbalance, a parameter μ is used to denotes the fraction of minority classes. μ is set to 0.5 [5] for all of the experiments. The parameter ρ of the constructed datasets denotes the imbalance ratio between the number of samples from the most frequent and the least frequent classes, i.e., $\rho = \max_i\{n_i\} / \min_j\{n_j\}$.

Imbalanced CINIC. The CINIC-10 dataset [8] is compiled by combining CIFAR-10 images with images downsampled from the ImageNet database. It contains 270,000 images and it splits train, validation, and test set equally. We only use the official split of training and validation. Using the CINIC-10 dataset helps us compare different methods better because it has 9000 training data per class which allows us to conduct extensive experiments with various imbalance ratios while making sure each class still preserves a certain number of data. This helps us focus more on the imbalance between classes rather than solving a few-shot classification problem for the minority classes.

iNaturalist 2018. The iNaturalist species classification dataset [13] is a real-world large-scale imbalanced dataset which has 437,513 training images of 8,142 classes. The dataset features many visually similar species which are extremely difficult to accurately classify without expert knowledge. We adopt the official training and validation splits for our experiments where the training datasets have a long-tailed label distribution and the validation set is designed to have a balanced distribution.

CIFAR and CINIC-10. For fair comparisons, we ported the official code of [5] into our codebase. We follow [5,7] use ResNet-32 for all CIFAR experiments and we use ResNet-18 for all CINIC experiments. We train 300 epochs and decay the learning rate 0.01 at 150, 225 epoch. We use stochastic gradient descent (SGD) with momentum 0.9 and weight decay 0.0002. For non-LDAM methods, We train it for 300 epochs with mini-batch size 128. We decay our learning rate by 0.1 at 150, 250 epoch. All CIFAR and CINIC experiment results are mean over 5 runs. Standard data augmentation is applied, which is the combination of random crop, random horizontal flip and normalization. If DRW or DRS are used for the training, we use re-weighting or re-sampling at the second learning rate decay. We set $\tau = 0.5$ and $\kappa = 3$ for all experiments. The choice of these two parameters can be found with simple grid search. Despite that one might want to use carefully-tuned parameters for different imbalance scenarios, we empirically found that setting $\tau = 0.5$ and $\kappa = 3$ is able to provide consistent improvements over the previous state-of-the-arts.

Table 1. Top-1 accuracy of on imbalanced CIFAR-10 and CIFAR-100. † denotes the results from the original paper.

Dataset	Imbalanced CIFAR-10				Imbalanced CIFAR-100			
Imbalance type	Long-tailed		Step		Long-tailed		Step	
Imbalance ratio	100	10	100	10	100	10	100	10
ERM	71.86	86.22	64.17	84.02	40.12	56.77	40.13	54.74
Focal [21] †	70.18	86.66	63.91	83.64	38.41	55.78	38.57	53.27
RW Focal [7] †	74.57	87.1	60.27	83.46	36.02	57.99	19.75	50.02
DRS [5] †	74.50	86.72	72.03	85.17	40.33	57.26	41.35	56.79
DRW [5] †	74.86	86.88	71.60	85.51	40.66	57.32	41.14	57.22
LDAM [5] †	73.35	86.96	66.58	85.00	39.60	56.91	39.58	56.27
LDAM-DRW [5]	76.57	86.7	75.94	86.52	42.64	57.18	45.40	57.09
Mixup [37]	73.09	88.00	65.80	85.20	40.83	58.37	39.64	54.46
Remix	75.36	88.15	68.98	86.34	41.94	59.36	39.96	57.06
BBN [38] †	**79.82**	88.32	–	–	42.56	59.12	–	–
Remix-LDAM-DRW	79.33	86.78	77.81	86.46	45.02	59.47	45.32	56.59
Remix-LDAM-DRS	79.45	87.16	78.00	86.91	45.66	59.21	45.74	56.19
Remix-RS	76.23	87.70	67.28	86.63	41.13	58.62	39.74	56.09
Remix-RW	75.1	87.91	68.74	86.38	33.51	57.65	17.42	54.45
Remix-DRS	79.53	88.85	77.46	88.16	46.53	60.52	**47.25**	**60.76**
Remix-DRW	79.76	**89.02**	**77.86**	**88.34**	**46.77**	**61.23**	46.78	60.44

iNaturalist 2018. We use ResNet-50 as the backbone network across all experiments for iNaturalist 2018. Each image is first resized to 256×256, and then a 224×224 crop is randomly sampled from an image or its horizontal flip. Then color jittering and lighting are applied. Follow [5,7], we train the network for 90 epochs with an initial learning rate of 0.1 and mini-batch size 256. We anneal the learning rate at epoch 30 and 60. For the longer training schedule, we train the network for 200 epochs with an initial learning rate of 0.1 and anneal the learning rate at epoch 75 and 150. Using the longer training schedule is necessary for Mixup-based regularizations to converge [36,37]. We set $\tau = 0.5$ and $\kappa = 3$.

Baseline Methods for Comparison. We compare our methods with vanilla training, state-of-the-art techniques and their combinations. (1) Empirical risk minimization (ERM): Standard training with no anti-imbalance techniques involved. (2) Focal: Use focal loss instead of cross entropy. (3) Re-weighting (RW): Re-weight each sample by the effective number which is defined as $E_n = (1 - \beta^n)/(1 - \beta)$, where $\beta = (N - 1)/N$. (4) Re-sampling (RS): Each example is sampled with probability proportional to the inverse of effective number. (5) Deferred re-weighting and deferred re-sampling (DRW, DRS): A deferred training procedure which first trains using ERM before annealing the learning

Table 2. Top-1 accuracy on imbalanced CINIC-10 using ResNet-18.

Imbalance type	Long-tailed				Step			
Imbalance ratio	200	100	50	10	200	100	50	10
ERM	56.16	61.82	72.34	77.06	51.64	55.64	68.35	74.16
RS [7]	53.71	59.11	71.28	75.99	50.65	53.82	65.54	71.33
RW [7]	54.84	60.87	72.62	76.88	50.47	55.91	69.24	74.81
DRW [5]	59.66	63.14	73.56	77.88	54.41	57.87	68.76	72.85
DRS [5]	57.98	62.16	73.14	77.39	52.67	57.41	69.52	75.89
Mixup [37]	57.93	62.06	74.55	79.28	53.47	56.91	69.74	75.59
Remix	58.86	63.21	75.07	79.02	54.22	57.57	70.21	76.37
LDAM-DRW [5]	60.80	65.51	74.94	77.90	54.93	61.17	72.26	76.12
Remix-DRS	61.64	65.95	75.34	79.17	60.12	66.53	75.47	**79.86**
Remix-DRW	**62.95**	**67.76**	**75.49**	**79.43**	**62.82**	**67.56**	**76.55**	79.36

rate, and then deploys RW or RS (6) LDAM: A label-distribution-aware margin loss which considers the trade-off between the class margins. (7) Mixup: Each batch of training samples are generated according to [37] (8) BBN [38] and LWS [17]: Two state-of-the-arts methods. We directly copy the results from the original paper.

5.2 Results on Imbalanced CIFAR and CINIC

In Table 1 and Table 2, we compare the previous state-of-the-arts with the pure Remix method and a variety of the Remix-integrated methods. Specifically, we first integrate Remix with basic re-weighting and re-sampling techniques with respect to the effective number [7], and we use the deferred version of them [5]. We also experiment with a variety that integrate our method with the LDAM loss [5]. We observe that Remix works particularly well with re-weighting and re-sampling. Among all the methods that we experiment with, the best performance was achieved by the method that integrates Remix with either the deferred re-sampling method (Remix-DRS) or the deferred re-weighting method (Remix-DRW).

Regarding the reason why Remix-DRS and Remix-DRW achieve the best performance, we provide an intuitive explanation as the following: We believe that the improvement of our Remix method comes from the imbalance-aware mixing equation for labels (Eq. 10), particularly when λ_y is set to either 0 or 1. The more often the conditions are satisfied to make λ_y be either 0 or 1, the more opportunity is given to the learning algorithm to adjust for the data imbalance. To increase the chance that those conditions are satisfied, we need to have more pairs of training samples where each pair is consisted of one sample of a majority class and one sample of a minority class. Since, by the definition of a minority class, there are not many samples from a minority class, the chance of forming such pairs of training data may not be very high.

Table 3. Validation errors on iNaturalist 2018 using ResNet-50.

Loss	Schedule	Top-1	Top-5
ERM	SGD	40.19	18.99
RW Focal	SGD	38.88	18.97
Mixup	SGD	39.69	17.88
Remix	SGD	38.69	17.70
LDAM	DRW	32.89	15.20
BBN [38] †	—	30.38	—
LWS (200 epoch) [17] †	—	30.5	—
LDAM (200 epoch)	DRW	31.42	14.68
Remix (200 epoch)	DRS	**29.26**	**12.55**
Remix (200 epoch)	DRW	**29.51**	**12.73**

Table 4. Top-1 accuracy of ResNet-18 trained with imbalanced CIFAR-10 with imbalance ratio $\rho = 100$

Imbalance type	Long-tailed	Step
Mixup-DRW	81.09	76.13
Remix-DRW	**81.60**	**79.35**
Mixup-DRS	80.40	75.58
Remix-DRS	**81.11**	**79.35**

When the re-sampling method is used with Remix, the re-sampling method increases the probability of having data of minority classes in the training batches. With more samples from minority classes in the training batches, the chance of forming sample pairs that satisfied the conditions is increased, thus allows Remix to provide better trade-off on the data imbalance among classes.

Likewise, although using Remix with re-weighting does not directly increase the probability of pairs that satisfy the conditions, the weights assigned to the minority classes will amplify the effect when a case of majority-minority sample pair is encountered, and thus, will also guide the classifier to have better trade-off on the data imbalance.

On the other hand, using the LDAM loss doesn't further improve the performance. We suspect that the trade-off LDAM intends to make is competing with the trade-off Remix intends to make. Therefore, for the rest of the experiments, we will focus more on the methods where DRS and DRW are integrated with Remix.

5.3   Results on iNaturalist 2018

In Table 3, we present the results in similar order as Table 2. The results show the same trends to the results on CIFAR and CINIC. Again, our proposed method, Remix, outperforms the original Mixup. Also, state-of-the-art results are achieved when Remix is used with re-weighting or re-sampling techniques. The model's performance is significantly better than the previous state-of-the-arts and outperforms the baseline (ERM) by a large margin. The improvement is significant and more importantly, the training cost remains almost the same in terms of training time.

5.4   Ablation Studies

When it comes to addressing the issue of data imbalance, one common approach is to simply combine the re-sampling or re-weighting method with Mixup. In Table 4, we show the comparison of Mixup and Remix when they are integrated with the re-sampling technique (DRS) or the re-weighting technique (DRW). Note that our methods still outperform Mixup-based methods. The results in Table 4 imply that, while Remix can be considered as over-sampling the minority classes in the label space, the performance gain from Remix does not completely overlap with the gains from the re-weighting or re-sampling techniques.

We also observe that the improvement is more significant on datasets with step imbalance, than it is on datasets with long-tailed imbalance. In the long-tailed setting, the distribution of the class sizes makes it less likely to have pairs of data samples that satisfy the conditions of Eq. 10 to make λ_y either 0 or 1. On the other hand, the conditions of Eq. 10 are relatively more likely to be satisfied, on a dataset with the step imbalance. In other words, there is less room for Remix to unleash its power on a dataset with long-tailed setting.

The proposed Remix method is general and can be applied with other Mixup-based regularizations, such as Manifold Mixup and CutMix. In Table 5, we show that, the performance of Manifold Mixup and CutMix increases when they employ the Remix regularization, significantly outperforming the performance of the vanilla Manifold Mixup or CutMix.

Moreover, we also observe that when the imbalance ratio is not very extreme ($\rho = 10$), using Mixup or Remix doesn't produce much difference. However, when the imbalance is extreme (e.g., $\rho = 100$), employing our proposed method is significantly better than the vanilla version of Manifold Mixup or CutMix.

Table 5. Top-1 accuracy of ResNet-32 on imbalanced CIFAR-10 and CIFAR-100.

Dataset	Imbalanced CIFAR-10				Imbalanced CIFAR-100			
Imbalance type	Long-tailed		Step		Long-tailed		Step	
Imbalance ratio	100	10	100	10	100	10	100	10
ERM	71.86	86.22	64.17	84.02	40.12	56.77	40.13	54.74
Mixup [37]	73.09	88.00	65.80	85.20	40.83	58.37	39.64	54.46
Remix	**75.36**	**88.15**	**68.98**	**86.34**	**41.94**	**59.36**	**39.96**	**57.06**
Manifold mixup [32]	73.47	87.78	66.13	85.22	41.19	58.55	39.52	53.72
Remix-MM	**77.07**	**88.70**	**69.78**	**87.39**	**44.12**	**60.76**	**40.22**	**58.01**
CutMix [36]	75.04	88.30	67.97	86.35	41.86	59.47	40.23	56.59
Remix-CM	**76.59**	87.96	**69.61**	**87.59**	**43.55**	**60.15**	**40.336**	**57.78**

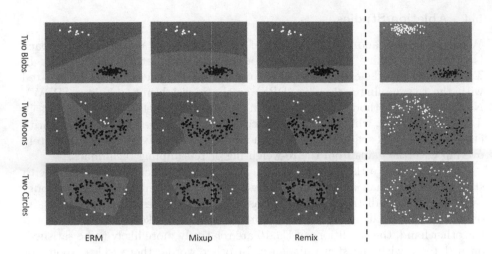

Fig. 4. A visualization of the decision boundary of the models learned by different methods. Remix creates tighter margin for the majority class, and compensates the effect from the data imbalance.

5.5   Qualitative Analysis

To further demonstrate the effect of Remix we present the results of Remix on the commonly used datasets, namely, "two blobs", "two moons", and "two circles" datasets from scikit-learn [26]. In Fig. 4, the original balanced datasets are shown at the rightmost column. The three columns at the left show the created imbalanced datasets with the imbalance ratio $\rho = 10$. The majority class is plotted in black and the minority class is plotted in white. The results in Fig. 4 show that Remix creates tighter margin for the majority class. In all three cases, we observe that even though our regularization sacrifices some training accuracy for the majority class (some black dots are misclassified), but it actually provides a better decision boundary for the minority class.

6   Conclusions and Future Work

In this paper, we redesigned the Mixup-based regularizations for imbalanced data, called Remix. It relaxes the mixing factor which results in pushing the decision boundaries towards majority classes. Our method is easy to implement, end-to-end trainable and computation efficient which are critical for training on large-scale imbalanced datasets. We also show that it can be easily used with existing techniques to achieve superior performance. Despite the ease of use and effectiveness, the current analysis is rather intuitive. Our future work is to dig into the mechanism and hopefully provide more in-depth analysis and theoretical guarantees which may shed lights on a more ideal form of Remix as the current one involves two hyper-parameters.

References

1. Berthelot, D., Carlini, N., Goodfellow, I.G., Papernot, N., Oliver, A., Raffel, C.: Mixmatch: a holistic approach to semi-supervised learning. In: NeurIPS (2019)
2. Bowyer, K.W., Chawla, N.V., Hall, L.O., Kegelmeyer, W.P.: Smote: synthetic minority over-sampling technique. J. Artif. Intell. Res. **16**, 321–357 (2002)
3. Bulò, S.R., Neuhold, G., Kontschieder, P.: Loss max-pooling for semantic image segmentation. In: 2017 IEEE Conference on Computer Vision and Pattern Recognition (CVPR), pp. 7082–7091 (2017)
4. Byrd, J., Lipton, Z.C.: What is the effect of importance weighting in deep learning? In: ICML (2018)
5. Cao, K., Wei, C., Gaidon, A., Arechiga, N., Ma, T.: Learning imbalanced datasets with label-distribution-aware margin loss. In: Advances in Neural Information Processing Systems (2019)
6. Chung, Y.A., Lin, H.T., Yang, S.W.: Cost-aware pre-training for multiclass cost-sensitive deep learning. In: IJCAI (2015)
7. Cui, Y., Jia, M., Lin, T.Y., Song, Y., Belongie, S.: Class-balanced loss based on effective number of samples. In: CVPR (2019)
8. Darlow, L.N., Crowley, E., Antoniou, A., Storkey, A.J.: CINIC-10 is not imagenet or CIFAR-10. ArXiv abs/1810.03505 (2018)
9. Devlin, J., Chang, M.W., Lee, K., Toutanova, K.: Bert: pre-training of deep bidirectional transformers for language understanding. In: NAACL-HLT (2019)
10. DeVries, T., Taylor, G.W.: Improved regularization of convolutional neural networks with cutout. arXiv preprint arXiv:1708.04552 (2017)
11. Elkan, C.: The foundations of cost-sensitive learning. In: IJCAI (2001)
12. He, H., Bai, Y., Garcia, E.A., Li, S.: ADASYN: adaptive synthetic sampling approach for imbalanced learning. In: 2008 IEEE International Joint Conference on Neural Networks (IEEE World Congress on Computational Intelligence), pp. 1322–1328 (2008)
13. Horn, G.V., et al.: The iNaturalist species classification and detection dataset. In: 2018 IEEE/CVF Conference on Computer Vision and Pattern Recognition, pp. 8769–8778 (2017)
14. Horn, G.V., Perona, P.: The devil is in the tails: Fine-grained classification in the wild. ArXiv abs/1709.01450 (2017)
15. Huang, C., Li, Y., Loy, C.C., Tang, X.: Learning deep representation for imbalanced classification. In: 2016 IEEE Conference on Computer Vision and Pattern Recognition (CVPR), pp. 5375–5384 (2016)
16. Kim, J., Jeong, J., Shin, J.: Imbalanced classification via adversarial minority over-sampling. In: OpenReview (2019). https://openreview.net/pdf?id=HJxaC1rKDS
17. Kang, B., et al.: Decoupling representation and classifier for long-tailed recognition. In: International Conference on Learning Representations (2020). https://openreview.net/forum?id=r1Ddp1-Rb
18. Lample, G., Ott, M., Conneau, A., Denoyer, L., Ranzato, M.: Phrase-based & neural unsupervised machine translation. In: EMNLP (2018)
19. Lan, Z.Z., Chen, M., Goodman, S., Gimpel, K., Sharma, P., Soricut, R.: Albert: a lite bert for self-supervised learning of language representations. ArXiv abs/1909.11942 (2019)
20. Li, Z., et al.: Learning the depths of moving people by watching frozen people. In: CVPR (2019)

21. Lin, T.Y., Goyal, P., Girshick, R.B., He, K., Dollár, P.: Focal loss for dense object detection. IEEE Trans. Pattern Anal. Mach. Intell. (2017)
22. Liu, W., Wen, Y., Yu, Z., Li, M., Raj, B., Song, L.: Sphereface: deep hypersphere embedding for face recognition. In: 2017 IEEE Conference on Computer Vision and Pattern Recognition (CVPR), pp. 6738–6746 (2017)
23. Liu, W., Wen, Y., Yu, Z., Yang, M.: Large-margin softmax loss for convolutional neural networks. In: ICML (2016)
24. Liu, Z., Miao, Z., Zhan, X., Wang, J., Gong, B., Yu, S.X.: Large-scale long-tailed recognition in an open world. In: CVPR (2019)
25. Mullick, S.S., Datta, S., Das, S.: Generative adversarial minority oversampling. In: The IEEE International Conference on Computer Vision (ICCV), October 2019
26. Pedregosa, F., et al.: Scikit-learn: machine learning in python. J. Mach. Learn. Res. **12**, 2825–2830 (2011)
27. Shaham, T.R., Dekel, T., Michaeli, T.: Singan: learning a generative model from a single natural image. In: The IEEE International Conference on Computer Vision (ICCV), October 2019
28. van Steenkiste, S., Greff, K., Schmidhuber, J.: A perspective on objects and systematic generalization in model-based RL. ArXiv abs/1906.01035 (2019)
29. Takahashi, R., Matsubara, T., Uehara, K.: Ricap: random image cropping and patching data augmentation for deep CNNs. In: Proceedings of The 10th Asian Conference on Machine Learning (2018)
30. Thulasidasan, S., Chennupati, G., Bilmes, J.A., Bhattacharya, T., Michalak, S.E.: On mixup training: improved calibration and predictive uncertainty for deep neural networks. In: NeurIPS (2019)
31. Tianyu Pang, K.X., Zhu, J.: Mixup inference: better exploiting mixup to defend adversarial attacks. In: ICLR (2020)
32. Verma, V., et al.: Manifold mixup: better representations by interpolating hidden states. In: Chaudhuri, K., Salakhutdinov, R. (eds.) Proceedings of the 36th International Conference on Machine Learning. Proceedings of Machine Learning Research, vol. 97, pp. 6438–6447. PMLR, Long Beach, California, USA, 09–15 Jun 2019. http://proceedings.mlr.press/v97/verma19a.html
33. Verma, V., Lamb, A., Kannala, J., Bengio, Y., Lopez-Paz, D.: Interpolation consistency training for semi-supervised learning. In: IJCAI (2019)
34. Wang, S., Liu, W., Wu, J., Cao, L., Meng, Q., Kennedy, P.J.: Training deep neural networks on imbalanced data sets. In: 2016 International Joint Conference on Neural Networks (IJCNN), pp. 4368–4374 (2016)
35. Wang, T.C., et al.: Video-to-video synthesis. In: NeurIPS (2018)
36. Yun, S., Han, D., Oh, S.J., Chun, S., Choe, J., Yoo, Y.: Cutmix: regularization strategy to train strong classifiers with localizable features. In: International Conference on Computer Vision (ICCV) (2019)
37. Zhang, H., Cissé, M., Dauphin, Y., Lopez-Paz, D.: mixup: Beyond empirical risk minimization. In: International Conference on Learning Representations (2018). https://openreview.net/forum?id=r1Ddp1-Rb
38. Zhou, B., Cui, Q., Wei, X.S., Chen, Z.M.: BBN: bilateral-branch network with cumulative learning for long-tailed visual recognition. In: IEEE/CVF Conference on Computer Vision and Pattern Recognition (CVPR), June 2020

Generalized Many-Way Few-Shot Video Classification

Yongqin Xian[1]([✉]), Bruno Korbar[2], Matthijs Douze[2], Bernt Schiele[1],
Zeynep Akata[1,3], and Lorenzo Torresani[2]

[1] Max Planck Institute for Informatics, Saarbrücken, Germany
yxian@mpi-inf.mpg.de
[2] Facebook AI, Menlo Park, USA
[3] University of Tübingen, Tübingen, Germany

Abstract. Few-shot learning methods operate in low data regimes. The aim is to learn with few training examples per class. Although significant progress has been made in few-shot image classification, few-shot video recognition is relatively unexplored and methods based on 2D CNNs are unable to learn temporal information. In this work we thus develop a simple 3D CNN baseline, surpassing existing methods by a large margin. To circumvent the need of labeled examples, we propose to leverage weakly-labeled videos from a large dataset using tag retrieval followed by selecting the best clips with visual similarities, yielding further improvement. Our results saturate current 5-way benchmarks for few-shot video classification and therefore we propose a new challenging benchmark involving more classes and a mixture of classes with varying supervision.

1 Introduction

In the video domain annotating data is time-consuming due to the additional time dimension. The lack of labeled training data is more prominent for fine-grained action classes at the "tail" of the skewed long-tail distribution (see Fig. 1), e.g., "arabesque ballet". It is thus important to study video classification in the limited labeled data regime. Visual recognition methods that operate in the few-shot learning setting aim to generalize a classifier trained on *base classes* with enough training data to *novel classes* with only a few labeled training examples. While considerable attention has been devoted to this scenario in the image domain [4,24,25,37], few-shot video classification is relatively unexplored.

Existing few-shot video classification approaches [2,43] are mostly based on frame-level features extracted from a 2D CNN, which essentially ignores the important temporal information. Although additional temporal modules have been added at the top of a pre-trained 2D CNN, necessary temporal cues may be lost when temporal information is learned on top of static image features. We argue that under-representing temporal cues may negatively impact the robustness of the classifier. In fact, in the few-shot scenario it may be risky for the model to rely exclusively on appearance and context cues extrapolated from

© Springer Nature Switzerland AG 2020
A. Bartoli and A. Fusiello (Eds.): ECCV 2020 Workshops, LNCS 12540, pp. 111–127, 2020.
https://doi.org/10.1007/978-3-030-65414-6_10

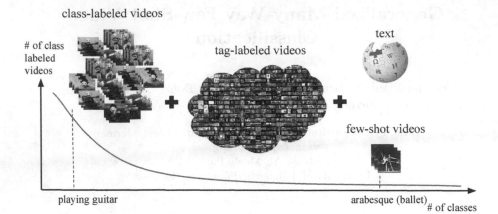

Fig. 1. Our 3D CNN approach combines a few class-labeled videos (time-consuming to obtain) and tag-labeled videos. It saturates existing benchmarks, so we move to a more challenging generalized many-way few-shot video classification task.

the few available examples. In order to make temporal information available we propose to represent the videos by means of a 3D CNN.

While obtaining labeled videos for target classes is time-consuming and challenging, there are many videos tagged by users available on the internet. For example, there are 400,000 tag-labeled videos in the YFCC100M [32] dataset. Our second goal is thus to leverage such tag-labeled videos (Fig. 1) to alleviate the lack of labeled training data.

Existing experimental settings for few-shot video classification [2,43] are limited. Predicting a label among just 5 novel classes in each testing episode is in fact relatively easy. Moreover, restricting the label space to only novel classes at test time, and ignoring the base classes is unrealistic. In real-world applications test videos are expected to belong to any class.

In this work, our goal is to push the progress of few-shot video classification in three ways: 1) To learn the temporal information, we revisit spatiotemporal CNNs in the few-shot video classification regime. We develop a 3D CNN baseline that maintains significant temporal information within short clips; 2) We propose to retrieve relevant videos annotated with tags from a large video dataset (YFCC100M) to circumvent the need for labeled videos of novel classes; 3) We extend current few-shot video classification evaluation settings by introducing two challenges. In our *generalized few-shot video classification* task, the label space has no restriction in terms of classes. In many-way few-shot video classification with, the number of classes goes well beyond five, and towards all available classes. Our extensive experimental results demonstrate that on existing settings spatiotemporal CNNs outperform the state-of-the-art by a large margin, and on our proposed settings weakly-labeled videos retrieved using tags successfully tackles both of our new few-shot video classification tasks.

2   Related Work

Low-Shot Learning Setup. The low-shot image classification [14,23,25] setting uses a large-scale fully labeled dataset for pre-training a DNN on the base classes, and a low-shot dataset with a small number of examples from a disjoint set of novel classes. The terminology "k-shot n-way classification" means that in the low-shot dataset there are n distinct classes and k examples per class for training. Evaluating with few examples (k small) is bound to be noisy. Therefore, the k training examples are often sampled several times and results are averaged [6,14]. Many authors focus on cases where the number of classes n is small as well, which amplifies the measurement noise. For that case [25] introduces the notion of "episodes". An episode is one sampling of n classes and k examples per class, and the accuracy measure is averaged over episodes. It is feasible to use distinct datasets for pre-training and low-shot evaluation. However, to avoid dataset bias [33] it is easier to split a large supervised dataset into disjoint sets of "base" and "novel" classes. The evaluation is often performed only on novel classes, except [14,28,41] who evaluate on the combination of base and novel classes.

Recently, a low-shot video classification setup has been proposed [7,43]. They use the same type of decomposition of the dataset as [25], with learning episodes and random sampling of low-shot classes. In this work, we follow and extend the evaluation protocol of [43].

Tackling Low-Shot Learning. The simplest low-shot learning approach is to extract embeddings from the images using the pre-trained DNN and train a linear classifier [1] or logistic regression [14] on these embeddings using the k available training examples. Another approach is to cast low-shot learning as a nearest-neighbor classifier [40]. The "imprinting" approach [24] builds a linear classifier from the embeddings of training examples, then fine-tunes it. As a complementary approach, [18] has looked into exploiting noisy labels to aid classification. In this work, we use videos from YFCC100M [32] retrieved by tags to augment and improve training of our few-shot classifier.

In a meta-learning setup, the low-shot classifier is assumed to have hyperparameters or parameters that must be adjusted before training. Thus, there is a preliminary meta-learning step that consists in training those parameters on simulated episodes sampled from the base classes. Both matching networks [37] and prototypical networks [30] employ metric learning to "meta-learn" deep features and adopt a nearest neighbor classifier. In MAML [11], the embedding classifier is meta-learned to adapt quickly and without overfitting to fine-tuning. Ren et al. [26] introduce a semi-supervised meta-learning approach that includes unlabeled examples in each training episode. While that method holds out a subset from the same target dataset as the unlabeled images, our retrieval-enhanced approach leverages weakly-labeled videos from another heterogeneous dataset which may have domain shift issues and a huge amount of distracting videos.

Recent works [4,40] suggest that state-of-the-art performance can be obtained without meta learning. In particular, Chen et al. [4] show that

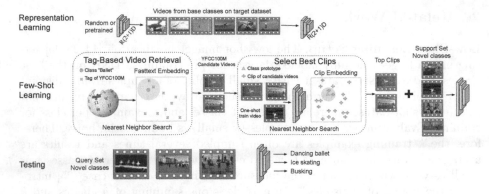

Fig. 2. Our approach comprises three steps: representation learning, few-shot learning and testing. In representation learning, we train a R(2+1)D CNN using the base classes of our target dataset starting from a random initialization or from a Sports1M-pretrained model. In few-shot learning, given few-shot support videos from novel classes, we first retrieve a list of candidate videos for each class from YFCC100M [32] using their tags, followed by selecting the best matching short clips from the retrieved videos using visual features. Those clips serve as additional training examples to learn classifiers that generalize to novel classes at test time.

meta-learning methods are less useful when the image descriptors are expressive enough, which is the case when they are from high-capacity networks trained on large datasets. Therefore, we focus on techniques that do not require a meta-learning stage.

Deep Descriptors for Videos. Moving from hand-designed descriptors [5,22, 27,38] to learned deep network based descriptors [9,10,20,29,34,39] has been enabled by labeled large-scale datasets [20,21], and parallel computing hardware. Deep descriptors are sometimes based on 2D-CNN models operating on a frame-by-frame basis with temporal aggregation [12,42]. More commonly they are 3D-CNN models that operate on short sequences of images that we refer to as video clips [34,36]. Recently, ever-more-powerful descriptors have been developed by leveraging two-stream architectures using additional modalities [10,29], factorized 3D convolutions [35,36], or multi-scale approaches [8].

3   Learning Spatiotemporal Features of Videos

In the few-shot learning setting [43], classes are split into two disjoint label sets, i.e., base classes (denoted as $\mathbf{C}_b$) that have a large number of training examples, and novel classes (denoted as $\mathbf{C}_n$) that have only a small set of training examples. Let $\mathbf{X}_b$ denote the training videos with labels from the base classes and $\mathbf{X}_n$ be the training videos with labels from the novel classes ($|\mathbf{X}_b| \gg |\mathbf{X}_n|$). Given the training data $\mathbf{X}_b$ and $\mathbf{X}_n$, the goal of the conventional few-shot video classification task (FSV) [2,43] is to learn a classifier which predicts labels among novel classes at test time. As the test-time label space is restricted to a few novel classes, the FSV setting is unrealistic. Thus, in this paper, we additionally study the generalized few-shot video classification (GFSV) which allows videos at test time to belong to any base or novel class.

3.1 3D CNN for FSV (3DFSV)

In this section, we introduce our spatiotemporal CNN baseline for few-shot video classification (3DFSV). Our approach in Fig. 2 consists of 1) a representation learning stage which trains a spatiotemporal CNN on the base classes, 2) a few-shot learning stage that trains a linear classifier for novel classes with few labeled videos, and 3) a testing stage which evaluates the model on unseen test videos. The details of each of these stages are given below.

Representation Learning. Our model adopts a 3D CNN [36] ϕ : $\mathbb{R}^{F \times 3 \times H \times W} \to \mathbb{R}^{d_v}$, encoding a short, fixed-length video clip of F RGB frames with spatial resolution $H \times W$ to a feature vector in a d_v-dimensional embedding space. On top of the feature extractor ϕ, we define a linear classifier $f(\bullet; W_b)$ parameterized by a weight matrix $W_b \in \mathbb{R}^{d_v \times |\mathbf{C}_b|}$, producing a probability distribution over the base classes. The objective is to jointly learn the network ϕ and the classifier W_b by minimizing the cross-entropy classification loss on video clips randomly sampled from training videos $\mathbf{X}_b$ of base classes. More specifically, given a training video $\mathbf{x} \in \mathbf{X}_b$ with a label $\mathbf{y} \in \mathbf{C}_b$, the loss for a video clip $\mathbf{x}_i \in \mathbb{R}^{F \times 3 \times H \times W}$ sampled from video $\mathbf{x}$ is defined as

$$\mathcal{L}(\mathbf{x}_i) = -\log \sigma(W_b^T \phi(\mathbf{x}_i))_{\mathbf{y}} \tag{1}$$

where σ denotes the softmax function that produces a probability distribution and $\sigma(\bullet)_{\mathbf{y}}$ is the probability at class $\mathbf{y}$. Following [4], we do not do meta-learning, so we can use all the base classes to learn the network ϕ.

Few-Shot Learning. This stage aims to adapt the learned network ϕ to recognize novel classes $\mathbf{C}_n$ with a few training videos $\mathbf{X}_n$. To reduce overfitting, we fix the network ϕ and learn a linear classifier $f(\bullet, W_n)$ by minimizing the cross-entropy loss on video clips randomly sampled from videos in $\mathbf{X}_n$, where $W_n \in \mathbb{R}^{d_v \times |\mathbf{C}_n|}$ is the weight matrix of the linear classifier. Similarly, we define the loss for a video clip $\mathbf{x}_i$ sampled from $\mathbf{x} \in \mathbf{X}_n$ with a label $\mathbf{y}$ as

$$\mathcal{L}(\mathbf{x}_i) = -\log \sigma(W_n^T \phi(\mathbf{x}_i))_{\mathbf{y}} \tag{2}$$

Testing. The spatiotemporal CNN operates on fixed-length video *clips* of F RGB frames and the classifiers make clip-level predictions. At test time, the model must predict the label of a test video $\mathbf{x} \in \mathbb{R}^{T \times 3 \times H \times W}$ with arbitrary time length T. We achieve this by randomly drawing a set L of clips $\{\mathbf{x}_i\}_{i=1}^L$ from video $\mathbf{x}$, where $\mathbf{x}_i \in \mathbb{R}^{F \times 3 \times H \times W}$. The video-level prediction is then obtained by averaging the prediction scores after the softmax function over those L clips. For few-shot video classification (FSV), this is:

$$\frac{1}{L} \sum_{i=1}^{L} f(\mathbf{x}_i; W_n).$$ (3)

For generalized few-shot video classification (GFSV), both base and novel classes are taken into account and we build a classifier by concatenating the base class weight W_b learned in the representation stage with the novel class weight W_n learned in the few-shot learning stage:

$$\frac{1}{L} \sum_{i=1}^{L} f(\mathbf{x}_i; [W_b; W_n]).$$ (4)

3.2   Retrieval-Enhanced 3DFSV (R-3DFSV)

During few-shot learning, fine-tuning the network ϕ or learning the classifier $f(\bullet; W_n)$ alone is prone to overfitting. Moreover, class-labeled videos to be used for fine-tuning are scarce. Instead, the hypothesis is that leveraging a massive collection of weakly-labeled real-world videos would improve our novel-class classifier. To this end, for each novel class, we propose to retrieve a subset of weakly-labeled videos, associate pseudo-labels to these retrieved videos and use them to expand the training set of novel classes. It is worth noting that those retrieved videos may be assigned with wrong labels and have domain shift issues as they belong to another heterogeneous dataset, making this idea challenging to implement. For efficiency and to reduce the label noise, we adopt the following two-step retrieval approach.

Tag-Based Video Retrieval. The YFCC100M dataset [32] includes about 800K videos collected from Flickr, with a total length of over 8000 h. Processing a large collection of videos has a high computational demand and a large portion of them are irrelevant to our target classes. Thus, we restrict ourselves to videos with tags related to those of the target class names and leverage information that is complementary to the actual video content to increase the visual diversity.

Given a video with user tags $\{t_i\}_{i=1}^{S}$ where $t_i \in \mathcal{T}$ is a word or phrase and S is the number of tags, we represent it with an average tag embedding $\frac{1}{S} \sum_{i=1}^{S} \varphi(t_i)$. The tag embedding $\varphi(.) : \mathcal{T} \to \mathbb{R}^{d_t}$ maps each tag to a d_t dimensional embedding space, e.g., Fasttext [17]. Similarly, we can represent each class by the text embedding of its class name and then for each novel class c, we compute its cosine similarity to all the video tags and retrieve the N most similar videos according to this distance.

Selecting Best Clips. The video tag retrieval selects a list of N candidate videos for each novel class. However, those videos are not yet suitable for training because the annotation may be erroneous, which can harm the performance. Besides, some weakly-labeled videos can last as long as an hour. We thus propose to select the best short clips of F frames from those candidate videos using the few-shot videos of novel classes.

Given a set of few-shot videos $\mathbf{X}_n^c$ from novel class c, we randomly sample L video clips from each video. We then extract features from those clips with the spatiotemporal CNN ϕ and compute the class prototype by averaging over clip features. Similarly, for each retrieved candidate video of novel class c, we also randomly draw L video clips and extract clip features from ϕ. Finally, we perform a nearest neighbour search with cosine distance to find the M best matching clips of the class prototype:

$$\max_{\mathbf{x}_j} \ \cos(p_c, \phi(\mathbf{x}_j)) \tag{5}$$

where p_c denotes the class prototype of class c, $\mathbf{x}_j$ is the clip belonging to the retrieved weakly-labeled videos. After repeating this process for each novel class, we obtain a collection of pseudo-labeled video clips $\mathbf{X}_p = \{\mathbf{X}_p^c\}_{c=1}^{|C_n|}$ where $\mathbf{X}_p^c$ indicates the best M video clips from YFCC100M for novel class c.

Batch Denoising. The retrieved video clips contribute to learning a better novel-class classifier $f(\bullet; W_n)$ in the few-shot learning stage by expanding the training set of novel classes from $\mathbf{X}_n$ to $\mathbf{X}_n \bigcup \mathbf{X}_p$. $\mathbf{X}_p$ may include video clips with wrong labels. During the optimization, we adopt a simple strategy to alleviate the noise: half of the video clips per batch come from $\mathbf{X}_n$ and another half from $\mathbf{X}_p$ at each iteration. The purpose is to reduce the gradient noise in each mini-batch by enforcing that half of the samples are trustworthy.

4   Experiments

In this section, we first describe the existing experimental settings and our proposed setting for few-shot video recognition. We then present the results comparing our approaches with the state-of-the-art methods in the existing setting on two datasets, the results of our approach in our proposed settings, model analysis and qualitative results.

4.1   Experimental Settings

Here we describe the four datasets we use, previous few-shot video classification protocols and our settings.

Datasets. Kinetics [21] is a large-scale video classification dataset which covers 400 human action classes including human-object and human-human interactions. Its videos are collected from Youtube and trimmed to include only one action class. The UCF101 [31] dataset is also collected from Youtube videos, consisting of 101 realistic human action classes, with one action label in each video. SomethingV2 [13] is a fine-grained human action recognition dataset, containing 174 action classes, in which each video shows a human performing a predefined basic action, such as "picking something up" and "pulling something from left to right". We use the second release of the dataset. YFCC100M [32] is the largest publicly available multimedia collection with about 99.2 million images and 800k videos from Flickr. Although none of these videos are annotated with a class label, half of them (400k) have at least one user tag. We use the tag-labeled videos of YFCC100M to improve the few-shot video classification.

Table 1. Statistics of our data splits on Kinetics, UCF101 and SomethingV2 datasets. We follow the train, val, and test class splits of [43] and [2] on Kinetics and SomethingV2 respectively. In addition, we add test videos (the second number under the second test column) from train classes for GFSV. We also introduce a new data split on UCF101 and for all datasets we propose 5-,10-,15-,24-way (the maximum number of test classes) and 1-,5-shot setting.

	# classes			# videos		
	Train	Val	Test	Train	Val	Test
Kinetics	64	12	24	6400	1200	2400+2288
UCF101	64	12	24	5891	443	971+1162
SomethingV2	64	12	24	67013	1926	2857+5243

Prior Setup. The existing practice of [2,43] indicates randomly selecting 100 classes on Kinetics and on SomethingV2 datasets respectively. Those 100 classes are then randomly divided into 64, 12, and 24 non-overlapping classes to construct the meta-training, meta-validation and meta-testing sets. The meta-training and meta-validation sets are used for training models and tuning hyper-parameters. In the testing phase of this meta-learning setting [2,43], each episode simulates a n-way, k-shot classification problem by randomly sampling a support set consisting of k samples from each of the n classes, and a query set consisting of one sample from each of the n classes. While the support set is used to adapt the model to recognize novel classes, the classification accuracy is computed at each episode on the query set and mean top-1 accuracy over 20,000 episodes constitutes the final accuracy.

Proposed Setup. The prior experimental setup is limited to $n = 5$ classes in each episode, even though there are 24 novel classes in the test set. As in this setting the performance saturates quickly, we extend it to 10-way, 15-way and 24-way settings. Similarly, the previous meta-learning setup assumes that test videos all come from novel classes. On the other hand, it is important in many real-world scenarios that the classifier does not forget about previously learned classes while learning novel classes. Thus, we propose the more challenging generalized few-shot video classification (GFSV) setting where the model needs to predict both base and novel classes.

To evaluate a n-way k-shot problem in GFSV, in addition to a support and a query set of novel classes, at each test episode we randomly draw an additional query set of 5 samples from each of the 64 base classes. We do not sample a support set for base classes because base class classifiers have been learned during the representation learning phase. We report the mean top-1 accuracy of both base and novel classes over 500 episodes.

Kinetics, UCF101 and SomethingV2 datasets are used as our few-shot video classification datasets with disjoint sets of train, validation and test classes (see Table 1 for details). Here we refer to base classes as train classes. Test classes include the classes we sample novel classes from in each testing episode. For Kinetics and SomethingV2, we follow the splits proposed by [2,43] respectively for a fair comparison. It is worth noting that 3 out of 24 test classes in Kinetics appear in Sports1M, which is used for pretraining our 3D ConvNet. But the performance drop is negligible if we replace those 3 classes with other 3 random kinetics classes that are not present in Sports1M (more details can be found in the supplementary material). Following the same convention, we randomly select 64, 12 and 24 non-overlapping classes as train, validation and test classes from UCF101 dataset, which is widely used for video action recognition. We ensure that in our splits the novel classes do not overlap with the classes of Sports1M. For the GFSV setting, in each dataset the test set includes samples from base classes coming from the validation split of the original dataset.

Implementation Details. Unless otherwise stated our backbone is a 34-layer R(2+1)D [36] pretrained on Sports1M [20] which takes as input video clips consisting of $F = 16$ RGB frames with spatial resolution of $H = 112 \times W = 112$. We extract clip features from the $d_v = 512$ dimensional top pooling units.

In the representation learning stage, we fine-tune the R(2+1)D with a constant learning rate 0.001 on all datasets and stop training when the validation accuracy of base classes saturates. We perform standard spatial data augmentation including random cropping and horizontal flipping. We also apply temporal data augmentation by randomly drawing 8 clips from a video in one epoch. In the few-shot learning stage, the same data augmentation is applied and the novel class classifier is learned with a constant learning rate 0.01 for 10 epochs on all the datasets. At test time, we randomly draw $L = 10$ clips from each video and average their predictions for a video-level prediction.

As for the retrieval approach, we use the 400 dimensional ($d_t = 400$) fast-text [16] embedding trained with GoogleNews. We first retrieve $N = 20$ candidate videos for each class with video tag retrieval and then select $M = 5$ best clips among those videos with visual similarities.

4.2   Comparing with the State-of-the-Art

In this section, we compare our model with the state-of-the-art in existing evaluation settings which mainly consider 1-shot, 5-way and 5-shot, 5-way problems and evaluate only on novel classes, i.e., FSV. The baselines CMN [43] and TAM [2] are considered as the state-of-the-art in few-shot video classification. CMN [43] proposes a multi-saliency embedding function to extract video descriptor, and few-shot classification is then done by the compound memory network [19]. TAM [2] proposes to leverage the long-range temporal ordering information in video data through temporal alignment. They additionally build a stronger CMN, namely CMN++, by using the few-shot learning practices from [4]. We use their reported numbers for fair comparison. The results are shown

Table 2. Comparing with the state-of-the-art few-shot video classification methods. We report top-1 accuracy on the novel classes of Kinetics and SomethingV2 for 1-shot and 5-shot tasks (both in 5-way). 3DFSV (ours, scratch): our R(2+1)D is trained from scratch; 3DFSV (ours, pretrained): our model is trained from the Sports1M-pretrained R(2+1)D. R-3DFSV (ours, pretrained): our model with retrieved videos.

	Kinetics		SomethingV2	
Method	1-shot	5-shot	1-shot	5-shot
CMN [43]	60.5	78.9	-	-
CMN++ [2]	65.4	78.8	34.4	43.8
TAM [2]	73.0	85.8	42.8	52.3
3DFSV (ours, scratch)	48.9	67.8	57.9	75.0
3DFSV (ours, pretrained)	92.5	**97.8**	**59.1**	**80.1**
R-3DFSV (ours, pretrained)	**95.3**	**97.8**	-	-

in Table 2. As the code from CMN [43] and TAM [2] is not available at the time of submission we do not include UCF101 results.

On Kinetics, we observe that our 3DFSV (pretrain) approach, i.e. without retrieval, outperforms the previous best results by a wide margin in both 1-shot and 5-shot cases. On SomethingV2, we would like to first highlight that our 3DFSV (scratch) significantly improves over TAM by 15.1% in 1-shot (42.8% of TAM vs 57.9% of ours) and by surprisingly 22.7% in 5-shot (52.3% of TAM vs 75.0% of ours). This is encouraging because the 2D CNN backbone of TAM is pretrained on ImageNet, while our R(2+1)D backbone is trained from random initialization. Our 3DFSV (pretrain) yields further improvement after using the Sports1M-pretrained R(2+1)D. We observe that the effect of the Sports1M-pretrained model on SomethingV2 is not as significant as on Kinetics because there is a large domain gap between Sports1M to SomethingV2 datasets. Those results show that a simple linear classifier on top of a pretrained 3D CNN, e.g. R(2+1)D [36], performs better than sophisticated methods with a pretrained 2D ConvNet as a backbone.

Although as shown in C3D [34], I3D [3], R(2+1)D [36], spatiotemporal CNNs have an edge over 2D spatial ConvNet [15] in the fully supervised video classification with enough annotated training data, we are the first to apply R(2+1)D in the few-shot video classification with limited labeled data. It is worth noting that our R(2+1)D is pretrained on the Sports1M while the 2D ResNet backbone of CMN [43] and TAM [2] is pretrained on ImageNet. A direct comparison between 3D CNNs and 2D CNNs is hard because they are designed for different input data. While it is standard to use an ImageNet-pretrained 2D CNN in image domains, it is common to apply a Sports1M-pretrained 3D CNN in video domains. One of our goals is to establish a strong few-shot video classification baseline with 3D CNNs. Intuitively, the temporal cue of the video is better preserved when clips are processed directly by a spatiotemporal CNN as opposed to processing them as images via a 2D ConvNet. Indeed, even though we

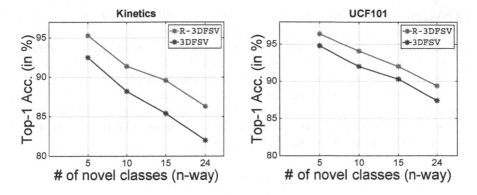

Fig. 3. Results of 3DFSV and R-3DFSV on both Kinetics and UCF101 in the one-shot video classification setting (FSV). In this experiment we go beyond the classical 5-way classification setting. We use 5, 10, 15 and 24 (all) of the novel classes in each testing episode. We report the top-1 accuracy of novel classes.

train our 3DFSV from the random initialization on SomethingV2 which requires strong temporal information, our results still remain promising. This confirms the importance of 3D CNNs for few-shot video classification.

Our R-3DFSV (pretrain) approach, i.e. with retrieved tag-labeled video clips, lead to further improvements in 1-shot case (3DFSV (pretrain) 92.5% vs R-3DFSV (pretrain) 95.3) on Kinetics dataset. This implies that tag-labeled videos retrieved from the YFCC100M dataset include discriminative cues for Kinetics tasks. In 5-shot, our R-3DFSV (pretrain) approach achieves similar performance as our 3DFSV (pretrain) approach however with an 97.8% this task is almost saturated. We do not retrieve any weakly-labeled videos for the SomethingV2 dataset because it is a fine-grained dataset of basic actions and it is unlikely that YFCC100M includes any relevant video for that dataset. As a summary, although 5-way classification setting is still challenging to those methods with 2D ConvNet backbone, the results saturate with the stronger spatiotemporal CNN backbone.

4.3   Increasing the Number of Classes in FSV

Although prior works evaluated few-shot video classification on 5-way, i.e. the number of novel classes at test time is 5, our 5-way results are already saturated. Hence, in this section, we go beyond 5-way classification and extensively evaluate our approach in the more challenging, i.e., 10-way, 15-way and 24-way few-shot video classification (FSV) setting. Note that from every class we use one sample per class during training, i.e. one-shot video classification.

As shown in Fig. 3, our R-3DFSV method exceeds 95% accuracy both in Kinetics and UCF101 datasets for 5-way classification. With the increasing number of novel classes, e.g. 10, 15 and 24, as expected, the performance degrades. Note that, our R-3DFSV approach with retrieval consistently outperforms our

3DFSV approach without retrieval and the more challenging the task becomes, e.g. from 5-way to 24-way, the larger improvement retrieval approach can achieve on Kinetics, i.e. our retrieval-based method is better than our baseline method by 2.8% in 5-way (ours 3DFSV 92.5% vs our R-3DFSV 95.3%) and the gap becomes 4.3% in 24-way (our 3DFSV 82.0% vs our R-3DFSV 86.3%).

The trend with a decreasing accuracy by going from 5-way to 24-way indicates that the more realistic task on few-shot video classification has not yet been solved even with a spatiotemporal CNN. We hope that these results will encourage more research in this challenging setting.

4.4 Evaluating Base and Novel Classes in GFSV

The FSV setting has a strong assumption that test videos all come from novel classes. In contrast to the FSV, GFSV is more realistic and requires models to predict both base and novel classes in each testing episode. In other words, 64 base classes become distracting classes when predicting novel classes which makes the task more challenging. Intuitively, distinguishing novel and base classes is a challenging task because there are severe imbalance issues between the base classes with a large number of training examples and the novel classes with only few-shot examples. In this section, we evaluate our methods in this more realistic and challenging generalized few-shot video classification (GFSV) setting.

Table 3. Generalized few-shot video classification (GFSV) results on Kinetics and UCF101 in 5-way 1-shot and 5-shot tasks. We report top-1 accuracy on both base and novel classes.

	Method	Kinetics		UCF101	
		Novel	Base	Novel	Base
1-shot	3DFSV	7.5	88.7	3.5	97.1
	R-3DFSV	13.7	88.7	4.9	97.1
5-shot	3DFSV	20.5	88.7	10.1	97.1
	R-3DFSV	22.3	88.7	10.4	97.1

Table 4. 5-way 1-shot results ablated on Kinetics. **PR**: pretraining on Sports1M; **RL**: representation learning on base classes; **VR**: retrieve videos with tags [32]; **BD**: batch denoising. **BC**: best clip selection.

PR	RL	VR	BD	BC	Acc
✓					27.1
	✓				48.9
✓	✓				92.5
✓	✓	✓			91.4
✓	✓	✓	✓		93.2
✓	✓	✓	✓	✓	95.3

In Table 3, on the Kinetics dataset, we observe a big performance gap between base and novel classes in both 1-shot and 5-shot cases. The reason is that predictions of novel classes are dominated by the base classes. Interestingly, our R-3DFSV improves 3DFSV on novel classes in both 1-shot and 5-shot cases, e.g., 7.5% of 3DFSV vs 13.7% of R-3DFSV in 1-shot. A similar trend can be observed on the UCF101 dataset. Those results indicate that our retrieval-enhanced approach can alleviate the imbalance issue to some extent. At the same time, we find that the GFSV setting, e.g. not restricting the test time search space only to novel classes but considering all of the classes, is still a challenging task and hope that this setting will attract interests of a wider community for future research.

4.5 Ablation Study and Retrieved Clips

In this section, we perform an ablation study to understand the importance of each component of our approach. After the ablation study, we evaluate the importance of the number of retrieved clips to the FSV performance.

Ablation Study. We ablate our model in the 1-shot, 5-way video classification task on Kinetics dataset with respect to five critical parts including pretraining R(2+1)D on Sports1M (**PR**), representation learning on base classes (**RL**), video retrieval with tags (**VR**), batch denoising (**BD**) and best clip selection (**BC**).

We start from a model with only a few-shot learning stage on novel classes. If a **PR** component is added to the model (the first result row in Table 4), the newly-obtained model can achieve only 27.1% accuracy It indicates that a pretrained 3D CNN alone is insufficient for a good performance. Adding **RL** component to the model (the second row) means to train representation on base classes from scratch, which results in a worse accuracy compared to our full model. The main reason is that optimizing the massive number of parameters of R(2+1)D is difficult on a small train set. Adding both **PR** and **RL** components (the third row) obtains an accuracy of 92.5% which significantly boosts having **PR** and **RL** alone.

Next, we study two critical components proposed in our retrieval approach. Comparing to our approach without retrieval (the third row), directly appending retrieved videos (**VR**) to the few-shot training set of novel classes (the fourth row) leads to 0.9% accuracy drop, while performing the batch denoising (**BD**, the fifth row) in addition to **VR** obtains 0.7% gain. This implies that noisy labels from retrieved videos may hurt the performance but our **BD** handles the noise well. Finally, adding the best clip selection (**BC**, the last row) after **VR**

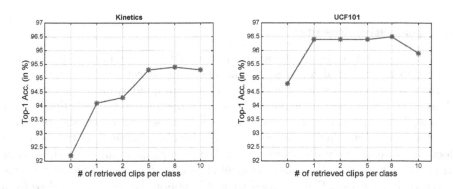

Fig. 4. The effect of increasing the number of retrieved clips, **left**: on Kinetics, **right**: on UCF101. Both experiments are conducted on the 1-shot, 5-way classification task.

and **BD** gets a big boost of 2.8% accuracy. In summary, those ablation studies demonstrate the effectiveness of the five different critical parts in our approach.

Influence of the Number of Retrieved Clips. When the number of retrieved clips increases, the retrieved videos become more diverse, but at the same time, the risk of obtaining negative videos becomes higher. We show the effectiveness of our R-3DFSV with the increasing number of retrieved clips in Fig. 4. On the Kinetics dataset (left of Fig. 4), without retrieving any videos, the performance is 92.5%. As we increase the number of retrieved video clips for each novel class, the performance keeps improving and saturates at retrieving 8 clips per class, reaching an accuracy of 95.4%. On the UCF101 dataset (right of Fig. 4), retrieving 1 clip gives us 1.6% gain. Retrieving more clips does not further improve the results, indicating more negative videos are retrieved. We observe a slight performance drop at retrieving 10 clips because the noise level becomes too high, i.e. there are 10 times more noisy labels than clean labels.

Fig. 5. Top-5 retrieved video clips from YFCC100M for 8 novel classes on Kinetics. The left column is the class name with its one-shot query video and the right column shows the retrieved 16-frame video clips (middle frame is visualized) together with their users tags. Negative retrievals are marked in red.

4.6   Qualitative Results

In Fig. 5, we visualize the top-5 video clips we retrieve from YFCC100M dataset with video tag retrieval followed by the best clips selection. We observe that the retrieved video clips of some classes are of high quality, meaning that those videos truly reveal the target novel classes. For instance, retrieved clips of class "Busking" are all correct because user tags of those videos consist of words like "buskers", "busking" that are close to the class name, and the best clip selection can effectively filter out the irrelevant clips. It is intuitive those clips can potentially help to learn better novel class classifiers by supplementing the limited training videos. Failure cases are also common. For example, videos from the class "Cutting watermelon" do not retrieve any positive videos. The reasons can be that there are no user tags of cutting watermelon or our tag embeddings are not good enough. Those negative videos might hurt the performance if we treat them equally, which is why the batch denoising is critical to reduce the effect of negative videos.

5   Conclusion

In this work, we point out that a spatiotemporal CNN trained on a large-scale video dataset saturates existing few-shot video classification benchmarks. Hence, we propose new more challenging experimental settings, namely generalized few-shot video classification (GFSV) and few-shot video classification with more ways than the classical 5-way setting. We further improve spatiotemporal CNNs by leveraging the weakly-labeled videos from YFCC100M using weak-labels such as tags for text-supported and video-based retrieval. Our results show that generalized more-way few-shot video classification is challenging and we encourage future research in this setting.

References

1. Akata, Z., Perronnin, F., Harchaoui, Z., Schmid, C.: Label-embedding for image classification. IEEE Trans. Pattern Anal. Mach. Intell. (TPAMI) (2016)
2. Cao, K., Ji, J., Cao, Z., Chang, C.Y., Niebles, J.C.: Few-shot video classification via temporal alignment. In: CVPR (2020)
3. Carreira, J., Zisserman, A.: Quo vadis, action recognition? A new model and the kinetics dataset. In: CVPR (2017)
4. Chen, W.Y., Liu, Y.C., Kira, Z., Wang, Y.C.F., Huang, J.B.: A closer look at few-shot classification. In: International Conference on Learning Representations (2019). https://openreview.net/forum?id=HkxLXnAcFQ
5. Dollár, P., Rabaud, V., Cottrell, G., Belongie, S.: Behavior recognition via sparse spatio-temporal features. In: Proceedings of the 14th International Conference on Computer Communications and Networks. VS-PETS Beijing, China (2005)
6. Douze, M., Szlam, A., Hariharan, B., Jégou, H.: Low-shot learning with large-scale diffusion. In: CVPR (2018)
7. Dwivedi, S.K., Gupta, V., Mitra, R., Ahmed, S., Jain, A.: Protogan: towards few shot learning for action recognition. arXiv preprint arXiv:1909.07945 (2019)
8. Feichtenhofer, C., Fan, H., Malik, J., He, K.: Slowfast networks for video recognition. In: Proceedings of the IEEE International Conference on Computer Vision, pp. 6202–6211 (2019)
9. Feichtenhofer, C., Pinz, A., Wildes, R.: Spatiotemporal residual networks for video action recognition. In: Advances in Neural Information Processing Systems, pp. 3468–3476 (2016)
10. Feichtenhofer, C., Pinz, A., Zisserman, A.: Convolutional two-stream network fusion for video action recognition. In: Proceedings of the IEEE Conference on Computer Vision and Pattern Recognition, pp. 1933–1941 (2016)
11. Finn, C., Abbeel, P., Levine, S.: Model-agnostic meta-learning for fast adaptation of deep networks. In: Proceedings of the 34th International Conference on Machine Learning, vol. 70, pp. 1126–1135. JMLR. org (2017)
12. Girdhar, R., Ramanan, D., Gupta, A., Sivic, J., Russell, B.: ActionVLAD: learning spatio-temporal aggregation for action classification. In: Proceedings of the IEEE Conference on Computer Vision and Pattern Recognition, pp. 971–980 (2017)

13. Goyal, R., et al.: The "something something" video database for learning and evaluating visual common sense. In: ICCV (2017)
14. Hariharan, B., Girshick, R.: Low-shot visual recognition by shrinking and hallucinating features. In: Proceedings of the IEEE International Conference on Computer Vision, pp. 3018–3027 (2017)
15. He, K., Zhang, X., Ren, S., Sun, J.: Deep residual learning for image recognition. In: CVPR (2016)
16. Joulin, A., Grave, E., Bojanowski, P., Douze, M., Jégou, H., Mikolov, T.: Fasttext. zip: compressing text classification models. arXiv preprint arXiv:1612.03651 (2016)
17. Joulin, A., Grave, E., Bojanowski, P., Mikolov, T.: Bag of tricks for efficient text classification. In: Proceedings of the 15th Conference of the European Chapter of the Association for Computational Linguistics. Short Papers, vol.2, pp. 427–431. Association for Computational Linguistics, April 2017
18. Joulin, A., van der Maaten, L., Jabri, A., Vasilache, N.: Learning visual features from large weakly supervised data. In: Leibe, B., Matas, J., Sebe, N., Welling, M. (eds.) ECCV 2016. LNCS, vol. 9911, pp. 67–84. Springer, Cham (2016). https://doi.org/10.1007/978-3-319-46478-7_5
19. Kaiser, L., Nachum, O., Roy, A., Bengio, S.: Learning to remember rare events. arXiv preprint arXiv:1703.03129 (2017)
20. Karpathy, A., Toderici, G., Shetty, S., Leung, T., Sukthankar, R., Fei-Fei, L.: Large-scale video classification with convolutional neural networks. In: CVPR (2014)
21. Kay, W., et al.: The kinetics human action video dataset. arXiv preprint arXiv:1705.06950 (2017)
22. Laptev, I.: On space-time interest points. Int. J. Comput. Vis. **64**(2–3), 107–123 (2005)
23. Mensink, T., Verbeek, J., Perronnin, F., Csurka, G.: Metric learning for large scale image classification: generalizing to new classes at near-zero cost. In: Fitzgibbon, A., Lazebnik, S., Perona, P., Sato, Y., Schmid, C. (eds.) ECCV 2012. LNCS, vol. 7573, pp. 488–501. Springer, Heidelberg (2012). https://doi.org/10.1007/978-3-642-33709-3_35
24. Qi, H., Brown, M., Lowe, D.G.: Low-shot learning with imprinted weights. In: Proceedings of the IEEE Conference on Computer Vision and Pattern Recognition, pp. 5822–5830 (2018)
25. Ravi, S., Larochelle, H.: Optimization as a model for few-shot learning. In: ICLR (2016)
26. Ren, M., et al.: Meta-learning for semi-supervised few-shot classification. In: ICLR (2018)
27. Sadanand, S., Corso, J.J.: Action bank: A high-level representation of activity in video. In: 2012 IEEE Conference on Computer Vision and Pattern Recognition, pp. 1234–1241. IEEE (2012)
28. Schoenfeld, E., Ebrahimi, S., Sinha, S., Darrell, T., Akata, Z.: Generalized zero- and few-shot learning via aligned variational autoencoders. In: IEEE Conference on Computer Vision and Pattern Recognition (CVPR) (2019)
29. Simonyan, K., Zisserman, A.: Two-stream convolutional networks for action recognition in videos. In: Advances in Neural Information Processing Systems, pp. 568–576 (2014)
30. Snell, J., Swersky, K., Zemel, R.: Prototypical networks for few-shot learning. In: NeurIPS (2017)
31. Soomro, K., Zamir, A.R., Shah, M.: Ucf101: a dataset of 101 human actions classes from videos in the wild. arXiv preprint arXiv:1212.0402 (2012)

32. Thomee, B., et al.: YFCC100M: the new data in multimedia research. arXiv preprint arXiv:1503.01817 (2015)
33. Torralba, A., Efros, A.A., et al.: Unbiased look at dataset bias. In: CVPR (2011)
34. Tran, D., Bourdev, L., Fergus, R., Torresani, L., Paluri, M.: Learning spatiotemporal features with 3D convolutional networks. In: CVPR (2015)
35. Tran, D., Wang, H., Torresani, L., Feiszli, M.: Video classification with channel-separated convolutional networks. arXiv preprint arXiv:1904.02811 (2019)
36. Tran, D., Wang, H., Torresani, L., Ray, J., LeCun, Y., Paluri, M.: A closer look at spatiotemporal convolutions for action recognition. In: CVPR (2018)
37. Vinyals, O., Blundell, C., Lillicrap, T., Wierstra, D., et al.: Matching networks for one shot learning. In: Advances in Neural Information Processing Systems, pp. 3630–3638 (2016)
38. Wang, H., Schmid, C.: Action recognition with improved trajectories. In: Proceedings of the IEEE International Conference on Computer Vision, pp. 3551–3558 (2013)
39. Wang, L., et al.: Temporal segment networks: towards good practices for deep action recognition. In: Leibe, B., Matas, J., Sebe, N., Welling, M. (eds.) ECCV 2016. LNCS, vol. 9912, pp. 20–36. Springer, Cham (2016). https://doi.org/10.1007/978-3-319-46484-8_2
40. Wang, Y., Chao, W.L., Weinberger, K.Q., van der Maaten, L.: Simpleshot: revisiting nearest-neighbor classification for few-shot learning. arXiv preprint arXiv:1911.04623 (2019)
41. Xian, Y., Sharma, S., Schiele, B., Akata, Z.: F-VAEGAN-D2: a feature generating framework for any-shot learning. In: IEEE Conference on Computer Vision and Pattern Recognition (CVPR) (2019)
42. Yue-Hei Ng, J., Hausknecht, M., Vijayanarasimhan, S., Vinyals, O., Monga, R., Toderici, G.: Beyond short snippets: deep networks for video classification. In: Proceedings of the IEEE Conference on Computer Vision and Pattern Recognition, pp. 4694–4702 (2015)
43. Zhu, L., Yang, Y.: Compound memory networks for few-shot video classification. In: Ferrari, V., Hebert, M., Sminchisescu, C., Weiss, Y. (eds.) ECCV 2018. LNCS, vol. 11211, pp. 782–797. Springer, Cham (2018). https://doi.org/10.1007/978-3-030-01234-2_46

GAN-Based Anomaly Detection
In Imbalance Problems

Junbong Kim[1], Kwanghee Jeong[2], Hyomin Choi[2], and Kisung Seo[2(✉)]

[1] IDIS, Seoul, Korea
[2] Department of Electronic Engineering, Seokyeong University, Seoul, Korea
ksseo@skuniv.ac.kr

Abstract. Imbalance pre one of the key issues that affect the performance greatly. Our focus in this work is to address an imbalance problem arising from defect detection in industrial inspections, including the different number of defect and non-defect dataset, the gap of distribution among defect classes, and various sizes of defects. To this end, we adopt the anomaly detection method that is to identify unusual patterns to address such challenging problems. Especially generative adversarial network (GAN) and autoencoder-based approaches have shown to be effective in this field. In this work, (1) we propose a novel GAN-based anomaly detection model which consists of an autoencoder as the generator and two separate discriminators for each of normal and anomaly input; and (2) we also explore a way to effectively optimize our model by proposing new loss functions: Patch loss and Anomaly adversarial loss, and further combining them to jointly train the model. In our experiment, we evaluate our model on conventional benchmark datasets such as MNIST, Fashion MNIST, CIFAR 10/100 data as well as on real-world industrial dataset – smartphone case defects. Finally, experimental results demonstrate the effectiveness of our approach by showing the results of outperforming the current State-Of-The-Art approaches in terms of the average area under the ROC curve (AUROC).

Keywords: Imbalance problems · Anomaly detection · GAN · Defects inspection · Patch loss · Anomaly adversarial loss

1 Introduction

The importance of the imbalance problems in machine learning is investigated widely and many researches have been trying to solve them [12,20,23,28,34]. For example, class imbalance in the dataset can dramatically skew the performance of classifiers, introducing a prediction bias for the majority class [23]. Not only class imbalance, but various imbalance problems exist in data science. A general overview of imbalance problems is investigated in the literature [12,20,23,28]. Specifically, the survey of various imbalance problems for object detection subject is described in the review paper [34].

J. Kim and K. Jeong—-Equal contribution.

A. Bartoli and A. Fusiello (Eds.): ECCV 2020 Workshops, LNCS 12540, pp. 128–145, 2020.
https://doi.org/10.1007/978-3-030-65414-6_11

We handle a couple of imbalance problems closely related to industrial defects detection in this paper. Surface defects of metal cases such as scratch, stamped, and stain are very unlikely to happen in the production process, thereby resulting in outstanding class imbalance. Besides, size of defects, loss scale, and discriminator distributional imbalances are covered as well. In order to prevent such imbalance problems, anomaly detection [8] approach is used. This method discards a small portion of the sample data and converts the problem into an anomaly detection framework. Considering the shortage and diversity of anomalous data, anomaly detection is usually modeled as a one-class classification problem, with the training dataset containing only normal data [40].

Reconstruction-based approaches [1, 41, 43] have been paid attention for anomaly detection. The idea behind this is that autoencoders can reconstruct normal data with small errors, while the reconstruction errors of anomalous data are usually much larger. Autoencoder [33] is adopted by most reconstruction-based methods which assume that normal and anomalous samples could lead to significantly different embeddings and thus differences in the corresponding reconstruction errors can be leveraged to differentiate the two types of samples [42]. Adversarial training is introduced by adding a discriminator after autoencoders to judge whether its original or reconstructed image [10, 41]. Schlegl et al. [43] hypothesize that the latent vector of a GAN represents the true distribution of the data and remap to the latent vector by optimizing a pre-trained GAN-based on the latent vector. The limitation is the enormous computational complexity of remapping to this latent vector space. In a follow-up study, Zenati et al. [52] train a BiGAN model [4], which maps from image space to latent space jointly, and report statistically and computationally superior results on the MNIST benchmark dataset. Based on [43, 52], GANomaly [1] proposes a generic anomaly detection architecture comprising an adversarial training framework that employs adversarial autoencoder within an encoder-decoder-encoder pipeline, capturing the training data distribution within both image and latent vector space. However, the studies mentioned above have much room for improvement on performance for benchmark datasets such as Fashion-MNIST, CIFAR-10, and CIFAR-100.

A novel GAN-based anomaly detection model by using a structurally separated framework for normal and anomaly data is proposed to improve the biased learning toward normal data. Also, new definitions of the patch loss and anomaly adversarial loss are introduced to enhance the efficiency for defect detection. First, this paper proves the validity of the proposed method for the benchmark data, and then expands it for the real-world data, the surface defects of the smartphone case. There are two types of data that are used in the experiments – classification benchmark datasets including MNIST, Fashion-MNIST, CIFAR10, CIFAR100, and a real-world dataset with the surface defects of the smartphone. The results of the experiments showed State-Of-The-Art performances in four benchmark dataset, and average accuracy of 99.03% in the real-world dataset of the smartphone case defects. To improve robustness and performance, we select the final model by conducting the ablation study. The result of the ablation study and the visualized images are described.

In summary, our method provides the methodological improvements over the recent competitive researches, GANomaly [1] and ABC [51], and overcome the State-Of-The-Art results from GeoTrans [13] and ARNet [18] with significant gap.

2 Related Works

Imbalance Problems

A general review of Imbalance problems in deep learning is provided in [5]. There are lots of class imbalance examples in various areas such as computer vision [3,19,25,48], medical diagnosis [16,30] and others [6,17,36,38] where this issue is highly significant and the frequency of one class can be much larger than another class. It has been well known that class imbalance can have a significant deleterious effect on deep learning [5]. The most straightforward and common approach is the use of sampling methods. Those methods operate on the data itself to increase its balance. Widely used and proven to be robust is oversampling [29]. The issue of class imbalance can be also tackled on the level of the classifier. In such a case, the learning algorithms are modified by introducing different weights to misclassification of examples from different classes [54] or explicitly adjusting prior class probabilities [26]. A systematic review on imbalance problems in object detection is presented in [34]. In here, total of eight different imbalance problems are identified and grouped four main types: class imbalance, scale imbalance, spatial imbalance, and objective imbalance. Problem based categorization of the methods used for imbalance problems is well organized also.

Anomaly Detection

For anomaly detection on images and videos, a large variety of methods have been developed in recent years [7,9,22,32,37,49,50,55]. In this paper, we focus on anomaly detection in still images. Reconstruction-based anomaly detection [2,10,43,44,46] is the most popular approach. The method compress normal samples into a lower-dimensional latent space and then reconstruct them to approximate the original input data. It assume that anomalous samples will be distinguished through relatively high reconstruction errors compared with normal samples.

Autoencoder and GAN-Based Anomaly Detection

Autoencoder is an unsupervised learning technique for neural networks that learns efficient data encoding by training the network to ignore signal noise [46]. Generative adversarial network (GAN) proposed by Goodfellow et al. [15] is the approach co-training a pair networks, generator and discriminator, to compete with each other to become more accurate in their predictions. As reviewed in

[34], adversarial training has also been adopted by recent work within anomaly detection. More recent attention in the literature has been focused on the provision of adversarial training. Sabokrou et al. [41] employs adversarial training to optimize the autoencoder and leveraged its discriminator to further enlarge the reconstruction error gap between normal and anomalous data. Furthermore, Akcay et al. [1] adds an extra encoder after autoencoders and leverages an extra MSE loss between the two different embeddings. Similarly, Wang et al. [45] employs adversarial training under a variational autoencoder framework with the assumption that normal and anomalous data follow different Gaussian distribution. Gong et al. [14] augments the autoencoder with a memory module and developed an improved autoencoder called memory-augmented autoencoder to strengthen reconstructed errors on anomalies. Perera et al. [35] applies two adversarial discriminators and a classifier on a denoising autoencoder. By adding constraint and forcing each randomly drawn latent code to reconstruct examples like the normal data, it obtained high reconstruction errors for the anomalous data.

3   Method

3.1   Model Structures

In order to implement anomaly detection, we propose a GAN-based generative model. The pipeline of the proposed architecture of training phase is shown in the Fig. 1. The network structure of the Generator follows that of an autoencoder, and the Discriminator consists of two identical structures to separately process the input data when it is normal or anomaly. In the training phase, the model learns to minimize reconstruction error when normal data is entered to the generator, and to maximize reconstruction error when anomaly data is entered. The loss used to minimize reconstruction error with normal image input is marked in blue color in four ways. Also, the loss used for maximizing the error with anomaly image input is marked in red color in two ways. In the inference phase, reconstruction error is used to detect anomalies as a criteria standard. The matrix maps in the right part of Fig. 1 show that each value of the output matrix represents the probability of whether the corresponding image patch is real or fake. The way is used in PatchGAN [11] and it is totally different from Patch Loss we proposed in this paper.

3.2   Imbalance Problems in Reconstruction-Based Anomaly Detection

In order to handle anomaly detection for defects inspection, the required imbalance characteristics are described. We define imbalance problems for defects as class imbalance, loss function scale imbalance, distributional bias on the learning model, and imbalance in image and object (anomaly area) sizes. Table 1 summarizes the types of imbalance problems and solutions.

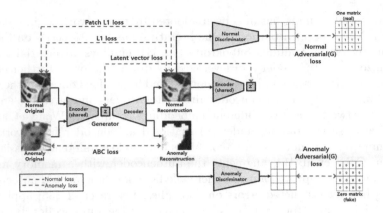

Fig. 1. Pipeline of the proposed approach for anomaly detection (Color figure online).

Table 1. Imbalance problems and solutions of the proposed method

Imbalance problems	Solutions
Class imbalance	Data sampling with k-means clustering (Sect. 3.5)
Loss scale imbalance	Loss function weight search (Sect. 3.4)
Discriminator distributional Bias	Two discriminator (Sect. 3.3)
Size imbalance between object(defect) and image	Reconstruction-based methods

Class Imbalance

Class imbalance is well known, and surface defects of metal cases such as scratch, stamped, and stain are very unlikely to happen in the production process, therefore resulting in outstanding class imbalance problems between normal and anomaly. Not only the number of normal and defective data is imbalanced, but also the frequency of occurrence differ among the types of defects such as scratch, stamped, and stain, so imbalance within each class exists in anomaly data. To resolve such class imbalance, data is partially sampled and used in training. Here, if the data is sampled randomly without considering its distribution, the entire data and the sampled data might not be balanced in their distribution. Therefore, in this paper, we use the method of dividing the entire data into several groups by k-means clustering, and then sample the same number of data within each group.

Loss Function Scale Imbalance

The proposed method uses the weighted sum of 6 types of loss functions to train the generator. The scale of the loss function used here is different, and even if the scale is the same, the effect on the learning is different. In addition, GAN contains a min-max problem that the generator and the discriminator learn by competing against each other, making the learning difficult and unstable. The loss scales of the generator and the discriminator should be sought at a similar rate to each other so that GAN is effectively trained. To handle such loss function scale imbalance problems, weights used in loss combination are explored by a grid search.

Discriminator Distributional Bias

The loss will be used to update the generator differently for normal and anomaly data. When the reconstruction data is given to the discriminator, the generator is trained to output 1 from normal data and 0 from anomaly data. Thus, when training from both normal and anomaly data, using a single discriminator results in training the model to classify only normal images well. Separating discriminator for normal data and for anomaly data is necessary to solve this problem. This method only increases the parameters or computations of the model in the training phase, but not those in inference phase. As a result, there is no overall increase in memory usage or latency at the final inferences.

Size Imbalance Between Object (defect) and Image

Industrial defect data exhibits smaller size of defect compared to the size of the entire image. Objects in such data occupy very small portion of the image, making it closer to object detection rather than classification, so it is difficult to expect fair performance with classification methods. To solve this, we propose a method generating images to make the total reconstruction error bigger not affected by the size of the defect and the size of the entire image which contains the defect.

3.3   Network Architecture

The proposed model is a GAN-based network structure consisting of a generator and a discriminator. The generator is in the form of an autoencoder to perform image to image translation. And a modified U-Net [39] structure is adopted, which has an effective delivery of features using a pyramid architecture. The discriminator is a general CNN network, and two discriminators are used only in the training phase.

The generator is a symmetric network that consists of four 4 x 4 convolutions with stride 2 followed by four transposed convolutions. The total parameters of generator is composed of a sum of 0.38 K, 2.08 K, 8.26 K. 32.9 K, 32.83 K, 16.42 K, 4.11 K, and 0.77 K, that 97.75 K totally. The discriminator is a general network that consists of three 4 x 4 convolutions with stride 2 followed by two 4 x 4 convolutions with stride 1. The total parameters of discriminator is composed of a sum of 0.38 K, 2.08 K, 8.26 K, 32.9 K, and 32.77 K, that is 76.39 K.

3.4   Loss Function

Total number of loss functions used in the proposed model is eight. Six losses for training of generator, one for normal discriminator and another for anomaly discriminator. The loss function for training of each discriminator is adopted from LSGAN [31] as shown in Eq. (1). It uses the a-b coding scheme for the discriminator, where a and b are the labels for fake data and real data, respectively.

$$\min_{D} V_{\text{LSGAN}}(D) = [(D(x) - b)^2] + [(D(G(x)) - a)^2] \tag{1}$$

Six kinds of loss functions, as shown in from Eqs. (2) to (8) are employed to train the generator. Among them, four losses are for normal images. First, L1 reconstruction error of generator for normal image is provided as shown in Eq. (2). It penalizes by measuring the L1 distance between the original x and the generated images ($\hat{x} = G(x)$) as defined in [1]:

$$\mathcal{L}_{\text{recon}} = \|x - G(x)\|_1 \tag{2}$$

Second, the patch loss is newly proposed in this paper as shown in Eq. (3). Divide a normal image and a generated image separately into M patches and select the average of the biggest n reconstruction errors among all the patches.

$$\mathcal{L}_{\text{patch}} = f_{avg}(n)(\|x_{patch(i)} - G(x_{patch(i)})\|_1), i = 1, 2, ..., m \tag{3}$$

Third, latent vector loss [1] is calculated as the difference between latent vectors of generator for normal image and latent vectors of cascaded encoder for reconstruction image as shown in Eq. (4)

$$\mathcal{L}_{\text{enc}} = \|G_E(x) - G_E(G(x))\|_1 \tag{4}$$

Equation (5) defines the proposed adversarial loss for the generator update use in LSGAN [31], where y denotes the value that G wants D to believe for fake data.

$$\min_{G} V_{\text{LSGAN}}(G) = [(D(G(x)) - y)^2] \tag{5}$$

Fourth, the adversarial loss used to update the generator is as shown in. Eq. (6). The loss function intends to output a real label of 1 when a reconstruction image (fake) is into the discriminator.

$$\min_{G} V_{\text{LSGAN}}(G) = [(D(G(x)) - 1)^2] \tag{6}$$

Two remaining losses for anomaly images are as follows. One is anomaly adversarial loss for updating generator and the other is ABC [51] loss. Unlike a general adversarial loss of Eq. (6), anomaly reconstruction image should be generated differently from real one to classify anomaly easily, the anomaly adversarial loss newly adopted in our work is as shown in Eq. (7).

$$\min_{G} V_{\text{LSGAN}}(G) = [(D(G(x)) - 0)^2] \tag{7}$$

ABC loss as shown Eq. (8) is used here to maximize L1 reconstruction error $\mathcal{L}_\theta(\cdot)$ for anomaly data. Because the difference between the reconstruction errors for normal and anomaly data is large, the equation is modified by adding the exponetial and log function to solve the scale imbalance.

$$\mathcal{L}_{\mathrm{ABC}} = -\log(1 - e^{-\mathcal{L}_\theta(x_i)}) \tag{8}$$

Total loss function consists of weighted sum of each loss. All losses for normal images are grouped together and same as for anomaly images. Those two group of losses are applied to update the weights of learning process randomly. The scale imbalances exist among the loss functions. Although the scale could be adjusted in same range, the effect might be different, so we explore the weight of each loss using the grid search. Because ABC loss can have the largest scale, the weighted sum of normal data is set more than twice as large as the weighted sum of anomaly data. In order to avoid huge and unnecessary search space, each weight of the loss functions is limited from 0.5–1.5 range. Then the grid search is executed the each weight adjusting by 0.5. Total possible cases for the grid search is 314. The final explored weights of loss function are shown in Table 2.

Table 2. Weight combination of loss functions obtained by Grid search

Normal reconstruction L1 loss	ABC loss	Normal adversarial loss	Normal reconstruction patch L1 loss	Normal latent vector loss	Anomaly adversarial loss
1.5	0.5	0.5	1.5	0.5	1.0

3.5   Data Sampling

As mentioned in Sect. 3.2, the experimental datasets include imbalance problems. For benchmark datasets such as MNIST, Fashion-MNIST, CIFAR-10, and CIFAR-100, have a class imbalance problems presenting imbalance of data sampling. The real-world dataset, surface defects of smartphone case is not only the number of normal and defective data are imbalanced, but also the frequency of occurrences differs among the types of defects. Also the size of image and object (defect) is imbalanced too. To solve those imbalance problem, k-means clustering-based data sampling is performed to make balanced distribution of data. In learning stage for benchmark datasets, all data is used for normal case. In case of anomaly, the same number is sampled for each class so that the total number of data is similar to normal. At this time, k-means clustering is performed on each class, and data is sampled from each cluster in a distribution similar to the entire dataset. For anomaly case of defect dataset, data is sampled using the same method as the benchmark, and a number of normal data is sampled, equal to the number of data combined with the three kinds of defects - scratch, stamped and stain. Detail number of data is described in Sect. 4.1

4   Experiments

In this section, we perform substantial experiments to validate the proposed method for anomaly detection. We first evaluate our method on commonly used benchmark datasets - MNIST, Fashion-MNIST, CIFAR-10, and CIFAR-100. Next, we conduct experiments on real-world anomaly detection dataset - smartphone case defect dataset. Then we present the respective effects of different designs of loss functions through ablation study.

4.1   Datasets

Datasets used in the experiments include of four standard image datasets: MNIST [27], Fashion-MNIST [47], CIFAR-10 [24], and CIFAR-100 [24]. Additionaly smartphone case defect data is added to evaluate the performance in real-world environments.

For MNIST, Fashion-MNIST, CIFAR-10 data set, each class is defined as normal and the rest of nine classes are defined as anomaly. Total 10 experiments are performed by defining each 10 class once as normal. With CIFAR-100 dataset, one class is defined as normal and the remaining 19 classes are defined as anomaly among 20 superclasses. Each superclass is defined as normal one by one, so in total 20 experiments were conducted. Also, in order to resolve imbalance in the number of normal data and anomaly data, and in distribution of sampled data, the method proposed in 3.5 is applied when sampling training data. 6000 normal and anomaly data images were used for training MNIST and Fashion-MNIST, and 5000 were used for CIFAR-10. Additionally, for MNIST and Fashion-MNIST, the images were resized into 32 x 32 so that the size of the feature can be the same when concatenating them in the network structure.

Smartphone case defect dataset consists of normal, scratch, stamped and stain classes. And there are two main types of data set. The first dataset contains patch images that are cropped into 100 x 100 from their original size of 2192 x 1000. The defective class is sampled in the same number as the class with the least number of data among defects by deploying the method from Sect. 3.5, and the normal data is sampled in a similar number to that of the detective data. 900 images of normal data and 906 images of anomaly data were used for training, and 600 images of normal data and 150 images of anomaly data were used for testing. In the experiments, images were resized from 100 x 100 to 128 x 128. The second dataset consists of patch images cropped into 548 x 500. The same method of sampling as the 100 x 100 patch images is used. 1800 images of normal data and 1814 images of anomaly data are used for training, and 2045 images of normal data and 453 images of normal data are used for testing.

4.2   Experimental Setups

Experimentation is performed using Intel Core i7-9700K @ 3.60 GHz and NVIDIA geforce GTX 1080ti with Tensorflow 1.14 deep learning framework. For augmentation on MNIST, the method of randomly cropping 0–2 pixels from the boundary and resizing again was used, while for Fashion-MNIST, images were vertically and horizontally flipped, randomly cropped 0–2 pixels from their boundary, and resized again. Then, the images were rotated 90, 180, or 270 degrees. For CIFAR-10 and CIFAR-100, on top of the augmentation method utilized for Fashion-MNIST, hue, saturation, brightness, and contrast are varied as additional augmentation. For defective dataset, vertical flipping and horizontal flipping are used along with rotation of 90, 180, or 270° vertical flipping an.

Hyperparameters and details on augmentation are as follows Table 3.

Table 3. Hyperparameters used for model training

Hyperparameters					Parameters of patch reconstruction error loss (Eq. (3))		
Epoch	Batch size	Learning rate init	Learning rate decay epoch	Learning rate decay factor	Patch size	Stride	Number of selected patch
300	1	0.0001	50	0.5	16	8	3

4.3 Benchmarks Results

In order to evaluate the performance of our proposed method, we conducted experiments on MNIST, Fashion-MNIST, CIFAR-10 and CIFAR-100. For estimating recognition rates of the trained model, AUROC(Area Under the Receiver Operating Characteristic) is used. Table 5 contains State-Of-The-Art studies and their recognition rates, and Fig. 2 compares recognition rates, model parameters, and FPS on CIFAR-10 among the representative studies.

According to Table 5, for the MNIST, the paper with highest performance among previous studies is ARNet, which presented average AUROC value of 98.3 for 10 classes, however our method obtains average AUROC of 99.7 with the proposed method. Also, ARNet have standard deviation of 1.78 while ours shows 0.16, therefore significantly reducing deviations among classes. For the Fashion-MNIST, ARNet previously performs the best with average AUROC of 93.9 and standard deviation of 4.7, but our method accomplishes much better results of average AUROC of 98.6 and standard deviation 1.20. For the CIFAR-10, the average AUROC ARNet, so far as the best, with is 86.6 with standard deviation 5.35, but our method achieves quite better result of average AUROC of 90.6 with standard deviation 3.14. Lastly, for CIFAR-100, compared to the results from ARNet, which are average AUROC of 78.8 and standard deviation 8.82, the proposed method shows outstanding improvement of average AUROC 87.4 and standard deviation 4.80. In summary, four different datasets were used for evaluating the performance, and we achieve highly improved results from the previous State-Of-The-Arts studies in terms of recognition rate and learning stability on all the tested datasets.

Figure 2 shows a graph that compares the performance of State-Of-The-Arts researches regarding for average AUROC, FPS, and the number of model parameters used in inference. FPS is calculated as averaging total 10 estimations of processing time for 10,000 CIFAR-10 images. In the graph, the x-axis stands for FPS, i.e. how many images are inferred per second, and the y-axis represents AUROC(%), the recognition rate. The area of the circle of each method indicates the number of model parameters used in inference. The graph shows that our method has AUROC of 90.6%, which is 4% higher than that of ARNet, 86.6%, known to be the highest in the field. Among the papers mentioned, Geo-Trans has the least number of model parameters, 1,216K. However, the model proposed in this paper 98 K parameters, which only 8% of those of GeoTrans therefore resulting in huge reduction. Finally in terms of FPS, ARNet was known to be the fastest with 270FPS, the proposed method was able to process with 532FPS, almost twice as fast.

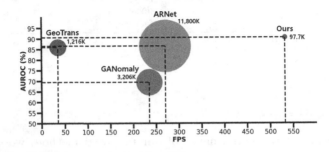

Fig. 2. Comparison of AUROC, FPS, and model parameters for CIFAR10 dataset.

Table 4. AUROC results for defect dataset (%)

Dataset	Train	Test
100 x 100 patch defect	99.73	99.23
548 x 500 patch defect	99.74	98.84

4.4 Defect Dataset Results

Table 4 shows the experiment results with smartphone detective dataset. Testing for 100 x 100 patch images, the performance is absolutely high with AUROC(%) of 99.23. The bigger size of image data with 548 x 500 patch, much more difficult than of 100 x 100 patch data, is tested and shows very similar results of 98.84 still. Therefore, the superiority of our method is proved even in real-world smartphone detective dataset.

Table 5. Comparison with the state-of-the-art literature in AUROC(%) for benchmark datasets

Dataset	Method	0	1	2	3	4	5	6	7	8	9	avg	SD
	AE	98.8	99.3	91.7	88.5	86.2	85.8	95.4	94.0	82.3	96.5	91.9	5.90
	VAE [21]	92.1	**99.9**	81.5	81.4	87.9	81.1	94.3	88.6	78.0	92.0	87.7	7.05
	AnoGAN [43]	99.0	99.8	88.8	91.3	94.4	91.2	92.5	96.4	88.3	95.8	93.7	4.00
	ADGAN [10]	99.5	**99.9**	93.6	92.1	94.9	93.6	96.7	96.8	85.4	95.7	94.7	4.15
MNIST	GANomaly [1]	97.2	99.6	85.1	90.6	94.5	94.9	97.1	93.9	79.7	95.4	92.8	6.12
	OCGAN [35]	99.8	**99.9**	94.2	96.3	97.5	98.0	99.1	98.1	93.9	98.1	97.5	2.10
	GeoTrans [13]	98.2	91.6	99.4	99.0	99.1	99.6	**99.9**	96.3	97.2	99.2	98.0	2.50
	ARNet [18]	98.6	**99.9**	99.0	99.1	98.1	98.1	99.7	99.0	93.6	97.8	98.3	1.78
	OURS	**99.9**	**99.9**	**99.7**	**99.8**	**99.7**	**99.8**	99.7	**99.7**	**99.5**	**99.4**	**99.7**	**0.16**
	AE	71.6	96.9	72.9	78.5	82.9	93.1	66.7	95.4	70.0	80.7	80.9	11.03
	DAGMM [53]	42.1	55.1	50.4	57.0	26.9	70.5	48.3	83.5	49.9	34.0	51.8	16.47
	DSEBM [55]	91.6	71.8	88.3	87.3	85.2	87.1	73.4	98.1	86.0	97.1	86.6	8.61
Fashion-	ADGAN [10]	89.9	81.9	87.6	91.2	86.5	89.6	74.3	97.2	89.0	97.1	88.4	6.75
MNIST	GANomaly [1]	80.3	83.0	75.9	87.2	71.4	92.7	81.0	88.3	69.3	80.3	80.9	7.37
	GeoTrans [13]	99.4	97.6	91.1	89.9	92.1	93.4	83.3	98.9	90.8	99.2	93.5	5.22
	ARNet [18]	92.7	99.3	89.1	93.6	90.8	93.1	85.0	98.4	**97.8**	98.4	93.9	4.70
	OURS	**99.5**	**99.6**	**98.2**	**98.6**	**98.1**	**99.5**	**95.9**	**99.4**	97.6	**99.6**	**98.6**	**1.20**
	AE	57.1	54.9	59.9	62.3	63.9	57.0	68.1	53.8	64.4	48.6	59.0	5.84
	VAE [21]	62.0	66.4	38.2	58.6	38.6	58.6	56.5	62.2	66.3	73.7	58.1	11.50
	DAGMM [53]	41.4	57.1	53.8	51.2	52.2	49.3	64.9	55.3	51.9	54.2	53.1	5.95
	DSEBM [55]	56.0	48.3	61.9	50.1	73.3	60.5	68.4	53.3	73.9	63.6	60.9	9.10
CIFAR-	AnoGAN [43]	61.0	56.5	64.8	52.8	67.0	59.2	62.5	57.6	72.3	58.2	61.2	5.68
10	ADGAN [10]	63.2	52.9	58.0	60.6	60.7	65.9	61.1	63.0	74.4	64.4	62.4	5.56
	GANomaly [1]	**93.5**	60.8	59.1	58.2	72.4	62.2	88.6	56.0	76.0	68.1	69.5	13.08
	OCGAN [35]	75.7	53.1	64.0	62.0	72.3	62.0	72.3	57.5	82.0	55.4	65.6	9.52
	GeoTrans [13]	74.7	**95.7**	78.1	72.4	87.8	**87.8**	83.4	**95.5**	93.3	91.3	86.0	8.52
	ARNet [18]	78.5	89.8	86.1	77.4	**90.5**	84.5	89.2	92.9	92.0	85.5	86.6	5.35
	OURS	92.6	93.6	**86.9**	85.4	89.5	87.8	**93.5**	91.0	**94.6**	91.7	90.6	3.14

Dataset	Method	0	1	2	3	4	5	6	7	8	9	10
	AE	66.7	55.4	41.4	49.2	44.9	40.6	50.2	48.1	66.1	63.0	52.7
	DAGMM [53]	43.4	49.5	66.1	52.6	56.9	52.4	55.0	52.8	53.2	42.5	52.7
	DSEBM [55]	64.0	47.9	53.7	48.4	59.7	46.6	51.7	54.8	66.7	71.2	78.3
	ADGAN [10]	63.1	54.9	41.3	50.0	40.6	42.8	51.1	55.4	59.2	62.7	79.8
	GANomaly [1]	57.9	51.9	36.0	46.5	46.6	42.9	53.7	59.4	63.7	68.0	75.6
	GeoTrans [13]	74.7	68.5	74.0	81.0	78.4	59.1	81.8	65.0	85.5	90.6	87.6
	ARNet [18]	77.5	70.0	62.4	76.2	77.7	64.0	86.9	65.6	82.7	90.2	85.9
CIFAR-	OURS	85.5	86.1	94.4	87.3	91.7	85.1	89.9	88.3	83.8	92.4	94.2
100												

Method	11	12	13	14	15	16	17	18	19	avg	SD
AE	62.1	59.6	49.8	48.1	56.4	57.6	47.2	47.1	41.5	52.4	8.11
DAGMM [53]	46.4	42.7	45.4	57.2	48.8	54.4	36.4	52.4	50.3	50.5	6.55
DSEBM [55]	62.7	66.8	52.6	44.0	56.8	63.1	73.0	57.7	55.5	58.8	9.36
ADGAN [10]	53.7	58.9	57.4	39.4	55.6	63.3	66.7	44.3	53.0	54.7	10.08
GANomaly [1]	57.6	58.7	59.9	43.9	59.9	64.4	71.8	54.9	56.8	56.5	9.94
GeoTrans [13]	83.9	83.2	58.0	**92.1**	68.3	73.5	93.8	**90.7**	85.0	78.7	10.76
ARNet [18]	83.5	**84.6**	67.6	84.2	74.1	**80.3**	91.0	85.3	85.4	78.8	8.82
OURS	**86.0**	83.7	**76.8**	89.5	**80.6**	80.2	**94.9**	89.7	**87.1**	**87.4**	**4.93**

4.5   Ablation Studies

To understand how each loss affect GAN-based anomaly learning, we defined CIFAR-10 bird class as normal and the rest of the classes as anomaly, and conducted an ablation study. The network structure and learning parameters are the same as the experiments with the benchmarks, and training data is also sampled based on k-means clustering to 5000 normal and 5000 anomaly images. All of the given testing data from the dataset is used for testing. The results shows that the performance improves as each loss gets added to the basic autoencoder. In the Table 6, No. 2 (only ABC added) and No.7 (all losses added) don't show much difference in AUROC, 85.44% and 86.76%. However, from Fig. 3 it can be found that reconstructed images exhibits significant difference. It visualizes the result of experiment No. 2 and No. 7 with normal data. In the case of ABC, the normal images could not be reconstructed similarly, so the center was made 0 due to loss from anomaly. On the other hand, experiment No. 7 which exploited combination of the proposed losses reconstructed the image similarly to the original. The average error regarding reconstruction error map also shows difference of 2–3 times (Fig. 4).

Table 6. Ablation study experiment results to confirm the impact of losses used in generator update (AUROC(%))

No.	Auto encoder	ABC	Generator normal	Generator anomaly	Normal patch	Normal latent	AUROC
1	✓						64.68
2	✓	✓					85.44
3	✓		✓				65.74
4	✓			✓			65.93
5	✓				✓		64.87
6	✓					✓	64.89
7	✓	✓	✓	✓	✓	✓	86.76

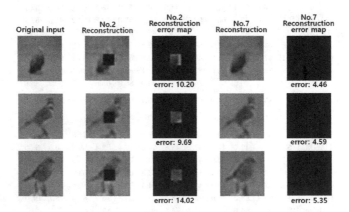

Fig. 3. Visualized results for No. 2 and No. 7 experiments in ablation study

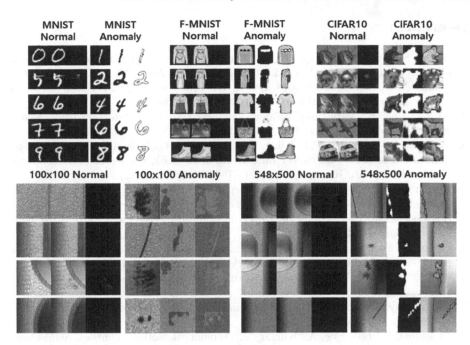

Fig. 4. Visualization of experimental results on the Benchmark dataset & defect data (Original, reconstruction, reconstruction error map order from the left of each image)

5   Conclusion

We proposed a novel GAN-based anomaly detection model by using a new framework, newly defined loss functions, and optimizing their combinations. The discriminators for normal and anomaly are structurally separated to improve the learning that has been biased toward normal data. Also, a new definition of patch loss and anomaly adversarial loss, which are effective for fault detection was introduced and combined with the major losses proposed from previous studies to perform joint learning. In order to systemize the proportion of each loss in the combination, the weight of each loss was explored using the grid search. The main results of our experiments successfully demonstrated that the proposed method much further improves the AUROC for CIFAR-10 data compared to the results of State-Of-The-Art including GANomaly [1], GeoTrans [13] and ARNet [18]. Especially, we applied our method to real-world data set with surface defects of smartphone case and validated outstanding superiority of anomaly detection of the defects. In the future, we will try to extend our approach to a hierarchical anomaly detection scheme from pixel level to video level.

Acknowledgement. This work was supported by National Research Foundation of Korea Grant funded by the Korea government (NRF-2019R1F1A1056135).

References

1. Akcay, S., Atapour-Abarghouei, A., Breckon, T.P.: GANomaly: semi-supervised anomaly detection via adversarial training. In: Jawahar, C.V., Li, H., Mori, G., Schindler, K. (eds.) ACCV 2018. LNCS, vol. 11363, pp. 622–637. Springer, Cham (2019). https://doi.org/10.1007/978-3-030-20893-6_39
2. An, J., Cho, S.: Variational autoencoder based anomaly detection using reconstruction probability. Spec. Lect. IE **2**(1), 1–18 (2015)
3. Beijbom, O., Edmunds, P.J., Kline, D.I., Mitchell, B.G., Kriegman, D.: Automated annotation of coral reef survey images. In: 2012 IEEE Conference on Computer Vision and Pattern Recognition, pp. 1170–1177. IEEE (2012)
4. Bergmann, P., Fauser, M., Sattlegger, D., Steger, C.: Uninformed students: Student-teacher anomaly detection with discriminative latent embeddings. In: Proceedings of the IEEE/CVF Conference on Computer Vision and Pattern Recognition, pp. 4183–4192 (2020)
5. Buda, M., Maki, A., Mazurowski, M.A.: A systematic study of the class imbalance problem in convolutional neural networks. Neural Netw. **106**, 249–259 (2018)
6. Cardie, C., Howe, N.: Improving minority class prediction using case-specific feature weights (1997)
7. Chalapathy, R., Chawla, S.: Deep learning for anomaly detection: a survey (2019). arXiv preprint: arXiv:1901.03407
8. Chandola, V., Banerjee, A., Kumar, V.: Anomaly detection: a survey. ACM Comput. Surv. (CSUR) **41**(3), 1–58 (2009)
9. Chu, W., Xue, H., Yao, C., Cai, D.: Sparse coding guided spatiotemporal feature learning for abnormal event detection in large videos. IEEE Transactions on Multimedia **21**(1), 246–255 (2018)
10. Deecke, L., Vandermeulen, R., Ruff, L., Mandt, S., Kloft, M.: Image anomaly detection with generative adversarial networks. In: Berlingerio, M., Bonchi, F., Gärtner, T., Hurley, N., Ifrim, G. (eds.) ECML PKDD 2018. LNCS (LNAI), vol. 11051, pp. 3–17. Springer, Cham (2019). https://doi.org/10.1007/978-3-030-10925-7_1
11. Demir, U., Unal, G.: Patch-based image inpainting with generative adversarial networks (2018). arXiv preprint arXiv:1803.07422
12. Fernández, A., García, S., Galar, M., Prati, R.C., Krawczyk, B., Herrera, F.: Learning From Imbalanced Data Sets. Springer, Heidelberg (2018)
13. Golan, I., El-Yaniv, R.: Deep anomaly detection using geometric transformations. In: Advances in Neural Information Processing Systems, pp. 9758–9769 (2018)
14. Gong, D., Liu, L., Le, V., Saha, B., Mansour, M.R., Venkatesh, S., Hengel, A.V.D.: Memorizing normality to detect anomaly: memory-augmented deep auto encoder for unsupervised anomaly detection. In: Proceedings of the IEEE International Conference on Computer Vision, pp. 1705–1714 (2019)
15. Goodfellow, I., Pouget-Abadie, J., Mirza, M., Xu, B., Warde-Farley, D., Ozair, S., Courville, A., Bengio, Y.: Generative adversarial nets. In: Advances in Neural Information Processing Systems, pp. 2672–2680 (2014)
16. Grzymala-Busse, J.W., Goodwin, L.K., Grzymala-Busse, W.J., Zheng, X.: An Approach to imbalanced data sets based on changing rule strength. In: Pal, S.K., Polkowski, L., Skowron, A. (eds.) Rough-Neural Computing. Cognitive Technologies. Springer, Heidelberg (2004)
17. Haixiang, G., Yijing, L., Shang, J., Mingyun, G., Yuanyue, H., Bing, G.: Learning from class-imbalanced data: review of methods and applications. Expert Syst. Appl. **73**, 220–239 (2017)

18. Huang, C., Cao, J., Ye, F., Li, M., Zhang, Y., Lu, C.: Inverse-transform autoencoder for anomaly detection (2019). arXiv preprint: arXiv:1911.10676

19. Johnson, B.A., Tateishi, R., Hoan, N.T.: A hybrid pansharpening approach and multiscale object-based image analysis for mapping diseased pine and oak trees. Int. J. Rem. Sens. **34**(20), 6969–6982 (2013)

20. Johnson, J.M., Khoshgoftaar, T.M.: Survey on deep learning with class imbalance. J. Big Data **6**(1), 27 (2019)

21. Kingma, D.P., Welling, M.: Auto-encoding variational bayes (2013). arXiv preprint: arXiv:1312.6114

22. Kiran, B.R., Thomas, D.M., Parakkal, R.: An overview of deep learning based methods for unsupervised and semi-supervised anomaly detection in videos. J. Imag. **4**(2), 36 (2018)

23. Krawczyk, B.: Learning from imbalanced data: open challenges and future directions. Prog. Artif. Intell. **5**(4), 221–232 (2016)

24. Krizhevsky, A., Hinton, G., et al.: Learning multiple layers of features from tiny images (2009)

25. Kubat, M., Holte, R.C., Matwin, S.: Machine learning for the detection of oil spills in satellite radar images. Mach. Learn. **30**(2–3), 195–215 (1998)

26. Lawrence, S., Burns, I., Back, A., Tsoi, A.C., Giles, C.L.: Neural network classification and prior class probabilities. In: Orr, G.B., Müller, K.-R. (eds.) Neural Networks: Tricks of the Trade. LNCS, vol. 1524, pp. 299–313. Springer, Heidelberg (1998). https://doi.org/10.1007/3-540-49430-8_15

27. LeCun, Y.: The mnist database of handwritten digits (1998). http://yann.lecun.com/exdb/mnist/

28. Leevy, J.L., Khoshgoftaar, T.M., Bauder, R.A., Seliya, N.: A survey on addressing high-class imbalance in big data. J. Big Data **5**(1), 42 (2018)

29. Ling, C.X., Li, C.: Data mining for direct marketing: problems and solutions. KDD **98**, 73–79 (1998)

30. Mac Namee, B., Cunningham, P., Byrne, S., Corrigan, O.I.: The problem of bias in training data in regression problems in medical decision support. Artif. Intell. Med. **24**(1), 51–70 (2002)

31. Mao, X., Li, Q., Xie, H., Lau, R.Y., Wang, Z., Paul Smolley, S.: Least squares generative adversarial networks. In: Proceedings of the IEEE International Conference On Computer Vision, pp. 2794–2802 (2017)

32. Markou, M., Singh, S.: Novelty detection: a review-part 2: neural network based approaches. Signal Process. **83**(12), 2499–2521 (2003)

33. Masci, J., Meier, U., Cireşan, D., Schmidhuber, J.: Stacked convolutional autoencoders for hierarchical feature extraction. In: Honkela, T., Duch, W., Girolami, M., Kaski, S. (eds.) ICANN 2011. LNCS, vol. 6791, pp. 52–59. Springer, Heidelberg (2011). https://doi.org/10.1007/978-3-642-21735-7_7

34. Oksuz, K., Cam, B.C., Kalkan, S., Akbas, E.: Imbalance problems in object detection: a review. IEEE Trans. Pattern Anal. Mach. Intell. (2020)

35. Perera, P., Nallapati, R., Xiang, B.: OCGAN: One-class novelty detection using GANs with constrained latent representations. In: Proceedings of the IEEE Conference on Computer Vision and Pattern Recognition, pp. 2898–2906 (2019)

36. Philip, K., Chan, S.: Toward scalable learning with non-uniform class and cost distributions: a case study in credit card fraud detection. In: Proceeding of the Fourth International Conference on Knowledge Discovery and Data Mining, pp. 164–168 (1998)

37. Pimentel, M.A., Clifton, D.A., Clifton, L., Tarassenko, L.: A review of novelty detection. Signal Process. **99**, 215–249 (2014)

38. Radivojac, P., Chawla, N.V., Dunker, A.K., Obradovic, Z.: Classification and knowledge discovery in protein databases. J. Biomed. Inform. **37**(4), 224–239 (2004)
39. Ronneberger, O., Fischer, P., Brox, T.: U-Net: convolutional networks for biomedical image segmentation. In: Navab, N., Hornegger, J., Wells, W.M., Frangi, A.F. (eds.) MICCAI 2015. LNCS, vol. 9351, pp. 234–241. Springer, Cham (2015). https://doi.org/10.1007/978-3-319-24574-4_28
40. Ruff, L., Vandermeulen, R., Goernitz, N., Deecke, L., Siddiqui, S.A., Binder, A., Müller, E., Kloft, M.: Deep one-class classification. In: International Conference on Machine Learning, pp. 4393–4402 (2018)
41. Sabokrou, M., Khalooei, M., Fathy, M., Adeli, E.: Adversarially learned one-class classifier for novelty detection. In: Proceedings of the IEEE Conference on Computer Vision and Pattern Recognition, pp. 3379–3388 (2018)
42. Sakurada, M., Yairi, T.: Anomaly detection using auto-encoders with nonlinear dimensionality reduction. In: Proceedings of the MLSDA 2014 2nd Workshop on Machine Learning for Sensory Data Analysis, pp. 4–11 (2014)
43. Schlegl, T., Seeböck, P., Waldstein, S.M., Schmidt-Erfurth, U., Langs, G.: Unsupervised anomaly detection with generative adversarial networks to guide marker discovery. In: Niethammer, M., et al. (eds.) IPMI 2017. Unsupervised anomaly detection with generative adversarial networks to guide marker discovery, vol. 10265, pp. 146–157. Springer, Cham (2017). https://doi.org/10.1007/978-3-319-59050-9_12
44. Schmidhuber, J.: Deep learning in neural networks: An overview. Neural networks **61**, 85–117 (2015)
45. Wang, X., Du, Y., Lin, S., Cui, P., Shen, Y., Yang, Y.: adVAE: a self-adversarial variational auto-encoder with Gaussian anomaly prior knowledge for anomaly detection. Knowl. Based Syst. **190**, 105187 (2020)
46. Xia, Y., Cao, X., Wen, F., Hua, G., Sun, J.: Learning discriminative reconstructions for unsupervised outlier removal. In: Proceedings of the IEEE International Conference on Computer Vision, pp. 1511–1519 (2015)
47. Xiao, H., Rasul, K., Vollgraf, R.: Fashion-mnist: a novel image dataset for benchmarking machine learning algorithms (2017). arXiv preprint: arXiv:1708.07747
48. Xiao, J., Hays, J., Ehinger, K.A., Oliva, A., Torralba, A.: Sun database: Large-scale scene recognition from abbey to zoo. In: 2010 IEEE Computer Society Conference on Computer Vision and Pattern Recognition, pp. 3485–3492. IEEE (2010)
49. Xu, K., Jiang, X., Sun, T.: Anomaly detection based on stacked sparse coding with intra-frame classification strategy. IEEE Trans. Multimed. **20**(5), 1062–1074 (2018)
50. Xu, K., Sun, T., Jiang, X.: Video anomaly detection and localization based on an adaptive intra-frame classification network. IEEE Trans. Multimed. **22**(2), 394–406 (2019)
51. Yamanaka, Y., Iwata, T., Takahashi, H., Yamada, M., Kanai, S.: Auto-encoding binary classifiers for supervised anomaly detection. In: Pacific Rim International Conference on Artificial Intelligence, pp. 647–659. Springer, Cham (2019)
52. Zenati, H., Foo, C.S., Lecouat, B., Manek, G., Chandrasekhar, V.R.: Efficient GAN-based anomaly detection (2018). arXiv preprint: arXiv:1802.06222
53. Zhai, S., Cheng, Y., Lu, W., Zhang, Z.: Deep structured energy based models for anomaly detection (2016). arXiv preprint: arXiv:1605.07717

54. Zhou, Z.H., Liu, X.Y.: Training cost-sensitive neural networks with methods addressing the class imbalance problem. IEEE Trans. knowl. Data Eng. **18**(1), 63–77 (2005)
55. Zong, B., et al.: Deep auto-encoding Gaussian mixture model for unsupervised anomaly detection. In: International Conference on Learning Representations (2018)

Active Class Incremental Learning
for Imbalanced Datasets

Eden Belouadah[1,2](✉) (iD), Adrian Popescu[1] (iD), Umang Aggarwal[1] (iD),
and Léo Saci[1] (iD)

[1] CEA, LIST, 91191 Gif-sur-Yvette, France
{eden.belouadah,adrian.popescu,umang.aggarwal,leo.saci}@cea.fr
[2] IMT Atlantique, Computer Science Department, 29238 Brest, France

Abstract. Incremental Learning (IL) allows AI systems to adapt to
streamed data. Most existing algorithms make two strong hypotheses
which reduce the realism of the incremental scenario: (1) new data are
assumed to be readily annotated when streamed and (2) tests are run
with balanced datasets while most real-life datasets are imbalanced.
These hypotheses are discarded and the resulting challenges are tackled with a combination of active and imbalanced learning. We introduce
sample acquisition functions which tackle imbalance and are compatible with IL constraints. We also consider IL as an imbalanced learning
problem instead of the established usage of knowledge distillation against
catastrophic forgetting. Here, imbalance effects are reduced during inference through class prediction scaling. Evaluation is done with four visual
datasets and compares existing and proposed sample acquisition functions. Results indicate that the proposed contributions have a positive
effect and reduce the gap between active and standard IL performance.

Keywords: Incremental learning · Active learning · Imbalanced
learning · Computer vision · Image classification.

1 Introduction

AI systems are often deployed in dynamic settings where data are not all available
at once [28]. Examples of applications include: (1) robotics - where the robot
evolves in a changing environment and needs to adapt to it, (2) news analysis -
where novel entities and events appear at a fast pace and should be processed
swiftly, and (3) medical document processing - where parts of the data might
not be available due to privacy constraints.

In such cases, incremental learning (IL) algorithms are needed to integrate
new data while also preserving the knowledge learned for past data. Following
a general trend in machine learning, recent IL algorithms are all built around
Deep Neural Networks (DNNs) [2,3,10,17,23,32]. The main challenge faced by
such algorithms is catastrophic interference or forgetting [24], a degradation of
performance for previously learned information when a model is updated with
new data.

© Springer Nature Switzerland AG 2020
A. Bartoli and A. Fusiello (Eds.): ECCV 2020 Workshops, LNCS 12540, pp. 146–162, 2020.
https://doi.org/10.1007/978-3-030-65414-6_12

IL algorithm design is an open research problem if computational complexity should remain bounded as new data are incorporated and/or if only a limited memory is available to store past data. These two conditions are difficult to satisfy simultaneously and existing approaches address one of them in priority. A first research direction allows for model complexity to grow as new data are added [2,3,23,34,40]. They focus on minimizing the number of parameters added for each incremental update and no memory of past data is allowed. Another research direction assumes that model complexity should be constant across incremental states and implements rehearsal over a bounded memory of past data to mitigate catastrophic forgetting [10,14,19,32,45]. Most existing IL algorithms assume that new data are readily labeled at the start of each incremental step. This assumption is a strong one since data labeling is a time consuming process, even with the availability of crowdsourcing platforms. Two notable exceptions are presented in [3] and [30] where the authors introduce algorithms for self-supervised face recognition. While interesting, these works are applicable only to a specific task and both exploit pretrained models to start the process. Also, a minimal degree of supervision is needed in order to associate a semantic meaning (i.e. person names) to the discovered identities. A second hypothesis made in incremental learning is that datasets are balanced or nearly so. In practice, imbalance occurs in a wide majority of real-life datasets but also in research datasets constructed in controlled conditions. Public datasets such as ImageNet [12], Open Images [20] or VGG-Face2 [9] are all imbalanced. However, most research works related to ImageNet report results with the ILSVRC subset [33] which is nearly balanced.

These two hypotheses limit the practical usability of existing IL algorithms. We replace them by two weaker assumptions to make the incremental learning scenario more realistic. First, full supervision of newly streamed data is replaced by the possibility to annotate only a small subset of these data. Second, no prior assumption is made regarding the balanced distribution of new data in classes. We combine active and imbalanced learning methods to tackle the challenges related to the resulting IL scenario.

The main contribution of this work is to adapt sample acquisition process, which is the core component of active learning (AL) methods, to incremental learning over potentially imbalanced datasets. A two phases procedure is devised to replace the classical acquisition process which uses a single acquisition function. A standard function is first applied to a subset of the active learning budget in order to learn an updated model which includes a suboptimal representation of new data. In the second phase, a balancing-driven acquisition function is used to favor samples which might be associated to minority classes (i.e. those having a low number of associated samples). The data distribution in classes is updated after each sample labeling to keep it up-to-date. Two balancing-driven acquisition functions which exploit the data distribution in the embedding space of the IL model are introduced here. The first consists of a modification of the core-set algorithm [36] to restrain the search for new samples to data points which are closer to minority classes than to majority ones. The second function prioritizes samples which are close to the poorest minority classes (i.e. those represented by the minimum number of samples) and far from any of the majority classes. The balancing-driven acquisition phase is repeated several times and new samples are successively added to the training set in order to enable an iterative active learning process [37].

A secondary contribution is the introduction of a backbone training procedure which considers incremental learning with memory as an instance of imbalanced learning. The widely used training with knowledge distillation [10,17,19,32,45] is consequently replaced by a simpler procedure which aims to reduce the prediction bias towards majority classes during inference [8]. Following the conclusions of this last work, initial predictions are rectified by using the prior class probabilities from the training set.

Four public datasets designed for different visual tasks are used for evaluation. The proposed balancing-driven sample acquisition process is compared with a standard acquisition process and results indicate that it has a positive effect for imbalanced datasets.

2   Related Works

We discuss existing works from incremental, imbalanced and active learning areas and focus on those which are most closely related to our contribution. Class incremental learning witnessed a regain of interest and all recent methods exploit DNNs. One influential class of IL methods build on the adaptation of fine tuning and exploit increasingly sophisticated knowledge distillation techniques to counter catastrophic forgetting [24]. *Learning without Forgetting (LwF)* [22] introduced this trend and is an inspiration for a wide majority of further IL works. *Incremental Classifier and Representation Learning (iCaRL)* [32] is one such work which uses *LwF* and also adds a bounded memory of the past to implement replay-based IL efficiently. *iCaRL* selects past class exemplars using a herding approach. The classification layer of the neural nets is replaced by a nearest-exemplars-mean, which adapts nearest-class-mean [25], to counter class imbalance. *End-to-end incremental learning* [10] uses a distillation component which adheres to the original definition of *LwF* from [16]. A balanced fine tuning step is added to counter imbalance. As a result, a consequent improvement over *iCaRL* is reported. *Learning a Unified Classifier Incrementally via Rebalancing (LUCIR)* [17] tackles incremental learning problem by combining cosine normalization in the classification layer, a less-forget constraint based on distillation and an inter-class separation to improve comparability between past and new classes. *Class Incremental Learning with Dual Memory (IL2M)* [4] and *Bias Correction (BiC)* [41] are recent approaches that add an extra layer to the network in order to remove the prediction bias towards new classes which are represented by more images than past classes.

Classical active learning is thoroughly reviewed in [37]. A first group of approaches exploits informativeness to select items for uncertain regions in the classification space. Uncertainty is often estimated with measures such as entropy [39], least confidence first [11] or min margin among top predictions [35]. Another group of approaches leverages sample representativeness computed in the geometric space defined by a feature extractor. Information density [38] was an early implementation of such an approach. *Core-set*, which rely on the classical *K-centers* algorithm to discover an optimal subset of the unlabeled dataset, was introduced in [36].

Recent active learning works build on the use of deep learning. The labeling effort is examined in [18] to progressively prune labels as labeling advances. An algorithm which learns a loss function specifically for AL was proposed in [43]. While very interesting, such an approach is difficult to exploit in incremental learning since the main challenge here is to counter data imbalance between new and past classes or among new classes. Another line of works proposes to exploit multiple network states to improve the AL process. *Monte Carlo Dropout* [13] uses softmax prediction from a model with random dropout masks. In [6], an ensemble approach which combines multiple snapshots of the same training process is introduced. These methods are not usable in our scenario because they increase the number of parameters due to the use of multiple models. We retain the use of the same deep model through incremental states to provide embeddings and propose a stronger role for them during the sample acquisition process. Recently, [1] proposed a method which focuses on single-stage AL for imbalanced datasets. They exploit a pretrained feature extractor and annotate the unlabeled samples so as to favor minority classes.

Ideally, incremental updates should be done in a fully unsupervised manner [42] in order to remove the need for manual labeling. However, unsupervised algorithms are not mature enough to capture dataset semantics with the same degree of refinement and performance as their supervised or semi-supervised counterparts. Closest to our work are the self-supervision approaches designed for incremental face recognition [3,30]. They are tightly related to unsupervised learning since no manual labeling is needed, except for naming the person. Compared to self-supervision, our approach requires manual labeling for a part of new data and has a higher cost. However, it can be applied to any class IL problem and not only to specific tasks such as face recognition as it is the case for [3,30].

A comprehensive review of imbalanced object-detection problems is provided in [26]. The authors group these problems in a taxonomy depending on their class imbalance, scale imbalance, spatial imbalance or objective imbalance. The study shows the increasing interest of the computer vision community in the imbalanced problems for their usefulness in real life situations.

3   Proposed Method

The proposed active learning adaptation to an incremental scenario is motivated by the following observations:

– Existing acquisition functions ($\mathcal{AF}$s) were designed and tested successfully for active learning over balanced datasets. However, a wide majority of real-life datasets are actually imbalanced. Here, no prior assumption is made regarding

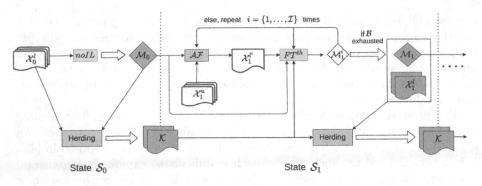

Fig. 1. Illustration of the proposed training process with one initial state $\mathcal{S}_0$, and one incremental state $\mathcal{S}_1$. The initial deep model $\mathcal{M}_0$ is trained from scratch on a fully-labeled dataset $\mathcal{X}_0^l$ using $noIL$ (a non-incremental learning). $\mathcal{M}_0$ and $\mathcal{X}_0^l$ are used to prepare the past class memory $\mathcal{K}$ using herding (a mechanism that selects the best representative past class images). State $\mathcal{S}_1$ starts with a sample acquisition function $\mathcal{AF}$ that takes the unlabeled set $\mathcal{X}_1^u$ and the model $\mathcal{M}_0$ as inputs, and provides a part of the budget $\mathcal{B}$ annotated as $\mathcal{X}_1^{l^i}$. The model $\mathcal{M}_0$ is then updated with data from $\mathcal{X}_1^{l^i} \cup \mathcal{K}$ using FT^{th} (a fine tuning followed by a threshold calibration). The updated model $\mathcal{M}_1^i$ is again fed into the acquisition function $\mathcal{AF}$ with the rest of unlabeled examples from $\mathcal{X}_1^u$ to further annotate a part of the budget $\mathcal{B}$ and the model is updated afterwards. This process is repeated $\mathcal{I}$ times until $\mathcal{B}$ is exhausted. The model $\mathcal{M}_1$ is then returned with the annotated dataset $\mathcal{X}_1^l$ and the memory $\mathcal{K}$ is updated by inserting exemplars of new classes from $\mathcal{X}_1^l$ and reducing exemplars of past classes in order to fit its maximum size. Note that the two blue arrows are applicable in the first AL iteration only (when $i = 1$). Best viewed in color.

the imbalanced or balanced character of the unlabeled data which is streamed in IL states. Unlike existing sample acquisition approaches which exploit a single $\mathcal{AF}$, we propose to split the process in two phases. The first phase uses a classical $\mathcal{AF}$ to kick-off the process. The second one implements an $\mathcal{AF}$ which is explicitly designed to target a balanced representation of labeled samples among classes.

– In IL, a single deep model can be stored throughout the process. This makes the application of recent ensemble methods [6] inapplicable. Following the usual AL pipeline, an iterative fine tuning of the model is implemented to incorporate labeled samples from the latest AL iteration.

– A memory $\mathcal{K}$ of past class samples is allowed and, following [4,17], we model IL as an instance of imbalanced learning. The distillation component, which is central to most existing class IL algorithms [10,19,32,41], is removed. Instead, we deal with imbalance by using a simple but efficient post-processing step which modifies class predictions based on their prior probabilities in the training set. The choice of this method is motivated by its superiority in deep imbalanced learning over a large array of other methods [8].

An illustration of the proposed learning process is provided in Fig. 1. In the next sections, we first formalize the proposed active incremental learning scenario. Then, we introduce the adapted sample acquisition process, with focus on the balancing-driven acquisition functions. Finally, we present the incremental learning backbone which is inspired from imbalanced learning [8].

3.1   Problem Formalization

The formalization of the problem is inspired by [10,32] for the incremental learning part and by [6] for the active learning part. We note $\mathcal{T}$ the number of states (including the first non-incremental state), $\mathcal{K}$ - the bounded memory for past classes, $\mathcal{B}$ - the labeling budget available for active learning, $\mathcal{AF}$ - an acquisition function designed to optimize sample selection in active learning, $\mathcal{I}$ the number of iterations done during active learning, $\mathcal{S}_t$ - an incremental state, N_t - the total number of classes recognizable in $\mathcal{S}_t$, $\mathcal{X}_t^u$ - the unlabeled dataset associated to $\mathcal{S}_t$, $\mathcal{X}_t^l$ - a manually labeled subset of $\mathcal{X}_t^u$, $\mathcal{M}_t$ - the deep model learned in $\mathcal{S}_t$. The initial state $\mathcal{S}_0$ includes a dataset $\mathcal{X}_0 = \{X_0^1, X_0^2, ..., X_0^j, ..., X_0^{P_0}\}$ with $N_0 = P_0$ classes. $X_t^j = \{x_1^j, x_2^j, ..., x_{n_j}^j\}$ is the set of n_j training examples for the jth class, p_t^j is its corresponding classification probability in the state $\mathcal{S}_t$.

We assume that all the samples are labeled in the first state. An initial non-incremental model $\mathcal{M}_0 : \mathcal{X}_0 \rightarrow \mathcal{C}_0$ is trained to recognize a set $\mathcal{C}_0$ containing N_0 classes using all their data from $\mathcal{X}_0$. P_t new classes need to be integrated in each incremental state $\mathcal{S}_t$, with $t > 0$. Each IL step updates the previous model $\mathcal{M}_{t-1}$ into the current model $\mathcal{M}_t$ which recognizes $N_t = P_0 + P_1 + ... + P_t$ classes in the incremental state $\mathcal{S}_t$. Active learning is deployed using $\mathcal{AF}(\mathcal{X}_t^u)$ to obtain $\mathcal{X}_t^l$, a labeled subset from $\mathcal{X}_t^u$.

$\mathcal{X}_t^l$ data of the P_t new classes are available but only a bounded exemplar memory $\mathcal{K}$ for the N_{t-1} past classes is allowed. $\mathcal{M}_t$, the model associated to the state $\mathcal{S}_t$ is trained over the $\mathcal{X}_t^l \cup \mathcal{K}$ training dataset. An iterative AL process is implemented to recognize a set of classes $\mathcal{C}_t = \{c_t^1, c_t^2, ..., c_t^{N_{t-1}}, c_t^{N_{t-1}+1}, ..., c_t^{N_t}\}$.

3.2   Active Learning in an Incremental Setting

We discuss the two phases of the adapted active learning process below. Classical sampling is followed by a phase which exploits the proposed balancing-driven acquisition functions.

Classical Sample Acquisition Phase. At the start of each IL state $\mathcal{S}_t$, an unlabeled dataset $\mathcal{X}_t^u$ is streamed into the system and classical AL acquisition functions are deployed to label $\mathcal{X}_t^l$, a part of $\mathcal{X}_t^u$, for inclusion in the training set. Due to IL constraints, the only model available at the beginning of $\mathcal{S}_t$ is $\mathcal{M}_{t-1}$, learned for past classes in the previous incremental step. It is used to extract the embeddings needed to implement acquisition functions. A number of acquisition functions were proposed to optimize the active learning process [37], with adaptations for deep learning in [6,13,36,46]. Based on their strong experimental performance [6,35–37], four $\mathcal{AF}$s are selected for the initial phase:

- **core-set sampling** [36] (*core* hereafter): whose objective is to extract a representative subset of the unlabeled dataset from the vectorial space defined by the deep embeddings. The method selects samples with:

$$x_{next} = \underset{x_u \in \mathcal{X}_t^u}{\operatorname{argmax}}\{ \min_{1 \leq k \leq n} \Delta(e(x_u), e(x_k))\} \tag{1}$$

where: x_{next} is the next sample to label, x_u is an unlabeled sample left, x_k is one of the n samples which were already labeled, $e()$ is the embedding extracted using $\mathcal{M}_{t-1}$ and Δ is the Euclidean distance between two embeddings.

- **random sampling** (*rand* hereafter) : a random selection of images for labeling. While basic, random selection remains a competitive baseline in active learning.

- **entropy sampling** [37] (*ent* hereafter): whose objective is to favor most uncertain samples as defined by the set of probabilities given by the model.

$$x_{next} = \underset{x_u \in \mathcal{X}_t^u}{\operatorname{argmax}}\{-\sum_{j=1}^{J}(p_t^j * \log(p_t^j))\} \tag{2}$$

where p_t^j is the prediction score of x_u for the class j and J is the number of detected classes so far by AL.

- **margin sampling** [35] (*marg* hereafter): selects the most uncertain samples based on their top-2 predictions of the model.

$$x_{next} = \underset{x_u \in \mathcal{X}_t^u}{\operatorname{argmax}}\{max(p_t^1, .., p_t^j, .., p_t^J) - max_2(p_t^1, .., p_t^j, .., p_t^J)\} \tag{3}$$

where $max(\cdot)$ and $max_2(\cdot)$ provide the top-2 predicted probabilities for the sample x_u. This $\mathcal{AF}$ favors samples that maximize the difference between their top two predictions.

This acquisition phase is launched once at the beginning of each incremental state to get an initial labeled subset of the new data. This step is necessary to include the samples for the new classes in the trained model and initiate the iterative AL process.

Balancing-Driven Sample Acquisition Phase. The second acquisition phase tries to label samples so as to tend toward a balanced distribution among new classes. The distribution of the number of samples per class is computed after each sample labeling to be kept up-to-date. The average number of samples per class is used to divide classes into minority and majority ones. These two sets of classes are noted $\mathcal{C}_t^{mnr}$ and $\mathcal{C}_t^{maj}$ for incremental state $\mathcal{S}_t$. Two functions are proposed to implement the balancing-driven acquisition:

- **balanced core-set sampling** ($b - core$ hereafter) is a modified version of *core* presented in Eq. 1. $b - core$ acts as a filter which keeps candidate samples for labeling only if they are closer to a minority class than to any majority class. We write the relative distance of an unlabeled image w.r.t. its closest minority and majority classes as:

$$\Delta_{\frac{mnr}{maj}}(x_u) = \min_{c_t^{mnr} \in \mathcal{C}_t^{mnr}} \Delta(e(x_u), \mu(c_t^{mnr})) - \min_{c_t^{maj} \in \mathcal{C}_t^{maj}} \Delta(e(x_u), \mu(c_t^{maj})) \quad (4)$$

where: x_u is an unlabeled sample, c_t^{mnr} and c_t^{maj} are classes from the minority and majority sets $\mathcal{C}_t^{mnr}$ and $\mathcal{C}_t^{maj}$ respectively, $e(x_u)$ is the embedding of x_u extracted from the latest deep model available, $\mu(c_t^{mnr})$ and $\mu(c_t^{maj})$ are the centroids of minority and majority classes c_t^{mnr} and c_t^{maj} computed over the embeddings of their labeled samples.

The next sample to label is chosen by using the core-set definition from Eq. 1 but after filtering remaining unlabeled samples with Eq. 4:

$$x_{next} = \operatorname*{argmax}_{x_u \in \mathcal{X}_t^u \ and \ \Delta_{\frac{mnr}{maj}}(x_u) < 0} \{ \min_{1 \leq k \leq n} \Delta(e(x_u), e(x_k)) \} \quad (5)$$

- **poorest class first sampling** (*poor*) is an acquisition function which gives priority to the class represented by the minimum number of labeled samples associated to it at a given moment during active learning. If there are several such classes, one of them is selected randomly. The method translates the hypothesis that samples which are close to a poor class and far from any majority class should be favored in order to achieve a more balanced distribution. The next candidate for labeling is selected with:

$$x_{next} = \operatorname*{argmin}_{x_u \in \mathcal{X}_t^u} \{ \Delta(e(x_u), \mu(c_t^{poor})) - \min_{\forall c_t^{maj} \in \mathcal{C}_t^{maj}} \Delta(e(x_u), \mu(c_t^{maj})) \} \quad (6)$$

where c_t^{poor} is a minority class from $\mathcal{C}_t^{mnr}$ which has the lowest number of samples in the current labeled subset.

poor is similar in spirit to $b - core$ but has a stronger drive towards balancing because an individual class with poorest representation is targeted instead of samples which are close to any minority class.

In an iterative active learning scenario, the balancing-driven acquisition can be repeated several time until the AL budget $\mathcal{B}$ is exhausted.

3.3 Imbalance-Driven Incremental Learning

The model update within each incremental state is inspired by a usual iterative AL approach [37] which includes a classical acquisition phase at the beginning and several balancing-driven iterations. For a total of $\mathcal{I}$ active learning iterations in each state $\mathcal{S}_t$, intermediate models $\mathcal{M}_t^1, ..., \mathcal{M}_t^i, ..., \mathcal{M}_t^{\mathcal{I}-1}$ are created while annotating $\mathcal{X}_t^{l^1}, ..., \mathcal{X}_t^{l^i}, ..., \mathcal{X}_t^{l^{\mathcal{I}-1}}$ during the first $\mathcal{I}-1$ iterations before obtaining the final $\mathcal{M}_t$. The number of iterations $\mathcal{I}$ and the size of each iteration can take different values. The choice of a particular setting is done empirically so as to: (1) have enough new samples in the initial iteration in order for the new classes to be trainable in $\mathcal{M}_t^1$, i.e. the model $\mathcal{M}_t$ in the first iteration, (2) have enough candidates left for the balancing-driven iterations and (3) do not repeat the fine tuning process too many times to keep the incremental update timely. $\mathcal{M}_{t-1}$ is used to extract embeddings if *core* is used in the initial AL iteration. Note that while iterative training increases the level of forgetting in IL [4,5], it is needed in AL to update model representation while annotating the images [37].

As we mentioned, we depart from the usual modeling of the IL problem [10,17,32,41] which exploits knowledge distillation to counter catastrophic forgetting. Following the recent observation that a simpler fine tuning based approach gives interesting results [4], we use an IL backbone inspired from imbalance learning results presented in [8]. This backbone is called fine tuning with thresholding (FT^{th} below), also known as threshold moving or post scaling [8]. Thresholding adjusts the decision threshold of the model. It consists in the addition of a calibration layer at the end of the model during inference to compensate the prediction bias in favor of majority classes. This layer rectifies the class prediction p_t^j of the jth class in the state $\mathcal{S}_t$ as follows:

$$p_t^{j'} = p_t^j \times \frac{|\mathcal{X}_t^l \cup \mathcal{K}|}{|X_t^j|} \tag{7}$$

where $|X_t^j|$ is the number of training examples for the jth class in the state $\mathcal{S}_t$ and $|\mathcal{X}_t^l \cup \mathcal{K}|$ is the total number of training examples in state $\mathcal{S}_t$.

FT^{th} boosts the scores of classes with a lower number of associated samples. The method has the interesting property of dealing with imbalance in IL in a uniform manner. It does not matter whether imbalance comes from the distribution of newly streamed data or from the fact that only a bounded memory of past classes is available. This stands in contrast with knowledge distillation which handles imbalance for past classes but not among new ones. FT^{th} is competitive against state-of-the-art algorithms. In a classical (i.e. fully supervised) IL setting, it has 59.59 top-1 accuracy for Cifar100, compared to $iCaRL$ [32] (57.35) and $LUCIR$ [17] (55.36). More results are provided in the next section.

4   Experiments

4.1   Datasets

Experiments are run with four public datasets, out of which three are imbalanced and one is balanced. We provide a brief description of the datasets below:

- **ImageNet100** - dataset for fine grained object recognition consisting of a subset of 100 randomly selected leaf classes from ImageNet [12] which have at least 50 training images and are not present in the ILSVRC subset [33].
- **Faces100** - face recognition dataset consisting of a subset of randomly selected 100 identities from VGG-Face2 [9] with at least 30 training images.

- **Food101** - dataset for fine-grained food recognition [7]. Since the initial dataset is perfectly balanced, an imbalance induction procedure was applied by removing a variable number of training samples keeping at least 25 images per class.
- **Cifar100** - dataset for object recognition used in its original version [21] which is perfectly balanced.

The main statistics of the experimental datasets are provided in Table 1. We provide the coefficient of variation $cv = \frac{\sigma}{\mu}$, with σ the standard deviation and μ the mean of the distribution of samples per class. cv provides information about the degree of imbalance associated to each dataset. The larger this value is, the more imbalanced the dataset will be.

Table 1. Dataset statistics. cv is the coefficient of variation defined as $cv = \frac{\sigma}{\mu}$.

Dataset	Classes	Train	Test	Mean train (μ)	Std train (σ)	cv
ImageNet100	100	50000	5K	500.0	376.17	0.7523
Faces100	100	23237	5K	232.37	167.68	0.7216
Food101	101	22374	10K	223.74	177.66	0.7940
Cifar100	100	50000	10K	500.0	0.0	0.0

4.2 Methodology

Incremental Learning Setting. We run the experiments with $\mathcal{T} = 10$ states for each dataset[1]. This setting is classically used in class incremental learning [10, 32]. A total of $\mathcal{K}$ images of past classes are kept at any time during incremental learning. $\mathcal{K}$ approximates 2% of the full training sets. Memory sizes are thus $\mathcal{K} = 1000$ for ImageNet100 and Cifar100, $\mathcal{K} = 465$ for Faces100 and $\mathcal{K} = 450$ for Food101. At the end of each incremental state, memory is updated by inserting exemplars of new classes and reducing exemplars of past classes in order to fit its maximum size. Note that since $\mathcal{K}$ is constant and the number of past classes grows, the imbalance in favor of new classes grows for later incremental states and the problem becomes more challenging. The exemplars are chosen using the herding mechanism introduced in [32]. The herding procedure consists in choosing the set of images that approximates the best the real mean of the class.

Active Learning Process. Three active learning budgets are tested covering $\mathcal{B} = \{20\%, 10\%, 5\%\}$ of the unlabeled dataset $\mathcal{X}_t^u$ streamed in state $\mathcal{S}_t$. These different values are used to get a comprehensive view of each configuration's

---

[1] The initial non-incremental state of Food101 includes 11 classes while the initial states for the other datasets include 10 classes each.

behavior. Active learning is implemented with a usual iterative approach [36,37] including $\mathcal{I} = 4$ iterations, 40% of $\mathcal{B}$ are used for classical acquisition and three times 20% of $\mathcal{B}$ for balancing-driven acquisition (values were experimentally chosen). Classical and balancing-driven acquisition phases are independent of one another and we test all their combinations. For completeness, we include results with a baseline in which both phases are implemented with random sampling. Note that the proposed acquisition functions are non-deterministic and experiments are run five times for each configuration in order to have a robust estimation of its performance. To improve comparability of configurations which use the same initial $\mathcal{AF}$, the same initial models are used for all subsequent balancing-driven $\mathcal{AF}$s.

Training Details. The experimental setup is inspired by the one proposed in *iCaRL* [32]. FT^{th} is implemented in Pytorch [29] using a ResNet-18 architecture [15] and an SGD optimizer. The first non-incremental state is run for 100 epochs with *batch size* = 128, *lr* = 0.1, *momentum* = 0.9, *weight decay* = 0.0005. The learning rate is divided by 10 when the error plateaus for 15 consecutive epochs. Fine tuning is run for 80 epochs, 20 epochs for each active learning iteration with *batch size* = 32, *lr* = 0.1, *momentum* = 0.9, *weight decay* = 0.0005. The learning rate is initialized at the beginning of the AL process and then divided by 10 when the error plateaus for 10 consecutive epochs.

Training images are preprocessed using randomly resized 224×224 crops and horizontal flipping and are normalized afterwards. While more advanced data augmentation is known to slightly improve performance [10], we did not apply other image transformations. For Faces100, face cropping is done with MTCNN [44] before further processing.

Upper Bound Methods. In addition to the active learning configurations, we present results with:

- sIL - usual supervised incremental learning in which all samples are labeled (equivalent to $\mathcal{B} = 100\%$).
- $noIL$ - classical non-incremental learning in which all samples are provided at once.

For comparability, sIL and $noIL$ are both trained using threshold calibration. sIL is an incremental upper bound for active learning configurations since it is fully supervised. $noIL$ is an upper bound for sIL since all the data are labeled and available at once. These upper bounds are useful insofar they provide information about the performance gap due to a partial labeling of streamed data.

4.3 Results and Discussion

FT^{th} **in supervised mode** - Instead of handling catastrophic forgetting [24] as previous works did [10,17,31,41], we address IL with bounded past memory as an imbalanced learning problem. We use threshold calibration [8] to rectify scores in order to give more chances to minority classes to be selected during inference. The comparison to recent IL methods in supervised mode from Table 2 indicates that FT^{th} is competitive. It clearly outperforms $iCaRL$ [32] and $IL2M$ [4] and is better than LUCIR [17] for three datasets out of four. We also provide the results of vanilla fine tuning before threshold calibration to underline the usefulness of thresholding. It has a positive effect for all four datasets, a finding which validates its usefulness in our scenario.

Table 2. Top-1 average supervised IL accuracy (%). Best results are in bold.

Dataset	FT	FT^{th} [8]	$LUCIR$ [17]	$iCaRL$ [32]	$IL2M$ [4]
Imagenet100	54.80	**61.42**	60.77	52.40	57.68
Faces100	69.11	73.26	**78.44**	60.48	70.33
Food101	30.21	**34.79**	25.70	21.99	32.20
Cifar100	50.98	**59.59**	55.36	57.35	54.24

Table 3. Top-1 average accuracy (%). Following [10], accuracy is averaged only for incremental states (i.e. excluding the initial, non-incremental state). Results are averaged over 5 runs for all AL configurations. sIL is the result obtained in a fully supervised IL scenario. $noIL$ is the non-incremental upper-bound performance obtained with all data available. *Best results for each active learning configuration (row) are in bold.*

Dataset	B	rand	poor	b-core	core	poor	b-core	ent	poor	b-core	marg	poor	b-core	sIL	noIL
Imagenet100	20%	57.48 ±0.50	**58.65** ±0.23	58.08 ±0.40	56.85 ±0.23	57.25 ±0.84	57.46 ±0.52	45.07 ±0.58	56.53 ±0.27	56.23 ±0.22	54.13 ±0.66	56.39 ±0.53	56.26 ±0.45	61.42	72.48
	10%	52.61 ±0.45	**54.89** ±0.53	53.40 ±0.26	52.09 ±0.41	53.55 ±1.21	52.22 ±1.13	42.15 ±0.43	51.91 ±0.51	51.81 ±0.76	46.26 ±1.24	51.40 ±0.89	50.61 ±0.36		
	5%	47.72 ±0.69	**48.71** ±0.97	48.18 ±0.56	46.01 ±0.51	47.39 ±0.85	46.45 ±0.30	37.95 ±0.61	45.10 ±1.46	44.70 ±1.02	36.74 ±1.05	44.18 ±1.09	43.93 ±0.93		
Faces100	20%	65.91 ±0.94	66.41 ±0.10	**67.24** ±0.36	66.41 ±0.66	66.46 ±0.46	66.94 ±1.37	48.62 ±0.95	63.30 ±0.33	64.99 ±0.80	59.51 ±1.17	62.85 ±1.75	65.27 ±0.53	73.26	93.62
	10%	58.40 ±0.71	**59.13** ±1.66	58.92 ±1.05	55.82 ±4.70	58.76 ±2.93	57.26 ±2.90	42.12 ±1.38	54.93 ±1.07	55.45 ±1.47	49.32 ±1.40	54.82 ±2.19	58.69 ±1.52		
	5%	48.38 +1.27	50.09 ±2.30	**50.12** ±0.96	48.74 ±1.21	47.71 ±1.28	50.04 ±2.04	35.61 ±0.51	45.79 ±0.83	45.39 ±1.13	38.37 ±1.02	45.90 ±2.31	45.64 ±1.56		
Food101	20%	28.67 ±0.42	**28.89** ±0.43	28.60 ±0.52	28.24 ±0.42	27.88 ±0.34	27.98 ±0.46	23.72 ±1.00	27.99 ±0.41	27.51 ±0.54	28.13 ±0.59	28.18 ±0.35	27.56 ±0.24	34.79	62.53
	10%	24.12 ±0.47	**24.17** ±0.56	24.07 ±0.68	22.91 ±0.63	23.46 ±0.12	23.07 ±0.31	19.41 ±0.66	22.32 ±0.73	22.25 ±0.64	23.35 ±0.64	22.68 ±0.81	22.84 ±0.52		
	5%	20.51 ±0.61	19.10 ±0.68	**20.63** ±0.46	19.22 ±0.36	19.17 ±0.58	18.79 ±0.64	16.80 ±0.75	18.66 ±0.31	18.57 ±0.47	18.62 ±0.48	18.79 ±0.75	18.41 ±0.82		
Cifar100	20%	**49.47** ±0.16	49.36 ±0.33	48.46 ±0.75	46.75 ±0.40	46.87 ±0.19	46.87 ±0.36	39.76 ±1.30	46.66 ±0.29	47.69 ±0.23	46.07 ±0.31	45.37 ±0.39	46.68 ±0.44	59.59	76.98
	10%	**45.49** ±0.61	45.23 ±1.17	44.83 ±0.29	41.76 ±0.54	42.60 ±0.77	42.04 ±0.77	34.87 ±0.66	42.64 ±0.50	43.76 ±0.55	39.92 ±0.43	40.94 ±0.31	41.82 ±0.35		
	5%	**41.58** ±0.29	40.69 ±0.23	39.69 ±0.46	35.23 ±0.64	37.70 ±0.46	35.72 ±0.67	31.74 ±0.74	37.68 ±0.34	38.02 ±0.66	31.88 ±0.58	35.96 ±0.42	35.69 ±0.64		

Active Learning - The experimental results obtained with FT^{th} for the proposed active incremental learning scenario are presented in Table 3. The comparison of classical $\mathcal{AF}$s ($rand$ -$rand$ and $core - core$ in Table 3) indicates that random sampling outperforms the $core - set$ sampling in a majority of cases. This result is at odds with the one reported in [36] but is in line with the findings of [6,13] that random sampling in AL is a strong baseline and is actually better than the recent core-set method from [34]. The authors of this last paper also report that random sampling has better performance for lower active learning budgets which are studied here. Consequently, improving over random sampling for imbalanced datasets is an interesting result.

The results from Table 3 indicate that the balancing-driven acquisition phase is useful for all three imbalanced datasets and active learning budgets tested. The gains for ImageNet100 and Faces100 are usually between 1 and 2 points compared to the classical acquisition processes implemented here ($rand$ - $rand$ or $core$ - $core$). The gains are low for Food101, the third imbalanced dataset tested. This is probably due to the fact that Food101 is a more difficult task, as shown by sIL. More labeled samples per class would probably be needed for an efficient training.

$poor$ strategy is better than $b - core$ for ImageNet100 while more mixed results are obtained for Faces100 and Food101 datasets. Interestingly, the best results are always obtained on top of a $rand$ initial sampling, even when $core$-$core$ baseline is better than a $rand$-$rand$ one, as it is the case for Faces100 with $B = 20\%$ and $B = 5\%$.

When applied without balancing, ent and $marg$ have poorer performance compared to that of $rand$ and $core$. Balancing significantly improves results for both of uncertainty-based methods, but their overall performance still lags behind that of random followed by balancing. This reinforces the finding that a random selection is a competitive acquisition function in our active incremental learning over imbalanced datasets scenario.

The performance drop between active learning configurations and fully supervised IL naturally grows as B is reduced. The drop between sIL and the best AL configuration is of 3, 6 and 5 points for $B = 20\%$ for ImageNet100, Faces100, and Food101 respectively. When the AL budget is reduced to only 5% of new data, the corresponding performance losses go to 12.5, 23 and 14 points. Even when as little as 5% of the new data are annotated, suboptimal models are trainable and usable if the IL system needs to be operational quickly.

While the focus is on imbalanced datasets, we also report results with Cifar100, a perfectly balanced dataset for completeness. In this case, the balancing-driven sampling has a slightly negative effect when applied over $rand$ and a slightly positive effect over $core$. It is however notable that $core$ lags consistently behind $rand$ for Cifar100. The best strategy for all B sizes is $rand$-$rand$, with $rand$ - $poor$ being a close second best configuration.

The gap between active IL and supervised IL is still notable, especially for smaller AL budgets. In practice, active IL is useful when the system needs to be operational quickly after new data are streamed but at the expense of suboptimal performance. If a longer delay is permitted, it is naturally preferable to annotate all new data before updating the incremental model. The gap is even higher between incremental and classical learning, even though FT^{th} has competitive performance compared to existing IL algorithms. Globally, our results provide further confirmation that the use of incremental learning vs. classical learning should be weighted depending on the time, memory and/or computation constraints associated to an AI system's operation.

5   Conclusion

We proposed a more realistic incremental learning scenario which does not assume that streamed data are readily annotated and that they are evenly distributed among classes. An adaptation of the active learning sampling process is proposed in order to obtain a more balanced labeled subset. This adaptation has a positive effect for imbalanced datasets and a slightly negative effect for the balanced dataset evaluated here. Both proposed acquisition functions improve results compared to a classical acquisition process. Also interesting, experiments show that the random baseline outperforms the $core - set$ function. The strong performance of random sampling indicates that this method should be consistently used as a baseline for future works in active incremental learning. As a secondary contribution, we introduce FT^{th}, a IL backbone which provides competitive results when compared to state-of-the-art methods. The code is publicly available to facilitate reproducibility[2].

The proposed approach brings the IL scenario closer to practical needs. It reduces the time needed for an IL system to become operational upon receiving new data. The obtained results are encouraging but further investigation is needed to reduce the gap between active and supervised IL. Future work aims to: (1) run experiments with semi-supervised learning methods to automatically expand the labeled dataset and improve overall performance. While appealing, not all semi-supervised methods prove efficient in practice [27] and their usefulness for imbalanced datasets needs to be studied. (2) complement the proposed balancing-driven acquisition functions with a component which pushes the sampling process towards a better coverage of the manifold of each modeled class. This could be done, for instance, by taking inspiration from the herding mechanism [32] already used to select past exemplars. (3) render the IL scenario even more realistic by testing incremental steps of variable size to account for the fact that data might arrive at variable pace and considering that newly streamed data might belong both to unseen and past classes.

Acknowledgements. This publication was made possible by the use of the FactoryIA supercomputer, financially supported by the Ile-de-France Regional Council.

References

1. Aggarwal, U., Popescu, A., Hudelot, C.: Active learning for imbalanced datasets. In: Proceedings of the IEEE/CVF Winter Conference on Applications of Computer Vision (WACV) (March 2020)

---

[2] https://github.com/EdenBelouadah/class-incremental-learning/.

2. Aljundi, R., Chakravarty, P., Tuytelaars, T.: Expert gate: Lifelong learning with a network of experts. In: Conference on Computer Vision and Pattern Recognition. CVPR (2017)
3. Aljundi, R., Kelchtermans, K., Tuytelaars, T.: Task-free continual learning. In: The IEEE Conference on Computer Vision and Pattern Recognition (CVPR) (June 2019)
4. Belouadah, E., Popescu, A.: Il2m: Class incremental learning with dual memory. In: Proceedings of the IEEE International Conference on Computer Vision, pp. 583–592 (2019)
5. Belouadah, E., Popescu, A.: Scail: Classifier weights scaling for class incremental learning. In: IEEE Winter Conference on Applications of Computer Vision, WACV 2020, Snowmass Village, CO, USA, March 2–5, 2020 (2020)
6. Beluch, W.H., Genewein, T., Nürnberger, A., Köhler, J.M.: The power of ensembles for active learning in image classification. In: 2018 IEEE Conference on Computer Vision and Pattern Recognition, CVPR 2018, Salt Lake City, UT, USA, June 18–22, 2018. pp. 9368–9377 (2018)
7. Bossard, L., Guillaumin, M., Van Gool, L.: Food-101 - mining discriminative components with random forests. In: European Conference on Computer Vision (2014)
8. Buda, M., Maki, A., Mazurowski, M.A.: A systematic study of the class imbalance problem in convolutional neural networks. Neural Networks **106**, 249–259 (2018)
9. Cao, Q., Shen, L., Xie, W., Parkhi, O.M., Zisserman, A.: Vggface2: A dataset for recognising faces across pose and age. In: 13th IEEE International Conference on Automatic Face & Gesture Recognition, FG 2018, Xi'an, China, May 15–19, 2018, pp. 67–74 (2018)
10. Castro, F.M., Marín-Jiménez, M.J., Guil, N., Schmid, C., Alahari, K.: End-to-end incremental learning. In: Proceedings of Computer Vision - ECCV 2018–15th European Conference, Munich, Germany, September 8–14, 2018, Part XII, pp. 241–257 (2018)
11. Culotta, A., McCallum, A.: Reducing labeling effort for structured prediction tasks. In: Proceedings, The Twentieth National Conference on Artificial Intelligence and the Seventeenth Innovative Applications of Artificial Intelligence Conference, July 9–13, 2005, Pittsburgh, Pennsylvania, USA, pp. 746–751 (2005)
12. Deng, J., Dong, W., Socher, R., Li, L., Li, K., Li, F.: Imagenet: a large-scale hierarchical image database. In: 2009 IEEE Computer Society Conference on Computer Vision and Pattern Recognition (CVPR 2009), 20–25 June 2009, Miami, Florida, USA, pp. 248–255 (2009)
13. Gal, Y., Islam, R., Ghahramani, Z.: Deep Bayesian active learning with image data. In: Proceedings of the 34th International Conference on Machine Learning, ICML 2017, Sydney, NSW, Australia, 6–11 August 2017, pp. 1183–1192 (2017)
14. He, C., Wang, R., Shan, S., Chen, X.: Exemplar-supported generative reproduction for class incremental learning. In: British Machine Vision Conference 2018, BMVC 2018, Northumbria University, Newcastle, UK, September 3–6, 2018, p. 98 (2018)
15. He, K., Zhang, X., Ren, S., Sun, J.: Deep residual learning for image recognition. In: Conference on Computer Vision and Pattern Recognition, CVPR (2016)
16. Hinton, G.E., Vinyals, O., Dean, J.: Distilling the knowledge in a neural network (2015). CoRR abs/1503.02531
17. Hou, S., Pan, X., Loy, C.C., Wang, Z., Lin, D.: Learning a unified classifier incrementally via rebalancing. In: IEEE Conference on Computer Vision and Pattern Recognition, CVPR 2019, Long Beach, CA, USA, June 16–20, 2019, pp. 831–839 (2019)

18. Hu, P., Lipton, Z.C., Anandkumar, A., Ramanan, D.: Active learning with partial feedback. In: 7th International Conference on Learning Representations, ICLR 2019, New Orleans, LA, USA, May 6–9, 2019 (2019)
19. Javed, K., Shafait, F.: Revisiting distillation and incremental classifier learning (2018). CoRR abs/1807.02802
20. Krasin, I., et al.: Openimages: a public dataset for large-scale multi-label and multi-class image classification (2017). Dataset: https://storage.oogleapis.com/openimages/web/index.html
21. Krizhevsky, A.: Learning multiple layers of features from tiny images, Technical report. University of Toronto (2009)
22. Li, Z., Hoiem, D.: Learning without forgetting. IEEE Trans. Pattern Anal. Mach. Intell. 40(12), 2935–2947 (2018)
23. Mallya, A., Lazebnik, S.: Packnet: Adding multiple tasks to a single network by iterative pruning. In: 2018 IEEE Conference on Computer Vision and Pattern Recognition, CVPR 2018, Salt Lake City, UT, USA, June 18–22, 2018, pp. 7765–7773 (2018)
24. Mccloskey, M., Cohen, N.J.: Catastrophic interference in connectionist networks: the sequential learning problem. Psychol. Learn. Motiv. 24, 104–169 (1989)
25. Mensink, T., Verbeek, J.J., Perronnin, F., Csurka, G.: Distance-based image classification: generalizing to new classes at near-zero cost. IEEE Trans. Pattern Anal. Mach. Intell. 35(11), 2624–2637 (2013)
26. Oksuz, K., Cam, B.C., Kalkan, S., Akbas, E.: Imbalance problems in object detection: a review (2019). CoRR abs/1909.00169
27. Oliver, A., Odena, A., Raffel, C.A., Cubuk, E.D., Goodfellow, I.: Realistic evaluation of deep semi-supervised learning algorithms. In: Bengio, S., Wallach, H., Larochelle, H., Grauman, K., Cesa-Bianchi, N., Garnett, R. (eds.) Advances in Neural Information Processing Systems, Vol. 31, pp. 3235–3246. Curran Associates, Inc. (2018)
28. Parisi, G.I., Kemker, R., Part, J.L., Kanan, C., Wermter, S.: Continual lifelong learning with neural networks: a review. Neural Netw. 113, 54–71 (2019)
29. Paszke, A., Gross, S., Chintala, S., Chanan, G., Yang, E., DeVito, Z., Lin, Z., Desmaison, A., Antiga, L., Lerer, A.: Automatic differentiation in pytorch. In: Advances in Neural Information Processing Systems Workshops. NIPS-W (2017)
30. Pernici, F., Bartoli, F., Bruni, M., Del Bimbo, A.: Memory based online learning of deep representations from video streams. In: The IEEE Conference on Computer Vision and Pattern Recognition (CVPR) (June 2018)
31. Rebuffi, S., Bilen, H., Vedaldi, A.: Learning multiple visual domains with residual adapters. In: Advances in Neural Information Processing Systems 30: Annual Conference on Neural Information Processing Systems 2017, 4–9 December 2017, Long Beach, CA, USA, pp. 506–516 (2017)
32. Rebuffi, S., Kolesnikov, A., Sperl, G., Lampert, C.H.: icarl: Incremental classifier and representation learning. In: Conference on Computer Vision and Pattern Recognition. CVPR (2017)
33. Russakovsky, O., et al.: Imagenet large scale visual recognition challenge. Int. J. Comput. Vis. 115(3), 211–252 (2015)
34. Rusu, A.A., et al.: Progressive neural networks (2016). CoRR abs/1606.04671
35. Scheffer, T., Decomain, C., Wrobel, S.: Mining the web with active hidden Markov models. In: Proceedings of the 2001 IEEE International Conference on Data Mining, 29 November - 2 December 2001, San Jose, California, USA, pp. 645–646 (2001)

36. Sener, O., Savarese, S.: Active learning for convolutional neural networks: A core-set approach. In: 6th International Conference on Learning Representations, ICLR 2018, Vancouver, BC, Canada, April 30 - May 3, 2018, Conference Track Proceedings (2018)

37. Settles, B.: Active learning literature survey, Technical report. University of Winsconsin (2010)

38. Settles, B., Craven, M.: An analysis of active learning strategies for sequence labeling tasks. In: 2008 Conference on Empirical Methods in Natural Language Processing, EMNLP 2008, Proceedings of the Conference, 25–27 October 2008, Honolulu, Hawaii, USA, A meeting of SIGDAT, a Special Interest Group of the ACL (2008)

39. Shannon, C.E.: A Mathematical Theory of Communication, pp. 379–423. University of Illinois Press, Champaign (1948)

40. Wang, Y., Ramanan, D., Hebert, M.: Growing a brain: Fine-tuning by increasing model capacity. In: Conference on Computer Vision and Pattern Recognition. CVPR (2017)

41. Wu, Y., et al.: Large scale incremental learning. In: IEEE Conference on Computer Vision and Pattern Recognition, CVPR 2019, Long Beach, CA, USA, June 16–20, 2019, pp. 374–382 (2019)

42. Yang, J., Parikh, D., Batra, D.: Joint unsupervised learning of deep representations and image clusters. In: The IEEE Conference on Computer Vision and Pattern Recognition (CVPR) (June 2016)

43. Yoo, D., Kweon, I.S.: Learning loss for active learning. In: The IEEE Conference on Computer Vision and Pattern Recognition (CVPR) (June 2019)

44. Zhang, K., Zhang, Z., Li, Z., Qiao, Y.: Joint face detection and alignment using multitask cascaded convolutional networks. IEEE Signal Process. Lett. **23**(10), 1499–1503 (2016)

45. Zhou, P., Mai, L., Zhang, J., Xu, N., Wu, Z., Davis, L.S.: M2KD: multi-model and multi-level knowledge distillation for incremental learning (2019). CoRR abs/1904.01769

46. Zhou, Z., Shin, J.Y., Zhang, L., Gurudu, S.R., Gotway, M.B., Liang, J.: Fine-tuning convolutional neural networks for biomedical image analysis: actively and incrementally. In: 2017 IEEE Conference on Computer Vision and Pattern Recognition, CVPR 2017, Honolulu, HI, USA, July 21–26, 2017, pp. 4761–4772 (2017)

Knowledge Distillation for Multi-task Learning

Wei-Hong Li$^{(\boxtimes)}$ and Hakan Bilen

VICO Group, University of Edinburgh, Edinburgh, UK
{w.h.li,hbilen}@ed.ac.uk

Abstract. Multi-task learning (MTL) is to learn one single model that performs multiple tasks for achieving good performance on all tasks and lower cost on computation. Learning such a model requires to jointly optimize losses of a set of tasks with different difficulty levels, magnitudes, and characteristics (*e.g.* cross-entropy, Euclidean loss), leading to the imbalance problem in multi-task learning. To address the imbalance problem, we propose a knowledge distillation based method in this work. We first learn a task-specific model for each task. We then learn the multi-task model for minimizing task-specific loss and for producing the same feature with task-specific models. As the task-specific network encodes different features, we introduce small task-specific adaptors to project multi-task features to the task-specific features. In this way, the adaptors align the task-specific feature and the multi-task feature, which enables a balanced parameter sharing across tasks. Extensive experimental results demonstrate that our method can optimize a multi-task learning model in a more balanced way and achieve better overall performance.

1 Introduction

The objective of multi-task learning (MTL) [3,26] is to develop methods that can tackle a large variety of tasks within a single model. MTL has multiple practical benefits. First, learning shared parameters across multiple tasks leads to representations that can be more data-efficient to train and also generalize better to unseen data. Second, sharing parameters and computations across tasks can significantly reduce both training and inference time over running multiple individual models, which is especially important in platforms with limited computational resources such as mobile devices. Therefore there is a growing interest in developing MTL methods and MTL has been successfully applied to machine learning problems in several fields including natural language processing [5], computer vision [2,14] and speech recognition [27].

There are at least two challenges to achieve better performance and efficiency with MTL. The first one is to design a multi-task deep neural network architecture that shares only the relevant parameters across the tasks and keeps the remaining ones task-specific. This is in contrast to the standard MTL methods that share all the layers except the last few ones across all the tasks. This

© Springer Nature Switzerland AG 2020
A. Bartoli and A. Fusiello (Eds.): ECCV 2020 Workshops, LNCS 12540, pp. 163–176, 2020.
https://doi.org/10.1007/978-3-030-65414-6_13

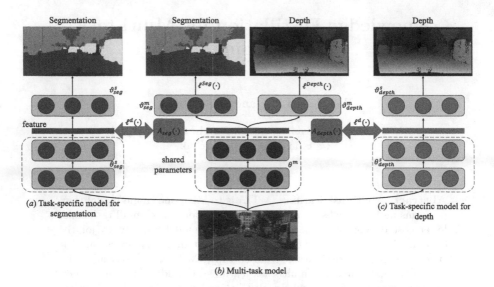

Fig. 1. Diagram of our method. We first train a task-specific model for each task in an offline stage and freeze their parameters (*i.e.* (a), (c)). We then optimize the parameters of the multi-task network for minimizing a sum of task-specific losses and also for producing similar features with the single-task networks (*i.e.* (b)). Best seen in color.

heuristic is possibly suboptimal when the tasks have different characteristics and goals (*e.g.* semantically low and high-level tasks), however, searching for an optimal architecture in an exponential configuration space is extremely expensive. The second one is to develop MTL training algorithms that achieve good performance not only in one of the tasks but in all of them. This problem is especially important when MTL involves jointly minimizing a set of loss functions for various problems with different difficulty levels, magnitudes, and characteristics (*e.g.* cross-entropy, Euclidean loss). Thus a naive strategy of uniformly weighing multiple losses can lead to sub-optimal performances and searching for optimal weights in a continuous hyperparameter space can be prohibitively expensive.

Concerned with the second problem, previous work [4,10,12,16,28] addresses the *unbalanced loss optimization* problem with balanced loss weighting and parameter updating strategies. Kendall *et al.* [12] weigh loss functions by considering the task-dependent uncertainty of the model at training time. Sener *et al.* [28] pose the MTL as a multiple objective optimization problem and propose an approximate Pareto optimization method that uses Frank-Wolfe algorithm to solve the constrained optimization. Yu *et al.* [32] project the gradients for each loss function to a space where conflicting gradient components are removed to eliminate the disturbance between the tasks. Although the previous work improves over the uniform weighing loss strategy in MTL, they still suffer from the problem of one task dominating the remaining ones and lower task performance than the single task models in standard multi-task benchmarks.

In this paper, we approach the unbalanced MTL problem from a different point and propose a knowledge distillation based method inspired from [11, 25]. As weighing the individual loss functions (*e.g.* [12]) or modifying the gradients for the loss functions by simple transformations (*e.g.* [4, 32]) provide a limited control on the learned parameters and are thus limited to prevent one task dominating the rest, we propose a more strict control on the parameters of the multi-task network. Given that single-task networks often perform well with sufficient training data, we hypothesize that the solution of the multi-task network should be close to the single task ones' and lie in the intersection of the single-task solutions. To this end, we first train a task-specific model for each task in an offline stage and freeze their parameters; then optimize the parameters of the multi-task network for minimizing a sum of task-specific losses and also for producing similar features with the single-task networks. As each task-specific network can compute different features, we introduce small task-specific adaptors that map multi-task features to the task-specific one's. The adaptors align the features of the single-task and multi-task networks, and enables a balanced parameter sharing across multiple tasks.

In the remainder of this paper, we first discuss how our method relates the previous MTL and data distillation methods in Sect. 2, formulate our method in Sect. 3, demonstrate that our method outperforms the state-of-the-art MTL methods in two standard benchmarks in Sect. 4 and conclude the paper with future remarks in Sect. 5.

2   Related Work

2.1   Multi-task Learning

Multi-task learning (MTL) is one of the long-standing problems in machine learning and has been used broadly [3, 7, 12, 14, 18, 18, 19, 23, 26, 33]. In computer vision, MTL has been used for image classification [23], facial landmark regression [33], segmentation and depth estimation [12] and so on. In this work, we specifically focus on tackling the unbalance in the optimization of multi-task networks to achieve good performance not only in a few tasks but in all tasks.

In recent years, several methods have been proposed for solving the imbalance problem in MTL by either designing loss weighting schemes [4, 10, 12, 16, 28] to weigh each task-specific loss or modifying parameter updates [32]. Chen *et al.* [4] develop a training strategy, namely GradNorm, that looks at the gradient's norm of each task and learns the weight to normalize each task's gradient so as to balance the losses for MTL. In [28], Sener *et al.* formulate the MTL as a multiple objectives optimization problem and proposed an approximation Pareto optimization method using Frank-Wolfe algorithm to learn weights of losses. Kendall *et al.* [12] propose to weigh multiple loss functions by considering the homoscedastic uncertainty of each task during training. To design the weighting scheme, Guo *et al.* [10] observe that the imbalances in task difficulty can lead to an unnecessary emphasis on easier tasks, thus neglecting and slowing progress on difficult tasks. Based on the observation, they introduce dynamic task prioritization for MTL, which allows the model to dynamically prioritize difficult tasks during training, where the difficulty is inversely proportional to performance. Rather than weighing the losses, Yu *et al.* [32] propose a form of gradient "surgery" that projects each task's gradient onto the normal plane of the gradient of any other task and modifies the gradients for each task so as to minimize negative conflict with other task gradients during the MTL optimization.

Unlike existing methods, we propose a knowledge distillation based MTL method to solve the unbalanced loss optimization problem from a different angle. To this end, we first train a task-specific model for each task in an offline stage and freeze their parameters. We then train the MTL network for minimizing task-specific loss and also for producing the same features with the task-specific networks. As the task-specific network encodes different features, we introduce small task-specific adaptors to project multi-task features to the task-specific features. In this way, the adaptors align the task-specific feature and the multi-task feature, which enables a balanced parameter sharing across tasks.

2.2   Knowledge Distillation

Our work is also related to knowledge distillation [11,17,22,25,31]. Hinton *et al.* [11] show that distilling the knowledge of the whole ensemble of models to a neural network can achieve better performance and avoid an expensive computation. Romero *et al.* [25] introduce the knowledge distillation to training a small student network to achieve better performance than the teacher network. Apart from the success in single-task learning, knowledge distillation has also been shown to be effective in MTL. Parisotto *et al.* [21] exploits the use of deep reinforcement learning and model compression techniques to train a single policy network that learns to perform in multiple tasks by using the guidance of several expert teachers. In the contrast, in [5], Clark *et al.* extends the Born-Again network [9] to MTL setting for NLP. More specifically, they apply the knowledge distillation loss proposed in [11] on each task's predictions and propose a weight annealing strategy to update the weight of the distillation losses and multiple tasks losses.

Different from these methods, we aim at solving the unbalanced loss optimization problem in MTL. Aligning the predictions from the multi-task network and the task-specific networks would still result in unbalance as the dimension of tasks' predictions is usually different and we need to use different loss functions for matching different tasks' predictions [5], *e.g.*, a kl-divergence loss for classification and l2-norm loss for regression. In this work, we first introduce a task-specific adaptor for each task to transform features from the multi-task network and we apply the same loss function to align the transformed multi-task feature and the task-specific networks' features. We train the MTL network for minimizing task-specific loss and for producing the same feature with task-specific networks. This enables the MTL to share the parameters in a balanced way.

3 Methodology

3.1 Single-Task Learning (STL)

Consider that we are given a dataset $\mathcal{D}$ that contains N training images $\boldsymbol{x}^i$ and their labels $\boldsymbol{y}_1^i, \ldots, \boldsymbol{y}_T^i$ for T tasks (e.g. semantic segmentation, depth estimation, surface normals). In case of the STL, we wish to learn T convolutional neural networks, one for each task, each maps the input $\boldsymbol{x}$ to the target label $\boldsymbol{y}_\tau$, i.e. $f(\boldsymbol{x}; \theta_\tau^s, \vartheta_\tau^s) = \boldsymbol{y}_\tau$ where the superscript s indicates the single-task, θ_τ^s and ϑ_τ^s are the parameters of the network. Each single-task network is composed of two parts: i) a feature encoder $\phi(\cdot; \theta_\tau^s)$ that takes in an image and outputs a high-dimensional encoding $\phi(\boldsymbol{x}; \theta_\tau^s) \in \mathbb{R}^{C \times H \times W}$ where C, H, W are the number channels, height and width of the feature map; ii) a predictor $\psi(\cdot; \vartheta_\tau^s)$ that takes in the encoding $\phi(\boldsymbol{x}; \theta_\tau^s)$ and predicts the output for the task τ, i.e. $\hat{\boldsymbol{y}}_\tau = \psi(\cdot; \vartheta_\tau^s) \circ \phi(\boldsymbol{x}; \theta_\tau^s)$ where θ_τ^s and ϑ_τ^s denote the parameters of the feature encoder and predictor respectively. The parameters for the network can be learned for each task independently by optimizing a task-specific loss function $\ell_\tau(\hat{\boldsymbol{y}}, \boldsymbol{y})$ (e.g. Cross-Entropy loss function for classification) over the training samples that measure the mismatch between the ground-truth label and prediction as following:

$$\min_{\theta_\tau^s, \vartheta_\tau^s} \sum_{\boldsymbol{x}, \boldsymbol{y}_\tau \in \mathcal{D}} \ell_\tau(\psi(\cdot; \vartheta_\tau^s) \circ \phi(\boldsymbol{x}; \theta_\tau^s), \boldsymbol{y}_\tau). \tag{1}$$

3.2 Multi-task Learning (MTL)

In the case of MTL, we would like to learn one network that shares the majority of its parameters across the tasks and solves all the tasks simultaneously. Similar to STL, the multi-task network can be decomposed into two parts: i) a feature encoder $\phi(\cdot; \theta^m)$ that encodes the input image into a high-dimensional encoding, now its parameters θ^m are shared across all the tasks; ii) a task-specific predictor $\psi(\cdot; \vartheta_\tau^m)$ for each task that takes in the shared encoding $\phi(\boldsymbol{x}; \theta^m)$ and outputs its prediction for task τ, i.e. $\psi(\cdot; \vartheta_\tau^m) \circ \phi(\boldsymbol{x}; \theta^m)$. Note that we use superscript m to denote MTL. The multi-task network can be learned by optimizing a linear combination of task-specific losses:

$$\min_{\theta^m, \vartheta_1^m, \ldots, \vartheta_T^m} \sum_{\tau=1}^{T} \sum_{\boldsymbol{x}, \boldsymbol{y}_\tau \in \mathcal{D}} w_\tau \ell_\tau(\psi(\cdot; \vartheta_\tau^m) \circ \phi(\boldsymbol{x}; \theta^m), \boldsymbol{y}_\tau) \tag{2}$$

where w_τ is a scaling hyperparameter for task τ that is used for balancing the loss functions among the tasks.

In contrast to the STL optimization in Eq. (1), optimizing Eq. (2) involves a joint learning of all the task-specific and shared parameters which is typically more challenging when the task-specific loss functions ℓ_τ have different characteristics such as their magnitude and dynamics (e.g. logarithmic, quadratic). One solution to balance the loss terms is to search for the best scaling hyperparameters w_τ by a cross-validation which has two shortcomings. First, the hyperparameter search in a continuous space is computationally expensive, especially when the number of tasks is large, as each validation step requires the training of the model. Second, even when the optimal hyperparameters can be found, it may be sub-optimal to use the same fixed ones throughout the optimization.

3.3   Knowledge Distillation for Multi-task Learning

Motivated by these challenges, the previous work [4,12,28] propose dynamic weighing strategies that can adjust them at each training iteration. Here we argue that these hyperparameters provide a limited control on the parameters of the network for preventing the unbalanced MTL and thus we propose a different view on this problem inspired by the knowledge distillation methods [11,25].

To this end, we first train a task-specific model $f(\cdot; \theta_\tau^s, \vartheta_\tau^s)$ for each task τ by optimizing Eq. (1) in an offline stage, freeze their parameters and use only their feature encoders $\phi(\cdot; \theta_\tau^s)$ to regulate the multi-task network at train time by minimizing the distance between the features of task-specific networks and multi-task network for given training samples (see Fig. 1). As the outputs of the task-specific encoders can differ significantly and the feature encoder of the multi-task network cannot match all of them simultaneously. Instead, we project the output of the multi-task feature encoder into each task-specific one via a task-specific adaptor $A_\tau : \mathbb{R}^{C \times H \times W} \to \mathbb{R}^{C \times H \times W}$ where H, W and C are the height, width and depth (number of channels) of the features. In our experiments, we use a linear layer that consists of a $1 \times 1 \times C \times C$ convolution for each adaptor. These adaptors are jointly learned along the parameters of the multi-task network to align its features with the single-task feature encoders.

$$\mathcal{L}^d = \sum_{\tau=1}^{T} \sum_{\boldsymbol{x}, \boldsymbol{y}_\tau \in \mathcal{D}} \ell^d(A_\tau(\phi(\boldsymbol{x}; \theta^m)), \phi(\boldsymbol{x}; \theta_\tau^s)) \tag{3}$$

where ℓ^d is the Euclidean distance function between the L2 normalized feature maps:

$$\ell^d(\boldsymbol{a}, \boldsymbol{b}) = \left\| \frac{\boldsymbol{a}}{||\boldsymbol{a}||_2} - \frac{\boldsymbol{b}}{||\boldsymbol{b}||_2} \right\|_2^2. \tag{4}$$

Now we can write the optimization formulation that is employed to learn the multi-task model as a linear combination of Eqs. (2) and (3):

$$\min_{\theta^m, \vartheta_1^m, \dots, \vartheta_T^m} \sum_{\tau=1}^{T} \sum_{\boldsymbol{x}, \boldsymbol{y}_\tau \in \mathcal{D}} w_\tau \ell_\tau(\psi(\cdot; \vartheta_\tau^m) \circ \phi(\boldsymbol{x}; \theta^m), \boldsymbol{y}_\tau) + \lambda_\tau \ell^d(A_\tau(\phi(\boldsymbol{x}; \theta^m)), \phi(\boldsymbol{x}; \theta_\tau^s))$$

$$\tag{5}$$

where λ_τ is the task-specific tradeoff hyperparameter.

Discussion. Alternatively, the inverse of each adaptor function can be thought as a mapping from each task-specific representation to a shared representation across all the tasks. The assumption here is that a large portion of encodings in the task-specific models is common to all the models up to a simple linear transformation. While the assumption of linear relations between the features of highly non-linear networks may be surprising, such linear relations have also been observed in multi-domain [24] and multi-task problems [30].

4   Experiments

4.1   Datasets

We evaluate our method on three multi-task computer vision benchmarks, including SVHN & Omniglot, NYU-v2, and Cityscapes.[1]

SVHN & Omniglot consists of two datasets, *i.e.* SVHN [20] and Omniglot [15] where SVHN is a dataset for digital number classification and Omniglot is the one for characters classification. Specifically, SVHN contains 47,217 training images and 26,040 validation images of 10 classes. Omniglot consists of 19,476 training and 6492 validation samples of 1623 categories. As the testing labels are not provided, we evaluate all methods on the validation images and report the accuracy of both tasks. Note that in contrast to the NYU-V2 and Cityscapes datasets where each image is associated with multiple labels, each image in this benchmark is labeled only for one task. Thus the goal is to learn a multi-task network that can learn both tasks from SVHN and Omniglot.

NYU-V2 [29] contains RGB-D indoor scene images, where we evaluate performances on 3 tasks, including 13-class semantic segmentation, depth estimation, and surface normals estimation. We use the true depth data recorded by the Microsoft Kinect and surface normals provided in [8] for depth estimation and surface normal estimation. All images are resized to 288×384 resolution as [16].

Cityscapes [6] consists of street-view images, each labeled for two tasks: 7-class semantic segmentation[2] and depth estimation. We resize the images to 128×256 to speed up the training.

4.2   Baselines

In this work, we use the hard parameters sharing architecture for all methods where the early layers of the network are shared across all tasks and the last layers are task-specific (See Fig. 1). We compare our method with two baselines:

- **STL** learns a task-specific model for each task.
- **Uniform**: This vanilla MTL model is trained by minimizing the uniformly weighted loss Eq. (2).

We also compare our method to the state-of-the-art MTL methods which are proposed for solving the unbalanced MTL, including Uncert [12], MGDA [28], GradNorm [4] and a knowledge distillation based method, namely BAM [5], that applies knowledge distillation to network's prediction. On NYU-v2 and Cityscapes, we also compare our method with Gradient Surgery (GS) [32] and

---

[1] The implementation of our method is available at https://weihonglee.github.io/Projects/KD-MTL/KD-MTL.htm.

[2] The original version of Cityscapes provides labels 19-class semantic segmentation. We follow the evaluation protocol in [16], we use labels of 7-class semantic segmentation. Please refer to [16] for more details.

Dynamic Weight Average (DWA) [16] with using different architectures, *i.e.* SegNet [1] and MTAN [16] which is the extension of SegNet by introducing task-specific attention modules for each task.

4.3   Comparison to the State-of-the-art

Results on SVHN & Omniglot. First, we evaluate all methods on SVHN & Omniglot. We extend the LeNet to MTL setting (See Fig. 2) and use the extended network for all methods. We set the batch size of the mini-batch as 512 where 256 samples from SVHN and 256 images from Omniglot. We use Adam [13] for optimizing the networks and adaptors. The learning rate of all task-specific adaptors is 0.01. We train all methods for 300 epochs in total where we scaled the learning rate by 0.85 every 15 epochs. In our method, weights of task-specific losses (*i.e.* w in Eq. (5)) are set uniformly. As a validation set for hyperparameter search (λ), we randomly pick 10% of training data. After the best hyperparameters are chosen, we retrain with the full training set and report the median validation accuracy of the last 20 epochs in Table 1. We search over the set $\lambda = \{1, 5, 10, 20\}$ of λ and we chose $\lambda = 10$.

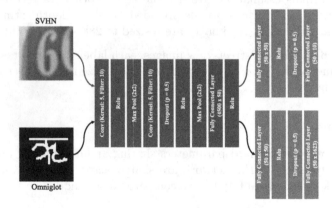

Fig. 2. Network architecture used in SVHN & Omniglot.

Table 1. Testing accuracy on SVHN & Omniglot.

Type	Methods	SVHN	Omniglot	avg
STL	- -	88.84	65.76	- -
MTL	Uniform	85.88	66.91	76.40
	Uncert [12]	85.43	64.1	74.77
	MGDA [28]	83.93	66.8	75.37
	GradNorm [4]	84.48	65.55	75.02
	BAM [5]	86.57	66.08	76.33
	Ours	**88.05**	**70.12**	**79.09**

Table 2. Testing results on NYU-v2. * Results of the 'GS' method are from [32].

Architecture	Type	Methods	Segmentation		Depth		Surface Normal				
			(Higher Better ↑)		(Lower Better ↓)		Angle Distance (Lower Better ↓)		Within $t°$ (Higher Better ↑)		
			mIoU	Pix Acc	Abs Err	Rel Err	Mean	Median	11.25	22.5	30
SegNet	STL	- -	17.32	55.70	0.6577	0.2828	29.99	23.81	24.31	48.06	60.05
	MTL	Uniform	18.14	55.82	0.5841	0.2490	31.89	26.81	20.61	42.95	55.33
		Uncert [12]	16.79	52.51	0.6183	0.2612	32.44	27.27	20.81	42.41	54.47
		MGDA [28]	15.57	49.05	0.6716	0.2710	29.90	24.13	23.81	47.58	59.75
		GradNorm [4]	18.08	55.76	0.5819	0.2495	30.65	25.38	22.64	45.34	57.59
		Ours	**18.75**	**58.02**	**0.5780**	**0.2467**	**29.40**	**23.71**	**24.33**	**48.22**	**60.45**
MTAN [16]	STL	- -	16.38	53.89	0.6792	0.2963	30.66	24.26	23.35	47.34	59.33
	MTL	Uniform	17.72	55.32	0.5906	0.2577	31.44	25.37	23.17	45.65	57.48
		DWA [16]	17.52	55.76	0.5869	0.2549	31.75	25.64	22.60	45.12	57.05
		Uncert [12]	17.67	55.61	0.5927	0.2592	31.25	25.57	22.99	45.83	57.67
		MGDA [28]	15.60	52.36	0.6215	0.2767	30.26	24.01	23.98	47.77	59.85
		GradNorm [4]	17.37	55.92	0.5924	0.2630	31.20	24.91	23.11	46.27	58.20
		GS* [32]	20.17	56.65	0.5904	0.2467	30.01	24.83	22.28	46.12	58.77
		Ours	**20.75**	**57.90**	**0.5816**	**0.2445**	**29.97**	**23.96**	**24.24**	**47.78**	**59.78**

Table 1 shows that MTL with uniform loss weights (Uniform) obtains worse results on SVHN while it achieves better performance on Omniglot than STL. The state-of-the-art methods which dynamically weigh the task-specific losses cannot achieve a good trade-off between these two tasks. More specifically, Uncert and GradNorm obtain worse overall performance than STL while MGDA improves the performance on Omniglot and obtains worse performance on SVHN. Though BAM obtains better overall performance on both tasks, the improvement is achieved mainly because of more informative information provided by the continuous predictions of the teacher network (task-specific models). The unbalanced problem in MTL when we use BAM is still unsolved as BAM applies knowledge distillation on network predictions, which would have similar problems with the vanilla MTL (Uniform). In contrast, our approach achieves significantly better performance than any other MTL methods *i.e.* our method obtains 88.05 % accuracy on SVHN and 70.12 % accuracy on Omniglot. Compared with STL, our method obtains comparable results on SVHN and significant gains on Omniglot. These results strongly verify that our method is able to alleviate the unbalanced problem in this benchmark and to outperform STL as it enables the MTL model to learn more informative features.

Results on NYU-V2. We follow the training and evaluation protocol in [16]. We use cross-entropy loss for semantic segmentation, l1-norm loss for depth estimation, and cosine similarity loss for surface normal estimation. We train all methods using Adam [13] with the learning rate initialized at 1e-4 and halved at the 100-th epoch for 200 epochs in total. The learning rate of all task-specific adaptors is 0.1 and the batch size is 2. In our method, weights of task-specific losses (*i.e.* w in Eq. (5)) are set uniformly. As a validation set for hyperparameter search (λ), we randomly pick 10% of training data. After the best hyperparameters are chosen, we retrain with the full training set and report the results of three

Table 3. Testing results on Cityscapes.

Architecture	Type	Methods	Segmentation (Higher Better ↑)		Depth (Lower Better ↓)	
			mIoU	Pix Acc	Abs Err	Rel Err
SegNet	STL	- -	51.85	91.08	0.0136	22.68
	MTL	Uniform	50.73	90.76	0.0151	40.81
		Uncert [12]	51.09	90.85	0.0143	27.66
		MGDA [28]	51.69	90.99	**0.0130**	**24.04**
		GradNorm [4]	50.06	90.82	0.0143	28.61
		Ours	**52.18**	**91.24**	0.0140	28.90
MTAN	STL	- -	51.24	91.16	0.0137	24.80
	MTL	Uniform	52.56	91.33	0.0152	24.64
		DWA [16]	51.95	91.33	0.0141	30.03
		Uncert [12]	50.37	91.11	0.0142	31.78
		MGDA [28]	52.32	**91.59**	**0.0138**	30.35
		GradNorm [4]	51.88	91.40	0.0148	31.43
		Ours	**52.71**	91.54	0.0139	**27.33**

tasks on the validation set in Table 2. We search over the set $\lambda = \{1, 2, 3, 4, 5, 6\}$ of λ and the distillation loss' weight of Segmentation, depth, and surface normal are set to 1, 1 and 2, respectively.

From the results shown in Table 2, we can see that it is possible to tackle multiple tasks within a network and achieve performance improvement on some tasks, *e.g.* when we use SegNet as the based network, the vanilla MTL (Uniform) achieves better performance on semantic segmentation and depth estimation though it causes a drop on surface normal estimation in comparison with STL. Though we see the benefits of using MTL, it is also clear that the unbalanced problem exists. We then apply existing methods that introduce loss weighting strategies for addressing the unbalanced loss optimization problem. From the results of using SegNet, GradNorm performs the best among all compared methods. However, it provides limited control on the learned parameters (*e.g.* it achieves similar results to the MTL model using a uniformly weighting scheme) and still suffers from lower task performance than the single task models.

In comparison with these methods, our method obtains significant gains over all tasks and achieves better results than single task learning models. This strongly verifies our hypothesis that the solution of the multi-task network should be close to the single task ones' and lie in the intersection of the single-task solutions and our method that applies stricter control on the parameters of the multi-task network can better address the unbalanced loss problem.

Table 4. Ablation study on NYU-v2. Here, '#layer' means which layers are selected for computing distillation loss. 'adaptors' indicates which kind of adaptors is used and '✗' means no adaptors are used.

Method			Segmentation (Higher Better ↑)		Depth (Lower Better ↓)		Surface Normal				
							Angle Distance (Lower Better ↓)		Within $t°$ (Higher Better ↑)		
backbone	#layers	adaptors	mIoU	Pix Acc	Abs Err	Rel Err	Mean	Median	11.25	22.5	30
SegNet	last	linear	18.66	57.78	0.5813	**0.2375**	30.17	24.74	22.96	46.44	58.76
	mid + last	linear	18.75	58.02	**0.5780**	0.2467	**29.40**	**23.71**	**24.33**	**48.22**	**60.45**
	mid + last	non-linear	**19.00**	**58.12**	0.5853	0.2398	29.74	24.25	23.06	47.16	59.53
	mid + last	✗	17.11	54.60	0.6060	0.2586	30.32	24.74	22.66	46.36	58.72
MTAN [16]	last	linear	18.52	56.81	**0.5756**	0.2489	31.13	24.93	23.52	46.29	58.18
	mid + last	linear	**20.75**	57.90	0.5816	**0.2445**	**29.97**	**23.96**	**24.24**	**47.78**	**59.78**
	mid + last	non-linear	20.42	56.93	0.5975	0.2589	30.32	24.86	22.98	46.21	58.55
	mid + last	✗	19.30	53.85	0.6064	0.2567	31.43	27.61	17.61	40.86	54.63

Results on Cityscapes. Similar to NYU-V2, we use cross-entropy loss for semantic segmentation and l1-norm loss for depth estimation in Cityscapes as in [16]. We train all methods using Adam [13] with the learning rate initialized at 1e-4 and halved at the 100-th epoch for 200 epochs in total. The learning rate of all task-specific adaptors is 0.1 and the batch size is 8. In our method, weights of task-specific losses (*i.e.* w in Eq. (5)) are set uniformly. As a validation set for hyperparameter search (λ), we randomly pick 10% of training data. After the best hyperparameters are chosen, we retrain with the full training set and report the results of two tasks on the validation set in Table 3. We search over the set $\lambda = \{1, 2, 3, 4, 5, 6\}$ of λ and the distillation loss' weight of Segmentation and depth are set to 2 and 6, respectively.

As shown in Table 3, in overall, MTL obtains worse performance than STL. It is clear that GradNorm obtains worse performance on semantic segmentation while it improves the performance on depth estimation. However, MGDA assigns much larger weight on depth estimation task and this enables the MTL model to achieve better performance on both tasks. Our method also achieves significant gains on both tasks. The results again demonstrate that our method is able to optimize MTL model in a more balanced way and to achieve better overall results.

4.4   Ablation Study

To better analyze the effect of the distillation loss, we conduct an ablation study on NYU-V2. We first evaluate the effect of applying distillation loss to more layer's features. On NYU-V2, we report the results of applying distillation loss to the last shared layer's feature only and the results of applying distillation loss to features of both the middle layer and the last layer. From the results presented in Table 4, adding more layers' features for computing distillation loss boosts the performance on NYU-V2 in general. However, results on SVHN & Omniglot and Cityscapes indicate that using the last layers obtains the best

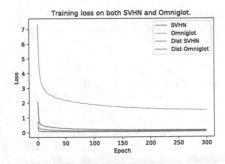

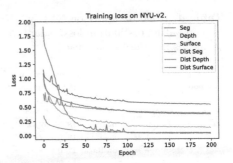

Fig. 3. Task-specific loss and distillation loss on training set of SVHN & Omniglot. Best view in color.

Fig. 4. Task-specific loss and distillation loss on training set of NYU-V2. Best view in color.

performance. We argue that adding more layers can enhance the distillation loss and a more strict control on the parameters of the multi-task network. This is not necessary for those tasks that use a small network, *e.g.* the network we used in SVHN & Omniglot and would be useful for tasks using a large network, *e.g.* the SegNet used in NYU-V2.

Analysis on Adaptors. We evaluate our method using different types of adaptors. As mentioned in Sect. 3, we use a linear layer that consists of a $1 \times 1 \times C \times C$ convolution for each adaptor and denote as 'linear' in Table 4. We also evaluate our method without any adaptors (*i.e.* indicated as '✗') and with 'non-linear 'adaptors (*i.e.* each adaptors consists of a $1 \times 1 \times C \times 2C$ convolution, a Relu activation layer and a $1 \times 1 \times 2C \times C$ convolution). From the results shown in Table 4, it is clear to see that the adaptors help to align the features of multi-task and single-task as each single-task model produce different features. Compared with our method using 'non-linear' adaptors, which have much larger capacity of mapping features, the results indicate that using 'linear' adaptors is sufficient as we discuss in Sect. 3.

Further Analysis. We also plot the task-specific loss and distillation loss on the training set of two benchmarks for analyzing our method. In both Fig. 3 and Fig. 4, it is clear that distillation loss is more balanced than task-specific loss. More specifically, the task-specific loss of SVHN and Omniglot converge at around 0.137 and 1.492, respectively. In contrast, the distillation loss of SVHN and Omniglot converge at around 0.089 and 0.041, respectively. On NYU-V2, the task-specific loss of semantic segmentation, depth estimation, and surface normal estimation end up at 0.027, 0.127, and 0.039 while the distillation loss ends up at around 0.534, 0.381, and 0.364. These results again verify that our method can optimize the MTL method in a more balanced way.

5   Conclusion

In this work, we proposed a knowledge distillation based multi-task method that learns to produce the same features with the single-task networks to address the unbalanced multi-task learning problem with the hypothesis that the solution of the multi-task network should be close to the single task ones' and lie in the intersection of the single solutions. We demonstrated that our method achieves significant performance gains over the state-of-the-art methods, on challenging benchmarks for image classification and scene understanding (semantic segmentation, depth estimation, and surface normal estimation). As future work, we plan to extend our method to multi-task network architecture searching.

References

1. Badrinarayanan, V., Kendall, A., Cipolla, R.: Segnet: a deep convolutional encoder-decoder architecture for image segmentation. Trans. Pattern Anal. Mach. Intell. **39**(12), 2481–2495 (2017)
2. Bilen, H., Vedaldi, A.: Integrated perception with recurrent multi-task neural networks. In: Advances in Neural Information Processing Systems, pp. 235–243 (2016)
3. Caruana, R.: Multitask learning. Mach. learn. **28**(1), 41–75 (1997)
4. Chen, Z., Badrinarayanan, V., Lee, C.Y., Rabinovich, A.: Gradnorm: Gradient normalization for adaptive loss balancing in deep multitask networks (2017). arXiv preprint: arXiv:1711.02257
5. Clark, K., Luong, M.T., Khandelwal, U., Manning, C.D., Le, Q.V.: Bam! born-again multi-task networks for natural language understanding (2019). arXiv preprint: arXiv:1907.04829
6. Cordts, M., et al.: The cityscapes dataset for semantic urban scene understanding. In: Computer Vision and Pattern Recognition. pp. 3213–3223 (2016)
7. Du, Y., Czarnecki, W.M., Jayakumar, S.M., Pascanu, R., Lakshminarayanan, B.: Adapting auxiliary losses using gradient similarity (2018). arXiv preprint arXiv:1812.02224
8. Eigen, D., Fergus, R.: Predicting depth, surface normals and semantic labels with a common multi-scale convolutional architecture. In: IEEE International Conference on Computer Vision, pp. 2650–2658 (2015)
9. Furlanello, T., Lipton, Z.C., Tschannen, M., Itti, L., Anandkumar, A.: Born again neural network (2018) arXiv preprint: arXiv:1805.04770
10. Guo, M., Haque, A., Huang, D.A., Yeung, S., Fei-Fei, L.: Dynamic task prioritization for multitask learning. In: European Conference on Computer Vision. pp. 270–287 (2018)
11. Hinton, G., Vinyals, O., Dean, J.: Distilling the knowledge in a neural network (2015). arXiv preprint: arXiv:1503.02531
12. Kendall, A., Gal, Y., Cipolla, R.: Multi-task learning using uncertainty to weigh losses for scene geometry and semantics. In: IEEE Conference on Computer Vision and Pattern Recognition, pp. 7482–7491 (2018)
13. Kingma, D.P., Ba, J.: Adam: A method for stochastic optimization (2014). arXiv preprint: arXiv:1412.6980
14. Kokkinos, I.: UberNet: training a universal convolutional neural network for low-, mid-, and high-level vision using diverse datasets and limited memory. In: IEEE Conference on Computer Vision and Pattern Recognition, pp. 6129–6138 (2017)

15. Lake, B.M., Salakhutdinov, R., Tenenbaum, J.B.: Human-level concept learning through probabilistic program induction. Science **350**(6266), 1332–1338 (2015)
16. Liu, S., Johns, E., Davison, A.J.: End-to-end multi-task learning with attention. In: IEEE Conference on Computer Vision and Pattern Recognition, pp. 1871–1880 (2019)
17. Ma, J., Mei, Q.: Graph representation learning via multi-task knowledge distillation (2019). arXiv preprint: arXiv:1911.05700
18. Meyerson, E., Miikkulainen, R.: Pseudo-task augmentation: From deep multitask learning to intratask sharing–and back (2018). arXiv preprint: arXiv:1803.04062
19. Misra, I., Shrivastava, A., Gupta, A., Hebert, M.: Cross-stitch networks for multi-task learning. In: IEEE Conference on Computer Vision and Pattern Recognition, pp. 3994–4003 (2016)
20. Netzer, Y., Wang, T., Coates, A., Bissacco, A., Wu, B., Ng, A.Y.: Reading digits in natural images with unsupervised feature learning. In: Advances in Neural Information Processing Systems Workshop on Deep Learning and Unsupervised Feature Learning (2011)
21. Parisotto, E., Ba, J.L., Salakhutdinov, R.: Actor-mimic: Deep multitask and transfer reinforcement learning (2015). arXiv preprint: arXiv:1511.06342
22. Phuong, M., Lampert, C.: Towards understanding knowledge distillation. In: International Conference on Machine Learning, pp. 5142–5151 (2019)
23. Rebuffi, S.A., Bilen, H., Vedaldi, A.: Learning multiple visual domains with residual adapters. In: Advances in Neural Information Processing Systems, pp. 506–516 (2017)
24. Rebuffi, S.A., Bilen, H., Vedaldi, A.: Efficient parametrization of multi-domain deep neural networks. In: IEEE Conference on Computer Vision and Pattern Recognition, pp. 8119–8127 (2018)
25. Romero, A., Ballas, N., Kahou, S.E., Chassang, A., Gatta, C., Bengio, Y.: Fitnets: Hints for thin deep nets (2014). arXiv preprint: arXiv:1412.6550
26. Ruder, S.: An overview of multi-task learning in deep neural networks (2017). arXiv preprint: arXiv:1706.05098
27. Seltzer, M.L., Droppo, J.: Multi-task learning in deep neural networks for improved phoneme recognition. In: IEEE International Conference on Acoustics, Speech and Signal Processing, pp. 6965–6969. IEEE (2013)
28. Sener, O., Koltun, V.: Multi-task learning as multi-objective optimization. In: Advances in Neural Information Processing Systems, pp. 527–538 (2018)
29. Silberman, N., Hoiem, D., Kohli, P., Fergus, R.: Indoor segmentation and support inference from rgbd images. In: European conference on computer vision. pp. 746–760. Springer (2012)
30. Stickland, A.C., Murray, I.: Bert and pals: Projected attention layers for efficient adaptation in multi-task learning (2019). arXiv preprint: arXiv:1902.02671
31. Tian, Y., Krishnan, D., Isola, P.: Contrastive representation distillation (2019). arXiv preprint: arXiv:1910.10699
32. Yu, T., Kumar, S., Gupta, A., Levine, S., Hausman, K., Finn, C.: Gradient surgery for multi-task learning (2020). arXiv preprint: arXiv:2001.06782
33. Zhang, Z., Luo, P., Loy, C.C., Tang, X.: Facial landmark detection by deep multi-task learning. In: Fleet, D., Pajdla, T., Schiele, B., Tuytelaars, T. (eds.) ECCV 2014. LNCS, vol. 8694, pp. 94–108. Springer, Cham (2014). https://doi.org/10.1007/978-3-319-10599-4_7

Mitigating Dataset Imbalance via Joint Generation and Classification

Aadarsh Sahoo[1], Ankit Singh[2], Rameswar Panda[3], Rogerio Feris[3],
and Abir Das[1(✉)] (iD)

[1] IIT Kharagpur, Kharagpur, West Bengal, India
[2] IIT Madras, Chennai, Tamil Nadu, India
[3] MIT-IBM Watson AI Lab, Cambridge, Massachusetts, USA

Abstract. Supervised deep learning methods are enjoying enormous success in many practical applications of computer vision and have the potential to revolutionize robotics. However, the marked performance degradation to biases and imbalanced data questions the reliability of these methods. In this work we address these questions from the perspective of dataset imbalance resulting out of severe under-representation of annotated training data for certain classes and its effect on both deep classification and generation methods. We introduce a joint dataset repairment strategy by combining a neural network classifier with Generative Adversarial Networks (GAN) that makes up for the deficit of training examples from the under-represented class by producing additional training examples. We show that the combined training helps to improve the robustness of both the classifier and the GAN against severe class imbalance. We show the effectiveness of our proposed approach on three very different datasets with different degrees of imbalance in them. The code is available at https://github.com/AadSah/ImbalanceCycleGAN.

1 Introduction

Deep neural networks (DNN) and large-scale datasets have been instrumental behind the remarkable progress in computer vision research. The data-driven feature and classifier learning enabled Convolutional Neural Networks (CNNs) to achieve superior performance than traditional machine learning methods. However, the over-reliance on data has brought with it new problems. One of them is the problem of becoming too adapted to the dataset by essentially memorizing all its idiosyncrasies [32,42]. Due to the presence of massive number of learnable parameters, most DNNs require huge amount of annotated examples for each class. This is resource consuming, expensive and often impractical too in cases

A. Sahoo and A. Singh—contributed equally. The work was done when Ankit Singh was at IIT Kharagpur.

Electronic supplementary material The online version of this chapter (https:// doi.org/10.1007/978-3-030-65414-6_14) contains supplementary material, which is available to authorized users.

A. Bartoli and A. Fusiello (Eds.): ECCV 2020 Workshops, LNCS 12540, pp. 177–193, 2020.
https://doi.org/10.1007/978-3-030-65414-6_14

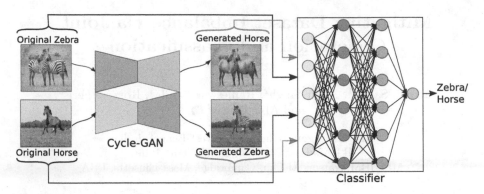

Fig. 1. We propose a joint generation and classification network to tackle severe dataset imbalance in classification scenario. The framework employs a Cycle-consistent GAN to generate new training examples from existing ones and feeds both to the classifier. The GAN and the classifier are trained jointly in an alternating fashion. Our model improves the classification performance especially when the dataset is highly imbalanced and the performance of the GAN is either at par with or experiences minor deterioration compared to the case when the classifier is not trained jointly.

where recognition involves rare classes. Such a situation is not rare where large number of annotated examples for one class and only a few for the other are available resulting in an imbalanced dataset.

Class imbalance is a classic problem in machine learning and is known to have detrimental effect [5,12,19]. Classifiers trained traditionally minimizes the average error across all training examples. Such a strategy often results in fitting to the majority class (*i.e.*, the class with large number of annotated training examples) only. Simply by virtue of their numbers, fitting to the majority class reduces the overall training error than fitting to the minority class.

Traditionally, dataset imbalance mitigation methods have tried to put all classes in the level playing field by focusing on to take away the number advantage enjoyed by the majority class. This includes oversampling minority population [6] or undersampling the majority population [8,12], providing more importance to errors from the minority class while training [49] or adjusting the predicted class probabilities during inference according to the prior class probabilities [34] *etc.* However, these approaches are limited to the data at hand in the sense that these can not generate additional unseen data that can help increase the much needed diversity for unbiased training. In this paper, we advocate for a generative approach to address dataset imbalance. A Generative Adversarial Network (GAN) is used to transform majority class examples to minority class and vice-versa. Starting from an imbalanced dataset, this generates training examples that are not only much broader and diverse in nature but also balanced in terms of the number of examples from the two different classes.

Our proposed approach explores a Cycle-consistent GAN or simply cycle-GAN [50] that is recently proposed to translate image from one domain (source) to another domain (target) (see Fig. 1). In addition to traditional generator and discriminator losses during training, a cycle-GAN employs a *cycle-consistency loss* that encourages the learned translations to be "cycle consistent", *i.e.*, if an image is translated from a source to a target domain and then it is translated back to the target domain again, original image and the doubly translated image should be same. We learn a joint architecture by extending cycle-GAN that feeds a classifier with images translated from a source class to a target class where the target class is also the minority class in the imbalanced dataset. Instead of using the cycle-GAN as only a component of the classification pipeline, we employ a joint training strategy that first trains the classifier using images generated by the cycle-GAN and then trains the cycle-GAN by back-propagating the classifier loss through it. The loss from the classifier acts as a multi-task supervision for the cycle-GAN in addition to the cycle-consistency and the traditional GAN loss. After few such iterations of alternate classifier and GAN training, the GAN learns to generate appropriate images for training a balanced classifier. Our key insight is that one can use a GAN to alleviate dataset imbalance effect on classifier and at the same time the gradients from the classifier loss can be propagated back to the GAN, to mitigate the same for the GAN.

We evaluate our proposed approach on three different datasets. The first one is the CelebA dataset [26] which is a largescale face dataset created from face images of celebrities, the second one is the Horse2Zebra dataset [50] which is a collection of natural images of horses and zebras downloaded from the internet and the third one is CUB-200-2011 dataset [43] consisting of fine-grained images of 200 bird species. Specifically, we demonstrate that in highly imbalanced scenarios, both classifier and the GAN achieves superior performances than traditional imbalance mitigation approaches.

2   Related Works

Class imbalance in machine learning has been studied long [19]. However, in the midst of present data revolution, the problem has become ever so important. While images corresponding to some concepts or classes are available in plenty and easy to annotate, examples corresponding to many concepts are either rare or might require expert annotators. As a result, many real world datasets are highly imbalanced. MS-COCO [22], SUN database [46], DeepFashion [25], Places [48], iNaturalist [16] etc. are all examples of datasets where the number of images in the most common class and the number of images in the least common class are hugely different. The community has commonly adopted two different kinds of approaches to address class imbalance: i) dataset level methods and ii) classifier or algorithmic level methods. A multidimensional taxonomy and categorization of the class imbalance problems can be obtained in the review papers [5,14].

The dataset level approaches consist of modification of imbalanced data in order to provide a balanced distribution with the aim that no change at the algorithm level is necessary and standard training algorithms would work. These are based on sampling either the majority or the minority classes differently. Basic random oversampling [6,9,45] involves replicating randomly selected training examples from the minority class such that the total number of minority and majority class examples that goes to train the classifier is same. However the same mechanism can be used to achieve varying degrees of class distribution balance by varying the amount of replication. Oversampling methods are simple and effective, however there are evidences that these can lead to overfitting [6,44]. To this end, various adaptive and informative oversampling methods have been proposed. SMOTE [6] creates synthetic examples of the minority class as a convex combination of existing examples and their nearest neighbors. Borderline-SMOTE [13] extends SMOTE by choosing more training examples near the class boundaries. Authors in [4] used a modified condensed nearest neighbor rule [41] to sample examples such that all minimally distanced nearest neighbor pairs are of same class and learned a classifier on these samples only. In CNNs, Shen *et al.* [38] used a class-aware sampling strategy where the authors first sample a class and then sample an image from that class to form minibatches that are uniform with respect to the classes.

Another variant of sampling is undersampling where training examples are removed randomly from the majority classes until the desired balance in the curated dataset is achieved. This approach, though counter-intuitive works better than oversampling in some situations [8]. One obvious disadvantage of undersampling is that discarding examples may cause the classifier to miss out on important variabilities in the dataset. EasyEnsemble and BalanceCascade [24] are two variants of informed undersampling that addresses this issue. EasyEnsemble creates an ensemble of classifiers by independently sampling several subsets from the minority class while BalanceCascade develops the ensemble classifier by systematically selecting the majority class samples. The one-sided selection [20] approach discards redundant examples especially close to the boundary to get a more informed reduced sample set than random undersampling of majority class. Though sampling based methods change the data distribution to be more balanced, they hardly change the data itself. Our proposed method on the other hand is not only able to change the data distribution to make it more balanced, but also adds variability as additional data is generated in the process.

Among the approaches that modify the learning algorithm without changing the dataset distribution, Zhou *et al.* [49] weigh misclassification of the minority class examples more than the majority class examples during training. A naive approach [21] is to weigh the predicted probabilities of different classes during inference according to the prior probability of occurrences of these classes in the training set. Recently Cui *et al.* [7] proposed a class balanced loss depending on the "effective" number of training examples per class based on the fact that the additional benefit diminishes with increasing number of examples. This generalizes the class specific loss balancing based on frequency of the samples [49] to their effective frequencies.

GAN [11] and its variants use a minimax game to model high dimensional distributions of visual data enabling them to be used for data augmentation [33] as well as for generating new images [2]. However, training with GAN generated minority class samples can lead to boundary distortion [36] resulting in performance drop for the majority class. Mullick *et al.* [30] proposed to generate minority class samples by convexly combining existing samples in an adversarial setting by fooling both the discriminator as well as the classifier. Our proposed approach, on the other hand, generates additional minority samples from images belonging to the majority class. In contrast to prior works that focus on only perturbing the whole image to generate a slightly different version of it, we use a translational generative model (Cycle-GAN) to perturb the *content* (object of interest) without changing the *context* much. Our conjecture is: the classifier gets two different objects in the same context which would likely help the classifier to focus more on the object of interest rather than on the context for the discriminative task. Training classifiers with such image pairs where the two objects are in the same common context not only alleviates the dataset imbalance problem but also creates a more robust classifier by learning object features rather than focusing on the context.

3 Approach

An overview of the training procedure is presented in Fig. 2. The goal is to learn a generator to translate images from majority class to images of minority class so that the generated images make the dataset balanced for training. We denote the majority class as A and minority class as B. In Fig. 2, horse images are assumed to be the majority class while the zebra images are assumed to be the minority class.

The proposed system uses a generator $(G_{A \to B})$ to translate a majority example (a) in the scene to an example (b_{gen}) belonging to the minority class. Following the cycle-GAN philosophy, another generator $(G_{B \to A})$ will translate b_{gen} to a majority class example (a_{cyc}). Since a_{cyc} is generated from the translated image b_{gen} which, in turn, is generated from the original majority class example a, a_{cyc} is known as the cyclic image of a. The generated minority class image $(G_{A \to B}(a))$ along with a real minority class image (b) constitute the GAN loss $(\mathcal{L}_{GAN})$ while the real majority class image (a) and the cyclic image (a_{cyc}) will constitute the cycle-consistency loss $(\mathcal{L}_{cyc})$ as,

$$
\begin{aligned}
\mathcal{L}_{GAN}(G_{A \to B}, D_B, A, B) = &-\mathbb{E}_{b \sim p_B(b)}\big[\log D_B(b)\big] \\
&- \mathbb{E}_{a \sim p_A(a)}\big[\log\big(1 - D_B\big(G_{A \to B}(a)\big)\big)\big] \\
\mathcal{L}_{cyc}(A) = \mathbb{E}_{a \sim p_A(a)}&\big[||G_{B \to A}\big(G_{A \to B}(a)\big) - a||_1\big]
\end{aligned}
\tag{1}
$$

where, D_B denotes the discriminator. Minimizing the GAN loss without the cycle-consistency constraint, can encourage the generator to map source domain images to a random permutation of the target domain images [50] as the mapped images still produce an identical target distribution. This is, especially, true with imbalanced dataset where the network capacity can be too large with regard to the size of the dataset, facilitating the generator to memorize the target domain. Cycle-consistency loss addresses this by putting further constraints on the generator by asking it to map back to the original domain if source-to-target

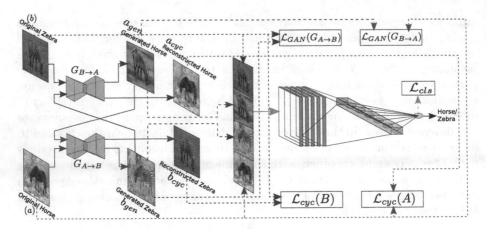

Fig. 2. Approach overview. We train a CNN to predict the presence of either horse or zebra in presence of high imbalance between the number of images of horses and zebras in the training data. In addition to images from the imbalanced dataset, the classifier uses images translated by cycle-GAN for training. What originally is a minority class becomes a majority class when translated. However, due to the obvious difference in quality between the generated and the real images they are weighed differently when used in training. The figure also shows the different losses along with the constituents contributing to them. Note that the identity loss for the cycle-GAN is not shown to make it less cluttered. The framework is trained by either keeping the GAN frozen throughout or alternatingly training the classifier and the GAN for a few epochs.

and target-to-source generators are composed sequentially. Following the original cycle-GAN formulation [50], the cycle-consistency loss in this work, is measured as the ℓ_1 distance between an image and its mapped back version.

A similar procedure provides the corresponding losses $(\mathcal{L}_{GAN}(G_{B \to A}, D_A, B, A)$ and $\mathcal{L}_{cyc}(B))$ for the scenario when we want to convert a visual scene containing majority class objects to a scene containing minority class objects.

$$\mathcal{L}_{GAN}(G_{B \to A}, D_A, B, A) = -\mathbb{E}_{a \sim p_A(a)}\big[\log D_A(a)\big]$$
$$- \mathbb{E}_{b \sim p_B(b)}\big[\log\big(1 - D_A\big(G_{B \to A}(b)\big)\big)\big]$$
$$\mathcal{L}_{cyc}(B) = \mathbb{E}_{b \sim p_B(b)}\big[||G_{A \to B}\big(G_{B \to A}(b)\big) - b||_1\big] \tag{2}$$

where, D_A denotes the discriminator.

In addition, following [40], an identity loss $(\mathcal{L}_{ide}(G_{A \to B}, G_{B \to A}))$ is used that encourages the generators to produce an identity image when a real sample of the target domain is passed through them.

$$\mathcal{L}_{ide}(G_{A \to B}, G_{B \to A}) = \mathbb{E}_{b \sim p_B(b)}\big[||G_{A \to B}(b) - b||_1\big] +$$
$$\mathbb{E}_{a \sim p_A(a)}\big[||G_{B \to A}(a) - a||_1\big] \tag{3}$$

Without this loss, translated images often come with additional tint. This is, especially, problematic for the proposed framework as systematically tinted images at training time can confuse the classifier to predict almost randomly during inference where images are without such systematic tints.

The real (but abundant) and the translated (but originally rare) images will constitute a more balanced training set for the classifier $z : A \cup B \rightarrow [0, 1]$, which provides a high value (close to 1) for a minority class image. Unlike the traditional classifiers, our binary cross entropy loss function comprises of losses from four different types of images namely, - Real images from minority class (b), Generated samples of the minority class (b_{gen}) and the same corresponding to the majority class (a and a_{gen} respectively). As the classifier provides the probability of an image to belong to the minority class, the loss coming from the minority class images is given by,

$$\mathcal{L}_{cls}^{B} = -\mathbb{E}_{b \sim p_B(b)} \left[\log z(b) \right] - \mathbb{E}_{a \sim p_A(a)} \left[\log z(G_{A \rightarrow B}(a)) \right] \tag{4}$$

Similarly the loss coming from the majority class images is given by,

$$\mathcal{L}_{cls}^{A} = -\mathbb{E}_{a \sim p_A(a)} \left[\log(1 - z(a)) \right] - \mathbb{E}_{b \sim p_B(b)} \left[\log(1 - z(G_{B \rightarrow A}(b))) \right] \tag{5}$$

The combined loss function for the classifier thus becomes,

$$\mathcal{L}_{cls} = \mathcal{L}_{cls}^{B} + \frac{1}{\gamma} \mathcal{L}_{cls}^{A} \tag{6}$$

where, γ is the imbalance ratio that determines the relative weight of the loss components coming from the minority and the majority classes. γ is defined to be the ratio of the number of training examples in the majority class to the same in the minority class. For a highly imbalanced scenario ($\gamma \gg 1$), the loss formulation penalizes the classifier much more when it misclassifies a minority class real image than when it misclassifies the same from the majority class. This is to encourage the classifier to be free from the bias induced by the majority class. The loss also penalizes the misclassification of images generated from the majority class more than the same generated from the minority class to help the classifier learn from the more diverse style captured by the GAN from the originally majority class images.

We propose two modes of training the proposed system - 1) **Augmented (AUG) Mode** 2) **Alternate (ALT) Mode**. For both the modes, the cycle-GAN is first trained on the imbalanced dataset. In augmented mode, the trained cycle-GAN is used throughout and acts as additional data generator for the classifier. In this mode, only the classifier loss $\mathcal{L}_{cls}$ (ref. Eq. (4)) is minimized. In alternate mode, we alternately train either the cycle-GAN or the classifier for a few epochs keeping the other fixed. The GAN is warm-started similar to the augmented mode of training while the classifier is trained from scratch. During the training of the classifier only the classifier loss $\mathcal{L}_{cls}$ is backpropagated to change the classifier weights and the GAN acts as additional data generator similar to the augmented mode. However, when the GAN is trained, the classifier acts as an additional teacher. The full objective in this case is given by,

$$\mathcal{L} = \mathcal{L}_{GAN}(G_{A \rightarrow B}, D_B, A, B) + \beta \mathcal{L}_{cyc}(A) + $$
$$\mathcal{L}_{GAN}(G_{B \rightarrow A}, D_A, B, A) + \beta \mathcal{L}_{cyc}(B) + $$
$$\alpha \mathcal{L}_{ide}(G_{A \rightarrow B}, G_{B \rightarrow A}) + \mathcal{L}_{cls}^{A} + \mathcal{L}_{cls}^{B} \tag{7}$$

where α and β are hyperparameters weighing the relative importance of the individual loss terms. We have also experimented with end-to-end training of the whole system which works good in a balanced data regime, but the performance is not satisfactory when there is high imbalance in data. This may be due to the fact that the naturally less stable GAN training [28,35] gets aggravated in presence of high imbalance in data. For example, with very few zebra and large number of horses, the cycle-GAN produces mostly the same image as the source horse with only a few faint stripes of zebra at different places of the image. Getting good quality input at the beginning is critical for a convnet classifier as the deficits due to the "noisy" inputs at the beginning of the learning process of the classifier can not be overcome later [1].

4 Experiments

The proposed approach is evaluated on three publicly available datasets namely CelebA [26], CUB-200-2011 [43], and Horse2Zebra [50]. Sections 4.1, 4.2 and 4.3 respectively provide the experimental details and evaluation results on them.

We evaluate the proposed approach in terms the improvement of performance of the classifier as well as that of the cycle-GAN in presence of high class imbalance in the training data. The performance of the classifier is measured in terms of the F_1 score of the two classes so that the evaluation is fair irrespective of the skewness, if any, in the test data. Following the common practice, we also report the classifier performance in terms of Average Class Specific Accuracy (ACSA) [17,30]. The best performing classifier on the validation set is used for reporting the test-set performances. Due to space constraints, we provide only the F_1 score corresponding to the minority class in the main paper while the others are included in the supplementary material. Although perceptual studies may be the gold standard for assessing GAN generated visual scenes, it requires fair amount of expert human effort. In absence of it, inception score [35] is a good proxy. However, for an image-to-image translation task as is done by a cycle-GAN, inception score may not be ideal [3]. This is especially true in low data regime as one of the requirements of evaluating using inception score is to evaluate using a large number of samples (*i.e.*, 50k) [35]. Thus, we have used an inception accuracy which measures how good an inception-v3 [39] model trained on real images can predict the true label of the generated samples. This is given by the accuracy of the inception-v3 model on the images translated by the cycle-GANs. A cycle-GAN providing higher accuracy to the same set of test images after translation, is better than one giving a lower accuracy. For this purpose, the ImageNet pretrained inception-v3 model is taken and the classification layer is replaced with a layer with two neurons. This modified network is then fine-tuned on balanced training data that is disjoint from the test set for all the three datasets. For each dataset the evaluation procedure is repeated 5 times while keeping the data fixed but using independent random initialization during runs. The average result of these 5 runs is reported.

4.1   Experiments on CelebA

Many datasets have contributed to the development and evaluation of GANs. However, the CelebA [26] dataset has been the most popular and canonical in this journey. CelebA contains faces of over 10,000 celebrities each of which has 20 images. We used images from this dataset to study the effect of imbalance for a binary classifier predicting the gender from the face images only. Gender bias due to imbalance in datasets [15,29] has started to gain attention recently and face images make the problem more challenging as many attributes can be common between males and females or can only be very finely discriminated.

Experimental Setup: We took a subset of 900 female images and considered this to be the majority class. The male images are varied in number from 100 to 900 in regular intervals of 100. In addition, we experimented with 50 male images too which makes the highest imbalance ratio (γ) for this dataset to be 18. We follow the same architecture of the cycle-GAN as that of Zhu *et al.* [50] where the generator consists of two stride-2 convolutions, 9 residual blocks, and 2 fractionally-strided convolutions. For discriminator, similarly, a 5 layer CNN is used for extracting features, following the PatchGAN [18] architecture. To avoid over-fitting in presence of less data, our binary classifier is a simple CNN without any bells and whistles. We used a classifier with same feature extraction architecture as that of the discriminator used above. The initial convolution layers are followed by two fully-connected layers of 1024 and 256 neurons along with a output layer consisting of two neurons. To handle over-fitting further, we used dropout with probability 0.5 of dropping out neurons from the last fully-connected layer. Following [50], we have used $\alpha = 5$ and $\beta = 10$ as the loss weights for the cycle-GAN throughout the experiments. The test set for this dataset comprises of 300 images of male and female each.

Results: Table 1 shows a comparative evaluation of the classifier performance of our proposed model for both AUG and ALT mode. In this table, we have provided the results for upto 500 male images due to space constraint and the rest are provided in the supplementary. We compared with some of the classic methods as well as methods particularly used with deep learning based classifiers. Namely they are random oversampling of minority class (OS), random undersampling of majority class (US), class frequency based cost sensitive weighing of loss (CS) [21], adjusting decision threshold by class frequency at prediction time (TS) and SMOTE [6] are some of the classical methods with which we compare our proposed method. We have also studied the effect of combining a few of the above classical methods and are reporting the two combinations - undersampling with cost sensitive learning (US+CS) and oversampling with cost sensitive learning (OS+CS). We have also compared with recent state-of-the-art approach using a class balanced loss (CBL) [7] for deep learning based classifiers.

We started with a highly imbalanced dataset of male and female faces where we randomly took 50 male and 900 female face images. Training the classifier without any data balancing strategy (we refer this classifier as the 'vanilla' classifier) misclassifies almost all of the minority class examples (male) as majority class examples (female) resulting in a accuracy of only 0.08 for males while the

Table 1. Comparison of the proposed method on the CelebA and CUB-200-2011 datasets using both AUG and ALT mode. The left half of the table shows the results for CelebA and the right half shows the results for the pairs of birds chosen from CUB-200-2011 dataset. The number of majority class training images (female) in CelebA is fixed to 900 while the CUB-200-2011 majority class (Sparrow) images 250. Minority class F_1 score for both are shown where the individual column headings indicate the number of minority class training images.

Dataset	CelebA						CUB-200-2011					
#Minority examples	50	100	200	300	400	500	12	25	50	75	100	125
Vanilla	0.1500	0.5220	0.7740	0.8460	0.9020	0.9160	0.0240	0.1180	0.2960	0.3700	0.5520	0.6160
TS	0.1560	0.6300	0.7880	0.8420	0.8960	0.9200	0.0240	0.1180	0.2880	0.3640	0.2567	0.6080
CS	0.7825	0.8012	0.8975	0.9137	0.9250	0.9244	0.3674	0.5007	0.5001	0.6384	0.6485	0.7212
US	0.8029	0.8491	0.9036	**0.9176**	0.9179	**0.9307**	0.2952	0.4263	0.6074	0.6893	**0.7169**	0.7080
OS	0.5805	0.7333	0.8749	0.9036	0.9181	0.9188	0.0602	0.1943	0.3910	0.5007	0.6295	0.6322
US+CS	0.8041	0.8463	0.9019	0.9163	0.9191	0.9220	0.5394	0.4760	0.6655	0.6647	0.6995	0.7201
OS+CS	0.7644	0.8249	0.8916	0.9065	0.9223	0.9260	0.3845	0.4225	0.5739	0.6147	0.6246	0.6913
SMOTE [6]	0.6208	0.7685	0.8807	0.8895	0.9167	0.9208	0.0586	0.4533	0.6375	0.5625	0.6708	0.6674
CBL [7] ($\beta = 0.9$)	0.6736	0.7771	0.8867	0.9061	0.9118	0.9206	0.1342	0.4006	0.5068	0.5624	0.6187	0.6890
CBL [7] ($\beta = 0.99$)	0.7012	0.8021	0.8938	0.9118	0.9178	0.9226	0.3259	0.5392	0.6001	0.6196	0.6369	0.6600
CBL [7] ($\beta = 0.999$)	0.7692	0.8250	0.8922	0.9122	0.9179	0.9220	0.3492	0.5256	0.6105	0.5400	0.5937	0.7212
CBL [7] ($\beta = 0.9999$)	0.7885	0.8099	0.8977	0.9127	0.9241	0.9226	0.3562	0.5950	0.5138	0.5933	0.6547	**0.7344**
(ours) ALT mode	**0.8240**	0.8520	0.8900	0.8880	0.8520	0.8920	0.5120	0.5120	0.5640	0.6340	0.6940	0.5960
(ours) AUG mode	0.8060	**0.8740**	**0.9140**	0.9160	**0.9340**	0.9220	**0.5940**	**0.6040**	**0.6680**	**0.7060**	0.7040	0.7180

accuracy for the female images is perfect (*i.e.*, 1). The F_1 score of the minority class for this case is 0.15. Our AUG mode of training improves the accuracy more than 5 fold to 0.8060 while the ALT mode of training improves it further to 0.8240. The fact that the classification is more balanced after the proposed dataset repairment is corroborated by the simultaneous increase of the precision of the majority class (female) from 0.5200 to 0.7720 (AUG mode) and 0.8260 (ALT mode). The male classification accuracy in AUG and ALT mode reaches up to 0.736 and 0.820 respectively. Table 1 shows the minority class F_1 score comparison with the state-of-the-arts. For 4 out of the 6 highly skewed regions (imbalance ratio ranging from 18 to 1.8) our proposed method significantly outperforms the rest. Out of the two cases, our augmentation mode is a close second (for 900:300 scenario). As the imbalance increases, our relative performance is better and better compared to other approaches. Performing better in highly skewed data distribution is very important as high imbalance implies more difficult case. This shows that our approach can deal with the hard cases in the challenging CelebA dataset.

We have also provided t-SNE [27] visualizations of the learned feature representation from the last fully-connected layer the classifier before and after the dataset repairment. Figure 3(a) and (b) show the visualization of the test set features for the vanilla classifier and the same trained in augmented mode respectively. This is for the case when the number of minority training examples is 50 while the majority training example is 900. It can be seen that the separation between the two classes is better after the repairment even in presence of such high imbalance. The same can be seen for the imbalanced training with 100 male and 900 female faces as shown in Fig. 3(c) (vanilla) and (d) (AUG).

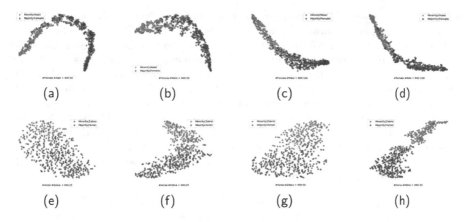

Fig. 3. t-SNE visualization for CelebA and Horse2Zebra dataset. Top row contains the t-SNE visualizations for 2 different imbalance ratio on CelebA dataset. Blue color indicates representation of majority class which is female while red color indicates the representation of the minority class i.e male. Figure 3(a) and (b) show the visualisation from the imbalanced (vanilla) model and augmented model for 900:50 imbalance ratio respectively. Figure 3(c) and (d) represent the 900:100 imbalance ratio. Similarly in the bottom row, Fig. 3(e) and (f) show the visualizations for imbalance ratio of 450:25 on Horse2Zebra dataset and Fig. 3(g) and (h) show the results for imbalance ratio of 450:50 respectively. Here, blue color indicates Horses (majority class) while the red color indicates Zebras (minority class). As can be seen from all the figures, a clear separation of the classes can be visualized. Best viewed in color (Color figure online).

Though in the high imbalance region (towards left of Table 1), both the modes of our proposed approach perform very good, the augmentation mode of training is consistently better than the alternate mode of training. The significance of the alternate mode of training, though, lies in the fact that it allows the cycle-GAN to improve itself by helping the classifier to discriminate better in presence of highly skewed training data. The improvement of performance of the GAN in terms of inception accuracy is shown in Table 2. Except for one scenario, the GAN performance improves for all the rest. Thus, the proposed approach provides a choice between going for bettering the GAN by trading off a little in the classifier performance (ALT mode) or going for improving the classifier without changing the GAN (AUG mode).

Visualizing Important Attributes for Classification: We aim to visually confirm that the proposed approach helps the classifier to look at the right evidences. For this purpose, we rely on two visual explanation techniques: Grad-CAM [37] and RISE [31]. They provide explanation heatmaps pointing to locations in the input image that are most important for a decision. Figure 4 shows visualizations for some representative minority images when they are passed through the vanilla classifier and the same after training in AUG and ALT mode. The top row shows a representative zebra image from the Horse2Zebra dataset. The vanilla model tends to concentrate on the surroundings (context) to classify

Table 2. Comparison of the inception accuracies of the Cycle-GAN before and after applying our proposed dataset repairment approach (in ALT mode). The table is divided into 3 parts from left to right corresponding to the three datasets on which the experimentations are performed. The dataset names are marked as headings of the parts. Each part is further divided into two halves where the left half shows the performance for the vanilla cycle-GAN and the right half shows the same after the cycle-GAN is trained in alternate mode of the proposed approach. The values shown are averaged over 5 runs for each of the cases. For two (CelebA and CUB-200-2011) out of the three datasets, the ALT mode of training improves the Cycle-GAN performance for most of the cases, while for Horse2Zebra dataset the performance is not increasing which may be due to the distinctive feature of the dataset where the attributes of the two classes are markedly different.

Celeb A dataset			CUB-200-2011 dataset			Horse2Zebra dataset		
# Male	CycleGAN	ALT mode	#Flycatcher	CycleGAN	ALT mode	#Zebra	CycleGAN	ALT mode
50	**0.4798**	0.4677	12	0.4149	**0.4554**			
100	0.5574	**0.6220**	25	**0.4238**	0.4059	25	**0.3532**	0.3480
200	0.7646	**0.8152**	50	0.4337	**0.4505**	50	**0.3880**	0.2680
300	0.7516	**0.8200**	75	0.4347	**0.4653**	75	**0.5000**	0.4440
400	0.8122	**0.8455**	100	**0.4317**	0.4307	100	**0.5492**	0.4780
500	0.8286	**0.8457**	125	0.4198	**0.4604**			

(incorrectly), while the proposed models accurately concentrate on the zebras (object) for correct classification. Similarly, the bottom row shows a flycatcher image from CUB dataset. When classifying the image, the vanilla model tends to scatter its attention while the proposed models base their prediction on right areas by learning to ignore inappropriate evidences.

4.2   Experiments on CUB

CUB-200-2011 birds dataset [43] is a popular benchmark dataset for fine-grained image classification. It contains 11,788 images of 200 different classes of North American bird species. Fine-grained classification is inherently difficult task as the members of the classes have very similar attributes with small inter-class variation requiring the classifier to focus on finer aspects of the images that are both descriptive and adequately discriminative. The challenges of fine-grained classification requires a lot of parameters to learn [10,23,47] which in turn needs a lot of training examples for proper training. Thus fine-grained classification provides us with an interesting testcase for low-data high-bias regime of classification.

Experimental Setup: For the experimentations, we have chosen two similar looking bird species (Flycatcher and Sparrow). We took all the images (226) of flycatcher that the dataset has and considered it to be the minority class. The 226 flycatcher images are split into 125 train, 50 val and 51 test images. The number of majority class (sparrow) images were kept constant at 250. We vary the number of flycatcher images from 25 to 125 in steps of 25 making the imbalance ratio vary from 10 to 5 in steps of 1. In addition, we also experimented

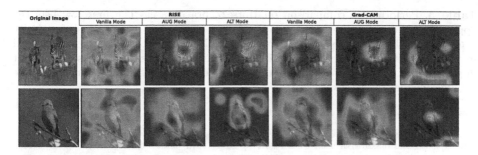

Fig. 4. Explanation heatmaps of images from Horse2Zebra (first row) and CUB (second row). Leftmost column shows the images while col 2–4 and 5–7 show RISE and Grad-CAM heatmaps respectively (importance increases from blue to red). The column sub-headings denote the corresponding classifier. Note that the heatmaps are generated for the predicted output by the classifiers. For vanilla mode, the prediction is wrong (horse and sparrow respectively) while for both AUG and ALT mode they are right. An interesting observation is revealed by the RISE generated heatmap (first row 4th col) showing that the evidence for ALT mode classifier is coming from a third zebra hidden in the grass. Best viewed in color (Color figure online).

with only 12 flycatcher images. The val and test set size of sparrow is kept same as those of the flycatcher class. The model architectures are kept same as those used for CelebA experiments (ref. Sect. 4.1).

Results: Comparative evaluations of the proposed approach for the pair of birds in the CUB-200-2011 data set is shown in the rightmost part of Table 1. The table provides the F_1 score of the minority class for the different approaches. It can be observed that our proposed method in augmentation mode is outperforming other approaches for highly skewed training data. The alternate mode of training the classifier does not perform as good as it does for the face images. However, the cycle-GAN inception accuracy gets improved for 4 out of 6 cases as can be seen in the middle part of Table 2 where for highly skewed training data, the alternate model betters the vanilla cycle-GAN.

4.3   Experiments on Horse2Zebra

Cycle-GAN [50] has been particularly successful in translating between images of horses and zebras. Unlike finegrained bird images, horses and zebras have distinctive features with marked difference in skin texture which allows a classifier to perform well even with very less data. To test with this end of the spectrum, our proposed approach was applied for a classification task between horse and zebra images in presence of high imbalance in training data.

Experimental Setup: The Horse2Zebra dataset [50] contains 1187 images of horses and 1474 images of zebras. We took the horse class as majority and used a total of 450 Horse images. The number of minority class (zebra) examples were varied from 25 to 100 in steps of 25 allowing us to experiment with low data

Table 3. Performance comparison on Horse2Zebra dataset. The first row lists the approaches against which we have compared our performance. The number of majority class training images (horses) is fixed at 450 while the number of zebra images used in training is shown as the row headings. The minority class F_1 score is seen to get improved with application of our proposed approach especially in the augmented mode for high imbalance regions.

#Zebra	Vanilla	TS	CS	US	OS	US+CS	OS+CS	SMOTE [6]	CBL [7]				Ours	
									$\beta = 0.9$	$\beta = 0.99$	$\beta = 0.999$	$\beta = 0.9999$	ALT	AUG
25	0.0500	0.0340	0.7925	0.1333	0.4750	02865	0.7366	0.2108	0.7613	0.8116	0.8263	0.8120	0.8040	**0.8500**
50	0.5580	0.6260	0.8329	0.7891	0.8012	0.7343	0.8522	0.7298	0.8347	0.8316	0.8222	0.8192	0.8320	**0.8620**
75	0.7180	0.7040	0.8699	0.8457	0.8370	0.8644	0.8680	0.7902	0.8562	0.8587	0.8524	0.8633	0.8020	**0.8780**
100	0.7680	0.7940	0.8684	0.8521	0.8628	**0.8711**	0.8735	0.8254	0.8446	0.8595	0.8683	0.8657	0.8220	0.8520

and low balance regime. The val and test set consists of 100 and 150 images respectively for each of the animal categories. For this case also, the model architectures are kept same as those used for CelebA experiments (ref. Sect. 4.1).

Results: Table 3 shows comparative evaluations of the proposed approach for this dataset in terms of the F_1 score of the minority class (zebra). The proposed method in AUG mode of training comprehensively outperforms the other approaches in the high imbalance region with the difference in F_1 score with the second best approach reaching almost 2.5% for the highest imbalance ratio with the number of training images for zebra being only 25. The inception accuracies of the cycle-GAN before and after training with ALT mode is provided in the rightmost part of Table 2. For this dataset we got consistently worse GANs while training in this mode. This can be due to the distinctive texture between the two animals that aids classification but imposes difficulty in translation especially in presence of the classifier loss coming from noisy data.

5   Conclusion

We propose a joint generation and classification network to handle severe data imbalance by transforming majority class examples to minority class and vice versa. Our approach explores a Cycle-consistent GANs that not only generates training samples that are much broader and diverse but also balanced in terms of number of examples. Experiments on three standard datasets demonstrate that both classifier and the GAN achieves superior performance than existing data imbalance mitigation approaches.

Acknowledgements. This work was partially supported by the IIT Kharagpur ISIRD program and the SERB Grant SRG/2019/001205.

References

1. Achille, A., Rovere, M., Soatto, S.: Critical learning periods in deep networks. In: International Conference on Learning Representations (2019)

2. Antoniou, A., Storkey, A., Edwards, H.: Data augmentation generative adversarial networks. arXiv preprint arXiv:1711.04340 (2017)
3. Barratt, S., Sharma, R.: A note on the inception score. arXiv preprint arXiv:1801.01973 (2018)
4. Batista, G.E., Prati, R.C., Monard, M.C.: A study of the behavior of several methods for balancing machine learning training data. SIGKDD Exp. Newsl 6(1), 20–29 (2004)
5. Buda, M., Maki, A., Mazurowski, M.A.: A systematic study of the class imbalance problem in convolutional neural networks. Neural Netw. 106, 249–259 (2018)
6. Chawla, N.V., Bowyer, K.W., Hall, L.O., Kegelmeyer, W.P.: SMOTE: synthetic minority over-sampling technique. J. Artif. Intell. Res. 16, 321–357 (2002)
7. Cui, Y., Jia, M., Lin, T.Y., Song, Y., Belongie, S.: Class-balanced loss based on effective number of samples. In: IEEE Conference on Computer Vision and Pattern Recognition, pp. 9268–9277 (2019)
8. Drummond, C., Holte, R.C.: C4.5, class imbalance, and cost sensitivity: why under-sampling beats over-sampling. In: Workshop on Learning from Imbalanced Datasets, vol. 11, pp. 1–8 (2003)
9. Estabrooks, A., Jo, T., Japkowicz, N.: A multiple resampling method for learning from imbalanced data sets. Comput. Intell. 20(1), 18–36 (2004)
10. Gao, Y., Beijbom, O., Zhang, N., Darrell, T.: Compact bilinear pooling. In: IEEE Conference on Computer Vision and Pattern Recognition, pp. 317–326 (2016)
11. Goodfellow, I., et al.: Generative adversarial nets. In: Advances in Neural Information Processing Systems, pp. 2672–2680 (2014)
12. Haixiang, G., Yijing, L., Shang, J., Mingyun, G., Yuanyue, H., Bing, G.: Learning from class-imbalanced data: review of methods and applications. Exp. Syst. Appl. 73, 220–239 (2017)
13. Han, H., Wang, W.-Y., Mao, B.-H.: Borderline-SMOTE: a new over-sampling method in imbalanced data sets learning. In: Huang, D.-S., Zhang, X.-P., Huang, G.-B. (eds.) ICIC 2005. LNCS, vol. 3644, pp. 878–887. Springer, Heidelberg (2005). https://doi.org/10.1007/11538059_91
14. He, H., Garcia, E.A.: Learning from imbalanced data. Trans. Knowl. Data Eng. 9, 1263–1284 (2008)
15. Hendricks, L.A., Burns, K., Saenko, K., Darrell, T., Rohrbach, A.: Women also snowboard: overcoming bias in captioning models. In: Ferrari, V., Hebert, M., Sminchisescu, C., Weiss, Y. (eds.) ECCV 2018. LNCS, vol. 11207, pp. 793–811. Springer, Cham (2018). https://doi.org/10.1007/978-3-030-01219-9_47
16. Horn, G.V., et al.: The iNaturalist species classification and detection dataset. In: IEEE Conference on Computer Vision and Pattern Recognition, pp. 8769–8778 (2018)
17. Huang, C., Li, Y., Loy, C.C., Tang, X.: Learning deep representation for imbalanced classification. In: IEEE Conference on Computer Vision and Pattern Recognition, pp. 5375–5384 (2016)
18. Isola, P., Zhu, J.Y., Zhou, T., Efros, A.A.: Image-to-image translation with conditional adversarial networks. In: IEEE Conference on Computer Vision and Pattern Recognition, pp. 1125–1134 (2017)
19. Japkowicz, N., Stephen, S.: The class imbalance problem: a systematic study. Intell. Data Anal. 6(5), 429–449 (2002)
20. Kubat, M., Matwin, S.: Addressing the curse of imbalanced training sets: one-sided selection. In: International Conference on Machine Learning, vol. 97, pp. 179–186 (1997)

21. Lawrence, S., Burns, I., Back, A., Tsoi, A.C., Giles, C.L.: Neural network classification and prior class probabilities. In: Montavon, G., Orr, G.B., Müller, K.-R. (eds.) Neural Networks: Tricks of the Trade. LNCS, vol. 7700, pp. 295–309. Springer, Heidelberg (2012). https://doi.org/10.1007/978-3-642-35289-8_19

22. Lin, T.-Y., et al.: Microsoft COCO: common objects in context. In: Fleet, D., Pajdla, T., Schiele, B., Tuytelaars, T. (eds.) ECCV 2014. LNCS, vol. 8693, pp. 740–755. Springer, Cham (2014). https://doi.org/10.1007/978-3-319-10602-1_48

23. Lin, T.Y., RoyChowdhury, A., Maji, S.: Bilinear CNN models for fine-grained visual recognition. In: IEEE International Conference on Computer Vision, pp. 1449–1457 (2015)

24. Liu, X.Y., Wu, J., Zhou, Z.H.: Exploratory undersampling for class-imbalance learning. Trans. Syst. Man Cybern. Part B (Cybern.) **39**(2), 539–550 (2008)

25. Liu, Z., Luo, P., Qiu, S., Wang, X., Tang, X.: DeepFashion: powering robust clothes recognition and retrieval with rich annotations. In: IEEE Computer Vision and Pattern Recognition, pp. 1096–1104 (2016)

26. Liu, Z., Luo, P., Wang, X., Tang, X.: Deep learning face attributes in the wild. In: IEEE International Conference on Computer Vision (December 2015)

27. van der Maaten, L., Hinton, G.: Visualizing data using t-SNE. J. Mach. Learn. Res. **9**, 2579–2605 (2008)

28. Arjovsky, M., Bottou, L.: Towards principled methods for training generative adversarial networks. In: International Conference on Learning Representations (2017)

29. van Miltenburg, E.: Stereotyping and bias in the Flickr30K dataset. In: Workshop on Multimodal Corpora: Computer Vision and Language Processing (May 2016)

30. Mullick, S.S., Datta, S., Das, S.: Generative adversarial minority oversampling. In: IEEE International Conference on Computer Vision (October 2019)

31. Petsiuk, V., Das, A., Saenko, K.: Rise: randomized input sampling for explanation of black-box models. In: British Machine Vision Conference (September 2018)

32. Ponce, J., et al.: Dataset issues in object recognition. In: Ponce, J., Hebert, M., Schmid, C., Zisserman, A. (eds.) Toward Category-Level Object Recognition. LNCS, vol. 4170, pp. 29–48. Springer, Heidelberg (2006). https://doi.org/10.1007/11957959_2

33. Ratner, A.J., Ehrenberg, H., Hussain, Z., Dunnmon, J., Ré, C.: Learning to compose domain-specific transformations for data augmentation. In: Advances in Neural Information Processing Systems, pp. 3236–3246 (2017)

34. Richard, M.D., Lippmann, R.P.: Neural network classifiers estimate Bayesian a posteriori probabilities. Neural Comput. **3**(4), 461–483 (1991)

35. Salimans, T., Goodfellow, I., Zaremba, W., Cheung, V., Radford, A., Chen, X.: Improved techniques for training GANs. In: Neural Information Processing Systems, pp. 2234–2242 (2016)

36. Santurkar, S., Schmidt, L., Madry, A.: A classification-based study of covariate shift in GAN distributions. In: International Conference on Machine Learning, pp. 4487–4496 (2018)

37. Selvaraju, R.R., Cogswell, M., Das, A., Vedantam, R., Parikh, D., Batra, D.: Grad-CAM: visual explanations from deep networks via gradient-based localization. In: IEEE International Conference on Computer Vision (October 2017)

38. Shen, L., Lin, Z., Huang, Q.: Relay backpropagation for effective learning of deep convolutional neural networks. In: Leibe, B., Matas, J., Sebe, N., Welling, M. (eds.) ECCV 2016. LNCS, vol. 9911, pp. 467–482. Springer, Cham (2016). https://doi.org/10.1007/978-3-319-46478-7_29

39. Szegedy, C., Vanhoucke, V., Ioffe, S., Shlens, J., Wojna, Z.: Rethinking the inception architecture for computer vision. In: IEEE Conference on Computer Vision and Pattern Recognition, pp. 2818–2826 (2016)
40. Taigman, Y., Polyak, A., Wolf, L.: Unsupervised cross-domain image generation. In: International Conference on Learning Representations (2017)
41. Tomek, I.: Two modifications of CNN. IEEE Trans. Syst. Man Cybern. **SMC– 6**(11), 769–772 (1976)
42. Torralba, A., Efros, A.A.: Unbiased look at dataset bias. In: IEEE CVPR, pp. 1521–1528 (June 2011)
43. Wah, C., Branson, S., Welinder, P., Perona, P., Belongie, S.: The Caltech-UCSD Birds-200-2011 Dataset. Technical report, CNS-TR-2011-001, California Institute of Technology (2011)
44. Wang, K.J., Makond, B., Chen, K.H., Wang, K.M.: A hybrid classifier combining SMOTE with PSO to estimate 5-year survivability of breast cancer patients. Appl. Soft Comput. **20**, 15–24 (2014)
45. Weiss, G.M., Provost, F.: The Effect of Class Distribution on Classifier Learning: An Empirical Study. Technical report, Technical Report ML-TR-43, Department of Computer Science, Rutgers University (2001)
46. Xiao, J., Hays, J., Ehinger, K.A., Oliva, A., Torralba, A.: SUN database: large-scale scene recognition from Abbey to Zoo. In: IEEE Computer Vision and Pattern Recognition, pp. 3485–3492 (2010)
47. Zhang, L., Huang, S., Liu, W., Tao, D.: Learning a mixture of granularity-specific experts for fine-grained categorization. In: IEEE International Conference on Computer Vision, pp. 8331–8340 (2019)
48. Zhou, B., Lapedriza, A., Xiao, J., Torralba, A., Oliva, A.: Learning deep features for scene recognition using places database. In: Neural Information Processing Systems, pp. 487–495 (2014)
49. Zhou, Z.H., Liu, X.Y.: Training cost-sensitive neural networks with methods addressing the class imbalance problem. IEEE Trans. Knowl. Data Eng. **1**, 63–77 (2006)
50. Zhu, J.Y., Park, T., Isola, P., Efros, A.A.: Unpaired image-to-image translation using cycle-consistent adversarial networks. In: IEEE International Conference on Computer Vision, pp. 2223–2232 (2017)

W40 - Computer Vision Problems
in Plant Phenotyping

W40 - Computer Vision Problems in Plant Phenotyping

This workshop has run since 2014 in conjunction with ECCV (2014, 2020), BMVC (2015, 2018), ICCV (2017), and CVPR (2019). Plant phenotyping is the identification of effects on the plant appearance and behavior as a result of genotype differences and the environment. Imaging has become common in phenotyping and is more and more integrated into regular agricultural procedures. These images are recorded by a range of capture devices, from small embedded camera systems to multi-million Euro smart-greenhouses, at scales ranging from microscopic images of cells, to entire fields captured by UAV imaging. These images need to be analyzed in a high throughput, robust, and accurate manner. UN-FAO statistics show that according to current population predictions, we will need to achieve a 70% increase in food productivity by 2050, to maintain current global agricultural demands. Large-scale measurement of plant traits is a key bottleneck and machine vision is ideally placed to help. The goal of this workshop was to not only present interesting computer vision solutions, but also to introduce challenging computer vision problems in the increasingly important plant phenotyping domain, accompanied with benchmark datasets and suitable performance evaluation methods. Together with the workshop, permanently open challenges were addressed, the Leaf Counting (LCC) and Leaf Segmentation Challenges (LSC), as well as a new one: the Global Wheat Detection Challenge with more than 2000 participating teams. Of the 17 full papers presented in CVPPP 2020, 16 responded to our open call's computer vision for plant phenotyping topics and 1 addresses the global wheat challenge. The challenge papers were processed as 'late breaking' papers due to the deadline of the kaggle challenge after the regular CVPPP submission deadline. All submissions, including 8 extended abstracts and the challenge papers and abstracts, were double-blind peer reviewed by at least three external reviewers. The committee then ranked papers and rejected those that did not receive sufficient scores of quality and priority as suggested by the reviewers. The acceptance rate for full papers was 64%, and 80% for extended abstracts. We did not distinguish between oral and poster presentations. More information, including the full workshop program, can be found on the CVPPP 2020 Website.

We would like to first and foremost thank all authors, reviewers, and members of the Program Committee for their contribution to this workshop. We would like to also thank the ECCV organizers and particularly the workshop chairs for their support. Finally, we thank our sponsor the Int. Plant Phenotyping Network.

August 2020

Hanno Scharr
Tony Pridmore
Sotirios A. Tsaftaris

Patch-Based CNN Evaluation for Bark Classification

Debaleena Misra, Carlos Crispim-Junior[✉][ID], and Laure Tougne[ID]

Univ Lyon, Lyon 2, LIRIS, 69676 Lyon, France
{debaleena.misra,carlos.crispim-junior,laure.tougne}@liris.cnrs.fr

Abstract. The identification of tree species from bark images is a challenging computer vision problem. However, even in the era of deep learning today, bark recognition continues to be explored by traditional methods using time-consuming handcrafted features, mainly due to the problem of limited data. In this work, we implement a patch-based convolutional neural network alternative for analyzing a challenging bark dataset Bark-101, comprising of 2587 images from 101 classes. We propose to apply image re-scaling during the patch extraction process to compensate for the lack of sufficient data. Individual patch-level predictions from fine-tuned CNNs are then combined by classical majority voting to obtain image-level decisions. Since ties can often occur in the voting process, we investigate various tie-breaking strategies from ensemble-based classifiers. Our study outperforms the classification accuracy achieved by traditional methods applied to Bark-101, thus demonstrating the feasibility of applying patch-based CNNs to such challenging datasets.

Keywords: Bark classification · Convolutional neural networks · Transfer learning · Patch-based CNNs · Image re-scaling · Bicubic interpolation · Super-resolution networks · Majority voting

1 Introduction

Automatic identification of tree species from images is an interesting and challenging problem in the computer vision community. As urbanization grows, our relationship with plants is fast evolving and plant recognition through digital tools provide an improved understanding of the natural environment around us. Reliable and automatic plant identification brings major benefits to many sectors, for example, in forestry inventory, agricultural automation [31], botany [26], taxonomy, medicinal plant research [3] and to the public, in general. In recent years, vision-based monitoring systems have gained importance in agricultural operations for improved productivity and efficiency [17]. Automated crop harvesting using agricultural robotics [2] for example, relies heavily on visual identification of crops from their images. Knowledge of trees can also provide landmarks in localization and mapping algorithms [30].

Although plants have various distinguishable physical features such as leaves, fruits or flowers, bark is the most consistent one. It is available round the year,

© Springer Nature Switzerland AG 2020
A. Bartoli and A. Fusiello (Eds.): ECCV 2020 Workshops, LNCS 12540, pp. 197–212, 2020.
https://doi.org/10.1007/978-3-030-65414-6_15

with no seasonal dependencies. The aging process is also a slow one, with visual features changing over longer periods of time while being consistent during shorter time frames. Even after trees have been felled, their bark remains an important identifier, which can be helpful for example, in autonomous timber assessment. Barks are also more easily visually accessible, contrary to higher-level leaves, fruits or flowers. However, due to the low inter-class variance and high intra-class variance for bark data, the differences are very subtle. Besides, bark texture properties are also impacted by the environment and plant diseases. Uncontrolled illumination alterations and branch shadow clutter can additionally affect image quality. Hence, tree identification from only bark images is a challenging task not only for machine learning approaches [5,7,8,24] but also for human experts [13].

Recent developments in deep neural networks have shown great progress in image recognition tasks, which can help automate manual recognition methods that are often laborious and time consuming. However, a major limitation of deep learning algorithms is that a huge amount of training data is required for attaining good performance. For example, the ImageNet dataset [11] has over 14 million images. Unfortunately, the publicly available bark datasets are very limited in size and variety. Recently released *BarkNet 1.0* dataset [8] with 23,000 images for 23 different species, is the largest in terms of number of instances, while *Bark-101* dataset [24] with 2587 images and 101 different classes, is the largest in terms of number of classes. The data deficiency of reliable bark datasets in literature presumably explains why majority of bark identification research has revolved around hand-crafted features and filters such as Gabor [4,18], SIFT [9,13], Local Binary Pattern (LBP) [6,7,24] and histogram analysis [5], which can be learned from lesser data.

In this context, we study the challenges of applying deep learning in bark recognition from limited data. The objective of this paper is to investigate patch-based convolutional neural networks for classifying the challenging Bark-101 dataset that has low inter-class variance, high intra-class variance and some classes with very few samples. To tackle the problem of insufficient data, we enlarge the training data by using patches cropped from original Bark-101 images. We propose a patch-extraction approach with image re-scaling prior to training, to avoid random re-sizing during image-loading. For re-scaling, we compare traditional bicubic interpolation [10] with a more recent advance of re-scaling by super-resolution convolutional neural networks [12]. After image re-scale and patch-extraction, we fine-tune pre-trained CNN models with these patches. We obtain patch-level predictions which are then combined in a majority voting fashion to attain image-level results. However, there can be ties, i.e. more than one class could get the largest count of votes, and it can be challenging when a considerable number of ties occur. In our study, we apply concepts of ensemble-based classifiers and investigate various tie-breaking strategies [21,22,25,33,34] of majority voting. We validated our approach on three pre-trained CNNs - Squeezenet [19], MobileNetV2 [27] and VGG16 [29], of which the first two are compact and light-weight models that could be used for applications on mobile devices in the future. Our study demonstrates the feasibility of using deep neural networks for challenging datasets and outperforms the classification accuracy achieved using traditional hand-crafted methods on Bark-101 in the original work [24].

The rest of the paper is organised as follows. Section 2 reviews existing approaches in bark classification. Then, Sect. 3 explains our methodology for patch-based CNNs. Section 4 describes the experimentation details. Our results and insights are presented in Sect. 5. Finally, Sect. 6 concludes the study with discussions on possible future work.

2   Related Work

Traditionally, bark recognition has been studied as a texture classification problem using statistical methods and hand-crafted features. Bark features from 160 images were extracted in [32] using textual analysis methods such as gray level run-length method (RLM), concurrence matrices (COMM) and histogram inspection. Additionally, the authors captured the color information by applying the grayscale methods individually to each of the 3 RGB channels and the overall performance significantly improved. Spectral methods using Gabor filters [4] and descriptors of points of interests like SURF or SIFT [9,13,16] have also been used for bark feature extraction. The AFF bark dataset, having 11 classes and 1082 bark images, was analysed by a bag of words model with an SVM classifier constructed from SIFT feature points achieving around 70% accuracy [13].

An earlier study [5] proposed a fusion of color hue and texture analysis for bark identification. First the bark structure and distribution of contours (scales, straps, cracks etc.) were described by two descriptive feature vectors computed from a map of Canny extracted edges intersected by a regular grid. Next, the color characteristics were captured by the hue histogram in HSV color space as it is indifferent to illumination conditions and covers the whole range of possible bark colors with a single channel. Finally, image filtering by Gabor wavelets was used to extract the orientation feature vector. An extended study on the resultant descriptor from concatenation of these four feature vectors, showed improved performance in tree identification when combined with leaves [4]. Several works have also been based on descriptors such as Local Binary Patterns (LBP) and LBP-like filters [6,7,24]. Late Statistics (LS) with two state-of-art LBP-like filters - Light Combination of Local Binary Patterns (LCoLBP) and Completed Local Binary Pattern (CLBP) were defined, along with bark priors on reduced histograms in the HSV space to capture color information [24]. This approach created computationally efficient, compact feature vectors and achieved state-of-art performance on 3 challenging datasets (BarkTex, AFF12, Bark-101) with SVM and KNN classifiers. Another LBP-inspired texture descriptor called Statistical Macro Binary Pattern (SMBP) attained improved performance in classifying 3 datasets (BarkTex, Trunk12, AFF) [7]. SMBP encodes macrostructure information with statistical description of intensity distribution which is rotation-invariant and applies an LBP-like encoding scheme, thus being invariant to monotonic gray scale changes.

Some early works [18] in bark research have interestingly been attempted using artificial neural networks (ANN) as classifiers. In 2006, Gabor wavelets were used to extract bark texture features and applied to a radial basis probabilistic neural network (RBPNN) for classification [18]. It achieved around 80% accuracy on a dataset of 300 bark images. GLCM features have also been used in combination with fractal dimension features to describe the complexity and self-similarity of varied scaled texture [18]. They used a 3-layer ANN classifier on a dataset of 360 images having 24 classes and obtained an accuracy of 91.67%. However, this was before the emergence of deep learning convolutional neural networks for image recognition.

Recently, there have been few attempts to identify trees from only bark information using deep-learning. LIDAR scans created depth images from point clouds, which were applied to AlexNet resulting in 90% accuracy, using two species only - Japanese Cedar and Japanese Cypress [23]. Closer to our study with RGB images, patches of bark images have been used to fine-tune pre-trained deep learning models [15]. With constraints on the minimum number of crops and projected size of tree on plane, they attained 96.7% accuracy, using more than 10,000 patches for 221 different species. However, the report lacked clarity on the CNN architecture used and the experiments were performed on private data provided by a company, therefore inaccessible for comparisons. Image patches were also used for transfer-learning with ResNets to identify species from the BarkNet dataset [8]. This work obtained an impressive accuracy of 93.88% for single crop and 97.81% using majority voting on multiple crops. However, BarkNet is a large dataset having 23,000 high-resolution images for 23 classes, which significantly reduces the challenges involved. We draw inspiration from these works and build on them to study an even more challenging dataset - Bark-101 [24].

3   Methodology

Our methodology presents a plan of actions consisting of four main components: *Image re-scaling, patch-extraction, fine-tuning pre-trained CNNs* and *majority voting analysis*. The following sections discuss our work-flow in detail.

3.1   Dataset

In our study, we chose the Bark-101 dataset. It was developed from PlantCLEF database, which is part of the ImageCLEF challenges for plant recognition [20]. Bark-101 consists of a total of 2587 images (split into 1292 train and 1295 test images) belonging to 101 different species. Two observations about Bark-101 explain the difficulty level of this dataset. Firstly, these images simulate real world conditions as PlantCLEF forms their database through crowdsourced initiatives (for example from mobile applications as Pl@ntnet [1]). Although the images have been manually segmented to remove unwanted background, Bark-101 still contains a lot of noise in form of mosses, shadows or due to lighting conditions. Moreover, no constraints were given for image sizes during Bark-101 preparation leading to a huge variability of size. This is expected in practical outdoor settings where tree trunk diameters fluctuate and users take pictures from varying distances. Secondly, Bark-101 has high intra-class variability and low inter-class variability which makes classification difficult. High intra-class variability can be attributed to high diversity in bark textures during the lifespan of a tree. Low inter-class variability is explained by the large number of classes (101) in the dataset, as a higher number of species imply higher number of visually alike textures. Therefore, Bark-101 can be considered a challenging dataset for bark recognition (Fig. 1).

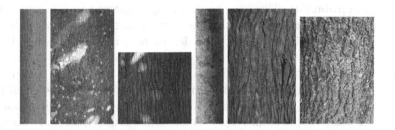

Fig. 1. Example images from Bark-101 dataset.

3.2 Patch Preparation

In texture recognition, local features can offer useful information to the classifier. Such local information can be obtained through extraction of patches, which means decomposing the original image into smaller crops or segments. Compared to semantic segmentation techniques that use single pixels as input, patches allow to capture neighbourhood local information as well as reduces execution time. These patches are then used for fine-tuning a pre-trained CNN model. Thus, the patch extraction process also helps to augment the available data for training CNNs and is particularly useful when the number of images per class is low, as is the case with Bark-101 dataset.

Our study focused on using a patch size of 224 × 224 pixels, following the default ImageNet size standards used in most CNN image recognition tasks today. However, when the range of image dimensions vary greatly within a dataset, it is difficult to extract a useful number of non-overlapping patches from all images. For example, in Bark-101, around 10% of the data is found to have insufficient pixels to allow even a single square patch of 224 × 224 size. Contrary to common data pre-processing for CNNs where images are randomly re-sized and cropped during data loading, we propose to prepare the patches beforehand to have better control in the patch extraction process. This also removes the risk of extracting highly deformed patches from low-dimension images. The original images are first upscaled by a given factor and then patches are extracted from them. In our experiments, we applied two different image re-scaling algorithms - traditional bicubic interpolation method [10] and a variant of super-resolution network, called efficient sub-pixel convolutional neural network (ESPCN) [28].

3.3   Image Re-scaling

Image re-scaling refers to creating a new version of an image with new dimensions by changing its pixel information. In our study, we apply *upscaling* which is the process of obtaining images with increased size. However, these operations are not loss-less and have a trade-off between efficiency, complexity, smoothness, sharpness and speed. We tested two different algorithms to obtain high-resolution representation of the corresponding low-resolution image (in our context, resolution referring to spatial resolution, i.e. size).

Bicubic Interpolation. This is a classical image upsampling algorithm involving geometrical transformation of 2D images with Lagrange polynomials, cubic splines or cubic convolutions [10]. In order to preserve sharpness and image details, new pixel values are approximated from the surrounding pixels in the original image. The output pixel value is computed as a weighted sum of pixels in the 4-by-4 pixel neighborhood. The convolution kernel is composed of piecewise cubic polynomials. Compared to bilinear or nearest-neighbor interpolation, bicubic takes longer time to process as more surrounding pixels are compared but the resampled images are smoother with fewer distortions. As the destination high-resolution image pixels are estimated using only local information in the corresponding low-resolution image, some details could be lost.

Super-Resolution Using Sub-pixel Convolutional Neural Network. Supervised machine learning methods can learn mapping functions from low-resolution images to their high-resolution representations. Recent advances in deep neural networks called Super-resolution CNN (SRCNN) [12] have shown promising improvements in terms of computational performances and reconstruction accuracy. These models are trained with low-resolution images as inputs and their high-resolution counterparts are the targets. The first convolutional layers of such neural networks perform feature extraction from the low-resolution images. The next layer maps these feature maps non-linearly to their corresponding high-resolution patches. The final layer combines the predictions within a spatial neighbourhood to produce the final high-resolution image (Fig. 2).

In our study, we focus on the efficient sub-pixel convolutional neural network (ESPCN) [28]. In this CNN architecture, feature maps are extracted from low-resolution space, instead of performing the super-resolution operation in the high-resolution space that has been upscaled by bicubic interpolation. Additionally, an efficient sub-pixel convolution layer is included that learns more complex upscaling filters (trained for each feature map) to the final low-resolution feature maps into the high-resolution output. This architecture is shown in Fig. 3, with two convolution layers for feature extraction, and one sub-pixel convolution layer which accumulates the feature maps from low-resolution space and creates the super-resolution image in a single step.

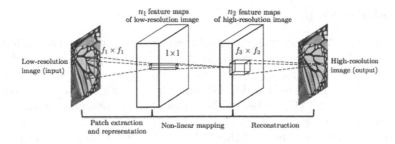

Fig. 2. SRCNN architecture [12].

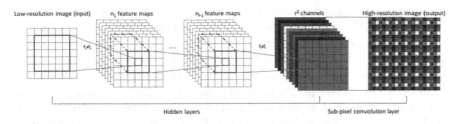

Fig. 3. Efficient sub-pixel convolutional neural network (ESPCN) [28]

3.4 CNN Classification

The CNN models used for image recognition in this work are 3 recent architectures: Squeezenet [19], MobileNetV2 [27] and VGG16 [29]. Since Bark-101 is a small dataset, we apply transfer-learning and use the pre-trained weights of the models trained on the large-scale image data of ImageNet [11]. The convolutional layers are kept frozen and only the last fully connected layer is replaced with a new one having random weights. We only train this layer with the dataset to make predictions specific to the bark data. We skip the detailed discussion of the architectures, since its beyond the scope of our study and can be found in the references for the original works.

3.5 Evaluation Metrics

In our study, we report two kinds of accuracy: *patch-level* and *absolute*.

- **Patch-level accuracy** - Describes performance at a patch-level, i.e. among all the patches possible (taken from 1295 test images), we record what percentage of patches has been correctly classified.
- **Absolute accuracy** - Refers to overall accuracy in the test data, i.e. how many among 1295 test images of Bark-101 could be correctly identified. In our patch-based CNN classification, we computed this by majority voting using 4 variants for resolving ties, described in the following Sect. 3.6.

3.6 Majority Voting

Majority voting [22, 34] is a popular label-fusion strategy used in ensemble-based classifiers [25]. We use this for combining the independent patch-level predictions into image-level results for our bark classification problem. In simple majority voting, the class that gets the largest number of votes among all patches is considered the overall class label of the image. Let us assume that an image I can be cropped into x parent patches and gets x_1, x_2, x_3 number of patches classified as the first, second and third classes. The final prediction class of the image is taken as the class that has $max(x_1, x_2, x_3)$ votes, i.e. the majority voted class.

However, there may be cases when more than one class gets the largest number of votes. There is no one major class and ties can be found, i.e. multiple classes can have the highest count of votes. In our study, we examine few tie-breaking strategies from existing literature in majority voting [21, 22, 25, 33, 34]. The most common one is *Random Selection* [22] of one of the tied classes, all of which are assumed to be equi-probable. Another trend of tie-breaking approaches relies on class priors and we tested two variants of class prior probability in our study. First, by *Class Proportions* strategy [33] that chooses the class having the higher number of training samples among the tied classes. Second, using *Class Accuracy* (given by F1-score), the tie goes in favor of the class having a better F1-score. Outside of these standard methods, more particular to neural network classifiers is the breaking of ties in ensembles by using Soft Max Accumulations [21]. Neural network classifiers output, by default, the class label prediction accompanied by a confidence of prediction (by using soft max function) and this information is leveraged to resolve ties in [21]. In case of a tie in the voting process, the confidences for the received votes of the tied classes, are summed up. Finally, the class that accumulates the *Maximum Confidence sum* is selected.

4   Experiments

4.1   Dataset

Pre-processing. As we used pre-trained models for fine-tuning, we resized and normalised all our images to the same format the network was originally trained on, i.e. ImageNet standards. For patch-based CNN experiments, no size transformations were done during training as the patches were already of size 224×224. For the benchmark experiments with whole images, the images were randomly re-sized and cropped to the size of 224×224 during image-loading. For data augmentation purposes, torchvision transforms were used to obtain random horizontal flips, rotations and color jittering, on all training images.

Patch Details. Non-overlapping patches of size 224×224 were extracted in a sliding window manner from non-scaled original and upscaled Bark-101 images (upscale factor of 4). Bark-101 originally has 1292 training images and 1295 test images. After patch-extraction by the two methods, we obtain a higher count of samples as shown in Table 1. In our study, 25% of the training data was kept for validation.

Table 1. Count of extracted unique patches from Bark-101.

Source image	Train	Validation	Test
Non-Scaled Bark-101	3156	1051	4164
Upscaled Bark-101	74799	24932	99107

4.2 Training Details

We used CNNs that have been pre-trained on ImageNet. Three architectures were selected - Squeezenet, MobileNetV2 and VGG16. We used Pytorch to fine-tune these networks with the Bark-101 dataset, starting with an initial learning rate of 0.001. Training was performed over 50 epochs with the Stochastic gradient descent (SGD) optimizer, reducing the learning rate by a factor of 0.1 every 7th epoch.

ESPCN was trained from scratch for a factor of 4, on the 1292 original training images of Bark-101, with a learning rate of 0.001 for 30 epochs.

5 Results and Discussions

We present our findings with the two kinds of accuracy described in Sect. 3.5. Compared to absolute accuracy, patch-level accuracy provides a more precise measure of how good the classifier model is. However, for our study, it is the absolute accuracy that is of greater importance as the final objective is to improve identification of bark species. It is important to note that Bark-101 is a challenging dataset and the highest accuracy obtained in the original work on Bark-101 [24] was 41.9% using Late Statistics on LBP-like filters and SVM classifiers. In our study, we note this as a benchmark value for comparing our performance using CNNs.

5.1 Using Whole Images

We begin by comparing the classification accuracy of different pre-trained CNNs fine-tuned with the non-scaled original Bark-101 data. Whole images were used and no explicit patches were formed prior to training. Thus, training was carried out on 1292 images and testing on 1295. Table 2 presents the results.

Table 2. Classification of whole images from Bark-101.

CNN	Absolute accuracy
Squeezenet	**43.7%**
VGG16	42.3%
MobilenetV2	34.2%

5.2 Using Patches

In this section, we compare patch-based CNN classification using patches obtained by the image re-scaling methods explained in Sect. 3.3.

In the following Tables 3, 4 and 5, the column for *Patch-level accuracy* gives the local performance, i.e how many of the test patches are correctly classified. This number of test patches vary across the two methods - patches from non-scaled original and those from upscaled Bark-101 (see Table 1). For *Absolute accuracy*, we calculate how many of the original 1295 Bark-101 test images are correctly identified, by majority voting (with 4 tie-breaking strategies) on patch-level predictions. Column *Random selection* gives results of arbitrarily breaking ties by randomly selecting one of the tied classes (averaged over 5 trials). In *Max confidence* column, the tied class having the highest soft-max accumulations is selected. The last two columns use class priors for tie-breaking. *Class proportions* selects the tied class that appears most frequently among the training samples (i.e. having highest proportion) while *Class F1-scores* resolves ties by selecting the tied class which has higher prediction accuracy (metric chosen here is F1-score). The best absolute accuracy among different tie-breaking methods for each CNN model, is highlighted in bold.

Patches from Non-scaled Original Images. Patches of size 224×224 were extracted from Bark-101 data, without any re-sizing of the original images. The wide variation in the sizes of Bark-101 images resulted in a minimum of zero and a maximum of 9 patches possible per image. We kept all possible crops, resulting in a total of 3156 train, 1051 validation and 4164 test patches. Although the number of training samples is higher than that for training with whole images (Sect. 5.1), only a total of 1169 original training images (belonging to 96 classes) allowed at least one square patch of size 224×224. Thus, there is data loss as the images from which not even a single patch could be extracted were excluded. Results obtained by this strategy are listed in Table 3.

Table 3. Classification of patches from non-scaled original Bark-101.

CNN model	Patch-level accuracy (%)	Absolute accuracy (%) by majority voting			
		Random selection	Max confidence	Class proportions	Class F1-scores
Squeezenet	47.84	43.47	**44.09**	43.17	43.63
VGG16	47.48	44.40	**45.25**	44.32	44.09
MobilenetV2	41.83	37.61	**38.69**	36.68	37.22

Fig. 4. An example pair of upscaled images and sample patches from them. The left-most image has been upscaled by ESPCN and the right-most one by bicubic interpolation. Between them, sample extracted patches are shown where the differences between the two methods of upscaling become visible.

Table 4. Classification of patches from Bark-101 upscaled by bicubic interpolation.

CNN model	Patch-level accuracy (%)	Absolute accuracy (%) by majority voting			
		Random	Max confidence	Class proportions	Class F1-scores
Squeezenet	35.69	48.32	**48.57**	48.11	48.19
VGG16	41.04	57.21	56.99	**57.22**	57.14
MobilenetV2	33.36	43.73	43.60	**43.83**	43.60

Table 5. Classification of patches from Bark-101 upscaled by ESPCN.

CNN model	Patch-level accuracy (%)	Absolute accuracy (%) by majority voting			
		Random	Max confidence	Class proportions	Class F1-scores
Squeezenet	34.85	49.40	49.38	**49.45**	49.38
VGG16	39.27	**55.86**	55.75	55.76	55.75
MobilenetV2	32.12	**42.19**	41.78	41.93	41.85

We observe that the patch-level accuracy is higher than absolute accuracy, which can possibly be explained due to the data-loss. From the original test data, only 1168 images (belonging to 96 classes) had dimensions that allowed at least one single patch of 224 × 224, resulting in a total of 4164 test patches. Around 127 test images were excluded and by default, classified as incorrect, therefore reducing absolute accuracy. For patch-level accuracy, we reported how many of the 4164 test patches were correctly classified.

Patches from Upscaled Images. The previous sub-section highlights the need for upscaling original images, so that none is excluded from patch-extraction.

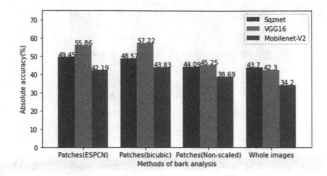

Fig. 5. Comparison of Bark-101 classification accuracy (absolute) using CNNs in this work. Best accuracy of 41.9% was obtained in the original work [24] using Late Statistics on LBP-like filters and SVM classifiers.

Here, we first upscaled all the original Bark-101 images by a factor of 4 and then extracted square patches of size 224×224 from them. Figure 4 shows an example pair of upscaled images and their corresponding patch samples.

We observe that among all our experiments, better absolute accuracy is obtained when patch-based CNN classification is performed on upscaled Bark-101 images and shows comparable performance between both methods of upscaling (bicubic or ESPCN). The best classifier performance in our study is **57.22%** from VGG16 fine-tuned by patches from Bark-101 upscaled by bicubic interpolation (Table 4). This is a promising improvement from both the original work [24] on Bark-101 (best accuracy of 41.9%) as well as the experiments using whole images (best accuracy of 43.7% by Squeezenet, from Table 2) (Fig. 5).

The comparison of tie-majority strategies shows that the differences are not substantial. This is because the variations can only be visible when many ties are encountered, which was not always the case for us. Table 6 lists the count of test images (whole) where ties were encountered. We observe that test images in the patch method with non-scaled original Bark-101, encounter 4–5 times more ties than when using upscaled images (bicubic or EPSCN). Our study thus corroborates that the differences among tie-breaking strategies are more considerable when several ties occur (Table 3), than when fewer ties are found (Tables 4 and 5). However, since the total number of test images in Bark-101 is 1295, the overall count of ties can still be considered low in our study. Nevertheless, we decided to include this comparison to demonstrate the difficulties of encountering ties in majority voting for patch-based CNN and investigate existing strategies to overcome this. It is interesting to observe (in Table 3) that for patches extracted from non-scaled original Bark-101 (where there is a higher number of ties), the best tie-breaking strategy is the *maximum confidence sum*, as affirmed in [21] where the authors had tested it on simpler datasets (having a maximum of 26 classes in the *Letter* dataset) taken from the UCI repository [14].

Table 6. Count of test images showing tied classes in majority voting.

Patch method	Squeezenet	VGG16	MobilenetV2
Non-Scaled Original	217	283	274
Upscaled by Bicubic	52	45	45
Upscaled by ESPCN	50	46	63

To summarise, we present few important insights. First, when the total count of training samples is low, patch-based image analysis can improve accuracy due to better learning of local information and also since the total count of training samples increases. Second, image re-scaling invariably introduces distortion and reduces the image quality, hence patches from upscaled images have a loss of feature information. As expected, patch-level accuracy is lower when using patches from upscaled images (Tables 4 and 5), compared to that of patches from non-scaled original images having more intact features (Table 3). However, we also observe that absolute accuracy falls sharply for patches taken from non-scaled original Bark-101. This is because several of the original images have such low image dimensions, that no patch formation was possible at all. Therefore, all such images (belonging to 5 classes, see Sect. 5.2 for details) were by default excluded from our consideration, resulting in low absolute accuracy across all the CNN models tested. Thus, we infer that for datasets having high diversity and variation of image dimensions, upscaling before patch-extraction can ensure better retention and representation of data. Finally, we also observe that it is useful to examine tie-breaking strategies in majority voting compared to relying on simple random selection. These strategies are particularly significant if a considerable number of ties are encountered.

6 Conclusion and Future Work

Our study demonstrates the potential of using deep learning for studying challenging datasets such as Bark-101. For a long time, bark recognition has been treated as a texture classification problem and traditionally solved using hand-crafted features and statistical analysis. A patch-based CNN classification approach can automate bark recognition greatly and reduce the efforts required by time-consuming traditional methods. Our study shows its effectiveness by outperforming accuracy on Bark-101 from traditional methods. An objective of our work was also to incorporate current trends in image re-scaling and ensemble-based classifiers in this bark analysis, to broaden perspectives in the plant vision community. Thus, we presented recent approaches in re-scaling by super-resolution networks and several tie-breaking strategies for majority voting and demonstrated their impact on performance. Super-resolution networks have promising characteristics to counter-balance the degradation introduced due to re-scaling. Although for our study with texture data as bark, its performance was comparable to traditional bicubic interpolation, we hope to investigate its effects on other plant data in future works. It would also be interesting to derive inspiration from patch-based image analysis in medical image segmentation where new label fusion methods are explored to integrate location information of patches for image-level decisions. In future works, we intend to accumulate new state-of-art methods and extend the proposed methodology to other plant organs and develop a multi-modal plant recognition tool for effectively identifying tree and shrub species. We will also examine its feasibility on mobile platforms, such as smart-phones, for use in real-world conditions.

Acknowledgements. This work has been conducted under the framework of the ReVeRIES project (Reconnaissance de Végétaux Récréative, Interactive et Educative sur Smartphone) supported by the French National Agency for Research with the reference ANR15-CE38-004-01.

References

1. Affouard, A., Goëau, H., Bonnet, P., Lombardo, J.C., Joly, A.: Pl@ntNet app in the era of deep learning. In: International Conference on Learning Representations (ICLR), Toulon, France (April 2017)
2. Barnea, E., Mairon, R., Ben-Shahar, O.: Colour-agnostic shape-based 3D fruit detection for crop harvesting robots. Biosyst. Eng. **146**, 57–70 (2016)
3. Begue, A., Kowlessur, V., Singh, U., Mahomoodally, F., Pudaruth, S.: Automatic recognition of medicinal plants using machine learning techniques. Int. J. Adv. Comput. Sci. Appl **8**(4), 166–175 (2017)
4. Bertrand, S., Ameur, R.B., Cerutti, G., Coquin, D., Valet, L., Tougne, L.: Bark and leaf fusion systems to improve automatic tree species recognition. Ecol. Inf. **46**, 57–73 (2018)
5. Bertrand, S., Cerutti, G., Tougne, L.: Bark recognition to improve leaf-based classification in didactic tree species identification. In: 12th International Conference on Computer Vision Theory and Applications, VISAPP 2017, Porto, Portugal (February 2017)
6. Boudra, S., Yahiaoui, I., Behloul, A.: A comparison of multi-scale local binary pattern variants for bark image retrieval. In: Battiato, S., Blanc-Talon, J., Gallo, G., Philips, W., Popescu, D., Scheunders, P. (eds.) ACIVS 2015. LNCS, vol. 9386, pp. 764–775. Springer, Cham (2015). https://doi.org/10.1007/978-3-319-25903-1_66
7. Boudra, S., Yahiaoui, I., Behloul, A.: Plant identification from bark: a texture description based on statistical macro binary pattern. In: 2018 24th International Conference on Pattern Recognition (ICPR), pp. 1530–1535. IEEE (2018)
8. Carpentier, M., Giguère, P., Gaudreault, J.: Tree species identification from bark images using convolutional neural networks. In: 2018 IEEE/RSJ International Conference on Intelligent Robots and Systems (IROS), pp. 1075–1081. IEEE (2018)
9. Cerutti, G., et al.: Late information fusion for multi-modality plant species identification. In: Conference and Labs of the Evaluation Forum, Valencia, Spain, pp. Working Notes (September 2013)
10. De Boor, C.: Bicubic spline interpolation. J. Math. Phys. **41**(1–4), 212–218 (1962)
11. Deng, J., Dong, W., Socher, R., Li, L.J., Li, K., Fei-Fei, L.: ImageNet: a large-scale hierarchical image database. In: 2009 IEEE Conference on Computer Vision and Pattern Recognition, pp. 248–255 (2009)

12. Dong, C., Loy, C.C., He, K., Tang, X.: Learning a deep convolutional network for image super-resolution. In: Fleet, D., Pajdla, T., Schiele, B., Tuytelaars, T. (eds.) ECCV 2014. LNCS, vol. 8692, pp. 184–199. Springer, Cham (2014). https://doi.org/10.1007/978-3-319-10593-2_13
13. Fiel, S., Sablatnig, R.: Automated identification of tree species from images of the bark, leaves or needles. In: 16th Computer Vision Winter Workshop, Mitterberg, Austria (February 2011)
14. Frank, A.: UCI machine learning repository (2010). http://archive.ics.uci.edu/ml
15. Ganschow, L., Thiele, T., Deckers, N., Reulke, R.: Classification of tree species on the basis of tree bark texture. Int. Arch. Photogram. Remote Sens. Spat. Inf. Sci. (ISPRS Arch.) **XLII–2/W13**, 1855–1859 (2019)
16. Goëau, H., et al.: LifeCLEF plant identification task 2014. In: Cappellato, L., Ferro, N., Halvey, M., Kraaij, W. (eds.) Conference and Labs of the Evaluation Forum (CLEF), Sheffield, United Kingdom, pp. 598–615. CEUR Workshop Proceedings (September 2014)
17. Hemming, J., Rath, T.: PA–precision agriculture: computer-vision-based weed identification under field conditions using controlled lighting. J. Agric. Eng. Res. **78**(3), 233–243 (2001)
18. Huang, Z.-K., Huang, D.-S., Du, J.-X., Quan, Z.-H., Guo, S.-B.: Bark classification based on Gabor filter features using RBPNN neural network. In: King, I., Wang, J., Chan, L.-W., Wang, D.L. (eds.) ICONIP 2006. LNCS, vol. 4233, pp. 80–87. Springer, Heidelberg (2006). https://doi.org/10.1007/11893257_9
19. Iandola, F.N., Han, S., Moskewicz, M.W., Ashraf, K., Dally, W.J., Keutzer, K.: SqueezeNet: AlexNet-level accuracy with 50x fewer parameters and <0.5 mb model size. arXiv preprint arXiv:1602.07360 (2016)
20. ImageCLEF: Plantclef 2017. https://www.imageclef.org/lifeclef/2017/plant. Accessed 15 Apr 2020
21. Kokkinos, Y., Margaritis, K.G.: Breaking ties of plurality voting in ensembles of distributed neural network classifiers using soft max accumulations. In: Iliadis, L., Maglogiannis, I., Papadopoulos, H. (eds.) AIAI 2014. IAICT, vol. 436, pp. 20–28. Springer, Heidelberg (2014). https://doi.org/10.1007/978-3-662-44654-6_2
22. Malmasi, S., Dras, M.: Native language identification using stacked generalization. arXiv preprint arXiv:1703.06541 (2017)
23. Mizoguchi, T., Ishii, A., Nakamura, H., Inoue, T., Takamatsu, H.: Lidar-based individual tree species classification using convolutional neural network. In: Videometrics, Range Imaging, and Applications XIV, vol. 10332, p. 1033200. International Society for Optics and Photonics (2017)
24. Ratajczak, R., Bertrand, S., Crispim Jr., C.F., Tougne, L.: Efficient bark recognition in the wild. In: International Conference on Computer Vision Theory and Applications, VISAPP 2019, Prague, Czech Republic (February 2019)
25. Rokach, L.: Ensemble-based classifiers. Artif. Intell. Rev. **33**(1–2), 1–39 (2010)
26. Sa Jr., J.J.M., Backes, A.R., Rossatto, D.R., Kolb, R.M., Bruno, O.M.: Measuring and analyzing color and texture information in anatomical leaf cross sections: an approach using computer vision to aid plant species identification. Botany **89**(7), 467–479 (2011)
27. Sandler, M., Howard, A., Zhu, M., Zhmoginov, A., Chen, L.C.: MobileNetV2: inverted residuals and linear bottlenecks. In: Proceedings of the IEEE Conference on Computer Vision and Pattern Recognition, pp. 4510–4520 (2018)
28. Shi, W., et al.: Real-time single image and video super-resolution using an efficient sub-pixel convolutional neural network. In: Proceedings of the IEEE Conference on Computer Vision and Pattern Recognition, pp. 1874–1883 (2016)

29. Simonyan, K., Zisserman, A.: Very deep convolutional networks for large-scale image recognition. In: Bengio, Y., LeCun, Y. (eds.) 3rd International Conference on Learning Representations, ICLR 2015, Conference Track Proceedings, San Diego, CA, USA, 7–9 May 2015 (2015)
30. Smolyanskiy, N., Kamenev, A., Smith, J., Birchfield, S.: Toward low-flying autonomous MAV trail navigation using deep neural networks for environmental awareness. In: 2017 IEEE/RSJ International Conference on Intelligent Robots and Systems (IROS), pp. 4241–4247. IEEE (2017)
31. Tian, H., Wang, T., Liu, Y., Qiao, X., Li, Y.: Computer vision technology in agricultural automation–a review. Inf. Process. Agric. **7**(1), 1–19 (2020)
32. Wan, Y.Y., et al.: Bark texture feature extraction based on statistical texture analysis. In: Proceedings of 2004 International Symposium on Intelligent Multimedia, Video and Speech Processing, pp. 482–485. IEEE (2004)
33. Woods, K., Kegelmeyer, W.P., Bowyer, K.: Combination of multiple classifiers using local accuracy estimates. IEEE Trans. Pattern Anal. Mach. Intell. **19**(4), 405–410 (1997)
34. Xu, L., Krzyzak, A., Suen, C.Y.: Methods of combining multiple classifiers and their applications to handwriting recognition. IEEE Trans. Syst. Man Cybern. **22**(3), 418–435 (1992)

Improving Pixel Embedding Learning Through Intermediate Distance Regression Supervision for Instance Segmentation

Yuli Wu, Long Chen$^{(\boxtimes)}$, and Dorit Merhof

Institute of Imaging and Computer Vision, RWTH Aachen, Aachen, Germany
yuli.wu@rwth-aachen.de, {long.chen,dorit.merhof}@lfb.rwth-aachen.de

Abstract. As a proposal-free approach, instance segmentation through pixel embedding learning and clustering is gaining more emphasis. Compared with bounding box refinement approaches, such as Mask R-CNN, it has potential advantages in handling complex shapes and dense objects. In this work, we propose a simple, yet highly effective, architecture for object-aware embedding learning. A distance regression module is incorporated into our architecture to generate seeds for fast clustering. At the same time, we show that the features learned by the distance regression module are able to promote the accuracy of learned object-aware embeddings significantly. By simply concatenating features of the distance regression module to the images as inputs of the embedding module, the mSBD scores on the CVPPP Leaf Segmentation Challenge can be further improved by more than 8% compared to the identical setup without concatenation, yielding the best overall result amongst the leaderboard at CodaLab.

Keywords: Instance segmentation · Pixel embedding · Distance regression

1 Introduction

Instance segmentation aims to label each individual object, which is critical to many biological and medical applications, such as plant phenotyping and cell quantification. Learning object-aware pixel embeddings is one of the trends in the field of instance segmentation. The embedding is essentially a high-dimensional representation of each pixel. To achieve instance segmentation, pixel embeddings of the same object should be located relatively close in the learned embedding space, while those of different objects should be discriminable.

The loss usually consists of two terms: the between-instance loss term $\mathcal{L}_{inter}$ and the within-instance loss term $\mathcal{L}_{intra}$. The former term $\mathcal{L}_{inter}$ encourages different-instance embeddings to be located far away from each other, while the latter term $\mathcal{L}_{intra}$ encourages same-instance embeddings to stay together.

© Springer Nature Switzerland AG 2020
A. Bartoli and A. Fusiello (Eds.): ECCV 2020 Workshops, LNCS 12540, pp. 213–227, 2020.
https://doi.org/10.1007/978-3-030-65414-6_16

Two most popular metrics used to describe the similarities of embeddings are Euclidean distance and cosine distance. Although the pixel embedding approaches have gained success in many datasets including CVPPP Leaf Segmentation Challenge [4,5,12,16], the trained embedding space is far from optimal.

Our idea was indirectly inspired by the "easy task first" concept behind curriculum learning [1]. Distance regression predicts the distance from a pixel to the object boundary and is used in [4,20], for example, as an auxiliary module. We have empirically found that the distance regression module is relatively easy to train on many datasets. Considering that the learned features by the distance regression module should be already recognizable for distinguishing instances, we prefix the embedding module with a distance regression module to promote the embedding learning process.

The main contributions of this paper are summarized as follows:

1. We propose an architecture to promote the pixel embedding learning by utilizing features learned from the distance regression module, which significantly improves the performance in the CVPPP Leaf Segmentation Challenge [19]. Our overall mean Symmetric Best Dice (mSBD) score is at the top position of the leaderboard with 0.879 by paper submission. Furthermore, the average of mSBD scores on Arabidopsis images (testing sets A1, A2, A4) outperforms the second best results from three different teams by over 3%, namely from 0.883 to 0.917;
2. We conduct a number of ablation experiments in terms of the stacked U-Net architecture, different types of concatenative layers and varied loss formats, to validate our architecture and also supplement some experimental vacancies in this field.

2   Related Work

We roughly categorize approaches of instance segmentation into two groups with respect to the overall pipeline: *instance-first* approaches and *one-stage* approaches. *Instance-first* approaches exploit the instance-level bounding boxes from the first-stage object detector. For example, Mask R-CNN [7] uses RPN [18], and recent methods like BlenderMask [3] and CenterMask [10] are based on the anchor-free detector FCOS [21]. Pixel-level segmentations are then produced through subjoined refinement modules. Mask R-CNN [7] constructs a lightweight segmentation network with consecutive convolutional layers, while the Blender Module and Spatial Attention-Guided Mask (SAG-Mask) are proposed in [3,10], respectively, for a more accurate segmentation.

In contrast, *one-stage* approaches predict the existence (*object-ness*) and mask of objects all at once. Masks are represented as polar coordinates in [20,25]. Specifically, the model regresses the distances to the boundary along a set of fixed directions at each location. To describe more complex shapes, masks are encoded with a linear projection in [27].

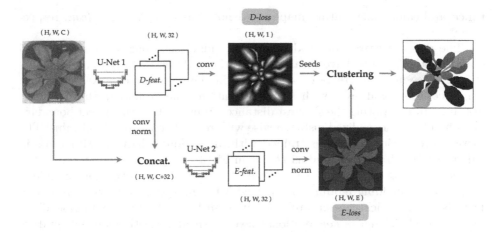

Fig. 1. Processing Pipeline. Distance regression features and distmaps are learned via distance module with U-Net 1. Concatenated distance regression features and images are fed into U-Net 2, from which the embeddings are learned. Final labels are generated based on seeds (thresholded maxima of distmaps) and embeddings via angular clustering. Denotations: H, W, C, E = Dimensions of Height, Width, Channel, Embedding.

Furthermore, the approaches based on pixel embedding learning, which also belong to *one-stage* approaches, are becoming a new trend. They share the general pipeline of *embedding and clustering*. Each pixel of input images is mapped to a high-dimensional vector (embedding), in which pixels of the same object are located closely. Then, clustering in the embedding space results in the final instance segmentation. De Brabandere and Neven [5,12] have proposed Euclidean distance based embedding loss for instance segmentation. Payer et al. [16] have demonstrated embedding loss which utilizes cosine similarity and recurrent stacked hourglass network [13]. Chen et al. [4] have introduced a U-Net based architecture of two heads, where the embeddings are trained with cosine embedding loss and local constraints. These two heads are distance regression head and embedding head. The distance regression head aims to provide seed candidates for clustering. Our proposed method inherits the fundamental modules from this work.

For current pixel embedding based approaches, clustering is an essential step. Mean Shift [6] and HDBSCAN [2] are used in [5,16] respectively. In [4,12], threshold based clustering is used with knowledge of the learned seeds.

3   Method

Our network consists of two cascaded parts (Fig. 1): the distance regression module and the embedding module. Each module uses a U-Net architecture with a 32-dimensional output feature map as the backbone network. The learned distance and embedding feature maps are denoted as *D-feat.* and *E-feat.*, respectively.

The distance regression module takes standardized images (by shifting and scaling each image to have mean 0 and variance 1) as the inputs and outputs the distance map (abbreviated as *distmap* in the following context) through a single convolutional layer with ReLU activation. The ground truth distmap is obtained by computing the shortest distances from pixels to the object boundary and then being normalized instance-wise with respect to the maximal value. The distance regression module is trained with Mean Squared Error (MSE) loss in this work, which is illustrated as *D-loss* in Fig. 1.

Distance feature map *D-feat.* learned by the distance regression module is fed to the embedding module together with the input image by concatenation. Details of the concatenation are introduced in Sect. 3.2. The final embeddings are obtained through a convolutional layer with linear activation, followed by L2 normalization. The embedding module is trained with the loss based on the cosine similarity and local constraints (Sect. 3.1), denoted as *E-loss* in Fig. 1.

The embedding space trained with loss in Eq. 1 has a comprehensive geometric interpretation: embedding vectors of neighboring objects tend to be orthogonal, which simplifies the complexity of clustering. The fast *angular clustering* can be effortlessly performed based on angles between embedding vectors. Firstly, seeds are obtained from distmaps by fetc.hing local maxima with a trivial threshold (selected as 70% of the global maximum in an image). After that, all neighboring pixels within the angular range δ_a of a seed are collected to form a cluster. In this work, we use $\delta_a = 45°$ for all experiments. At last, the labels outside of the officially provided ground truth foreground masks are omitted.

3.1   Cosine Embedding Loss with Local Constraints

For the embedding module training, we build upon the loss format from [4]. The training loss, denoted as *E-loss* in Fig. 1, is defined based on the cosine similarity $\mathcal{S}_{cos}(\mathbf{e_1}, \mathbf{e_2}) = \mathbf{e_1}^T \mathbf{e_2} / (\|\mathbf{e_1}\| \|\mathbf{e_2}\|)$ and is formularized as:

$$\mathcal{L}_{emb} = \lambda \cdot \mathcal{L}_{inter} + \mathcal{L}_{intra}$$

$$\mathcal{L}_{inter} = \frac{1}{C} \sum_{c_A=1}^{C} \frac{1}{|\mathbf{N}_{c_A}|} \sum_{c_B \in \mathbf{N}_{c_A}} \mathcal{S}_{cos}(\boldsymbol{\mu}_{c_A}, \boldsymbol{\mu}_{c_B})$$

$$\mathcal{L}_{intra} = \frac{1}{C} \sum_{c=1}^{C} \frac{1}{E_c} \sum_{i=1}^{E_c} \left[1 - \mathcal{S}_{cos}(\boldsymbol{e}_i, \boldsymbol{\mu}_c) \right],$$

(1)

where the embedding loss is defined as the weighted sum of the between-instance loss term $\mathcal{L}_{inter}$ and within-instance loss term $\mathcal{L}_{intra}$ with the weighting factor λ. $\boldsymbol{e}$ and $\boldsymbol{\mu}$ represents the pixel embedding vector and the mean embedding of an object, respectively. C denotes the number of objects, while the number of pixels of a single object c is denoted as E_c. $\mathbf{N}_{c_A}$ represents the set of neighboring objects around the object c_A and $|\mathbf{N}_{c_A}|$ is the number of neighbors.

The between-instance loss term $\mathcal{L}_{inter}$ encourages the embeddings of different object to be separated, while the within-instance loss term $\mathcal{L}_{intra}$ punishes the case where pixel embeddings of the same object diverge from the mean. In addition, the local constraints of this loss only force neighboring objects to form separable clusters in the embedding space. The benefits of local constraints and the comparison with the global constraint are demonstrated in Sect. 4.3.

3.2 Feature Concatenative Layer

The feature map *D-feat.* learned by the distance regression module is firstly transformed to the desired dimensions (shown with an example of 32 in Fig. 1) via a convolutional layer and then L2 normalized pixel-wise along through the feature channels before being concatenated to the images. Our experiment shows that the feature map normalization is critical to a stable training process.

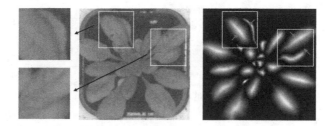

Fig. 2. Ambiguity between Leaf Boundary and Leaf Midvein. Although the embedding space learned with U-Net often fails at such locations, distmap (right) is able to distinguish them well: lower values (darker areas) indicate boundaries and higher values (brighter areas) indicate midveins.

As illustrated in Fig. 2, the difference between leaf boundary and leaf midvein (primary vein) is ambiguous. The learned embeddings by the U-Net architecture [4] often fail at those locations. However, the distmaps are able to tell the difference with lower values representing leaf boundaries and higher values representing leaf midveins. From another perspective, the distmap, which gives an approximate outline of objects, can be interpreted as a *object-ness* score, the pixel-wise probability about existence of object. In addition, as proposed by [14], mixing convolutional operations with the pixel location helps constructing dense pixel embeddings that can separate object instances. From this perspective, the distance regression features can indirectly provide location information to the subsequent module.

To this end, we construct a two-stage architecture, as depicted in Fig. 1, by forwarding the distance regression features to the embedding module. And the concatenation of the distance regression features and images can bring in best performance in the experiments. We term the distance features as concatenative layer in between the stacked U-Nets as *intermediate distance regression supervision*.

In the experiments, other different features have also been tested to forward: the 1-dimensional distmap, 8-dimensional distance features, 32-dimensional distance features, 32-dimensional embedding features, concatenated 16-dimensional distance features and 16-dimensional embedding features. Inspired by [12,14], we have also evaluated the performance of augmenting the input image with x- and y-coordinates.

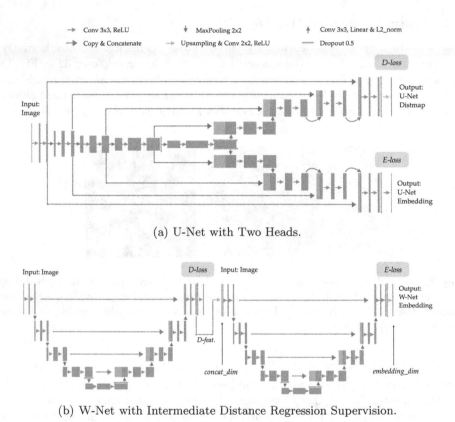

(a) U-Net with Two Heads.

(b) W-Net with Intermediate Distance Regression Supervision.

Fig. 3. Network Architectures of U-Net and W-Net.

3.3 From U-Net to W-Net

We abbreviate the proposed network as W-Net to differ from the existing U-Net with two heads, although the novelty and characteristic are not fully represented: the distance regression features as intermediate supervision and the cosine embedding loss with local constraints.

In Fig. 3, the detailed architectures of U-Net with two heads and W-Net with intermediate distance regression supervision are illustrated. The parallel distance and embedding heads of U-Net (Fig. 3a) are modified towards the serial distance and embedding modules in W-Net (Fig. 3b). Apart from the types of concatenative layer as discussed previously, we have also investigated the final dimensions of embeddings as another hyper-parameter, denoted as *embedding_dim* in Fig. 3b. The corresponding ablation experiments can be found in Sect. 4.4.

4   Experiments

Ablation experiments are conducted with U-Net and W-Net, as depicted in Fig. 3. The training loss is the sum of the distance regression loss (ReLU + MSE) and the cosine embedding loss with local constraints (Eq. 1). The latest CodaLab dataset of CVPPP2017 LSC is used as training set without augmentation. Model parameters are initialized by He Normal [8] and optimized by Adam [9]. The initial learning rate is set to 0.0001 and scheduled with exponential decay, with the decay period being set to 5000 steps and the decay rate 0.9. The batch size is set to 4 in most experiments, or 2 if high embedding dimensions are used. The maximal training epochs are set to 500. We show mSBD scores of testing set from CodaLab as the evaluation metric.

Fig. 4. Learned Embeddings with U-Net and W-Net. Numbered leaves are treated as one object by U-Net, while they are successfully separated in the embedding space learned with W-Net.

4.1   U-Net vs. W-Net

Firstly, we illustrate the performance improvement from U-Net with two heads to the proposed W-Net. In Fig. 4, two representative cases are demonstrated, where the U-Net fails to separate closely located leaves. In contrast, the W-Net has successfully distinguished the numbered leaves in Fig. 4.

Quantitatively, W-Net surpasses U-Net on overall mSBD by approximately 8% from 0.794 to 0.879 with the best set-ups for W-Net, as shown in Table 1. Under different settings of embedding dimensions (Fig. 6a) and loss weights (Fig. 6b), the performance gap between U-Net and W-Net can be continuously observed and remain about 8%.

Table 1. Comparison of Different Types of Concatenative Layers. Denotation: dfeat.16 + efeat.16 = concatenated distance features of 16 dim and embedding features of 16 dim. Others can be analogously educed.

Concatenative layer	Net	mSBD
none (baseline)	U-Net	.794
coordinate	U-Net	.798
distmap	W-Net	.824
dfeat.8	W-Net	.864
dfeat.32	W-Net	**.879**
efeat.32	W-Net	.847
dfeat.16 + efeat.16	W-Net	.873

Table 2. Comparison of Local/Global Constraints, Network and Clustering. Denotations: Local = local constraints, otherwise global; 64d = 64 dim embeddings, otherwise 8 dim; AC = Angular clustering; MWS = Mutex Watershed [24].

Local	Net	Clustering	mSBD
✓	W-Net	AC	**.879**
✓	W-Net 64d	AC	.854
	W-Net	AC	.835
	W-Net 64d	AC	.823
✓	U-Net	MWS	.719
✓	W-Net	MWS	.771
✓	U-Net	MeanShift	.679
✓	W-Net	MeanShift	.733
✓	U-Net	HDBSCAN	.631
✓	W-Net	HDBSCAN	.681

4.2   Concatenative Layer

We compare the effects of different types of concatenative layer. Firstly, the distmap (1-dimensional) can be directly forwarded. Alternatively, the distance regression features instead of the distmap can be utilized. Before concatenation, we convert the 32-channel *D-feat.* into 8 and 32 dimensions (denoted as *dfeat.8* and *dfeat.32* in Table 1) through a single convolutional layer.

Meanwhile, the case of using embedding loss as the intermediate supervision (*efeat.32*) has also been tested. Specifically, the embedding features from the first U-Net are concatenated with the images as the inputs of the second embedding module. Furthermore, the concatenated distance regression features and embedding features (*dfeat.16 + efeat.16*) are also investigated. At last, augmenting the input image with coordinates is tested. As proposed in [14], constructing dense object-aware pixel embeddings cannot be easily achieved using convolutions and the situation can be improved by incorporating information about the pixel location. In this work, we augment the input image with two coordinate channels for the normalized x- and y-coordinates, respectively.

Experimental results are summarized in Table 1. First of all, forwarding distmaps is not as effective as forwarding feature maps, including the distance regression features and the embedding features. The embedding features (*efeat.32*) can also boost the performance, but not as significantly as the distance regression features. This is verified by the fact that *efeat.32* is worse than *dfeat.32* and the mixed feature map *dfeat.16 + efeat.16*. For the distance regression feature itself, higher dimensions of 32 are preferred. Finally, augmenting images with coordinates does not show apparent differences in our experiments. The effects could be further studied. For example, augmenting each intermediate feature map with coordinates is also worth being investigated.

4.3 Local vs. Global Constraints

Local constraints make it possible to exploit lower-dimensional embedding space more efficiently, as in this case, different labels only have to be distributed to the neighboring objects. In contrast, the global constraints have to thoroughly give each single object in the images a different label, which requires larger receptive fields and more redundant embedding space. The combination of local constraints and cosine embeddings utilizes the embedding space further comprehensively, as the push force imposed by loss expects orthogonal embedding clusters for neighboring instances.

This is confirmed qualitatively by examples showcased in Fig. 5. In Fig. 5c, 8-dimensional embeddings are trained with global constraints. Not surprisingly, there are exactly 8 colors in the image, indicating 8 orthogonal clusters in the embedding space. Apparently, the global constraint will fail when the embedding dimensions are fewer than the number of objects. In contrast, the local constraints (Fig. 5a–5b) can distribute labels alternately between objects, with the same labels appearing multiple times for non-adjacent objects. This makes it possible to utilize a lower-dimensional embedding space. Quantitatively, the W-Net trained with local constraints surpasses the one trained with global constraints by more than 4% on overall mSBD, as listed in Table 2.

Intuitively, a higher-dimensional embedding space is able to provide a higher degree of freedom, i.e. we could simply use higher-dimensional embeddings to alleviate the problem of global constraints. At least the embedding vector does not have to be restricted to low dimensions. However, from the results in Fig. 6a,

| (a) local 8 | (b) local 64 | (c) global 8 | (d) global 64 |

Fig. 5. Learned Embeddings for Combined Cases of Local/Global Constraints and 8/64-dimensional Embeddings. (a,b) vs. (c,d): Local constraints ensure the effective utilization of embedding space, as same embeddings appear alternately for non-adjacent objects. (a) vs. (b): Higher-dimensional embeddings are redundant in the local constraint case. (c) vs. (d): Lower-dimensional embeddings with global constraints are not sufficient to distinguish all objects. This problem is slightly mitigated via higher-dimensional embeddings, still not as effective as local constraints.

we find that higher-dimensional embeddings produce worse results. This makes the capability of using lower-dimensional embedding space particularly important.

4.4   Dimensions of Embeddings

As discussed previously, the local constraints make the use of lower-dimensional embedding possible. It is thus worth investigating the influence of different embedding dimensions on the overall performance. The mSBD scores of both U-Net and W-Net for {4, 8, 16, 32, 64}-dimensional embeddings are plotted in Fig. 6a. For 32 and 64 dimensions, the batch size is set to 2, instead of 4 as in other cases, to fit the memory of a single GPU.

Our experiments show that the 8-dimensional embedding brings in the best result. First of all, merely 4 dimensions are incompetent to separate all adjacent objects, since it is common that one object has more than 4 neighbors. Although higher dimensions may not bring in more labels under local constraints, comparing Fig. 5a to 5b, increasing the embedding dimensions should not degrade the performance hypothetically. However, the mSBD score decreases slightly as the dimensions increase. Therefore we believe, under the premise that the dimensions are sufficient for all objects to fulfill the local constraints, higher-dimensional embedding space is more difficult to train.

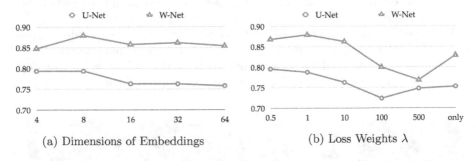

(a) Dimensions of Embeddings   (b) Loss Weights λ

Fig. 6. mSBD w.r.t. Dimensions of Embeddings and Loss Weight λ in $\mathcal{L}_{emb} = \lambda \cdot \mathcal{L}_{inter} + \mathcal{L}_{intra}$ using U-Net and W-Net. In (b), *only* denotes $\mathcal{L}_{emb} = \mathcal{L}_{inter}$. W-Net surpasses U-Net generally. Best overall performance of W-Net can be obtained with 8-dimensional embeddings and $\lambda = 1$.

4.5   Loss Weights

During the experiments, we find that the values of between-instance loss term $\mathcal{L}_{inter}$ are approximately 10 times greater than the values of within-instance loss term $\mathcal{L}_{intra}$. This is consistent with the fact that pixel embeddings of the same object converge tightly, but adjacent objects are not correctly segmented occasionally. The larger weighting factor λ of between-instance loss term $\mathcal{L}_{inter}$

might be helpful to emphasize the significance of it by amplification of its gradient. We set λ as $\{0.5, 1, 10, 100, 500\}$, and moreover, we omit the within-instance loss, denoted as *only* in Fig. 6b. The experiments are preformed for both U-Net and W-Net under identical main set-ups: 32-dimensional distance features as concatenative layer, local constraints and 8-dimensional embeddings.

From the experiments, we find that larger weighting factor of the between-instance loss term does not further help to encourage the network to separate the confused objects when λ is larger than 1, but reduces the consistency of embeddings in the same object. Figure 7 showcases the trade-off between the discrimination of adjacent objects (larger λ) and the consistency of individual object (smaller λ). The experiments show that $\lambda = 1$ brings in best overall performance, as shown in Fig. 6b. Besides, one surprising conclusion is that training the network with just the between-instance loss term can also, to some extent, form clusters in the embedding space (Fig. 7d).

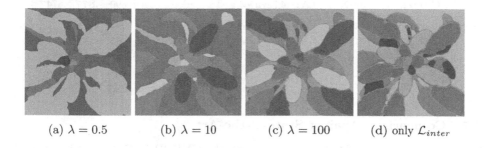

(a) $\lambda = 0.5$ (b) $\lambda = 10$ (c) $\lambda = 100$ (d) only $\mathcal{L}_{inter}$

Fig. 7. Learned Embeddings with Different Weights λ as in $\mathcal{L}_{emb} = \lambda \cdot \mathcal{L}_{inter} + \mathcal{L}_{intra}$. With ascending λ, overall segmentation performance becomes worse (Fig. 6) with the decreased consistency of embeddings in the same object. It is worth noting that training with just the between-instance loss term can also to some extent form clusters in the embedding space.

4.6   Clustering

Apart from the default angular clustering used along through the experiments, other three clustering techniques have been tested based on the predicted embeddings of the best results: Mutex Watershed [24], Mean Shift [6] and HDB-SCAN [2]. On the one hand, this provides a reference for the performance of different clustering methods on the embeddings trained with cosine similarity based loss. On the other hand, it can also indirectly reflect the quality of embeddings generated by U-Net and W-Net. Results are shown in Table 2.

In conclusion, the angular clustering has advantages in terms of performance and speed. Nevertheless, it should be noted that this method is only applicable to the case, where seeds are available and clusters are orthogonal in the embedding space. Additionally, all clustering approaches produce better results with embeddings predicted from the W-Net, which again confirms the improvement of our proposed method.

Table 3. Comparison of results. Abbreviations: Aug. = Data augmentation; Emb. = Metric of embedding similarity; Fg. = Ground truth foreground masks are used; syn = Synthetic images are used for training; HG = Stacked Hourglass network; Lb. = Results shown in the leaderboard of CodaLab.

Method	Backbone	Train	Aug.	Emb.	Fg.	Lb.	mSBD		
							A1	A1-3	A1-5
IPK [15,19]	-	A1−3				✓	.791	.782	-
Nottingham [19]	-	A1−3				✓	.710	.686	-
MSU [19,26]	-	A1−3				✓	.785	.780	-
Wageningen [19]	-	A1−3				✓	.773	.769	-
MRCNN [4,7]	ResNet	A1−3					-	.797	-
Stardist [4,20]	U-Net	A1−3					-	.802	-
IS-RA [17]	FCN	A1					.849	-	-
Ward [23]	ResNet	A1−4+syn	✓				.900	.740	.810
UPGen [22]	ResNet	A1−4+syn	✓				.890	**.877**	.874
DiscLoss [5]	ResNet	A1	✓	euc	✓		.842	-	-
CE-RH [16]	HG	A1	✓	cos			.845	-	-
E-LC [4]	U-Net	A1−3		cos			-	.831	.823
W-Net (ours)	U-Net	A1−4		cos	✓	✓	**.919**	.870	**.879**

4.7 Comparison Against State-of-the-Art

Comparison of state-of-the-art methods on the CVPPP LSC dataset is quantitatively shown in Table 3. It is clear that the learning based methods (denoted with backbones) can achieve better results than the first four classical methods. The last four methods are based on pixel embedding learning. Roughly speaking, they bring in promising results. Our overall result mSBD for A1-5 outperforms all others. In the leaderboard, our overall result is at the 1. Position by paper submission. Furthermore, the average of mSBD scores for Arabidopsis images (A1, A2, A4) outperforms the second best results from three different users, respectively, by over 3%, namely 0.883 to 0.917. Due to the extremely imbalanced training images on Arabidopsis (783 images) and Tobacco (27 images), our result on testing set A3 are not as good as others, with mSBD of 0.77. Compared to this, the current 1. Place mSBD of A3 in the leaderboard reaches 0.89. It implies that the sufficient number of training images is critical in our proposed method. We leave this room for improvement in the future. One thing worth mentioning is that the authors tend to not submit their results to the leaderboard of CodaLab, which makes the consistent comparison and review rather difficult.

4.8 Application to Human U2OS Cells

Our method has also been tested on the image set BBBC006v1 of human U2OS cells from the Broad Bioimage Benchmark Collection [11]. Totally 754 images are randomly separated into two equally distributed training and testing set with 377 images respectively. Other set-ups are identical to previously introduced ones. We use U-Net and W-Net with distance concatenative layer to show results in mSBD and mean Average Precision with IoU = {0.5, 0.55, 0.6, ..., 0.9} (mAP). The mSBD has increased from 0.896 to 0.915 and the mAP from 0.577 to 0.664. We showcase two examples of final labels in Fig. 8. As reported in [4], some embeddings around boundaries might be incomplete, which leads to incomplete segmentations. This problem has been mainly solved, as showcased in Fig. 8.

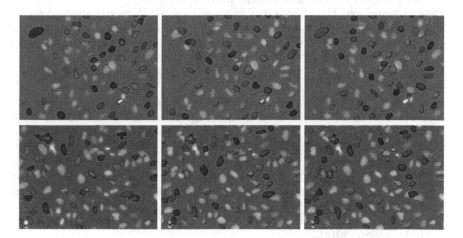

Fig. 8. Cell Segmentation Results of the BBBC006v1 Data: Ground Truth (left), U-Net (middle) and W-Net (right); Each row illustrates one example. The improvement from U-Net to W-Net is salient. The mSBD and mAP scores have increased from 0.896 to 0.915 and from 0.577 to 0.664, respectively.

5   Conclusion

In this work we propose a novel W-Net, which forwards the distance regression features learned by the first-stage U-Net to the subsequent embedding learning module. The intermediate distance regression supervision effectively promotes the accuracy of learned pixel embedding space, with the mSBD score on the CVPPP LSC dataset increased by more than 8% compared to the identical set-up without supervision of distance regression features. We have also conducted a number of experiments to investigate the characteristics of the pixel embedding learning with the cosine similarity based loss, involving the embedding dimensions, the weighting factor of the within-instance loss term and the between-instance loss term. We are looking forward to applying this method to more datasets in the future.

Acknowledgments. This work was supported by the German Research Foundation (DFG) Research Training Group 2416 MultiSenses-MultiScales.

References

1. Bengio, Y., Louradour, J., Collobert, R., Weston, J.: Curriculum learning. In: Annual International Conference on Machine Learning, pp. 41–48 (2009)
2. Campello, R.J., Moulavi, D., Zimek, A., Sander, J.: Hierarchical density estimates for data clustering, visualization, and outlier detection. ACM Trans. Knowl. Discov. Data (TKDD) **10**(1), 1–51 (2015)
3. Chen, H., Sun, K., Tian, Z., Shen, C., Huang, Y., Yan, Y.: BlendMask: top-down meets bottom-up for instance segmentation. In: Proceedings of the IEEE Conference on Computer Vision and Pattern Recognition (2020)
4. Chen, L., Strauch, M., Merhof, D.: Instance segmentation of biomedical images with an object-aware embedding learned with local constraints. In: Shen, D., et al. (eds.) MICCAI 2019. LNCS, vol. 11764, pp. 451–459. Springer, Cham (2019). https://doi.org/10.1007/978-3-030-32239-7_50
5. De Brabandere, B., Neven, D., Van Gool, L.: Semantic instance segmentation with a discriminative loss function. arXiv preprint arXiv:1708.02551 (2017)
6. Fukunaga, K., Hostetler, L.: The estimation of the gradient of a density function, with applications in pattern recognition. IEEE Trans. Inf. Theory **21**(1), 32–40 (1975)
7. He, K., Gkioxari, G., Dollár, P., Girshick, R.: Mask R-CNN. In: Proceedings of the IEEE International Conference on Computer Vision, pp. 2961–2969 (2017)
8. He, K., Zhang, X., Ren, S., Sun, J.: Delving deep into rectifiers: Surpassing human-level performance on imagenet classification. In: The IEEE International Conference on Computer Vision (ICCV), December 2015
9. Kingma, D.P., Ba, J.: Adam: a method for stochastic optimization. arXiv preprint arXiv:1412.6980 (2014)
10. Lee, Y., Park, J.: Centermask: real-time anchor-free instance segmentation. In: Proceedings of the IEEE Conference on Computer Vision and Pattern Recognition (2020)
11. Ljosa, V., Sokolnicki, K.L., Carpenter, A.E.: Annotated high-throughput microscopy image sets for validation. Nat. Method **9**(7), 637–637 (2012)
12. Neven, D., Brabandere, B.D., Proesmans, M., Gool, L.V.: Instance segmentation by jointly optimizing spatial embeddings and clustering bandwidth. In: Proceedings of the IEEE Conference on Computer Vision and Pattern Recognition, pp. 8837–8845 (2019)
13. Newell, A., Yang, K., Deng, J.: Stacked hourglass networks for human pose estimation. In: Leibe, B., Matas, J., Sebe, N., Welling, M. (eds.) ECCV 2016. LNCS, vol. 9912, pp. 483–499. Springer, Cham (2016). https://doi.org/10.1007/978-3-319-46484-8_29
14. Novotny, D., Albanie, S., Larlus, D., Vedaldi, A.: Semi-convolutional operators for instance segmentation. In: Ferrari, V., Hebert, M., Sminchisescu, C., Weiss, Y. (eds.) ECCV 2018. LNCS, vol. 11205, pp. 89–105. Springer, Cham (2018). https://doi.org/10.1007/978-3-030-01246-5_6
15. Pape, J.-M., Klukas, C.: 3-D histogram-based segmentation and leaf detection for rosette plants. In: Agapito, L., Bronstein, M.M., Rother, C. (eds.) ECCV 2014. LNCS, vol. 8928, pp. 61–74. Springer, Cham (2015). https://doi.org/10.1007/978-3-319-16220-1_5

16. Payer, C., Štern, D., Neff, T., Bischof, H., Urschler, M.: Instance segmentation and tracking with cosine embeddings and recurrent hourglass networks. In: Frangi, A.F., Schnabel, J.A., Davatzikos, C., Alberola-López, C., Fichtinger, G. (eds.) MICCAI 2018. LNCS, vol. 11071, pp. 3–11. Springer, Cham (2018). https://doi.org/10.1007/978-3-030-00934-2_1
17. Ren, M., Zemel, R.S.: End-to-end instance segmentation with recurrent attention. In: Proceedings of the IEEE Conference on Computer Vision and Pattern Recognition, pp. 6656–6664 (2017)
18. Ren, S., He, K., Girshick, R., Sun, J.: Faster R-CNN: towards real-time object detection with region proposal networks. In: Advances in Neural Information Processing Systems, pp. 91–99 (2015)
19. Scharr, H., et al.: Leaf segmentation in plant phenotyping: a collation study. Mach. Vis. Appl. **27**(4), 585–606 (2015). https://doi.org/10.1007/s00138-015-0737-3
20. Schmidt, U., Weigert, M., Broaddus, C., Myers, G.: Cell detection with star-convex polygons. In: Frangi, A.F., Schnabel, J.A., Davatzikos, C., Alberola-López, C., Fichtinger, G. (eds.) MICCAI 2018. LNCS, vol. 11071, pp. 265–273. Springer, Cham (2018). https://doi.org/10.1007/978-3-030-00934-2_30
21. Tian, Z., Shen, C., Chen, H., He, T.: FCOS: fully convolutional one-stage object detection. In: Proceedings of the IEEE International Conference on Computer Vision, pp. 9627–9636 (2019)
22. Ward, D., Moghadam, P.: Scalable learning for bridging the species gap in image-based plant phenotyping. arXiv preprint arXiv:2003.10757 (2020)
23. Ward, D., Moghadam, P., Hudson, N.: Deep leaf segmentation using synthetic data. arXiv preprint arXiv:1807.10931 (2018)
24. Wolf, S., et al.: The mutex watershed: efficient, parameter-free image partitioning. In: Ferrari, V., Hebert, M., Sminchisescu, C., Weiss, Y. (eds.) ECCV 2018. LNCS, vol. 11208, pp. 571–587. Springer, Cham (2018). https://doi.org/10.1007/978-3-030-01225-0_34
25. Xie, E., et al.: Polarmask: single shot instance segmentation with polar representation. In: Proceedings of the IEEE Conference on Computer Vision and Pattern Recognition (2020)
26. Yin, X., Liu, X., Chen, J., Kramer, D.M.: Multi-leaf tracking from fluorescence plant videos. In: 2014 IEEE International Conference on Image Processing (ICIP), pp. 408–412. IEEE (2014)
27. Zhang, R., Tian, Z., Shen, C., You, M., Yan, Y.: Mask encoding for single shot instance segmentation. In: Proceedings of the IEEE Conference on Computer Vision and Pattern Recognition (2020)

Time Series Modeling for Phenotypic Prediction and Phenotype-Genotype Mapping Using Neural Networks

Sruti Das Choudhury$^{(\boxtimes)}$ (iD)

School of Natural Resources and Department of Computer Science & Engineering,
University of Nebraska-Lincoln, Lincoln, NE, USA
S.D.Choudhury@unl.edu

Abstract. Image-based high throughput plant phenotyping refers to the process of computing phenotypes non-destructively by analyzing images of plants captured at regular time intervals. The non-invasive measurements of phenotypes at multiple timestamps during a plant's life cycle provides the motivation to extend the application of time series modeling in the field of phenomic research to (1) predict phenotypes for missing imaging days or for a time in the future based on analyzing past measurements; (2) predict a derived or composite phenotype from its one or more constituents and (3) bridge the phenotype-genotype gap to contribute in the study of improved crop breeding and understanding the genetic regulation of temporal variation of phenotypes. The paper uses long short-term memory, a variant of recurrent neural networks, for phenotype-genotype mapping, while autoregressive neural networks, autoregressive neural network with exogenous input and non-linear input output neural networks are used for phenotypic prediction. The experimental analyses on the benchmark dataset called Phenoseries dataset show the efficacy and future prospects of this foundational study.

Keywords: Time series modeling · Neural networks · Phenotypic prediction · Phenotype-genotype mapping · Long short-term memory

1 Introduction

A time series is an ordered sequence of values of a variable measured at successive points in time often at regular time intervals, e.g., sales data, stock prices, exchange rates in finance, weather forecast, biomedical measurements (e.g., blood pressure and temperature), biometrics and locations in particle tracking in physics. Time series analysis is carried out to understand the underlying structure that produces the observed data for forecasting, monitoring, and even feedback and feedforward control [15]. It can be mathematically represented by a vector, i.e., $v(t)$, where $t = 0, 1, 2, ... t$, represents different timestamps. The variable $v(t)$ is treated as a random variable, and its measurements are arranged in chronological order. If the observations of a time series are measured continuously over time, it is termed as continuous, e.g., readings of a temperature and

© Springer Nature Switzerland AG 2020
A. Bartoli and A. Fusiello (Eds.): ECCV 2020 Workshops, LNCS 12540, pp. 228–243, 2020.
https://doi.org/10.1007/978-3-030-65414-6_17

flow of water in a river. On the other hand, a discrete time series is characterized by recordings of observations at equally spaced intervals, e.g., daily, weekly or yearly. Since, in high throughput plant phenotyping platforms (HTP3) plants are imaged at regular intervals to capture salient information about a plant's development, a phenotypic time series is considered as a discrete time series.

Image based plant phenotyping refers to the proximal sensing and quantification of plant's traits resulting from complex interaction between the genotype and its environement non-destructively, by analyzing image sequences that obviate the need for time-consuming physical human labor. Image-based high throughput plant phenotyping is an interdisciplinary research field involving computer science, plant science, statistics, data science and genomics in the effort to link intricate plant phenotypes to genetic expression in order to meet current and emerging issues in agriculture relating to future food security under dwindling natural resources and projected climate variability and change [3]. Since plant's growth and development are regulated by the genetic composition and plant's interaction with the environment, different genotypes will likely produce different phenotypes, and plants of the same genotype might have different phenotypes under different environmental conditions. Furthermore, it is also likely that plants belonging to different genotypes show similar characteristics in terms of certain phenotypic traits. Thus, the phenotype-genotype mapping is many to many in nature, and its study has been an active research field for improved crop breeding. It also helps in the understanding of the genetic regulation of phenotypes over time. In addition, the study of phenotypic predictions for the following cases will significantly contribute in the breakthrough of the phenomics research. These cases are: (a) prediction of phenotypes for the missing imaging days to account for mechanical breakdown of HTP3; (b) prediction of phenotypes for a time in the future based on analyzing past measurements; and (c) prediction of derived phenotypes from one or few of its constituent primary phenotypes [5]. This paper uses artificial neural networks based time series modeling for phenotype-genotype mapping and phenotypic predictions.

We define four types of time series in the context of plant phenotyping:

- Linear: The tendency of a phenotypic time series to increase, decrease, or stagnate over time is referred to as a linear time series. For example, the total leaf area of the plant increases with time under normal growth conditions; however, it often starts to decrease as the leaves experience curling or shedding due to application of any kind of stress, e.g., drought, thermal, and salinity. Note that for many cereal crops, e.g., maize, the height increases monotonically with time and then reaches a stagnant condition on completion of the vegetative stage.
- Recovery: The growth of a plant is significantly affected by the application of a stress, and if the stress condition is reverted, e.g., re-watering of a drought stressed plant or adjusting temperature for a plant under thermal stress, the plant may resume its normal growth, if the stress is below a threshold. Hence, the study of speed of stress recovery or determination of a plant's adaptability to stress regulated by genotypes could be analyzed by time series modeling.

- Seasonal: Plants can sense change in seasons, and respond to climatic variations by changing leaf colors, shedding leaves, blooming and generating new leaves. Thus, a time series representing total number of leaves present at any time in a year will show seasonal effect.
- Catastrophic: The catastrophic effect on phenotyping time series is caused by unprecedented incidents, e.g., floods, storms and earthquakes, which do not follow any particular pattern but significantly affect a plant's development.

A time series can either be a univariate or a multivariate. A univariate time series is a series with a single time-dependent variable, e.g., a height of a plant, a primary phenotype, computed over an image sequence for a significant period of a plant's life cycle is a univariate time series. A derived phenotype is a composite of two or more primary phenotypes [2]. The time series for plant's aspect ratio (dependent variable) is an example of a multivariate phenotype as it is a ratio of two variables, i.e., plant's height and plant's width. Similarly, the areal density of a plant (which is a ratio of the total number of plant pixels of a side view image to the area of the convex-hull enclosing the plant at the same view) as a function of time is a multivariate time series.

The aim of time series modeling is to study the past observations of a time series to develop an appropriate model that describes its underlying structure for making predictions. Efforts have been made by researchers over the past several decades to develop efficient models for improved prediction accuracy. The popular time series modeling techniques are based on stochastic approaches (e.g., autoregressive moving average [1], autoregressive integrated moving average [1,20] and seasonal autoregressive integrated moving average [1]), support vector machines (SVM) (e.g., least-square SVM, dynamic least-square SVM ([7,17])), and artificial neural networks (ANNs) (e.g., time lagged neural networks, seasonal artificial neural networks).

With enormous growth of processing power of the computers over the years, ANNs have gained immense popularity as a time series prediction technique whose basic objective is mimicking the intelligence of human brain into machine. It is characterized by a structure that consists of multiple layers, and in each layer, neurons are able to implicitly recognize regularities and patterns from the input data which propagates to the next layer for improved learning experience and then provide generalized results based on the previously known knowledge. ANNs are data-driven and self-adaptive in nature which does not require any a priori assumption about the statistical distribution of the data, and hence, well-suited for many practical applications. ANNs are inherently non-linear and universal function approximaters that can efficiently model any complex time-dependent functions to the desired level of accuracy.

It is important to evaluate the performance of the prediction. The commonly used prediction performance measures are mean absolute error, mean absolute percentage error, mean squared error, sum of squared error, root mean squared error, normalized mean squared error and Theil's U-statistics. In this research, we used mean squared error (MSE) and the R value as the performance evaluation measures of the time series modeling. MSE is a measure of average squared deviation of predicted values from actual observations, and is computed as follows [11,20]:

$$MSE = \frac{1}{n} \sum_{t=1}^{n} (y_t - p_t)^2, \tag{1}$$

where y_t is the actual value, p_t is the predicted value and n is the total number of samples in the test set. R value measures the correlation between the actual value and the predicted value for a regression analysis. An R value of 1 means a close relationship while a value 0 implies a random relationship.

2   Image-Based Phenotypic Time Series Computation

The phenotypes used in this study are computed based on image sequence analysis. Figure 1 shows the image processing pipeline for computing phenotypes by considering the whole plant as a single object (referred to as holistic phenotypes [5]). First, a frame differencing technique is used to extract the foreground, i.e., the plant, by subtracting the background image from the original image. The resulting foreground thus obtained, is often associated with undesirable soil pixels and the part of the background due to illumination variations. In order to remove the noisy pixels, the pixels of the extracted foreground that correspond to green color of the original image, are assigned green color. A color-based thresholding in HSV (Hue, Saturation and Value) color space is applied on this image using the following ranges: hue (range 0.051–0.503), saturation (range: 0.102-0.804) and value (range 0.000–0.786), which results the noise-free binary image [2,4]. The resulting binary image is enclosed by geometrical shapes (e.g., bounding rectangle, convex-hull) to compute holistic phenotypes.

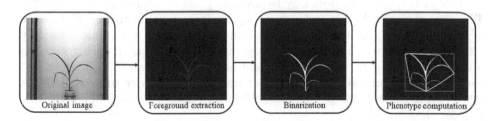

Fig. 1. Image processing pipeline to compute holistic phenotypes.

The row-1 and row-2 of Fig. 2 show sample side-view images from an image sequence of a maize plant enclosed by convex-hull and bounding rectangle, which respectively contribute to the measurements of plant's biomass and height. Row-3 of the figure shows the top-view images of the same samples each enclosed by a minimum enclosing circle, the diameter of which contributes to the computation of plant's width. The Fig. 3 shows the temporal variation of height, width and area of convex-hull computed by analyzing the image sequence (consisting of

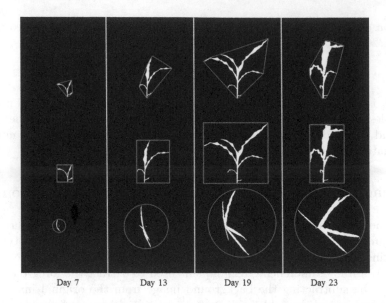

Day 7   Day 13   Day 19   Day 23

Fig. 2. Demonstration of image-based holistic plant phenotyping analysis: side view images of a maize plant enclosed by convex-hull (row-1), bounding rectangle (row-2), and top view images of the same plant enclosed by minimum bounding circle for Day 7, Day 13, Day 19 and Day 23.

images captured daily from Day 3 to Day 26) of the plant shown in Fig. 2, to demonstrate linear phenotypic time series. Note that Day 1 correspond to the first imaging day of the plants, which is two days after germination. The Fig. 3 shows that initially the height and width of the plant have similar growth characteristics, however, from Day 16 onwards, plant's width surpasses its height. The area of convex-hull is scaled down by a factor of 100 in this graph for ease of visualization. Note that the convex-hull area increases with time until Day 22 with the increase in the size of the plant. However, a sharp decline trend is observed after Day 22, although the plant has actually grown. This is attributed to the fact that maize plants alter leaf positioning in order to optimize light interception through the phytochrome [6], which affects the area of the convex-hull for the same view.

3   Dataset

For algorithm development and performance evaluation based on experimental analysis, a benchmark dataset is indispensable. Thus, we introduce a dataset called Phenoseries dataset with an aim to facilitate the study of time series modeling of structural phenotypes which facilitates in the understanding of morphological attributes of a plant. The dataset contains a set of primary and derived structural phenotypes, namely, height, width, plant aspect ratio, convex-hull area, and plant areal density, computed by analyzing the image sequences of

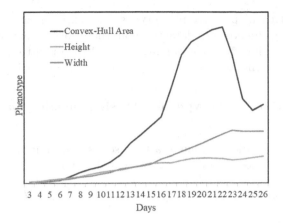

Fig. 3. Illustration of linear time series: the temporal variation of height, width and convex-hull area of a plant over consecutive days.

maize plants captured by using a visible light camera (once daily for 22 days) in LemnaTec Scanalyzer 3D HTP3 at the University of Nebraska-Lincoln, USA. The plants are placed in a metallic carrier (dimension: 236 mm × 236 mm × 142 mm) on a movable conveyor belt that moves the plants from the greenhouse to the imaging cabinet for capturing images. We selected nine genotypes of maize plants and there are eight plants in each genotypic category. Thus, the dataset contains 72 phenotypic time series corresponding to 72 plants belonging to nine different genotypes. Table 1 shows the genotypes of maize plants that are used in this study[1].

Table 1. The genotypes of maize plants used in the Phenoseries dataset.

G_{ID}	G_{name}	G_{ID}	G_{name}	G_{ID}	G_{name}
1	740	4	C103	7	PHG47
2	A619	5	PH207	8	PHV63
3	B73	6	PHG39	9	PHW52

4 Time Series Modeling for Phenotypic Prediction

In this section, the phenotypic predictions are performed based on (1) non-linear autoregressive neural network (NAR) for prediction of phenotypes for missing imaging days; (2) non-linear autoregressive neural network with exogenous input (NARX) to predict phenotypes based on dependency relationship; and (3) non-linear input-output neural network (N-IO) to predict derived phenotypes from its

---

[1] The dataset can be freely downloaded from https://plantvision.unl.edu/dataset.

one or more costituent primary phenotypes. The experimental demonstrations are provided using Phenoseries dataset. In all these cases, the networks use 70% of the data for training, 15% for validation, and 15% for testing. The training is performed by using the Levenberg-Marquardt algorithm [13,14].

4.1   Prediction of Phenotypes for Missing Imaging Days Using NAR:

A NAR is used to predict a value of a time series y at time t, i.e., $y(t)$, using d past values of the series, and can be represented as follows:

$$y(t) = f(y(t-1), y(t-2), ..., y(t-d)) + \epsilon(t). \tag{2}$$

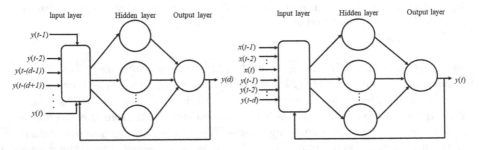

Fig. 4. A typical architecture for non-linear autoregressive neural network (NAR) (left) and a non-linear autoregressive neural network with exogenous input (NARX) (right).

A typical architecture for NAR is shown in Fig. 4(left). In HTP3, the process of image capturing using cameras in different modalities is time consuming, and the required time increases with the increase in the number of viewing angles. Thus, to allow imaging slot to all plants in a large greenhouse, we often need to negotiate with the interval of imaging, i.e., instead of daily, we might need to image the plants every alternate day in case of a HTP3 with a capacity of hosting large number of plants. In addition, unpredictable mechanical breakdown in a HTP3 also causes missing imaging days. Figure 5(left) shows the prediction performance of height of a plant for missing imaging days using NAR. In this experiment, we used four plants of genotype C103. The error histogram (see Fig. 5(top-right)) and autocorrelation (see Fig. 5(bottom-right)) are also presented. Note that training multiple times will generate different results due to different initial conditions and sampling. Thus, we trained the model five times and reported the average regression R values for training, validation and testing as 0.87, 0.79 and 0.89, respectively. The results show the efficiency of NAR to account for phenotypes over missing imaging days.

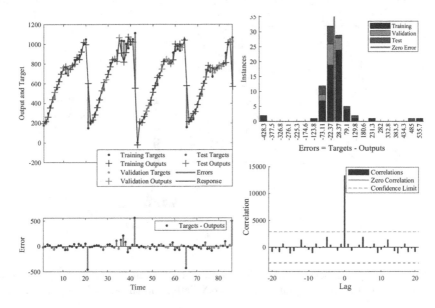

Fig. 5. Prediction of height for missing imaging days of a plant using NAR.

4.2 Prediction of a Future Phenotype Based on Dependency Relationship Using NARX:

A NARX is used to predict a value of a time series y at time t, i.e., $y(t)$, using d past values of the series and an another series $x(t)$. It can be represented as follows:

$$y(t) = f(y(t-1), y(t-2), ...y(t-d)) + f(x(t-1), x(t-2), ...x(t)) + \epsilon(t). \quad (3)$$

Figure 4(right) shows a typical architecture for NARX. Note that there are phenotypes which are related to each other, and hence, one can be used as a predictor for another. For example, the total number of plant pixels of a plant at a view is proportional to the area of the convex-hull of that plant at the same view, and vice versa. Thus, NARX can be successfully deployed to predict phenotypes of a time in future based on its past measurements and a related time series. Figure 6(left) shows the prediction response of the total number of plant pixels at any time in the future by using area of the convex-hull enclosing the plant at the same view and the past measurements of total number of plant pixels. Figure 6(right) shows the error histogram. Similarly, Fig. 7(left) shows the prediction response of area of the convex-hull enclosing the plant at any time in the future by using the total number of plant pixels at the same view and the past measurements of convex-hull area. Figure 7(right) shows the corresponding error histogram. We trained the model five times. The average R values computed for

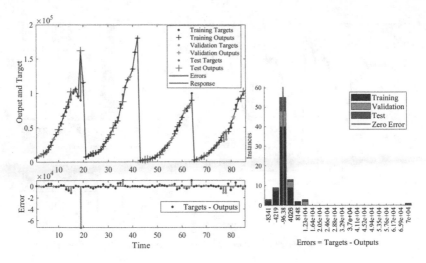

Fig. 6. The prediction performance of total number of plant pixels (target) based on convex-hull area of a plant (input) using NARX (left); and the error histogram (right).

training, validation and testing are 0.98, 0.84 and 0.89 respectively. The results show the efficacy of NARX for phenotypic prediction based on dependency relationship.

4.3   Prediction of Derived Phenotypes from Primary Phenotypes Using N-IO:

A N-IO is used to predict a time series y at time t, i.e., $y(t)$, using d past values of the series $x(t)$. It can be represented as follows:

$$y(t) = f(x(t-1), x(t-2), ..., x(t-d)) + \epsilon(t). \qquad (4)$$

Figure 8(a) shows a typical architecture for a N-IO. N-IO can be used to predict a derived phenotypic time series (target) by using one of its components as the input to the network. We use a derived phenotype called areal density (PAD), a potential measure of field planting density, as the experimental demonstration. It is defined as [5]

$$PAD = \frac{Plant_{Tpx} \text{ at a side view}}{Area_{CH} \text{ at the same view}}, \qquad (5)$$

where $Plant_{Tpx}$ denotes the total number of plant pixels at a viewing angle and $Area_{CH}$ denotes the area of the convex-hull of the plant at the same view. Figure 8(b) shows the prediction performance of plant areal density using the area of convex-hull enclosing the plant in the same view. We trained the model five times and computed the average MSE values for training, validation and

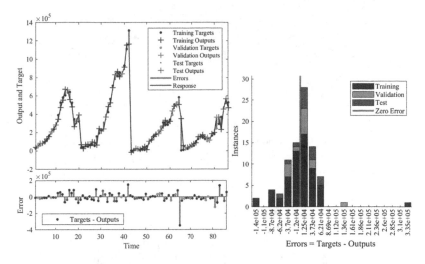

Fig. 7. The prediction performance (left) and the error histogram (right) of convex-hull area (target) based on total number of plant pixels (input) using NARX.

testing as 0.0002, 0.3858 and 0.0036, respectively. The error histogram and the cross-correlation shown in Fig. 8(c) and (d) respectively, demonstrate the efficacy of N-IO as a predictor of areal density from convex-hull area.

5   Time Series Modeling for Phenotype-Genotype Mapping

A long short-term memory (LSTM) network, a special type of recurrent neural networks, is capable of making predictions based on individual time steps of a sequence fed as the input to the network [12]. In recent times, LSTM has been proven effective in a variety of real-world applications, e.g., speech recognition, hand writing recognition, short-term traffic forecast, human action recognition, robot control and sign language translation [8,9,21]. The method by [18] is composed of a CNN-based deep visual descriptor and a LSTM network that uses temporal growth dynamics of Arabidopsis plants for genotypic classification. In this paper, we exploit the efficiency of LSTM in time series classification to achieve the novel task of bridging phenotype-genotype gap, i.e., to predict a maize plant's genotype based on analyzing its phenotypic time series.

5.1   Experimental Analysis

We split the Phenoseries dataset into a training set and a test set for a open set phenotype-genotype mapping, i.e., the plants in the test set are not present in the training set. The training set contains phenotypic time series corresponding to five plants and test set contains phenotypic time series of the remaining three plants for each of nine genotypes.

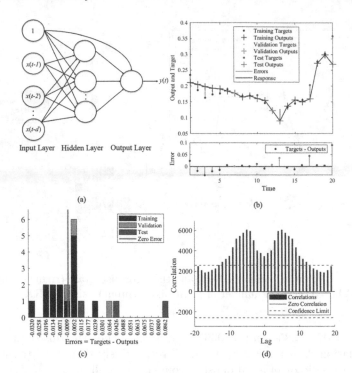

Fig. 8. (a) An architecture for N-IO; (b) Prediction performance of areal density (target) using convex-hull area of a sample maize plant as the input; (c)-(d) Performance analysis of N-IO as a predictor of areal density from convex-hull area using error histogram and cross-correlation, respectively.

Deep learning usually benefits from large training sets. However, in many real-world applications, the size of training set is limited, which penalizes the network accuracy by overfitting. For example, while HTP3 are quite efficient at acquiring and managing abundant and high-quality images of the target plants, the setup and maintenance cost of HTP3 infrastructure and imaging facilities are extremely high. Hence, to optimize space and reduce cost of experiment, it is a common practice to carry out multiple experiments simultaneously in an automated greenhouse that limit the number of plants participating in each experiment. A widely used solution to increase the size of the training dataset is data augmentation that aims at generating synthetic data by transforming original labeled samples so as to enable the model better learn the intra-class invariances. While the application of various geometric and photometric transformations to the original samples have been proven effective for images, these data augmentation techniques do not generalize well to time series [19].

In our method, we use window warping, scaling and averaging techniques for augmenting the phenotypic time series [10,19]. While window warping refers to warping a randomly selected slice of a time series by speeding it up or down [10], scaling is obtained by multiplying each observation of a time series by a fixed

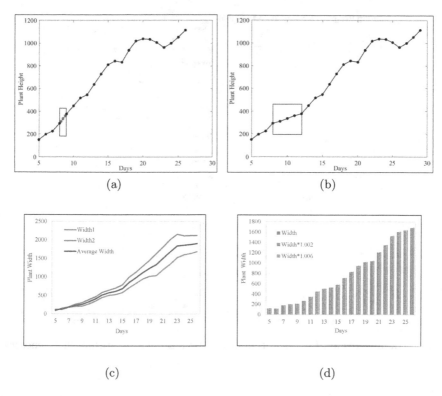

Fig. 9. Illustration of time series augmentation techniques: (a)–(b) window warping; (c) averaging; and (d) scaling.

fraction. The averaging technique of data augmentation is obtained by averaging different combinations of two or more time series of the same genotype. Figure 9 shows different techniques of time series augmentation, i.e., window warping, averaging and scaling. In window warping, we randomly select a region in the time series, and insert few additional equally spaced timestamps between any two consecutive days in the chosen region. Then, we use linear interpolation to determine the values of phenotypes for those timestamps (marked in red in Fig. 9(a)). These newly inserted timestamps are treated as days, and as a result, the time series is stretched (see Fig. 9(b)). Figure 9(c) shows the averaging technique of time series, where, the width time series of two plants belonging to same genotype are averaged to create a new time series. Figure 9(d) shows an example of scaling technique of time series augmentation, where a width time series of a plant is multiplied by factors 1.002 and 1.006 to generate two more time series. Note that test phenotypic time series are supposed to be representative of the originals without any alteration, and hence, are not subjected to augmentation for unbiased evaluation.

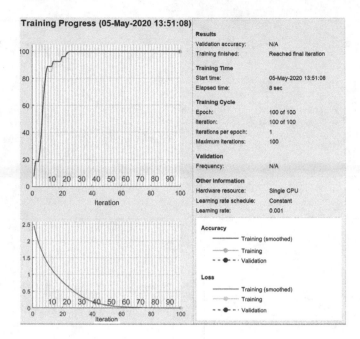

Fig. 10. Training results for phenotype-genotype mapping algorithm on Phenoseries dataset.

Time series augmentation is applied to the training set. The training set thus created, consists of a cell array of size 72 (number of phenotypic time series) and a categorical vector of genotypes for each phenotypic time series. Each entry of the cell array is a matrix of dimension 3 (total number of phenotypes, i.e., height, width and aspect ratio) × 22 (total number of days). The classification accuracy of genotypic prediction is measured by

$$\text{accuracy} = \frac{\text{Total number of correctly classified plants}}{\text{Total number of plants}} \times 100. \qquad (6)$$

Figure 10 demonstrates the training phase of the phenotype-genotype mapping algorithm using LSTM network. A bidirectional LSTM layer with 100 hidden units is used. The mini-batch size is set to 27, and the maximum number of epochs is 100. Table 2 presents the confusion matrix to visualize the performance of the phenotype-genotype mapping algorithm. The columns of a confusion matrix represent instances in a predicted class, while the rows represent the instances in the actual class [16]. The classification accuracy is 81.4%.

Table 2. Confusion matrix for showing the classification performance of the phenotype-genotype mapping algorithm.

Actual class	Predicted class								
	1	2	3	4	5	6	7	8	9
1	3	0	0	0	0	0	0	0	0
2	0	2	0	1	0	0	0	0	0
3	0	0	3	0	0	0	0	0	0
4	0	0	0	3	0	0	0	0	0
5	1	0	0	0	2	0	0	0	0
6	0	0	0	0	0	3	0	0	0
7	0	0	1	0	1	0	1	0	0
8	0	0	0	1	0	0	0	2	0
9	0	0	0	0	0	0	0	0	3

6 Conclusions

The paper introduces a novel application of time series modeling and prediction in the field of image based plant phenotyping to address the current and emerging issues related to agriculture for global food security. For any foundational study, a benchmark dataset is indispensable for algorithm development and evaluation. Thus, we introduce a benchmark dataset, first of its kind, called Phenoseries dataset, which consists of phenotypic traits of maize plants grouped into nine different genotypes. The phenotypes are computed based on analyzing image sequences captured once daily for 22 days. We described three time series augmentation techniques, i.e., window warping, averaging and scaling, to increase the size of the training set for improved network accuracy. Experimental analyses using the Phenoseries dataset show that neural network based time series modeling using NAR, NARX and N-IO hold great promise to achieve important tasks, i.e., prediction of phenotypes for missing imaging days or for a time in the future based on analyzing past measurements and prediction of derived phenotypes based on primary phenotypes. The paper uses LSTM, a variant of sequential neural networks, for the study of bridging phenotype-genotype gap. The ground-laying study presented in this paper has lot opportunities for future exploration. In the future work, we will consider a large-scale experimental setup with various plant species belonging to a wide variety of genotypes for both controlled environment and field phenotyping platforms.

Acknowledgment. This research is partially supported by the Nebraska Agricultural Experiment Station with funding from the Hatch Act Capacity Program (Accession Number 1011130) from the USDA National Institute of Food and Agriculture.

References

1. Box, G.E.P., Jenkins, G.M., Reinsel, G.C., Ljung, G.M.: Time Series Analysis: Forecasting and Control. Wiley, New York (2015)
2. Das Choudhury, S., Bashyam, S., Qiu, Y., Samal, A., Awada, T.: Holistic and component plant phenotyping using temporal image sequence. Plant Methods14(35) (2018)
3. Das Choudhury, S., Samal, A., Awada, T.: Leveraging image analysis for high-throughput plant phenotyping. Front. Plant Sci. **10**(508) (2019)
4. Das Choudhury, S., Goswami, S., Bashyam, S., Samal, A., Awada, T.: Automated stem angle determination for temporal plant phenotyping analysis. In: The IEEE International Conference on Computer Vision (ICCV) Workshop on Computer Vision Problems in Plant Phenotyping, pp. 2022–2029, October 2017
5. Das Choudhury, S., Stoerger, V., Samal, A., Schnable, J., Liang, Z., Yu, J.G.: Automated vegetative stage phenotyping analysis of maize plants using visible light images. In: Knowledge Discovery and Data Mining Workshop on Data Science for Food, Energy and Water, August 2016
6. Dubois, P.G., Olsefski, G.T., Flint-Garcia, S., Setter, T.L., Hoekenga, O.A., Brutnell, T.P.: Physiological and genetic characterization of end-of-day far-red light response in maize seedlings. Plant Physiol. **154**(1), 173–186 (2010). https://doi.org/10.1104/pp.110.159830
7. Fan, Y., Li, P., Song, Z.: Dynamic least square support vector machine. In: Proceedings of the 6th European Conference on Computer Systems, pp. 4886–4889. ACM (2006). https://doi.org/10.1109/WCICA.2006.1713313
8. Graves, A., Schmidhuber, J.: Offline handwriting recognition with multidimensional recurrent neural networks. In: Koller, D., Schuurmans, D., Bengio, Y., Bottou, L. (eds.) Advances in Neural Information Processing Systems, vol. 21, pp. 545–552. Curran Associates, Inc. (2009)
9. Graves, A., Schmidhuber, J.: Framewise phoneme classification with bidirectional LSTM and other neural network architectures. Neural Netw. **18**(5), 602–610 (2005)
10. Guennec, A.L., Malinowski, S., Tavenard, R.: Data augmentation for time series classification using convolutional neural networks. In: ECML/PKDD Workshop on Advanced Analytics and Learning on Temporal Data. Riva Del Garda, Italy (2016). https://doi.org/10.1109/WCICA.2006.1713313
11. Hamzaçebi, C.: Improving artificial neural networks' performance in seasonal time series forecasting. Inf. Sci. **178**(23), 4550–4559 (2008). https://doi.org/10.1016/j.ins.2008.07.024, http://www.sciencedirect.com/science/article/pii/S0020025508002958, including Special Section: Genetic and Evolutionary Computing
12. Hochreiter, S., Schmidhuber, J.: Long short-term memory. Neural Comput. **9**(8), 1735–1780 (1997). https://doi.org/10.1162/neco.1997.9.8.1735
13. Levenberg, K.: A method for the solution of certain non-linear problems in least squares. Quart. Appl. Math. **2**(2), 164–168 (1944)
14. Marquardt, D.W.: An algorithm for least-squares estimation of nonlinear parameters. J. Soc. Ind. Appl. Math. **11**(2), 431–441 (1963)
15. Montgomery, D.C., Jennings, C.L., Kulahci, M.: Introduction to Time Series Analysis and Forecasting. Wiley, New York (2015)
16. Stehman, S.V.: Selecting and interpreting measures of thematic classification accuracy. Remote Sens. Environ. **62**(1), 77–89 (1997)

17. Suykens, J., Vandewalle, J.: Recurrent least squares support vector machines. IEEE Trans. Circuits Syst. I Fundam. Theory Appl. **47**, 1109–1114 (2000). https://doi.org/10.1109/81.855471

18. Taghavi, S., Esmaeilzadeh, M., Najafi, M., Brown, T., Borevitz, J.: Deep phenotyping: deep learning for temporal phenotype/genotype classification. Plant Methods **14**, 66 (2018). https://doi.org/10.1186/s13007-018-0333-4

19. Um, T.T., et al.: Data augmentation of wearable sensor data for parkinson's disease monitoring using convolutional neural networks. In: ACM International Conference on Multimodal Interaction, pp. 216–220 (2017). http://arxiv.org/abs/1706.00527

20. Zhang, G.P.: Time series forecasting using a hybrid ARIMA and neural network model. Neurocomputing **50**(17), 159–175 (2003)

21. Zhao, Z., Chen, W., Wu, X., Chen, P.C.Y., Liu, J.: LSTM network: a deep learning approach for short-term traffic forecast. IET Intell. Transp. Syst. **11**(2), 68–75 (2017)

3D Plant Phenotyping: All You Need is Labelled Point Cloud Data

Ayan Chaudhury[1,2]([⊠]) [iD], Frédéric Boudon[3,4] [iD], and Christophe Godin[1,2] [iD]

[1] INRIA Grenoble Rhône-Alpes, Team MOSAIC, Montbonnot-Saint-Martin, France
{ayan.chaudhury,christophe.godin}@inria.fr
[2] Laboratoire Reproduction et Développement des Plantes, Univ Lyon,
ENS de Lyon, UCB Lyon 1, CNRS, INRA, Lyon, France
[3] CIRAD, UMR AGAP, 34098 Montpellier, France
frederic.boudon@cirad.fr
[4] AGAP, Univ Montpellier, CIRAD, INRAE, Institut Agro, Montpellier, France

Abstract. In the realm of modern digital phenotyping technological advancements, the demand of annotated datasets is increasing for either training machine learning algorithms or evaluating 3D phenotyping systems. While a few 2D datasets have been proposed in the community in last few years, very little attention has been paid to the construction of annotated 3D point cloud datasets. There are several challenges associated with the creation of such annotated datasets. Acquiring the data requires instruments having good precision and accuracy levels. Reconstruction of full 3D model from multiple views is a challenging task considering plant architecture complexity and plasticity, as well as occlusion and missing data problems. In addition, manual annotation of the data is a cumbersome task that cannot easily be automated. In this context, the design of synthetic datasets can play an important role. In this paper, we propose an idea of automatic generation of synthetic point cloud data using virtual plant models. Our approach leverages the strength of the classical procedural approach (like L-systems) to generate the virtual models of plants, and then perform point sampling on the surface of the models. By applying stochasticity in the procedural model, we are able to generate large number of diverse plant models and the corresponding point cloud data in a fully automatic manner. The goal of this paper is to present a general strategy to generate annotated 3D point cloud datasets from virtual models. The code (along with some generated point cloud models) are available at: https://gitlab.inria.fr/mosaic/publications/lpy2pc.

Keywords: Procedural model · L-System · Virtual plant · Point cloud · Labelled synthetic data

1 Introduction

With the recent breakthrough of the advancements on plant phenotyping research, botanical and agronomic studies have been revolutionized to enter in

A. Bartoli and A. Fusiello (Eds.): ECCV 2020 Workshops, LNCS 12540, pp. 244–260, 2020.
https://doi.org/10.1007/978-3-030-65414-6_18

a new stage of development [35]. In recent years, 3D computer vision based systems have been widely developed in different types of phenotyping applications including robotic branch pruning [5], automated growth analysis [7,8], agricultural automation tasks [38], etc. Point cloud based approach has been an integral part of machine learning aspects of plant phenotyping [47]. However, many of these techniques require ground truth data for training the algorithms and for validation of the results in a quantitative manner. For example, methods performing plant organ segmentation [22,27] require labelled point cloud data for quantifying the segmentation accuracy. Similarly, in-field 3D segmentation [34] or plant structure classification tasks [10] also demand annotated datasets for validation. In addition, to exploit the recent success of deep learning technologies for point cloud data [30] in agricultural applications, large amount of annotated data are in extreme demand. Unfortunately at the current moment, there is a scarcity of available annotated point cloud dataset in the plant phenotyping community. Only a handful of datasets of certain plant species are proposed in the literature. Among the currently available 3D datasets, Cruz et al. [9] created a multimodal database of top view depth images of Arabidopsis and bean plants. Wen et al. [44] proposed a database of 3D geometric plant models which are based on the measurements of in-situ morphological data. Different types of species in different growth stages are modelled in the framework. However the data is limited to the organ scale of the plants, and not on the full 3D model. Bernotas et al. [1] created a photometric stereo dataset along with manually annotated leaf masks of Arabidopsis rosettes. The Komatsuna plant dataset [37] contains RGB and depth images with annotated leaf labels. However, none of the above mentioned datasets are built on full 3D point cloud model of plants. Recently, the ROSE-X dataset [11] is created with 3D models of rosebush plants acquired through X-ray tomography method. Manual annotation is performed to label voxels and point clouds belonging to different organs of the plant. For remote sensing applications of vegetation studies, Wang [41] simulated a synthetic Terrestrial Laser Scanning (TLS) data of forest scenes. The generated point cloud contains simulated data of large trees having structurally complex crowns with labels belonging to leaf and wood class.

In contrast to the scarcity of 3D point cloud datasets, there has been quite a few 2D datasets available for public use, as well as software tool to access different types of existing solutions according to the experimental need [24]. These have revealed extremely useful for training machine learning algorithms and performance validations tasks for 2D image based plant phenotyping applications. The dataset proposed in [25] is one of the first kind, where annotated top-view color images of rosette plants at different growth stages are available. Shadrin et al. [31] proposed a manually annotated dataset of raw salad images acquired over a period of several days for plant growth dynamics assessment. Among other recent datasets, the Oil Radish dataset [26] and Grass clover dataset [32] have been proposed. In last few years, leaf segmentation and counting of rosette plants have been an active area of interest. Deep learning techniques have been very successful in solving this type of problem, which demand lot of annotated data for training the network. In order to overcome the scarcity of available datasets, different types of data augmentation strategies have been proposed to generate realistic synthetic data. Giuffrida et al. [13] proposed a Generative Adversarial Network (GAN) model to artificially create Arabidopsis rosette images. The model allows the user to create plants with desired number of leaves in a realistic manner. With a similar type of motivation, Ward et al. [43] generated synthetic leaf images inspired by domain randomization strategy using a Mask-RCNN deep learning model. Kuznichov et al. [21] proposed a data augmentation strategy which preserves the geometric structure of the generated leaves with the real data. Different types of synthetic images are generated by applying some heuristic rules on the leaf shape and growth and recently, parametric L-systems have been used to perform data augmentation in generating synthetic leaf models [36]. Various leaf databases have also been proposed [20,33,46] which are useful for species identification and leaf architecture analysis [6,40,45].

Some of the difficulties associated with creating labelled 3D point cloud datasets include the expensive scanning devices, cumbersome and error prone manual annotation of large number of points, and reconstructing full 3D structure of complex plants in the presence of occlusion and noise. Also, growing varieties of plant species and acquiring data at different growth stages require lot of manpower and constant monitoring of the plants. In this regard, synthetic (or virtual) plant models can play an important role. Recently Ward et al. [42] demonstrated that synthetic data can be extremely useful in training deep learning models, which can be reliably used for measurements of different types of phenotypic traits in the general case. Modelling the geometry of complex plant structures have been a center of attention for mathematicians and biologists for decades [14,15,23,29]. Virtual models can be extremely useful in agricultural studies [3,12], as well as have great potential for mechanical simulation of plant behavior [4]. In generating the virtual plant models, L-system based modelling technique has been quite successful [28]. Different types of platforms have been developed to simulate the L-system rules, e.g., L+C modelling language [18], L-Py framework [2], etc. Functional Structural Plant Models (FSPM) [15] have emerged as powerful tool to construct 3D models of plant functioning and growth. These models can play an important role in mechanical simulation in crop science research [39]. Although generation of synthetic plant models have been a well studied area, the virtual plant models have not been exploited in generating synthetic 3D point cloud data for plant phenotyping applications. Compared to the synthetic artery data simulation software like Vascusynth [16] for medical imaging research, there is no such tool available in the plant phenotyping community.

Inspired by the necessity of synthetic plant point cloud datasets for 3D plant phenotyping applications, we propose a general approach to sample points on the surface of virtual plant models to generate annotated 3D point cloud data in a fully automatic manner. The framework is general in the sense that it can be plugged into any procedural plant model, and surface point cloud can be sampled from the virtual plant. Another motivation of the work comes from a computer graphics perspective. Among different types of representations of geometric shapes in computer graphics, two types are widely used: triangular mesh and point based representation [19]. These two types are complementary to each other in terms of information richness and model complexity, and our framework builds a bridge between the two paradigms for the varieties of applications of the geometric plant models in phenotyping problems.

Initially we represent the virtual model as a collection of geometric primitives, each having their own label belonging to different organs of a plant. Each primitive is then tessellated into triangles to obtain a basic coarse mesh model. By a random sampling strategy using the idea of barycentric coordinates, we are able to generate uniformly distributed point cloud on the surface of the virtual model. Points residing inside any primitive are removed by a geometric filtering technique. We also model some basic characteristics of real world scanning devices by incorporating different sampling strategies, which can be adopted according to the need of the application. We demonstrate the efficiency of the approach by generating realistic synthetic point cloud model of Arabidopsis plants, as well as plants with different types of architecture. These datasets are generated by the stochasticity inherent to the procedural model and by a second level of stochasticity at the level of point cloud sampling. We also show some examples on other synthetic plants having different types of geometry structures.

In the next section we briefly describe how a virtual plant model should be specified to allow our algorithm to construct annotated point clouds from it. Then the basic random point cloud sampling strategy is explained, followed by the insideness testing process. Then we discuss the sampling strategy to obtain device specific point cloud models, and experimental results are demonstrated. Finally the scope of the work and directions for further improvements are discussed.

2   Virtual Plant Models

Virtual plants are computer generated synthetic models mimicking the development and growth of real plants. The models are typically developed by the recursive rules of L-systems [23]. One of the main attribute of L-system based plant models is that, complex plant structures can be generated by successive application of simple recursive rules. The recursive rules help in modelling natural phenomena like plant growth, creation and destruction of plant organs over time, change of shape (e.g. bending, spiralling), etc. This helps us to model a desired plant species in different growth stages by incorporating biological knowledge about the plant in the L-system rules.

In a L-system framework, a plant is defined by a string of symbols, called L-string, representing the different organs of the plant, A for Apex, I for internode, L for leaf for example. Each symbol is given a set of arguments describing variables attached to the corresponding organ. For example an internode can be modeled as a cylinder of height h and radius r, a leaf can be modeled as a polygon of size s, an apex can have an age t, etc. Square brackets are used to delineate branches within the L-string. An initial L-string, called the axiom, defines the initial state of the plant, before growth starts. For example, the plant may initially be composed of an apex A(t). This is represented as:

```
Axiom: A(0)   # meaning that the L-string initially contains a
              single apex at time 0
```

A typical L-system model, then consists of a set of rules that make it possible to evolve the initial L-string. For instance the following rule typically says that at each time step *dt* the apex produces an internode I, a branching apex A and a leaf L, and an apical apex A:

```
A(t) --> I(0.01,0.1) [+(angle)[L(0.1)]A(t+dt)] A(t+dt)
```

where 'angle' is a variable (that can be a constant for example). This rule can be complemented by rules that express how internodes and leaf should change in time:

```
I(r,h) --> I(r',h')
    with r' = r + a*dt and h' = r' + b*dt
L(s) --> L(s')
    with s' = s + c*dt
```

In general, these so-called production rules do not describe how the symbols are associated with geometric models. For this, a way to interpret geometrically the symbols of the L-string should be defined. This is done using a new set of rules called interpretation rules. Here they could for instance look like, e.g.:

```
A(t) --> Sphere(radius(t))
    with radius(t) = 0.1 (constant radius in time)
I(r,h) --> Cylinder(r,h)
L(s) --> Polygon(size)
```

To use our algorithm, one would need to attach a unique annotation for each class of organ that one would like to recognize in the application. This is done by adding an extra 'label' argument to the symbols of interest.

```
A(t) --> I(0.01,0.1,8) [+(angle)[L(0.1,9)]A(t+dt)] A(t+dt)
I(r,h,label_I) --> I(r',h',Label_I)
    with r' = r + a*dt and h' = r' + b*dt
L(size,label_L) --> L(s',label_L)
    with s' = s + c*dt
```

where label = 8 is attached to internodes and label = 9 to leaves. The interpretation rules should be modified accordingly:

```
A(t) --> Sphere(radius(t))
    with radius(t) = 0.1 (constant radius in time)
I(r,h,label_I) --> Cylinder(r,h,label_I)
L(size,label_L) --> Polygon(size,label_L)
```

In this way, the geometric models will be tagged with a label corresponding to their category (here Internode or Leaf), and will make it possible to identify the triangles (which are used for rendering the model) that belong respectively to internodes and leaves in the 3D scene. Basically we transfer the label of an organ to the triangles while performing the tessellation, as discussed next.

3   Surface Point Cloud Generation

After creating the virtual model, we aim at generating points on the surface of the model where each point has a label of the plant organ it belongs to. In order to achieve this, we follow a series of steps. Given a parametric L-system model with unique label associated with each organ of the plant, we extract the set of primitives (along with their label of the organ they belong to in the procedural model) which are used to display the virtual plant model as a 3D object. In general, the following types of primitives are typically used in constructing the 3D object: cylinder, frustum, sphere, Bézier curve, NURBS patch, etc. Each primitive is then locally tessellated into a set of triangles (these triangles are used to render the 3D object), and the triangles are assigned the label of the primitive. Finally a global list of labelled triangle vertices are created to obtain a coarse mesh model of the 3D structure of the plant. Now the idea is to sample a dense set of labelled point cloud on the triangle surfaces, with a controlled density to create a realistic point cloud model of the plant. In the next section, we describe a simple and effective random point sampling strategy that can generate desired number of labelled points on the surface of the model.

3.1   Random Point Sampling

In order to generate labelled points on the surface of the virtual model, we propose two types of sampling strategies. The first type of sampling aims at generating uniform density of points on the model surface, and the second type of sampling is motivated by the effect of real world scanning devices, for which point density depends on various sensor-related parameters. In this section, we present the basic sampling strategy to generate point cloud of uniform density.

As described earlier, at this stage we represent the virtual model as a set of triangles of different sizes in a 3D scene. Sampling of points on the triangle surface is performed by random selection of triangle in each iteration and generating a point at random location inside the triangle. If we want to generate a point cloud of size m, we repeat the process of random triangle selection m times, and each time we generate a random point inside the selected triangle. However since the triangle sizes are not uniform, the part of the model having higher number of small triangles will likely to have higher point density than the part having smaller number of large triangles. In order to handle this problem, we use the triangle area as a selection probability while performing the sampling, thus resulting in a list of triangles sampled in a uniform manner.

Next, in order to perform point sampling on each triangle surface, we exploit barycentric coordinates, which are widely used in ray tracing applications of computer graphics. The barycentric coordinates can be used to define the position of any point located inside a triangle. In the general case for a simplex in an affine space, let the vertices are denoted as $(\mathbf{v}_1, \cdots, \mathbf{v}_n)$. If the following condition holds true for some point $\mathbf{P}$:

$$(\alpha_1, \cdots, \alpha_n)\mathbf{P} = \alpha_1\mathbf{v}_1 + \cdots + \alpha_n\mathbf{v}_n, \tag{1}$$

and at least one of the α_i's are not zero, then the coefficients $(\alpha_1, \cdots, \alpha_n)$ are the barycentric coordinates of $\mathbf{P}$ with respect to $(\mathbf{v}_1, \cdots, \mathbf{v}_n)$. For the case of triangles, any point inside a triangle can be described by a convex combination of its vertices. Let the vertices of a triangle are denoted as $\mathbf{v}_1, \mathbf{v}_2, \mathbf{v}_3$. Then any random point $\mathbf{Q}$ inside the triangle can be expressed as,

$$\mathbf{Q} = \sum_{i=1}^{3} \alpha_i\mathbf{v}_i, \tag{2}$$

where $\alpha_1, \alpha_2, \alpha_3 \geq 0$ are three scalars (the barycentric coordinates of the point $\mathbf{Q}$) such that the following condition holds true:

$$\alpha_1 + \alpha_2 + \alpha_3 = 1. \tag{3}$$

For all points inside the triangle, there exists a unique triple $(\alpha_1, \alpha_2, \alpha_3)$ such that the condition of Eq. 2 holds true. Hence, due to the property in Eq. 3, any point can be defined by any two of the barycentric coordinates.

Using barycentric coordinates, random points can be generated easily on the triangle surfaces. For each randomly selected triangle, we generate two random numbers α_1 and α_2. Then α_3 is computed as, $\alpha_3 = 1 - (\alpha_1 + \alpha_2)$ (using Eq. 3). This gives us a random point $\mathbf{Q}$ inside the triangle. By performing this process for m number of times, a uniformly spaced point cloud of size m is obtained on the surface of the virtual model.

3.2   Insideness Testing

Although the point sampling is performed on the surface of the primitives, there can be points remaining inside the volumetric primitives (cylinder, frustum, and sphere). As shown in Fig. 1, for the case of a simple branching structure approximated by cylinders and frustums, part of a branch remains inside the other. This is physically inconsistent, and we need to filter out the points inside any volumetric primitive. Hence for each point, we perform a geometric testing for different types of primitives as follows.

Cylinder Primitive. Let an arbitraily oriented cylinder has radius r and two endpoints $(\overrightarrow{p_i}, \overrightarrow{p_j})$ on the central axis. Given a random point $\overrightarrow{q}$, we aim at determining if the point is inside the cylinder or not (left of Fig. 2). First, we

Fig. 1. In case of volumetric primitives, joining of two (or more) branches results in one primitive to retain inside another primitive.

test if $\vec{q}$ lies between the planes of the two circular facets of the cylinder. This is determined by the following two conditions:

$$(\vec{q} - \vec{p_i}) \cdot (\vec{p_j} - \vec{p_i}) \geq 0,$$
$$(\vec{q} - \vec{p_j}) \cdot (\vec{p_j} - \vec{p_i}) \leq 0. \tag{4}$$

If the above conditions are true, we compute the perpendicular distance of the point $\vec{q}$ from the central axis of the cylinder. If the distance is less than the radius of the cylinder, then the point lies inside. Mathematically the condition is defined as,

$$\frac{|(\vec{q} - \vec{p_i}) \cdot (\vec{p_j} - \vec{p_i})|}{|(\vec{p_j} - \vec{p_i})|} \leq r. \tag{5}$$

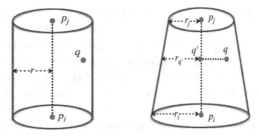

Fig. 2. Insideness testing for the case of cylinder and frustum.

Frustum Primitive. Let an arbitraily oriented frustum has base radius r_i and top radius r_j. The two endpoints on the central axis are defined as $(\vec{p_i}, \vec{p_j})$. Given a random point $\vec{q}$, we aim at determining if the point is inside the frustum or not (right of Fig. 2).

Similar to the case of cylinder, first we check if the point lies between the planes of the two circular facets of the frustum using Eq. 4. Then we project the point $\vec{q}$ on the line joining $\vec{p_i}$ and $\vec{p_j}$ (the central axis of the frustum) as,

$$\vec{q}' = \vec{p_i} + \left[\frac{(\vec{q} - \vec{p_i}) \cdot (\vec{p_j} - \vec{p_i})}{(\vec{p_j} - \vec{p_i}) \cdot (\vec{p_j} - \vec{p_i})} \right] * (\vec{p_j} - \vec{p_i}). \tag{6}$$

The distance between $\vec{q}$ and $\vec{q}'$ is computed as, $d = \text{dist}(\vec{q}, \vec{q}')$. Next we compute the frustum radius $r_{q'}$ at the point $\vec{q}'$. This can be computed as,

$$r_{q'} = \frac{\text{dist}(\vec{p_i}, \vec{q}')}{\text{dist}(\vec{p_i}, \vec{p_j})} * r_i + \frac{\text{dist}(\vec{p_j}, \vec{q}')}{\text{dist}(\vec{p_i}, \vec{p_j})} * r_j. \tag{7}$$

If the distance $d < r_{q'}$, then the point lies inside the frustum.

Sphere Primitive. Let a sphere centered at point $\vec{p} = (x_c, y_c, z_c)$ having radius r. A point $\vec{q} = (x_i, y_i, z_i)$ is inside the sphere if the following condition holds true:

$$\sqrt{(x_c - x_i)^2 + (y_c - y_i)^2 + (z_c - z_i)^2} \leq r. \tag{8}$$

General Primitives. For more general geometric models that represent more complex plant organs, we use general computer graphics techniques like ray casting [17] that make it possible to test whether a particular point lies inside or outside a closed geometric envelop.

The step by step procedure to generate the point cloud using the basic point sampling strategy is outlined in Algorithm 1.

3.3   Sensor Specific Distance Effect

The point cloud sampling described so far, is based on the assumption that the point density is uniform in the whole plant model. However in real scanning devices, this is usually not the case. While there are different types of sensor devices having different types of specific traits associated with the generated surface point cloud, one of the most common is the effect of distance from the scanner on the point density. A typical effect of this type yields higher point density near the scanner, and the density decreases as the distance of different parts of the object from the scanner increases. In order to simulate the effect of these type of real scanning devices, we implemented a sampling strategy which can be adapted to different types of cases according to the need of the application.

Instead of random sampling of triangles in every iteration as in the case of naive sampling technique described before, we compute the number of points to be sampled on each triangle based on a distance factor. First we compute the total surface area $\mathcal{A}_T$ of all the triangles as, $\mathcal{A}_T = \sum_i \mathcal{A}_{T_i}$. Then we compute the point density ρ_u per unit area of the mesh as, $\rho_u = n/\mathcal{A}_T$, where n is the maximum number of points we wish to sample. Basically ρ_u gives the point density for the case of uniform sampling: given a random triangle $\mathcal{T}_i$, the number of points on the triangle surface can be approximated as $(\rho_u * \mathcal{A}_{\mathcal{T}_i})$.

Next we compute the distance d_i of each triangle centroid from the camera position, and normalize the value in the range $[0, 1]$. Now the idea is that, as the distance of a triangle increases from the camera position, the point density ρ_u is decreased by some function of d_i. For i-th triangle, the number of points to be sampled is computed as,

$$n_i = \rho_u \, \mathcal{A}_{\mathcal{T}_i} \, \gamma(d_i), \tag{9}$$

where $\gamma(\cdot)$ is a real valued function that controls the effect of distance on the point sampling. To demonstrate the effect of $\gamma(\cdot)$, we show an example of a simple branching structure in Fig. 3, where we used $\gamma(d_i) = 1$ (which is the uniform sampling), and $\gamma(d_i) = d_i^2$. The Figure shows how the point density decreases as the distance of the object part increases from the camera position, compared to the case of uniform sampling.

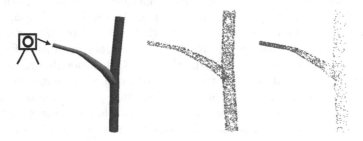

Fig. 3. Simulation of the effect of distance from the scanner position. From left to right: the original model along with camera position, using $\gamma(d_i) = 1$ (uniform point sampling), and $\gamma(d_i) = d_i^2$.

3.4   Effect of Noise

Presence of noise is a very common effect of scanning devices. While the type of noise can vary among different types of devices, we simulate a few common effects of noise in our framework. Some examples are shown in Fig. 4 on a simple branching structure as in Fig. 3. On the left of the figure, we show the effect of uniform noise, where each point is generated by a Gaussian having the surface position of the point as mean and a standard deviation of $\sigma = 0.06$. We also check the insideness testing while applying the noise, so that the noisy points do not remain inside any branch. In the middle of the figure, the noise is modelled with the effect of distance from the camera, as described in the previous section. Points closer to the scanner has almost zero noise, while the noise increases as the distance of the point increases from the position of the scanner. Finally at the right of the figure, the distance effect is shown considering the decrease of point density as described before.

Fig. 4. Simulation of noise effect. Left: uniform distribution of noise, Middle: Effect of noise on the distance from the scanner, Right: Effect of noise on the distance along with point density factor.

4 Results

We have implemented the model in L-Py framework [2]. However, other types of frameworks (e.g. L+C [18]) can also be used. In the left of Fig. 5, we show the synthetic model of an Arabidopsis plant that is generated by parametric L-system using stochastic parameters to compute different instances of the plant architecture. Using the stochasticity in the model, it is possible to generate a large variety of plants along with their labelled point cloud data. In the same figure we show some of the varieties of the generated point cloud models of the Arabidopsis plant. In all examples, we have used 3 labels (flowers, internodes and leaves) and about 100k points, although the number of labels and points can be controlled according to the need of the application. In Fig. 6, we show more detailed result of the Arabidopsis plant.

Fig. 5. From left to right: a virtual model of Arabidopsis plant, followed by labelled point cloud representation of the plant model in different growth stages.

We also show some results on different types of artificial plants in Fig. 7. This demonstrates the portability of the framework which allows us to use it in any procedural plant model. In the last example (rightmost in the same figure), we

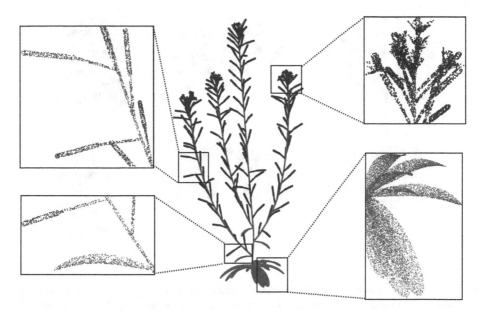

Fig. 6. An example of a point cloud model generated by the approach using 3 labels and about 100k points. Different parts are shown in details by the zoom-in.

demonstrate that the number of labels can be controlled according to the need of the application. In this case, we have used different labels for each flower of the Lilac plant (denoted by different colors).

5 Discussion

There are many ways the framework can be improved and adapted to different types of applications. For example, incorporating back face culling in the framework can be useful for applications where occlusion needs to be modelled. While we have implemented a basic sampling strategy to mimic the sensor specific distance effect, there can be several other types of effects in real world scanning devices, that yield a research avenue by themselves. Incorporating these effects in the point sampler will make the system more versatile and robust, and increase its application range. As we have simulated the effect of distance on the point density by considering the distance of the triangle centroids from the scanner, the point density remains same in all areas of one triangle. For large triangles where one part of the triangle is closer to the scanner than the other part, ideally the point density should vary accordingly inside the same triangle. However in the current framework, we have not modelled this effect.

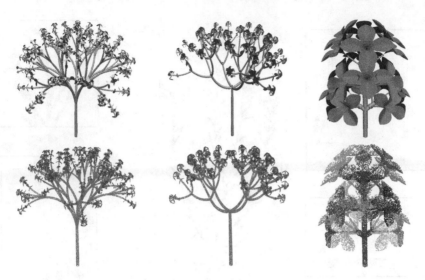

Fig. 7. Some examples of artificial plant models (top row), and the corresponding labelled point cloud data (bottom row). In the first two examples, 2 and 3 labels are used, respectively. In the last example, each flower is assigned a different label denoted by different colors. (Color figure online)

In order to model the sensor specific distance effect, the desired number of points should be significantly higher than the total number of triangles in the model. In the current implementation, we estimate the number of points per triangle based on its area, without taking into consideration the number of sampled points on the neighboring triangles. While the number of estimated points per triangle can be a fractional number according to Eq. 9, we need to round off the value to a nearest integer. Hence the part(s) of the model having large number of tiny triangles will contain at least one point inside each triangle, even when the number of estimated point for the triangle is less than 1, resulting in higher point density in these areas. Ideally the sampling should be performed based on the point density of the neighboring triangles, although this is not considered in the framework.

The current implementation is not optimized, and thus is slow in terms of computation time. For example, in order to generate 100k points in the Arabidopsis model, it takes about 1.5 h to run the basic sampler. While we have decomposed each shape into triangles to perform point sampling, the points can also be sampled directly on the surface of the geometrical shape. However, we use the triangle representation since it is simple and straightforward.

Algorithm 1: Basic Point Sampling Algorithm

Input: Procedural Plant Model
Output: Surface point cloud of the model $\mathcal{P} = \{\mathbf{p}_1, \cdots, \mathbf{p}_n\}$, label list $\boldsymbol{\ell}$
begin

    $\mathcal{P} \leftarrow \{\Phi\}, \boldsymbol{\ell} \leftarrow \{\Phi\}$
    Assign label $\ell_i, i \in \{1, \cdots, r\}$ to r organs in the procedural model
    Extract all primitives with organ labels from the model $\{\mathcal{C}_1^{\ell_i}, \cdots, \mathcal{C}_m^{\ell_i}\}$
    Extract the set of geometry parameters of the primitives $\{\lambda_1, \cdots, \lambda_m\}$
    Tessellate each primitive into set of triangles $\mathcal{T} = \{\{\mathcal{T}_1^{\ell_i}\}, \cdots, \{\mathcal{T}_m^{\ell_i}\}\}$
    Create global list of triangle vertices along with their corresponding labels
    while *Total number of Points* $! = n$ **do**

        Randomly select a triangle $\mathcal{T}_j^{\ell_i} \in \mathcal{T}$ with area probability
        $\{\mathbf{v}_1^j, \mathbf{v}_2^j, \mathbf{v}_3^j\} \leftarrow vertex(\mathcal{T}_j^{\ell_i})$ // `Extract triangle vertices`
        $\alpha_1 \leftarrow rand(0,1), \alpha_2 \leftarrow rand(0,1), \alpha_3 \leftarrow rand(0,1)$
        $\alpha_1 \leftarrow \alpha_1/\sum_{i=1}^3 \alpha_i, \alpha_2 \leftarrow \alpha_2/\sum_{i=1}^3 \alpha_i, \alpha_3 \leftarrow \alpha_3/\sum_{i=1}^3 \alpha_i$
        $\mathbf{p} \leftarrow \alpha_1 \mathbf{v}_1^j + \alpha_2 \mathbf{v}_2^j + \alpha_3 \mathbf{v}_3^j$
        insidenessFlag $\leftarrow$ *False*
        // `Loop over all primitives`
        for $k \leftarrow 1$ **to** m **do**

            if $\mathcal{C}_k^{\ell_i}$ *is a Cylinder* **then**
                insidenessFlag $\leftarrow$ isInsideCylinder$(\lambda_k, \mathbf{p})$// `using Eqn.4,5`
            else if $\mathcal{C}_k^{\ell_i}$ *is a Frustum* **then**
                insidenessFlag $\leftarrow$ isInsideFrsutum$(\lambda_k, \mathbf{p})$// `using Eqn.4,6,7`
            else if $\mathcal{C}_k^{\ell_i}$ *is a Sphere* **then**
                insidenessFlag $\leftarrow$ isInsideSphere$(\lambda_k, \mathbf{p})$// `using Eqn.8`

        if *insidenessFlag = False* **then**
            $\mathcal{P} \leftarrow \mathcal{P} \cup \mathbf{p}$
            $\boldsymbol{\ell} \leftarrow \boldsymbol{\ell} \cup \ell_i$

---

6 Conclusion

In this paper we have proposed a generalized approach to generate labelled 3D point cloud data from procedural plant models. As a preliminary work, we have presented results on a simulated Arabidopsis plant along with some artificial plant models. The point sampler is portable enough to use in different types of plant models, and corresponding point cloud data can be generated. This can be beneficial to various types of plant phenotyping research problems in performing quantitative evaluation of point cloud based algorithms, as well as in training the machine learning models. It will be worth investigating how the synthetic models perform in training deep neural networks. In future, we plan to model different types of plant species and more complex plants in the framework. While currently there is no such method available to generate synthetic labelled point cloud data, our approach is the first step towards this in promoting 3D point cloud based plant phenotyping research.

Acknowledgment. This work is supported by Robotics for Microfarms (ROMI) European project.

References

1. Bernotas, G., et al.: A photometric stereo-based 3D imaging system using computer vision and deep learning for tracking plant growth. GigaScience **8**(5), giz056 (2019)
2. Boudon, F., Pradal, C., Cokelaer, T., Prusinkiewicz, P., Godin, C.: L-Py: an L-system simulation framework for modeling plant architecture development based on a dynamic language. Front. Plant Sci. **3**, 76 (2012)
3. Buck-Sorlin, G., Delaire, M.: Meeting present and future challenges in sustainable horticulture using virtual plants. Front. Plant Sci. **4**, 443 (2013)
4. Bucksch, A., et al.: Morphological plant modeling: unleashing geometric and topological potential within the plant sciences. Front. Plant Sci. **8**, 900 (2017)
5. Chattopadhyay, S., Akbar, S.A., Elfiky, N.M., Medeiros, H., Kak, A.: Measuring and modeling apple trees using time-of-flight data for automation of dormant pruning applications. In: IEEE Winter Conference on Applications of Computer Vision (WACV), pp. 1–9 (2016)
6. Chaudhury, A., Barron, J.L.: Plant species identification from occluded leaf images. IEEE/ACM Trans. Comput. Biol. Bioinf. **17**, 1042–1055 (2018)
7. Chaudhury, A., et al.: Machine vision system for 3D plant phenotyping. IEEE/ACM Trans. Comput. Biol. Bioinf. **16**(6), 2009–2022 (2019)
8. Chaudhury, A., et al.: Computer vision based autonomous robotic system for 3D plant growth measurement. In: Proceedings of the 12th Conference on Computer and Robot Vision (CRV), pp. 290–296 (2015)
9. Cruz, J.A., et al.: Multi-modality imagery database for plant phenotyping. Mach. Vis. Appl. **27**(5), 735–749 (2015). https://doi.org/10.1007/s00138-015-0734-6
10. Dey, D., Mummert, L., Sukthankar, R.: Classification of plant structures from uncalibrated image sequences. In: Proceedings of the IEEE Workshop on the Applications of Computer Vision (WACV 2012), pp. 329–336 (2012)
11. Dutagaci, H., Rasti, P., Galopin, G., Rousseau, D.: Rose-X: an annotated data set for evaluation of 3D plant organ segmentation methods. Plant Methods **16**(1), 1–14 (2020)
12. Evers, J.B., Vos, J.: Modeling branching in cereals. Front. Plant Sci. **4**, 399 (2013)
13. Giuffrida, M.V., Scharr, H., Tsaftaris, S.A.: ARIGAN: synthetic Arabidopsis plants using generative adversarial network. In: Proceedings of ICCV Workshop on Computer Vision Problems in Plant Phenotyping, pp. 2064–2071 (2017)
14. Godin, C., Costes, E., Sinoquet, H.: Plant architecture modelling - virtual plants and complex systems. In: Turnbull, C.G.N. (ed.) Plant Architecture and Its Manipulation, chap. 9. Blackwell Publishing (2005)
15. Godin, C., Sinoquet, H.: Functional-structural plant modelling. New Phytol. **166**(3), 705–708 (2005)
16. Hamarneh, G., Jassi, P.: VascuSynth: Simulating vascular trees for generating volumetric image data with ground-truth segmentation and tree analysis. Comput. Med. Imaging Graph. **34**(8), 605–616 (2010)

17. Horvat, D., Zalik, B.: Ray-casting point-in-polyhedron test. In: Proceedings of the CESCG 2012: The 16th Central European Seminar on Computer Graphics (2012)
18. Karwowski, R., Prusinkiewicz, P.: Design and implementation of the L+C modeling language. Electron. Notes Theor. Comput. Sci. **86**(2), 134–152 (2003)
19. Kobbelt, L., Botsch, M.: A survey of point-based techniques in computer graphics. Comput. Graph. **28**(6), 801–814 (2004)
20. Kumar, N., et al.: Leafsnap: a computer vision system for automatic plant species identification. In: Fitzgibbon, A., Lazebnik, S., Perona, P., Sato, Y., Schmid, C. (eds.) ECCV 2012. LNCS, vol. 7573, pp. 502–516. Springer, Heidelberg (2012). https://doi.org/10.1007/978-3-642-33709-3_36
21. Kuznichov, D., Zvirin, A., Honen, Y., Kimmel, R.: Data augmentation for leaf segmentation and counting tasks in rosette plants. In: Proceedings of CVPR Workshop on Computer Vision Problems in Plant Phenotyping (2019)
22. Li, Y., Fan, X., Mitra, N.J., Chamovitz, D., Cohen-Or, D., Chen, B.: Analyzing growing plants from 4D point cloud data. ACM Trans. Graph. **32**(6), 1–10 (2013)
23. Lindenmayer, A., Prusinkiewicz, P.: The Algorithmic Beauty of Plants, vol. 1. Springer-Verlag, New York (1990)
24. Lobet, G., Draye, X., Périlleux, C.: An online database for plant image analysis software tools. Plant Methods **9**(1), 38 (2013)
25. Minervini, M., Fischbach, A., Scharr, H., Tsaftaris, S.A.: Finely-grained annotated datasets for image-based plant phenotyping. Pattern Recogn. Lett. **81**, 80–89 (2016)
26. Mortensen, A.K., Skovsen, S., Karstoft, H., Gislum, R.: The oil radish growth dataset for semantic segmentation and yield estimation. In: Proceedings of CVPR Workshop on Computer Vision Problems in Plant Phenotyping (2019)
27. Paulus, S., Dupuis, J., Mahlein, A., Kuhlmann, H.: Surface feature based classification of plant organs from 3D laserscanned point clouds for plant phenotyping. BMC Bioinf. **14**(1), 238 (2013)
28. Prusinkiewicz, P., Mündermann, L., Karwowski, R., Lane, B.: The use of positional information in the modeling of plants. In: Proceedings of SIGGRAPH, pp. 289–300 (2001)
29. Prusinkiewicz, P., Runions, A.: Computational models of plant development and form. New Phytol. **193**(3), 549–569 (2012)
30. Qi, C.R., Su, H., Mo, K., Guibas, L.J.: Pointnet: Deep learning on point sets for 3D classification and segmentation. In: Proceedings of the IEEE Conference on Computer Vision and Pattern Recognition, pp. 652–660 (2017)
31. Shadrin, D., Kulikov, V., Fedorov, M.: Instance segmentation for assessment of plant growth dynamics in artificial soilless conditions. In: Proceedings of BMVC Workshop on Computer Vision Problems in Plant Phenotyping (2018)
32. Skovsen, S., et al.: The GrassClover image dataset for semantic and hierarchical species understanding in agriculture. In: Proceedings of CVPR Workshop on Computer Vision Problems in Plant Phenotyping (2019)
33. Soderkvist, O.J.O.: Computer vision classification of leaves from Swedish trees. Masters thesis, Linkoping University, Sweden (2001)
34. Sodhi, P., Vijayarangan, S., Wettergreen, D.: In-field segmentation and identification of plant structures using 3D imaging. In: Proceedings of the IEEE/RSJ International Conference on Intelligent Robots and Systems (IROS 2017), pp. 5180–5187 (2017)
35. Tardieu, F., Cabrera-Bosquet, L., Pridmore, T., Bennett, M.: Plant phenomics, from sensors to knowledge. Curr. Biol. **27**(15), R770–R783 (2017)

36. Ubbens, J., Cieslak, M., Prusinkiewicz, P., Stavness, I.: The use of plant models in deep learning: an application to leaf counting in rosette plants. Plant Methods **14**(1), 6 (2018)
37. Uchiyama, H., et al.: An easy-to-setup 3d phenotyping platform for komatsuna dataset. In: Proceedings of ICCV Workshop on Computer Vision Problems in Plant Phenotyping, pp. 2038–2045 (2017)
38. Vázquez-Arellano, M., Griepentrog, H.W., Reiser, D., Paraforos, D.S.: 3-D imaging systems for agricultural applications - a review. Sensors **16**(5), 1039 (2016)
39. Vos, J., Evers, J.B., Buck-Sorlin, J.H., Andrieu, B., Chelle, M., Visser, P.H.B.D.: Functional-structural plant modelling: a new versatile tool in crop science. J. Exp. Bot. **61**(8), 2101–2115 (2010)
40. Wang, B., Gao, Y., Sun, C., Blumenstein, M., Salle, L.J.: Can walking and measuring along chord bunches better describe leaf shapes? In: Proceedings of the IEEE Conference on Computer Vision and Pattern Recognition, pp. 6119–6128 (2017)
41. Wang, D.: Unsupervised semantic and instance segmentation of forest point clouds. ISPRS J. Photogrammetry Remote Sens. **165**, 86–97 (2020)
42. Ward, D., Moghadam, P.: Scalable learning for bridging the species gap in image-based plant phenotyping. Comput. Vis. Image Underst., 103009 (2020)
43. Ward, D., Moghadam, P., Hudson, N.: Deep leaf segmentation using synthetic data. In: Proceedings of BMVC Workshop on Computer Vision Problems in Plant Phenotyping (2018)
44. Wen, W., Guo, X., Wang, Y., Zhao, C., Liao, W.: Constructing a three-dimensional resource database of plants using measured in situ morphological data. Appl. Eng. Agric. **33**(6), 747–756 (2017)
45. Wilf, P., Zhang, S., Chikkerur, S., Little, S.A., Wing, S.L., Serre, T.: Computer vision cracks the leaf code. Proc. Nat. Acad. Sci. **113**(12), 3305–3310 (2016)
46. Wu, S.G., Bao, F.S., Xu, E.Y., Wang, Y., Chang, Y., Xiang, Q.: A leaf recognition algorithm for plant classification using probabilistic neural network. In: IEEE International Symposium on Signal Processing and Information Technology, pp. 11–16 (2007)
47. Ziamtsov, I., Navlakha, S.: Machine learning approaches to improve three basic plant phenotyping tasks using three-dimensional point clouds. Plant Physiol. **181**(4), 1425–1440 (2019)

Phenotyping Problems
of Parts-per-Object Count

Faina Khoroshevsky[1]($\boxtimes$), Stanislav Khoroshevsky[1], Oshry Markovich[2],
Orit Granitz[3], and Aharon Bar-Hillel[1]

[1] Ben-Gurion University of the Negev, Be'er Sheva, Israel
{bordezki,khoroshe,barhille}@bgu.ac.il
[2] Rahan Meristem (1998) Ltd., 22825 Kibbutz Rosh Hanikra,
Western Galilee, Israel
oshrym@rahan.co.il
[3] Evogene, 13 Gad Feinstein Street, Park Rehovot, Rehovot, Israel
orit.granitz@evogene.com

Abstract. The need to count the number of parts per object arises in many yield estimation problems, like counting the number of bananas in a bunch, or the number of spikelets in a wheat spike. We propose a two-stage detection and counting approach for such tasks, operating in field conditions with multiple objects per image. The approach is implemented as a single network, tested on the two mentioned problems. Experiments were conducted to find the optimal counting architecture and the most suitable training configuration. In both problems, the approach showed promising results, achieving a mean relative deviation in range of 11%–12% of the total visible count. For wheat, the method was tested in estimating the average count in an image, and was shown to be preferable to a simpler alternative. For bananas, estimation of the actual physical bunch count was tested, yielding mean relative deviation of 12.4%.

Keywords: Deep neural networks · Object detection · Part counting · Yield estimation

1 Introduction

This work handled a specific kind of phenotyping problems: visual object's part counting, done by first detecting the objects in the image, and then counting their constituting parts. Such problems repeatedly arise in field phenotyping contexts. Examples include counting the number of bananas in banana bunches [48], the number of spikelets in wheat spikes [3,37], the number of flowers on apple trees [10,13], or the number of leaves in potted plants [11,21]. While such problems involve (object) detection and (part) counting, they should not be confused with single-stage counting or detection problems, and require a non-trivial combination of them.

© Springer Nature Switzerland AG 2020
A. Bartoli and A. Fusiello (Eds.): ECCV 2020 Workshops, LNCS 12540, pp. 261–278, 2020.
https://doi.org/10.1007/978-3-030-65414-6_19

Fig. 1. Example images with ground truth annotations. **Left:** Bananas in banana bunches. **Right:** Wheat spikelets in spikes. Blue bounding boxes are placed around measurable objects. Black dots are placed on countable parts. The target count number (number of visible parts) is stated above each bounding box. (Color figure online)

Object detection is a basic step in several types of agricultural applications. One important kind is robotic manipulation, like harvesting [4,43], plant's diseases and pests recognition [15] or autonomous selective spraying [8]. In another type of applications, object detection is the first stage for measuring object properties like height, width and length [5,49], cluster size estimation [24], or object classification [18,25,54]. In a third type of applications object detection is used as part of an objects (not parts) counting task [35]. In recent years the detection task is performed mainly with deep Convolutional Neural Networks (CNNs) [28,41,45], that enabled dramatically increased accuracy.

Object counting is also a common phenotyping need for tasks such as yield estimation [29,35,52], blossom level estimation for fruit thinning decisions [10,13, 51], or plant's leaf counting for growth stage and health estimation [7,11,21,34]. Different CNN architectures are currently the state of the art for counting as well [9,36] and for other phenotyping tasks [22].

Specifically, we address two part-counting problems in this work: banana count in a bunch, and spikelet count in a wheat spike. Image examples with annotations for each task are shown in Fig. 1. In both cases, the part count is an important yield indicator, and measuring it automatically is important for yield prediction and breed selection [17]. Spikelets counting can assist in crop management as it can be seen as a form of yield quantification for wheat crops [3] and the number of bananas in a bunch is one of the properties that is related to bunch weight and thus productivity [50]. Since these two very different problems are handled here with the same algorithm, we believe the proposed method is general and can handle other part-counting problems with minor adjustments.

One may wonder why the 'parts-counting' task (which is essentially a 'detect-then-count' procedure) cannot be reduced into plain counting, by simply taking pictures containing a single object in each image. Indeed, in principal this is possible. In some of the problems mentioned above the two-stage part-object nature was often bypassed by using images of a single centralized object [7, 13, 34]. However, solving a part counting problem by taking images of single objects has significant disadvantages. The counting procedure needs to be automated, rather than being done by tedious manual approach. If a human has to detect the objects and take single-object images, the process is partially manual with human intervention. It is hence slower, costs more, and less scalable than a fully automated system. Another option is to have a robotic system first detecting the objects, than moving to a close position to each object for taking a second image, isolating it from a wider scene. While this does not involve human intervention, such a system is much more complex to develop, and it also is likely to be slower than a system based on high resolution images containing multiple objects. Hence for automatic and flexible field phenotyping, the solution of keeping one object per image is significantly less scalable.

While parts counting is essentially a two-stage 'detect then count' task, it nevertheless raises some unique difficulties. First, parts are much smaller than their containing objects. Therefore high initial resolution and a re-scaling mechanism are required. Second, the system has to filter which detected objects are passed to the counting stage, and this involves a complex array of considerations. On the one hand not all objects are 'countable': some objects are too far, out of focus, occluded, or not well-detected. On the other hand one can usually afford measuring only a subset of the objects, and sometimes it is better to have good measurements for few than inaccurate measurement for many. While the exact policy may be task-dependent, the filtering creates dependency between the detector and the data distribution seen by the counter. As will be demonstrated in Sect. 4.2, this has implications w.r.t the preferred training procedure.

A third issue arises since counting itself is not a solved problem: there are several competing approaches, and the better choice seems to depend on data distribution and quantity, and on availability of annotation. Some counting networks work by explicit detection of the objects to count [6, 17, 33], some use density estimation [26, 44], and others directly regress the count locally or globally, without intermediate stages [32, 52].

This work explores the design space of possible solutions for parts-counting, and finds a useful and relatively simple solution. We proposed a single network combining stages for the object detector, cropping and scaling of the objects, and then a counting stage with two optional implementations. One uses a counting module which is a direct regressor, hence part position annotation is not necessary for training. The other does require the objects' parts annotations, but in addition to counting, it also provides parts localization. Following experiments, we recommend choosing between the two possible solutions based on computational efficiency and the need for part localization considerations, since both solution provide good and comparable results. The training procedure of the entire network is also discussed, showing that it can be done jointly or separately, as long as proper noise is introduced during learning.

We benchmark the proposed solutions on the two problems using several metrics. The most intuitive count measure is the (average) relative deviation, stating the average deviation of the count from its true value in percentage. Statistically, an important measure is the fraction of explained variance in the data. In relative deviation terms, the networks achieve 10.1% and 10.8% for banana and spikelet count respectively. Statistically, we are able to explain 0.58 and 0.75 of the count variance for the two problems respectively.

While the obtained accuracy is encouraging, it is obtained for estimation of the visible number of parts. For bananas, which are round and with significant volume, the visible number and the actual number of bananas in a bunch may be far from each other, as approximately half the bananas cannot be seen. We have collected a separate dataset of bunches with actual counts made manually, and were able to establish a strong linear connection between visible and actual banana count. Hence fairly accurate actual counts can also be obtained.

Our contribution is hence twofold: developing a successful (in terms of relative deviation and the fraction of explained variance) general algorithmic framework for the parts-per-object counting problem, and showing its merit on two important phenotyping cases. The rest of the paper is organized as follows: we review related work in Sect. 2, describe the algorithm in Sect. 3, present results in Sect. 4, and discuss further work in Sect. 5.

2   Related Work

Since our network is composed of two sub-networks for detection and counting, we discuss how these tasks are approached in general and in the agricultural domain, than focus on the works closest to our specific tasks.

Visual object detection had significantly advanced in recent years by incorporating deep neural networks extending CNN architectures. This enables tackling more intricate tasks as we do here. One line of influential networks includes two-staged networks like Faster R-CNN [41] or Mask R-CNN [19]. These networks include a stage of Region Proposal Network (RPN) which finds initial object candidates, and a second sub-network which accepts the candidates and handles final classification and bounding-box fine-tuning. Object candidates are passed between the two stages by sampling the feature maps produced by the RPN at Regions of Interest (RoIs). Technically, this is done with a non-differentiable layer termed RoI Pooling in Faster-R-CNN [16,41], and RoI Align in Mask-R-CNN [19], which samples feature map regions into small tensors with a fixed spatial extent. In this work we used the RoI Align layer to pass detected objects from the detector into the counting sub-network.

Another line of work includes fully differentiable one-stage architectures. Early one-stage networks as YOLO [40] and SSD [31] presented faster detectors, but with accuracy lower by 10–40% relative to two-stage methods [28]. Later one-stage networks like RetinaNet [28] and more recently EfficientDet [47] were able to remove the accuracy gap. Specifically RetinaNet introduced the usage of a Feature Pyramid Network (FPN) [27] for handling larger scale variability of objects, and a 'focal loss' variant which well balances between positive examples (which are few in a detection task) and negative examples.

In the agricultural context, object detection has numerous usages as it is a basic building block in automated harvesting [4,43], yield estimation [13,52], and phenotyping [5,49]. Several detection benchmarks [6,42,54] have shown that current detectors obtain high F_1 and AP scores in agricultural settings, hence encouraging two-staged tasks as pursued here. However, in [42,54] the datasets included mostly images taken very near to the crop of interest, hence including only few (typically less than 5) large objects. These images are significantly less challenging than images taken in our wheat spikelet counting task. In another work, images were taken in field conditions from a significant distance, with the target of fruit yield estimation [6]. To overcome the problem of object occlusions, it was assumed that on average there is a constant ratio of visible to occluded fruits, and detection outputs were calibrated with ground-truth counts. We use a similar assumption in this work.

Counting tasks in agriculture contexts were approached with traditional computer vision techniques [1,2,30], but has advanced in recent years by the incorporation of CNNs. Basically, there are two main approaches for building counting networks. The first is based on some form of explicit object localization: objects are first detected, then counted. The detection task can be defined as detection of the object's centers, and so the output is a 'density estimation' heat map, a two dimensional map showing where 'objecthood probability' is high [21,44]. Alternatively localization can be based on bounding box detection [17,33,35] or segmentation [14,38,55]. These methods require object location annotations like center point annotations (for density estimation), bounding boxes (for detection) or surrounding polygons (for segmentation). The second approach is via a global [11,21,39] or local direct regression model [32,52]. In global regression, the model implements a function mapping the entire image (or region) to a single number, which is the predicted count. In local regression the image (or region) is divided into local regions and each of them is mapped to a local predicted count. The local counts are then summed to get the global count.

The literature contains contrasting evidence regarding the relative accuracy of localization-based versus direct regression methods. In [33] a detection based method was found superior to direct regression and it is claimed to be more robust to train-test domain changes. Contrary, at [32] direct regression provides higher accuracy than density estimation. At [21] the two approaches show similar performance. An advantage of detection based methods is in being more "debuggable" and "explainable", as it allows to visually inspect where in the image the network finds objects. It is also able to provide the objects location information when it is needed for further processing. However, more computation effort is required at test time, and an additional annotating effort at train time.

While a lot of previous work has addressed wheat spike counting [2,14,17,32, 33,52,55] we are not familiar with work attempting to solve the spikelet-count-per-spike or the banana-per-bunch problems we face here. Global spikelet count was addressed in the recent papers [3,37]. In [37] both spike and spikelets were counted in glassdoor conditions, where background is suppressed and each image contains a few large spikes. A density estimation approach was used, with very good results of relative error <1% for spikelets and <5% for spikes. While in principal the average spikelet-per-spike statistic can be estimated from the two global counts, we show in our experimental results that this method provides inferior accuracy in field images. In [3] only spikelets are counted with density estimation, but in challenging field conditions as we address in our problem.

3 Method

The suggested network is comprised of several components. First is a detection section responsible for detecting the objects of interest. These are passed to the RoI-Align module that crops and resizes the detected objects from the original image. The crops are passed to the counting section, in which each crop is treated as an independent input image. It is assumed that each crop contains a single detected object, and the counting section outputs the part's count for that object. The assumption's correctness naturally depend on the performance of the object detection section. The basic composition of the detection and counting network is demonstrated in Fig. 2.

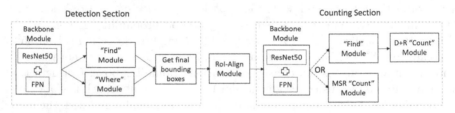

Fig. 2. The detection and counting network. The network includes two separate sections for object detection and part counting, connected by an object re-sampling layer

The detection section is a re-implementation of the RetinaNet architecture, composed of "Backbone", "Find" and "Where" modules, whose details are explained in Sect. 3.1. The RoI-Align module receives as input the original image and a set of rectangle coordinates of the detected objects. By re-sampling the image according to the given coordinates, it outputs a tensor of size $s \times s \times n_d$ of image crops, where s is the new object size (empirically set to 640) and n_d is the number of detected objects. Such resizing allows handling the significant variation in object size of the different datasets without the need of special configuration. We use the RoI-Align implementation of Mask-R-CNN [19].

The counting section includes re-implementations of the counters proposed by [21], composed using "Backbone", "Find" and "Count" modules. Modules reused in different sections ("Backbone" and "Find" appear in both the detector and counter sections) share code and architecture, but not parameters. The counters and their composition using the modules are described at Sect. 3.2.

Both the detection and the counting sub-networks have their own separate losses, which are learned separately. The training methodology of the whole network is discussed at Sect. 4.2.

3.1   The Detection Section

The relevant modules for the detection section of the network are the following:

"Backbone": A module used for creating a dense feature-rich representation of the image. It is based on applying ResNet-50 [20] as a dense convolutional network followed by FPN [27] on top of it. The FPN architecture generates a rich 5-scale feature pyramid (P_3–P_7) representation of the input image. These tensors include similar representations of the original image in multiple octaves, with P_i twice smaller than P_{i-1} in its spatial dimensions. Therefore they enable object detection at multiple scales. In all experiments we have used the ResNet-50 network with weights pre-trained on ImageNet and fine-tuned them.

"Find" (For Detection): The module includes 4 convolutional layers and it is trained for spatially locating objects in an input tensor representation. It supports two output modes with the first used for the detection section, the second for the counting section. For the detection pipeline, it implements the fully convolutional classification sub-network of RetinaNet that predicts the probability of object presence at each spatial position of an input tensor. At each position, this probability is estimated for nine possible anchor rectangles of pre-defined sizes and aspect ratios. The output tensor is thus with the same spatial dimensions as the input, but with nine score maps. The "Find" module processes independently all the five pyramid tensors produced by the Backbone's FPN. It is trained with the Focal Loss, designed to address possible imbalance between foreground and background classes during training [28].

"Where": This module implements the bounding box regression sub-network of RetinaNet. Accepting a representation tensor as input, it predicts for every position and possible anchor (among the nine considered) a 4-dimensional bounding box refinement vector (so the output includes 36 maps overall). The vector includes corrections required for the bounding box to better match the object in its $(x, y, width, height)$ parameters. It is trained by propagating a smooth-$L1$-regression loss matching the predicted values to ground-truth rectangles - only for anchors with relevant objects. Like the "Find" module, the module processes all the pyramid tensors P_3–P_7, and produces 5 output tensors.

"Get Final Bounding Boxes": This function accepts the output tensors of the "Find" and "Where" modules and creates a refined list of predicted boxes containing presumed objects. It filters boxes with predicted object probability higher than 0.7 from the "Find" module, and decodes the refined boxes for them according to the "Where" module. The predictions from all pyramid levels are merged and a non-maximum suppression procedure with a threshold of 0.3 is applied to yield the final detections list.

3.2 The Counting Section

Given an input image with a single object, the counting section outputs the count value of its parts. It has two variations, following the proposed architectures by [21]. The first variation uses the implementation of the Multiple Scale Regression (MSR) algorithm, that directly regresses the count value. The second variation is a re-implementation of the Detection+Regression (D+R) algorithm which predicts heat maps of the parts centers in addition to counting them, and thus requires parts location annotations in training. In both cases, the counting section starts with a second "Backbone" module receiving as input detected region objects.

MSR "Count": In this module, the representation tensors P_3–P_7 generated by the "Backbone", or a subset of them, are each sent to a direct regression module, so multiple count estimations are produced based on different resolutions. This regression module includes several convolutional and fully connected layers culminating in a two outputs neurons: the first predicts the expected count, and the second estimates the variance of the error expected in prediction, using the loss function suggested in [23]. The count estimates are then fused as a weighted average with weights based on the variance values.

For the D+R variation, the count section includes a "Find" module variation described next, and then a different "Count" module.

"Find" (For Counting): This module has similar architecture to the "Find" module used in detection, but it is used to "Find" center points instead of bounding box rectangles. The module operates only on the high-resolution pyramid scale (P_3), and a single output map of the same resolution is created, stating the probability of object presence at each location. To learn this network, a ground truth heat map is created in training, with a Gaussian kernel placed around each part's center. Like the detection "Find" module, the module contains four convolutional layers, but after that a single final map is predicted. It is trained to mimic the ground truth heat map using a dense L_1 regression loss.

Following [21], we tested the option of guiding internal layers by predicting an intermediate heat map after each convolutional layer, and forcing it to mimic the ground truth heat map with additional regression losses. When this is done, the intermediate guiding heat maps are created with decreasing kernel size, so the heat map regression task is cruder and simpler at initial layers.

D+R "Count": This module gets from the "Find" module the predicted heat map and the feature tensor preceding it, and provides a count estimate based on them. It is an implementation of the "Counting sub-network" of the D+R pipeline of [21]. The heat map is subjected to a smooth non-maxima suppression operation, keeping activity of the most active points close to ones while all others are zeroed. These remaining active points are object center predictions, and a global sum operation now gives an initial detection-based estimator of the count.

In addition to this path, the tensor preceding the map is globally summed to extract additional useful features, and the final count estimate is computed as linear regression from these features and the detection-based estimate.

4   Experiments and Results

We describe the datasets used and performance evaluation methodology in Sect. 4.1. Results in the part counting tasks and experiments measuring the relation between visible and actual number of parts for banana bunches are presented in Sect. 4.2.

Table 1. Datasets sizes

Dataset	Train (Images/Objects/Parts)	Validation (Images/Objects/Parts)	Test (Images/Objects/Parts)
Wheat	44/223/3523	27/102/1779	30/135/2347
Banana	72/82/6822	34/41/3372	35/43/3432

4.1   Experimental Setup

Wheat Dataset: The collected dataset includes 101 RGB images taken in an agricultural facility in the Central District of Israel. High resolution, 6000 × 4000, images of several wheat varieties in field conditions were taken using a commercial DSLR camera. Since the original images included hundreds of wheat spikes, several regions (usually 4–5, 1020 × 830 crops) in focus were handpicked in each image. These areas were cropped and treated as separate images that were passed for annotation of well defined, measurable spikes.

Bananas Dataset: 141 RGB images were captured in a facility in the Northern District of Israel. Images included banana bunches of different varieties and stages of the reproductive phase. These were taken using commercial 9MP-12MP mobile device cameras and a digital point-and-shoot cameras in field conditions. The common resolutions were 2340 × 4160 and 3024 × 4032.

All images were annotated with bounding boxes for countable objects, and with center dot annotation for each object part. Countable objects were defined as "objects for which a human annotator can visually count the parts". At least for wheat, the distinction between countable and non-countable spikes is slightly vague, and depends to some extent on the annotators' judgement. The number of images, objects and parts used is listed in Table 1.

Detection performance is measured by the Average Precision (AP) metric [12]. An object is considered to be found if its Intersection over Union (IoU) is at least 0.5 with a ground truth object annotation. For counting, denote the set of ground truth counts of the test set by $\{y_i\}_{i=1}^{N}$ and the set of count prediction of the model by $\{\hat{y}_i\}_{i=1}^{N}$. We estimate counting accuracy using the Mean Relative Deviation (MRD) and $1 - FVU$ statistics defined as follows:

$$MRD = \frac{1}{N} \sum_{i=1}^{N} [\frac{|\hat{y}_i - y_i|}{y_i}], \quad 1 - FVU = 1 - \frac{\sum_{i=1}^{N}(y_i - \hat{y}_i)^2}{\sum_{i=1}^{N}(y_i - \bar{y})^2} \qquad (1)$$

Where $\bar{y} = \frac{1}{N} \sum_{i=1}^{N} y_i$ is the mean ground truth count. The relative deviation states the count deviation as a fraction of the total count (the count deviation percentage). We believe this metric is more informative for our datasets than MSE or MAE since our task requires estimation of large numbers - there are dozens of bananas per bunch and spikelets per spike. $1 - FVU$, known as the fraction of explained variance, checks if the quality of the predictor we use (its mean squared error in the nominator) is significantly better than the trivial predictor of guessing that y_i is always equal to the mean $\bar{y}$ (the mean squared error of this predictor is the denominator – it is also the variance of y).

For each dataset, counting networks were trained for 300 epochs. The best epoch was chosen as the one with the lowest mean relative deviation, measured on the validation set, and averaged across IoU thresholds of 0.3, 0.5 and 0.7. This model was then tested on the test set images, providing the reported results.

4.2   Results

Based on the stages order of the suggested method, we start by describing object detection results. Then we report part counting results obtained with various counters, and consider the effect of training procedure on the obtained accuracy. Next, we compare the results to an alternative approach, in which objects and parts are counted independently, and report experiments predicting the actual physical count for banana bunches.

Detection Results: A confidence threshold of 0.7 were chosen for the trained object detectors, to provide high precision. Table 2 shows the recall and precision of the detectors used on the test set. As can be observed from the recall statistics, for wheat the detection problem is more difficult, due to the fine distinction required between countable and non-countable spikes.

Table 2. Test set recall and precision of the object detectors.

Dataset	Number of objects	Recall	Precision
Wheat spike	135	45.9%	74.7%
Banana bunch	43	81.4%	83.3%

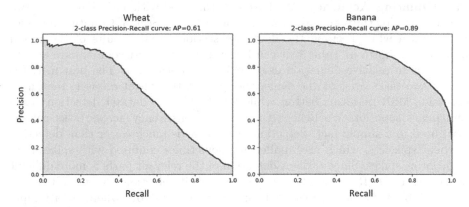

Fig. 3. Precision-Recall curves for parts detection, obtained by the D+R method. Unlike the direct regression, this method provides explicit localization of the detected parts, evaluated in these curves. In order to determine if a part detection is a hit or a miss we use the Percentage of Correct Keypoints (PCK) criterion, introduced in [53] and applied as in [21]. **Left:** Wheat spikelets. **Right:** Banana fruits in a bunch.

Fig. 4. Example of the network's stages performance with the D+R pipeline. **Top Row:** Banana, **Bottom Row:** Wheat. **(1).** Object detection (blue-ground truth, red-predictions),**(2).** Getting image crops of the detected objects, **(3).** Ground truth Gaussian heat map of the object's parts, **(4).** The final predicted heat map of the part's centers. White rectangle shows the detections of unannotated parts, found by the predicted map.

Part Counting Accuracy: We have experimented with variations of the two counting architectures suggested in [21], described in Sect. 3. Counting modules were trained while keeping the detection section fixed. Results of the tested counters are shown in Table 3. It can be seen that the best counters are able to achieve mean relative deviation close to 11% in both tasks. The best methods for the two tasks are not the same, though. On the wheat dataset, the direct regression MSR performs better, while for the banana dataset detection based counting is superior. We believe this lack of consistency across tasks can be attributed to a simple fact: detection of bananas is much easier than detection of wheat spikelets, which are smaller and sometimes confused with spikelets of neighboring spikes. Since explicit detection of the relevant parts is more difficult for wheat, the D+R method based on such explicit detection is weaker. It can be seen in Table 3 that for the wheat task, the results improve when less attempts are made to "educate" the network toward explicit detection (as we move from the original D+R method toward complete lack of intermediate detection losses).

The part detection results of the D+R method, presented at Fig. 3, further support the above mentioned view. It can be seen that banana fruits are much easier to detect, with AP of 0.89 obtained, while for spikelets it is only 0.61. A demonstration of the networks D+R pipeline operation can be seen in Fig. 4.

Table 3. Part counting results for several counter ablations. The first three rows show the results obtained by the MSR approach, when integrating a single resolution (P_3), 3 resolutions (P_3–P_5), or the originally proposed 5 resolutions (P_3–P_7). The last three rows show results of the D+R method. The original D+R method is compared to a variation in which the learned heat maps are generated with fixed kernel size (same radii), and to a variation in which no intermediate losses are used.

Counting approach	Wheat		Banana	
	Mean relative deviation	1-FVU	Mean relative deviation	1-FVU
MSR (P_3)	11.4%	0.718	13.5%	0.341
MSR (P_3–P_5)	10.8%	0.745	13.4%	0.349
MSR (P_3–P_7)	11.6%	0.652	12.7%	0.375
D+R	14.1%	0.641	12.1%	0.461
D+R (same radii)	12.9%	0.684	11.5%	0.581
D+R (no intermediate losses)	11.0%	0.759	13.1%	0.426

Training Methodology. In the training procedure described above, the detector is kept fixed while the counter is training. However, several other options are possible. One appealing option is to use a 'perfect' detector in training, which provides the counter with ground truth object bounding boxes, thus enabling it to train in 'ideal' conditions. An opposite intuition is that the counter should train in far from ideal, noisy conditions, so training should include detection noise. We have experimented with several alternatives using the best counter found for the wheat dataset (MSR using resolutions P_3–P_5), and the results are shown in Table 4. The main trend visible is that training with a perfect oracle detector provides inferior results – the counter needs to see imperfect bounding boxes while training. However, obtaining such imperfect bounding-boxes can be done either by using the imperfect detector, or/and by adding explicit detection noise. Training the detector simultaneously with the counter reduced accuracy a bit for the wheat problem, but it can be considered if training time is an issue (as training simultaneously is faster than stage-wise training). Moreover using a similar simultaneous training approach helped us to improve the relative deviation of the D+R method on the banana dataset to 10.1% (with 1-FVU = 0.584).

Table 4. Results of the MSR with P_3–P_5 counter ("fusing" three resolutions) trained with several procedures varying w.r.t the training detector. All methods were tested under the same conditions as in previous experiment, i.e. using the wheat object detector reported in Table 2. In bounding box noise addition, boxes provided by the detector were translated horizontally and vertically and scaled by random factors, drawn from the uniform distributions over $[-10 : 10]$, $[0.9 : 1.1]$ respectively (translation offsets are in pixels).

Training regime	Description	Mean relative deviation	1-FVU
T1	Fixed test detector	10.8%	0.745
T1 w. ST	Fixed test detector + scale and translation augmentation	10.7%	0.649
T1+	Test detector at $t = 0$ + subsequent learning	11.4%	0.588
TP	Perfect detector	13.8%	0.479
TP w. ST	Perfect detector + scale and translation augmentation	10.6%	0.637
L	Learning detector together with counter	11.6%	0.600

Image Average Counts and Comparison to Independent Object and Part Counting. In the wheat task, the statistic of interest for the breeder is not the spikelet count per specific spikes, but the *average* of this count over the entire piece of land captured in the image. Hence an alternative approach seems reasonable: counting the number of objects in the image (spike count), independently count the parts (spikelets), and then divide the two counts to get the average of interest. While this approach is simple, it suffers from two disadvantages relative to the two staged approach. First, it does not include a resize mechanism enlarging the relevant objects, hence detection of the small

Fig. 5. Ground truth (blue) and predicted (red) bounding boxes for spike and spikelets detected independently. The lack of object-part correspondence is clearly visible.

Table 5. Accuracy comparison between the two-stage and the independent detection methods.

Method	Mean relative deviation	Pearson correlation coefficient
Two stage	11.50%	0.937
Independent	17.82%	0.565

Table 6. Prediction of actual banana fruits count per bunch based on visual ground truth count, and based on network predictions.

Predicting from	Mean relative deviation	1-FVU
Visual ground truth	12.42%	0.554
Network predictions	12.42%	0.541

spikelets is more difficult. Second, not all spikes are 'countable' (due to scale and occlusion), and in many spikes only part of the spikelets can be observed. Independent counting of spikelets is therefore likely to count spikelets which do not belong to countable spikes, producing an over-estimation.

We empirically tested the subject by training a detector for independent detection of spikes and spikelets. An output example of this detector is presented in Fig. 5. For each image, an estimated value $\hat{y}_d$ of the average spikelets per spike was computed as the ratio of the detected spikelets count and the detected spikes count. To enable fair comparison with the two-stage method, which has internal ability for linear regression, a regressor of the form $\hat{y} = a\hat{y}_d + b$ was trained. The coefficients a, b were chosen to minimize the squared error expectation over the validation set $E_v[\hat{y} - y_{gt})^2]$. For the two stage method, an estimate of the average (over image) spikelets-per-spike was computed by simple averaging of the spikelet count over all detected spikes. Comparison between the two methods, presented in Table 5, shows that the 2-stage method is significantly more accurate.

Physical Counting. While for countable wheat spikes the vast majority of the spikelets are visible, banana bunches are round objects, where typically at least half of the bananas are occluded. An additional dataset was collected to investigate the relation between actual and visual banana count, containing 30 banana bunches with known actual count, photographed from 3 viewpoints each. Linear regression deviation was then estimated using a Leave-one-out cross validation procedure. This was done for inference of the actual count from the visual ground truth count, and (separately) from the network predicted count. The accuracy obtained is reported in Table 6.

5   Conclusions and Future Work

We presented a two-staged network that successfully performs per-object part counting in terms of mean relative count deviation and 1-FVU measurements. It was demonstrated on two real world agricultural problems, counting banana fruits in a bunch and spikelets in a spike, with images taken in field conditions. Several technical improvements should be considered in future work, including replacing the non-differentiable RoI-Align module with a differentiable one, or replacing the ResNet based backbone with an improved alternative such as EfficientNet [46]. Beyond these, it would be interesting to test the suggested approach in additional part counting tasks.

Acknowledgements. This research was supported by the Generic technological R&D program of the Israel innovation authority, the Phenomics consortium and the Ministry of Science & Technology, Israel.

References

1. Aich, S., et al.: Deepwheat: Estimating phenotypic traits from crop images with deep learning. In: Proceedings of the IEEE Winter Conference on Applications of Computer Vision (WACV 2018), pp. 323–332. IEEE (2018)
2. Alharbi, N., Zhou, J., Wang, W.: Automatic counting of wheat spikes from wheat growth images (2018)
3. Alkhudaydi, T., Zhou, J., De La lglesia, B.: SpikeletFCN: counting spikelets from infield wheat crop images using fully convolutional networks. In: Rutkowski, L., Scherer, R., Korytkowski, M., Pedrycz, W., Tadeusiewicz, R., Zurada, J.M. (eds.) ICAISC 2019. LNCS (LNAI), vol. 11508, pp. 3–13. Springer, Cham (2019). https://doi.org/10.1007/978-3-030-20912-4_1
4. Arad, B., et al.: Development of a sweet pepper harvesting robot. J. Field Robot. **37**, 1027–1039 (2020)
5. Baharav, T., Bariya, M., Zakhor, A.: In situ height and width estimation of sorghum plants from 2.5 d infrared images. Electron. Imaging **2017**(17), 122–135 (2017)
6. Bargoti, S., Underwood, J.P.: Image segmentation for fruit detection and yield estimation in apple orchards. J. Field Robot. **34**(6), 1039–1060 (2017)
7. Bell, J., Dee, H.: Aberystwyth leaf evaluation dataset. (17–36), 2 (2016). https://doi.org/10.5281/zenodo.168158
8. Berenstein, R., Shahar, O.B., Shapiro, A., Edan, Y.: Grape clusters and foliage detection algorithms for autonomous selective vineyard sprayer. Intell. Serv. Robot. **3**(4), 233–243 (2010)
9. Cholakkal, H., Sun, G., Khan, F.S., Shao, L.: Object counting and instance segmentation with image-level supervision. In: The IEEE Conference on Computer Vision and Pattern Recognition (CVPR) (2019)
10. Dias, P.A., Tabb, A., Medeiros, H.: Apple flower detection using deep convolutional networks. Comput. Ind. **99**, 17–28 (2018)
11. Dobrescu, A., Valerio Giuffrida, M., Tsaftaris, S.A.: Leveraging multiple datasets for deep leaf counting. In: Proceedings of the IEEE International Conference on Computer Vision Workshops, pp. 2072–2079 (2017)

12. Everingham, M., Van Gool, L., Williams, C.K.I., Winn, J., Zisserman, A.: The Pascal visual object classes (VOC) challenge. Int. J. Comput. Vision **88**(2), 303–338 (2010)
13. Farjon, G., Krikeb, O., Hillel, A.B., Alchanatis, V.: Detection and counting of flowers on apple trees for better chemical thinning decisions. Precis. Agric. **21**, 1–19 (2019)
14. Fernandez-Gallego, J.A., Kefauver, S.C., Gutiérrez, N.A., Nieto-Taladriz, M.T., Araus, J.L.: Wheat ear counting in-field conditions: high throughput and low-cost approach using RGB images. Plant Methods **14**(1), 22 (2018)
15. Fuentes, A., Yoon, S., Kim, S.C., Park, D.S.: A robust deep-learning-based detector for real-time tomato plant diseases and pests recognition. Sensors **17**(9), 2022 (2017)
16. Girshick, R.: Fast R-CNN. In: Proceedings of the IEEE International Conference on Computer Vision, pp. 1440–1448 (2015)
17. Hasan, M.M., Chopin, J.P., Laga, H., Miklavcic, S.J.: Detection and analysis of wheat spikes using convolutional neural networks. Plant Methods **14**(1), 100 (2018)
18. Haug, S., Ostermann, J.: A crop/weed field image dataset for the evaluation of computer vision based precision agriculture tasks. In: Agapito, L., Bronstein, M.M., Rother, C. (eds.) ECCV 2014. LNCS, vol. 8928, pp. 105–116. Springer, Cham (2015). https://doi.org/10.1007/978-3-319-16220-1_8
19. He, K., Gkioxari, G., Dollár, P., Girshick, R.: Mask R-CNN. In: Proceedings of the IEEE International Conference on Computer Vision, pp. 2961–2969 (2017)
20. He, K., Zhang, X., Ren, S., Sun, J.: Deep residual learning for image recognition. In: Proceedings of the IEEE Conference on Computer Vision and Pattern Recognition, pp. 770–778 (2016)
21. Itzhaky, Y., Farjon, G., Khoroshevsky, F., Shpigler, A., Bar-Hillel, A.: Leaf counting: multiple scale regression and detection using deep CNNs. In: BMVC, p. 328 (2018)
22. Kamilaris, A., Prenafeta-Boldú, F.X.: A review of the use of convolutional neural networks in agriculture. J. Agric. Sci. **156**(3), 312–322 (2018)
23. Kendall, A., Gal, Y.: What uncertainties do we need in Bayesian deep learning for computer vision? In: Advances in Neural Information Processing Systems, pp. 5574–5584 (2017)
24. Kurtser, P., Ringdahl, O., Rotstein, N., Berenstein, R., Edan, Y.: In-field grape cluster size assessment for vine yield estimation using a mobile robot and a consumer level RGB-D camera. IEEE Robot. Autom. Lett. **5**(2), 2031–2038 (2020)
25. Le, T.T., Lin, C.Y., et al.: Deep learning for noninvasive classification of clustered horticultural crops – a case for banana fruit tiers. Postharvest Biol. Technol. **156**, 110922 (2019)
26. Lempitsky, V., Zisserman, A.: Learning to count objects in images. In: Advances in Neural Information Processing Systems, pp. 1324–1332 (2010)
27. Lin, T.Y., Dollár, P., Girshick, R., He, K., Hariharan, B., Belongie, S.: Feature pyramid networks for object detection. In: Proceedings of the IEEE Conference on Computer Vision and Pattern Recognition, pp. 2117–2125 (2017)
28. Lin, T.Y., Goyal, P., Girshick, R., He, K., Dollár, P.: Focal loss for dense object detection. In: Proceedings of the IEEE International Conference on Computer Vision, pp. 2980–2988 (2017)
29. Linker, R.: A procedure for estimating the number of green mature apples in night-time orchard images using light distribution and its application to yield estimation. Precis. Agric. **18**(1), 59–75 (2017)

30. Liu, T., Wu, W., Chen, W., Sun, C., Zhu, X., Guo, W.: Automated image-processing for counting seedlings in a wheat field. Precis. Agric. **17**(4), 392–406 (2016)
31. Liu, W., et al.: SSD: single shot multiBox detector. In: Leibe, B., Matas, J., Sebe, N., Welling, M. (eds.) ECCV 2016. LNCS, vol. 9905, pp. 21–37. Springer, Cham (2016). https://doi.org/10.1007/978-3-319-46448-0_2
32. Lu, H., Cao, Z., Xiao, Y., Zhuang, B., Shen, C.: TasselNet: counting maize tassels in the wild via local counts regression network. Plant Methods **13**(1), 79 (2017)
33. Madec, S., et al.: Ear density estimation from high resolution RGB imagery using deep learning technique. Agric. For. Meteorol. **264**, 225–234 (2019)
34. Minervini, M., Fischbach, A., Scharr, H., Tsaftaris, S.A.: Finely-grained annotated datasets for image-based plant phenotyping. Pattern Recogn. Lett. **81**, 80–89 (2016)
35. Neupane, B., Horanont, T., Hung, N.D.: Deep learning based banana plant detection and counting using high-resolution red-green-blue (RGB) images collected from unmanned aerial vehicle (UAV). PLoS ONE **14**(10), e0223906 (2019)
36. Paul Cohen, J., Boucher, G., Glastonbury, C.A., Lo, H.Z., Bengio, Y.: Countception: Counting by fully convolutional redundant counting. In: Proceedings of the IEEE International Conference on Computer Vision Workshops, pp. 18–26 (2017)
37. Pound, M.P., Atkinson, J.A., Wells, D.M., Pridmore, T.P., French, A.P.: Deep learning for multi-task plant phenotyping. In: Proceedings of the IEEE International Conference on Computer Vision Workshops, pp. 2055–2063 (2017)
38. Qiongyan, L., Cai, J., Berger, B., Okamoto, M., Miklavcic, S.J.: Detecting spikes of wheat plants using neural networks with laws texture energy. Plant Methods **13**(1), 83 (2017)
39. Rahnemoonfar, M., Sheppard, C.: Deep count: fruit counting based on deep simulated learning. Sensors **17**(4), 905 (2017)
40. Redmon, J., Divvala, S., Girshick, R., Farhadi, A.: You only look once: unified, real-time object detection. In: Proceedings of the IEEE Conference on Computer Vision and Pattern Recognition, pp. 779–788 (2016)
41. Ren, S., He, K., Girshick, R., Sun, J.: Faster R-CNN: towards real-time object detection with region proposal networks. In: Advances in Neural Information Processing Systems, pp. 91–99 (2015)
42. Sa, I., Ge, Z., Dayoub, F., Upcroft, B., Perez, T., McCool, C.: Deepfruits: a fruit detection system using deep neural networks. Sensors **16**(8), 1222 (2016)
43. Santos, T.T., de Souza, L.L., dos Santos, A.A., Avila, S.: Grape detection, segmentation, and tracking using deep neural networks and three-dimensional association. Comput. Electron. Agric. **170**, 105247 (2020)
44. Sindagi, V.A., Patel, V.M.: Generating high-quality crowd density maps using contextual pyramid CNNs. In: Proceedings of the IEEE International Conference on Computer Vision, pp. 1861–1870 (2017)
45. Tan, M., Pang, R., Le, Q.: Efficientdet: Scalable and efficient object detection. arXiv preprint arXiv:1911.09070 (2019)
46. Tan, M., Le, Q.V.: Efficientnet: Rethinking model scaling for convolutional neural networks. arXiv preprint arXiv:1905.11946 (2019)
47. Tan, M., Pang, R., Le, Q.V.: EfficientDet: scalable and efficient object detection. arXiv preprint arXiv:1911.09070 (2019)
48. Turner, D., Mulder, J., Daniells, J.: Fruit numbers on bunches of bananas can be estimated rapidly. Sci. Hortic. **34**(3–4), 265–274 (1988)

49. Vit, A., Shani, G., Bar-Hillel, A.: Length phenotyping with interest point detection. In: Proceedings of the IEEE Conference on Computer Vision and Pattern Recognition Workshops (2019)
50. Wairegi, L., Van Asten, P., Tenywa, M., Bekunda, M.: Quantifying bunch weights of the east African highland bananas (Musa spp. AAA-EA) using non-destructive field observations. Sci. Hortic. **121**(1), 63–72 (2009)
51. Wang, Z., Underwood, J., Walsh, K.B.: Machine vision assessment of mango orchard flowering. Comput. Electron. Agric. **151**, 501–511 (2018)
52. Xiong, H., Cao, Z., Lu, H., Madec, S., Liu, L., Shen, C.: TasselNetv2: in-field counting of wheat spikes with context-augmented local regression networks. Plant Methods **15**(1), 150 (2019)
53. Yang, Y., Ramanan, D.: Articulated human detection with flexible mixtures of parts. IEEE Trans. Pattern Anal. Mach. Intell. **35**(12), 2878–2890 (2012)
54. Zheng, Y.Y., Kong, J.L., Jin, X.B., Wang, X.Y., Su, T.L., Zuo, M.: CropDeep: the crop vision dataset for deep-learning-based classification and detection in precision agriculture. Sensors **19**(5), 1058 (2019)
55. Zhou, C., Liang, D., Yang, X., Yang, H., Yue, J., Yang, G.: Wheat ears counting in field conditions based on multi-feature optimization and TWSVM. Front. Plant Sci. **9**, 1024 (2018)

Abiotic Stress Prediction from RGB-T Images of Banana Plantlets

Sagi Levanon[1], Oshry Markovich[2], Itamar Gozlan[1], Ortal Bakhshian[2],
Alon Zvirin[1($\boxtimes$)], Yaron Honen[1], and Ron Kimmel[1]

[1] Computer Science Department, Technion – Israel Institute of Technology,
Haifa, Israel
salz@cs.technion.ac.il
[2] Rahan Meristem (1998) Ltd., Kibbutz Rosh Hanikra, Israel
http://www.cs.technion.ac.il
http://www.rahan.co.il

Abstract. Prediction of stress conditions is important for monitoring plant growth stages, disease detection, and assessment of crop yields. Multi-modal data, acquired from a variety of sensors, offers diverse perspectives and is expected to benefit the prediction process. We present several methods and strategies for abiotic stress prediction in banana plantlets, on a dataset acquired during a two and a half weeks period, of plantlets subject to four separate water and fertilizer treatments. The dataset consists of RGB and thermal images, taken once daily of each plant. Results are encouraging, in the sense that neural networks exhibit high prediction rates (over 90% amongst four classes), in cases where there are hardly any noticeable features distinguishing the treatments, much higher than field experts can supply.

Keywords: Water stress · Thermal images · Neural networks

1 Introduction

Stress conditions in plants are usually divided into two categories - biotic, induced by biological factors such as fungi or bacteria, and abiotic, caused by climatic conditions such as heat or drought. In this paper we deal with prediction of abiotic stress, by designing an experiment for creating and analysing four separate treatments of water and fertilizer conditions for banana plantlets in a greenhouse. The specific species is known as *Musa acuminata*, considered one of the earliest domesticated plants [11,13,35,38,46], cultivated and hybridized extensively in recent centuries for mass production of human edible clones [18,30,52]. Banana is the fourth largest fruit crop in the world, considered a staple food in many countries [56]. However, one of the major limitations to it's productivity is water stress. Sensitivity to reduction in soil moisture is reflected in dwindled growth due to reduced stomatal conductance and leaf size which eventually may lead to plant death [27].

© Springer Nature Switzerland AG 2020
A. Bartoli and A. Fusiello (Eds.): ECCV 2020 Workshops, LNCS 12540, pp. 279–295, 2020.
https://doi.org/10.1007/978-3-030-65414-6_20

The experiment was aimed at analysis of banana plantlets undergoing four levels of water stress, intended for detection and early prediction of drought conditions. The plantlets were grown in a greenhouse in northern Israel by Rahan-Meristem [48]. Each treatment category included 30 plants, followed daily for 17 consecutive days, and photographed once daily with a high resolution RGB camera and a low resolution thermal (infra-red) sensor. All plants were cloned from the same original, and the experiment was conducted during September 2018. All images were manually annotated with plant contours. Image backgrounds were deliberately discarded, since they contain obstructive objects such as irrigation equipment, tagging labels, ground and cement.

It should be noted that neither laymen nor experts can notice any difference of treatment in the images, and hardly from a sequence of images. The single cue from the agricultural growers is that the appearance rate of the newest leaf in the plant is lower as the plant gets a decreased amount of water. In general, the RGB images display no visible difference, except appearance of new leaves once every 4 to 6 days. The average temperature of the plants, extracted from the thermal images, does show a noticeable range variation, about 5.0° between the extreme conditions.

Our main contributions include (1) creation of a RGB-T dataset in a controlled environment, planned specifically for analysis of plant structure and development, (2) proposing and comparing several methods for stress detection and prediction, (3) exploiting temporal image sequences to leverage the accuracy rates, and (4) employing shallow neural networks capable of feature extraction relevant to the issue at hand. The following sections present related efforts in the field, description of the experiment and data acquisition, methods and strategies employed, results and conclusions.

2   Related Efforts

RGB and Thermal Images. Several papers document the combined usage of RGB images, capturing light in the human visible range, and thermal images captured by infra-red (IR) or near infra-red (NIR) sensors. Naturally, papers dealing with autonomous driving, motion tracking, and human pose estimation attract much attention in the computer vision community, hence the existence of RGB-T datasets in these fields is more common [36,37,45,53,58]. In the agricultural domain, much research often rely on Unmanned Aerial Vehicles (UAV) and satellites collecting IR and NIR data for terrain mapping [19,40,42], vegetation indexing of crop fields [6,7,20,47], and canopy chlorophyll content [12,21,54]. Surveys related to UAV imagery all point to the importance of collecting and analyzing thermal and hyper spectral data for agricultural enterprises [1–3,5,31,39,41]. Simultaneous acquisition from multi-modal sensor systems is also indispensable for diagnosing plant stress at short-range [8,25].

Thermal images are used in various agricultural applications – plant monitoring, yield estimation, crop maturity evaluation, irrigation scheduling, soil salinity detection, disease and pathogen detection, and bruise detection [26,29]. Recently, several new RGB-T datasets were collected, specifically intended for applying computer vision methods to handle agricultural tasks, for example semantic segmentation of field patterns [9] and estimation of weed growth stages [16]. Multi-modal data (LIDAR, radar, RGB, and thermal) was collected and studied for the purpose of obstacle avoidance in agricultural fields [32,33]. Other examples relying on thermal imagery include detection of tea disease [63], detection of apple scab disease [43], irrigation monitoring [49], and estimation of leaf water potential [10].

Neural Networks for Plant Phenotyping and Stress Prediction. Research in recent years has produced an abundance of practical implementations applying Convolutional Neural Networks (CNN) for many vision based tasks in the advancing field of precision farming and automated agriculture, usually fine-tuned for specific goals related to detection of crop traits and early stress prediction. CNNs are by far the most popular computational tool for feature extraction relevant to crop phenotypes. To list just a few CNN based examples, RGB-T is used for wheat ear detection [22], RGB and depth sensors for object detection intended for measurement of width and height of banana leaves, and height of banana trees [61], salt stress due to soil salinity using hyperspectral reflectance [17], and leaf area, count, and genotype classification [14]. The importance of leaf segmentation for practically every plant phenotyping task is frequently mentioned [51,59,60], leading to various tailored augmentation methods for accurate leaf segmentation, especially of Arabidopsis-thaliana [34,50,62,64]. Reflecting on current trends, phenotyping capabilities are already upgraded by research into the structure and optimization of neural networks. A discussion on reducing network size for flower segmentation is presented in [4]. Lately, Dobrescu et al. presented a deep attempt to understand how CNNs reach decisions in leaf counting schemes [15]. Future studies will most likely continue to adapt neural networks, acquire larger amounts of multi-modal data, and aim at further automation of data collection and analysis.

3   Experiment Outline and Data Acquisition

A Stress Detection Venture. The experiment was aimed at analysis of banana plantlets undergoing four levels of water treatment, and planned specifically for detection and prediction of drought conditions. One-month old plantlets were grown in a $1L$ pot in a commercial greenhouse. The plantlets were positioned according to the scheme in Fig. 1. The four different treatments were labeled A, B, C, D for convenience, and each plant tagged with an ID, from 01 to 30. During the first three days all plants received the same treatment; different water and fertilizer amounts were administered starting on the fourth day. The plants in treatment A were watered and fertilized every day according to the normal commercial growing conditions (100%), and plants undergoing treatments B, C and D were watered every day with lower amounts, 80%, 60% and 40%, respectively, of the normal conditions. On the 17th and last day of the experiment, irrigation of all plants returned to normal conditions, to support their recovery and verify that they are in a productive state for further usage. Data collection time was between 07:00–10:00, and watering of the plants at 13:00.

Data Collection. The RGB camera in use was a mobile Samsung SM-G930F, with pixel resolution 4032 × 3024, focal length 4 mm, and exposure time ∼3.6 ms. The IR sensor was a Opgal-ThermApp, pixel resolution 384 × 288, focal length 35 mm, and wavelength region 7.5–14 μm [44], outputting 16 bits per pixel, each value an integer representing 100×°C. All images are top views, taken approximately 1.2 m from the top of plants. The temperature inside the greenhouse was measured once daily at a fixed position, being between 34–39°C. In addition, a log was kept, recording dates of leaf buddings in all plants. These manual measurements are displayed in Fig. 2. Observing the leaf buddings turned out to be crucial; as will be explained later, our findings suggest a strong correlation between the rate of new leaf appearance and our stress prediction.

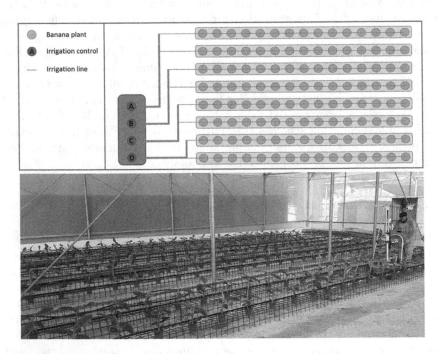

Fig. 1. Schematic representation of the experiment layout (top) and aerial view inside the greenhouse (bottom).

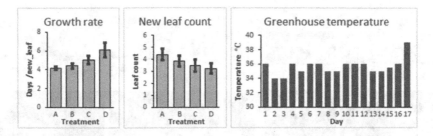

Fig. 2. Manual measurements during the experiment: Number of days between emergence of new leaves (left), count of new leaves during the entire experiment (middle), mean and standard deviation for each treatment, and measured temperature inside the greenhouse (right).

4   Methods and Strategies

Pre-processing. All images were manually annotated with plant contours. A custom built annotation tool was used, and contour points were marked on all images. Image backgrounds were deliberately discarded, since they contain obstructive objects such as irrigation equipment, iron support rails, tagging labels, ground and cement. Figure 3 shows a sample of raw input images, RGB and thermal, one from each treatment category, taken at the eighth day of the experiment. Furthermore, the thermal images were passed through an erosion phase in order to discard any parts of the background that remained after the segmentation. A similar erosion in the RGB images was not deemed necessary, simply due to finer contour annotations, performed on the original high resolution images.

Thermal analysis commenced with measuring average temperatures on the plants, then comparing them with the average temperature of the plant contours. Several neural network architectures were then employed to use RGB and thermal images, separately, then combining both modalities. The main boost to accuracy rates was obtained by inspecting consecutive images of the same plant from several days, three or more.

Of the 30 plants from each category, 26 were used for training, the remaining used solely for testing. Plants with ID's $05, 10, 15, 20$ served for test purposes, and in no way were seen by the training algorithms, neither RGB nor thermal. This separation of train/test was consistent throughout all the methods and prediction strategies.

RGB Data. Several network architectures were attempted, among them Resnet-50 [23], GoogLeNet [57], MobileNet [24], and VGG16 [55], or compacted versions of them, all widely used in classification problems. A Cifar-10 [28] based CNN architecture, trained from scratch, proved best. The deep pre-trained models resulted in performance far worse than training from scratch. This leads to

Fig. 3. Sample of original and segmented RGB images (top rows), original and segmented thermal images (bottom rows), from the eighth day of the experiment; A, B, C, D treatments (left to right), respectively.

the conjecture that general-purpose datasets are not well suited, in a transfer learning sense, to particular agricultural data. Our dataset is fairly small, so in order to reduce over-fitting, we used Keras built-in augmentation in our training process. The augmentations we used include horizontal and vertical flips, width and height shifts, and rotations.

Thermal Data. Similar to the RGB data, we first tried to fine-tune pre-built models like the GoogLeNet and Resnet-50 architectures, but here too, the results were far inferior to results from models we trained from scratch. We used a fairly similar model for this data compared to the RGB data model, but changed the order of convolutions because it proved to produce slightly better results. Augmentations were applied during this learning process as well, horizontal and vertical flips, width and height shifts, and rotations.

Next, we tried training the same model with image sequences from 3 consecutive days. Each sequence of 3 images was concatenated and fed to our model as a single 288×1152 pixel image to which we refer to as a *triplet*. We define $triplet_n$ as constructed from images taken at days n-2, n-1, and n. If n-1 < 1 or n-2 < 1 we simply duplicate the same image, so overall, we have the same amount of triplets as single images. An example of such a triplet is shown in

Fig. 4. According to our field observations, there is a strong correlation between the plant's drought stress and the rate in which new leaves appear. We believe that by using these triplets, our machine could deduce temporal features like leaf growth and size.

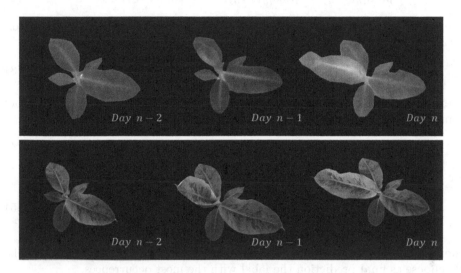

Fig. 4. Concatenation of images from 3 consecutive days (treatment A, plant ID 14, days 13–15). Notice the appearance of a new leaf bud at day n-2 (a small cigar-looking leaf), its rapid growth between days n-2 and n-1, and its final opening on day n.

Combination of RGB and Thermal Data. Because the thermal and RGB images are not aligned and have different resolutions, a model which trains on both types of images simultaneously may not perform well. For that reason, a unique model was trained for each type of images separately.

Prediction Model. To classify a plant x, we pass N consecutive thermal images (one for each day) through the thermal data model, and the corresponding RGB images through the RGB data model. For every image prediction, we save P labels that match the thermal data model prediction, and Q labels that match the RGB data model prediction, where $\frac{P}{Q}$ is the ratio between the first model accuracy and the second model accuracy. For instance, if the thermal model's accuracy is 60% and the RGB model's accuracy is 90% then $\frac{P}{Q} = \frac{60}{90} = \frac{2}{3} \rightarrow P = 2, Q = 3$.

To get the final prediction, we use a variation of the rolling average prediction method, meaning that our final prediction will be the highest probability (the one which appears the most times). In a formal manner, the prediction of plant x is obtained as follows. For each RGB image, we predict the label by passing it to the RGB model. We then duplicate the prediction Q times,

$$\text{pred}^*_{\text{rgb}}(RI_n) = (\text{pred}_{\text{rgb}}(RI_n), \cdots, \text{pred}_{\text{rgb}}(RI_n)) \in \mathbb{N}^Q, \qquad (1)$$

where RI_n is the RGB image of plant x at day n. Similarly, each thermal image is fed to the thermal model to obtain the prediction. We then duplicate the prediction P times,

$$\text{pred}_{th}^*(TI_n) = (\text{pred}_{th}(TI_n), \cdots, \text{pred}_{th}(TI_n)) \in \mathbb{N}^P, \tag{2}$$

where TI_n is the thermal image of plant x at day n. The combined prediction of this image at day n is a concatenation of both label vectors

$$\text{pred}^*(n) = \text{pred}_{th}^*(TI_n) \parallel \text{pred}_{\text{rgb}}^*(RI_n) \in \mathbb{N}^{(P+Q)}, \tag{3}$$

where $\parallel$ is the concatenation operator. Then, we combine all prediction vectors from all N images of the sequence to form a $1 \times N(P+Q)$ buffer of labels,

$$\text{buffer} = \text{pred}^*(1) \parallel \text{pred}^*(2) \parallel \cdots \parallel \text{pred}^*(N) \in \mathbb{N}^{N(P+Q)}. \tag{4}$$

For each possible label we sum the number of times it appears in the vector,

$$C(l) = \sum_{l' \in \text{buffer}} I\{l' = l\}, \tag{5}$$

where l' is the predicted label and l an element in the set of 4 treatment labels. We choose as final prediction the label with the most occurrences,

$$\text{pred}(x) = \text{argmax}_{l \in \text{labels}}(C(l)). \tag{6}$$

We also tried to apply this prediction method on each model separately and save only one label per prediction in the buffer. This simple method enhanced the prediction accuracy on both models as can be seen in the results section. Since the available data has a temporal quality, a Recurrent Neural Network (RNN) approach was also attempted but produced less satisfactory results.

5   Results

Preliminary Analysis. Thermal analysis commenced with extraction of average temperatures on the plants compared with the average temperature of the plant contours. To this end, the contour temperature is defined as a narrow band of pixels between the original and slightly eroded image mask. Prediction accuracy by average plant temperature only, and difference between plant and contour temperature, resulted in $\sim$50% and $\sim$60%, respectively, amongst the four categories. In this preliminary analysis, a simple prediction scheme was employed: First, plant temperatures where obtained per treatment and per day from each thermal image, then averaged for each category. Next, the predicted label of each image from the test set was defined as the treatment label (amongst A, B, C, and D) with average temperature closest to the test image. The prediction is considered true if the predicted label is the same as the ground truth label, and false otherwise. Finally, average accuracies were computed for each day and each treatment category. Average plant temperatures and prediction accuracy at this stage are presented in Fig. 5. Subtraction of the temperature measured once daily inside the greenhouse affected these results only slightly. A note should be made that on the 17th and last day of the experiment, irrigation of all plants was returned to normal conditions for all treatments, and the measured temperature on this day was significantly high (see Fig. 2), so these facts may explain the convergence of average temperatures on this day, as seen in Fig. 5(left). The temperature drop of treatment D on the 13th day may have been caused by a fault in the irrigation equipment. This step already demonstrates that classification amongst the treatments can be achieved, to some degree, correlating the well known fact that plants receiving less water have a higher temperature. By using images from three consecutive days, the prediction accuracy was improved to ~70%.

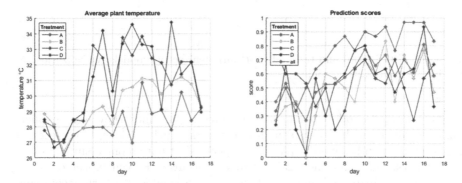

Fig. 5. Average daily plant temperatures for each treatment (left), and prediction accuracy rates obtained by plant and contour temperatures (right).

These results, moderately satisfactory, motivated us to apply deeper machine learning models, and integrate the temporal sequencing. The basic scheme is a network architecture composed of two convolution and max-pooling layers, followed by two fully connected layers, supplemented by 0.5 dropout, with 32 feature maps in the first convolution layer, and 64 in the second. The input is the original thermal images *as-is*, the RGB down-scaled to the same 384 × 288 resolution. The RGB and thermal network architectures are presented in Fig. 6. Again, implementation experience demonstrated that shallow networks trained from scratch performed better than deep pre-trained networks.

Prediction by RGB Images Only. In this experiment we used a batch size of 24 and trained for 100 epochs using an ADAM optimizer. As seen in Table 1, augmentation improved the results significantly, mainly because the dataset is rather small. Also, using a sequence of 3 images from consecutive days achieved slightly better results compared to a single image. One should not be surprised that the

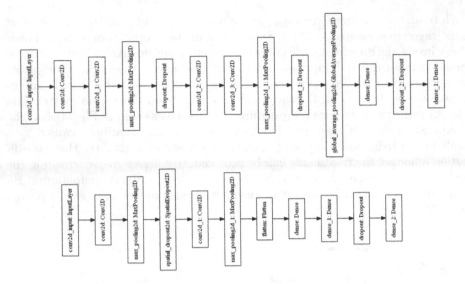

Fig. 6. The RGB (top) and thermal (bottom) architecture models; visualizations created by Keras visualising tool.

deep pre-trained models result in poor prediction, since they were explicitly trained to distinguish among semantic categories such as dogs, birds, cars, airplanes, etc., possibly different plant sub-species, but not amongst similar looking plant images. That said, the ∼25% prediction accuracy is even expected a-priory. In contrast, with limited amount of raw data, a shallow network accompanied with proper augmentation is expected to learn prominent features beneficial to the specific prediction task at hand, while deep training from scratch is an overkill.

Prediction by Thermal Images Only. In this experiment we used a batch size of 32 and trained for 1000 epochs using a SGD optimizer. Here too, augmentation improved the results significantly and applying a sequence of 3 images from consecutive days achieved better results than using a single image by a considerable margin. However, results are far inferior to prediction by RGB images only. We

Table 1. Treatment prediction accuracy using RGB and thermal images, separately; comparison of network architectures.

	GoogLeNet RGB	MobileNet RGB	Our model RGB	Our model thermal
Without augmentation	25	36	72	40
With augmentation	-	-	82	62.5
Triplets with augmentation	25	32	**84**	**72**

believe there is more than one reason for that. Firstly, the thermal data is not very consistent; there were variations in the daily measured temperature inside the greenhouse, and the manufacturer specifies a ±3.0 °C accuracy at 25 °C [44]. These inconsistencies can be partially seen in Fig. 5. Secondly, the visual information gathered from a plant may have more discriminative information than its temperature. RGB contains 3 channels, has much higher resolution, and also, today, low cost RGB cameras capture high quality info, while *quality-vs-cost* IR is still lagging behind. For example it is easier to isolate each leaf in a RGB image than in its thermal image. Due to these mentioned reasons, as well as less practical experience on learning from thermal images, the thermal network was trained for more epochs than the RGB network. We found that 400 epochs proved enough for converging on an asymptotic accuracy rate, and kept the process running in order to observe stable results before overfitting creeps in.

Integration of RGB and Thermal Data. In this experiment we integrated the models that performed best on their respective thermal and RGB data. The entire architecture is depicted in Fig. 7. We tried to combine the triplet models with the single images models and check which combination produces the best results. We also experimented with the sequence length N to see how early it is possible to predict abiotic stress with a fairly good accuracy. Table 2 shows the results of these experiments. Each row represents a single or combined model, and each column represents the number of consecutive images fed to the models. For example, column 8 shows the mean prediction accuracy of the rolling average method over every sequence of 8 consecutive images of all plants. Figure 8 is a graph visualization of prediction accuracy per treatment, obtained from one of the models, corresponding to the single RGB + single thermal line in Table 2. Among all scenarios, best results were obtained by the single RGB + triplet thermal model.

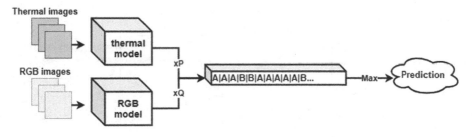

Fig. 7. The complete architecture. x^P and x^Q are P and Q prediction labels (thermal and RGB, respectively), possibly different for the same image, because of the dropout layer. These are concatenated into the label vector, which is *arg-maxed* to obtain the final prediction.

Table 2. Accuracy of treatment prediction by rolling average, comparison of methods and sequence length.

RGB and thermal rolling average prediction accuracy																
Sequence length N	1	2	3	4	5	6	7	8	9	10	11	12	13	14	15	16	17
Single RGB	80.7	82.8	81.6	80.2	80.6	82.3	83.6	86	85.8	88.3	88.4	91.1	92	93.6	93.5	93.3	93.3
Single thermal	69.4	71.1	67.7	65.7	68.6	72.5	78	81.8	82.6	83.8	87.3	83.5	87.3	85.1	87	86.6	80
RGB triplets	78	78.2	80.2	80.6	81.6	82.8	83.6	84.6	85	86.5	87.3	87.3	87.3	87.2	87	86.6	86.6
Thermal triplets	75.3	76.5	79.8	80.2	80.1	80.5	80.5	81.1	84.2	85.5	87.3	87.3	92	93.6	96.7	100	100
Single RGB + Single thermal	84.3	84.9	84.7	84.5	85.8	88.5	88.6	92.3	92.9	91.9	92.6	93.6	93.6	93.6	93.5	93.3	93.3
Triplet RGB + Single thermal	78.9	79.6	80.3	82	82.6	83.2	84	84.9	85.3	86.6	87.4	87.3	87.3	87.3	87.2	87	86.6
Single RGB + Triplet thermal	**89**	**89.5**	**91.9**	**91.8**	**91**	**91.4**	**91.2**	**92.3**	**92.9**	**94.6**	**94.7**	**100**	**100**	**100**	**100**	**100**	**100**
Triplet RGB + Triplet thermal	78.8	80.3	82.5	84		84.8	85.7	86.1	87.4	87.4	87.4	87.3	87.3	87.2	87.1	86.6	86.6

Discussion. One may wonder about the seemingly *too-good-to-be-true* results presented in Table 2, especially columns 12–17 for the single RGB + triplet thermal architecture, and some explanation must be provided. It is important to notice that the whole test set is fairly small, with only 16 plants, 4 from each of the 4 categories (although the entire test set consists of 272 thermal and 272 RGB images, 4 plants, 4 treatments, 17 days). Furthermore, as mentioned in the preliminary analysis, the naïve *average-per-treatment* prediction already reaches roughly 50% accuracy, as opposed to random classification of 25%, amongst 4 classes. Another thing worth mentioning is that the models were trained on all images from all days (except the test images). This means that the network can already be equipped to use this information for prediction while in some

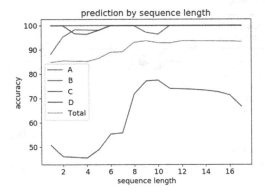

Fig. 8. Prediction accuracy by sequence length of the single RGB + single thermal framework.

realistic situations we would not be able to train on *future* images. As mentioned earlier, the *new-leaf-appearance-rate* is certainly highly correlated with plant stress. The triplets help capture this dependency after training the thermal and RGB models. In addition, the input to the complete architecture, which is a sequence of N images from consecutive days, further increases its ability to detect high *new-leaf-appearance-rates*. In some cases, as seen in Table 2, the *single-image* scheme outperformed the *triplets*. We believe the reason for this is an inherent restriction on augmentations in the triplet arrangement; although a convolution based architecture is expected to be invariant to flips and rotations, the very arrangement of the triplets defies usage of horizontal flips, and rotations are also very limited, thus reducing generality of the training set.

Two other test scenarios were addressed, a binary normal *vs.* deficit stress, and capability of day-by-day stress prediction. Table 3 presents an example of a binary dichotomy, normal treatment (A) *vs.* all others (B,C,D) suffering more or less from water deficit. No surprise here, the healthy plants are easily distinguished from the stressed. In order to address the early prediction issue, we also supply the prediction accuracy per day, as opposed to any sequence of the last N consecutive days. Results of day-by-day prediction, obtained from combinations of single and triplet models, are presented in Fig. 9.

Table 3. Binary prediction accuracy of normal condition (treatment A), *vs.* all other water deficit treatments.

RGB and thermal rolling average prediction accuracy																
Sequence length N	1	2	3	4	5	6	7	8	9	10	11	12	13	14	15	16	17
Single RGB + Single thermal	97.0	98.8	99.6	99.6	99.5	100	100	100	99.3	98.4	99.1	100	100	100	100	100	100

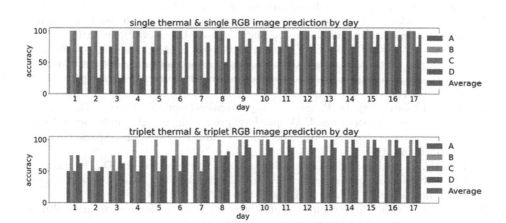

Fig. 9. Bar graph showing the prediction accuracy of each class as well as the average of all classes. Notice the total accuracy (in purple) rising as the day number goes up. (Color figure online)

6 Conclusion

We have shown that machine learning methods and neural networks with comparatively small architecture have the capability to detect and predict stress conditions in plants. Combination of RGB and thermal images has improved the results, and a temporal sequence of images has proven to obtain the best results. Our results lead to the suggestion of combining multi-modal data, and acquisition of a controlled time-lapsed sequences of images. We hope this modest contribution will serve the plant phenotyping community, by encouraging the ongoing efforts of carefully planned data collection, applying up-to-date computer vision methods, and analysis of stress detection and prediction.

Acknowledgment. This research was partly supported by the Israel Innovation Authority, the Phenomics Consortium.

References

1. Aasen, H., Honkavaara, E., Lucieer, A., Zarco-Tejada, P.J.: Quantitative remote sensing at ultra-high resolution with UAV spectroscopy: a review of sensor technology, measurement procedures, and data correction workflows. Remote Sens. **10**(7), 1091 (2018)
2. Abdullahi, H.S., Mahieddine, F., Sheriff, R.E.: Technology impact on agricultural productivity: a review of precision agriculture using unmanned aerial vehicles. In: Pillai, P., Hu, Y.F., Otung, I., Giambene, G. (eds.) WiSATS 2015. LNICST, vol. 154, pp. 388–400. Springer, Cham (2015). https://doi.org/10.1007/978-3-319-25479-1_29
3. Adão, T., et al.: Hyperspectral imaging: a review on UAV-based sensors, data processing and applications for agriculture and forestry. Remote Sens. **9**(11), 1110 (2017)
4. Atanbori, J., Chen, F., French, A.P., Pridmore, T.P.: Towards low-cost image-based plant phenotyping using reduced-parameter CNN (2018)
5. Barbedo, J.G.A.: A review on the use of unmanned aerial vehicles and imaging sensors for monitoring and assessing plant stresses. Drones **3**(2), 40 (2019)
6. Bellvert, J., Zarco-Tejada, P.J., Girona, J., Fereres, E.: Mapping crop water stress index in a 'Pinot-noir' vineyard: comparing ground measurements with thermal remote sensing imagery from an unmanned aerial vehicle. Precis. Agric. **15**(4), 361–376 (2014)
7. Berni, J., Zarco-Tejada, P., Suárez, L., González-Dugo, V., Fereres, E.: Remote sensing of vegetation from UAV platforms using lightweight multispectral and thermal imaging sensors. Int. Arch. Photogramm Remote Sens. Spatial Inform. Sci. **38**(6), 6 (2009)
8. Chaerle, L., Lenk, S., Leinonen, I., Jones, H.G., Van Der Straeten, D., Buschmann, C.: Multi-sensor plant imaging: towards the development of a stress-catalogue. Biotechnol. J. Healthc. Nutr. Technol. **4**(8), 1152–1167 (2009)
9. Chiu, M.T., et al.: Agriculture-vision: a large aerial image database for agricultural pattern analysis. arXiv preprint arXiv:2001.01306 (2020)
10. Cohen, Y., et al.: Crop water status estimation using thermography: multi-year model development using ground-based thermal images. Precis. Agric. **16**(3), 311–329 (2015)

11. De Langhe, E., Vrydaghs, L., De Maret, P., Perrier, X., Denham, T.: Why bananas matter: an introduction to the history of banana domestication. Ethnobotany Res. Appl. **7**, 165–177 (2009)
12. Delloye, C., Weiss, M., Defourny, P.: Retrieval of the canopy chlorophyll content from sentinel-2 spectral bands to estimate nitrogen uptake in intensive winter wheat cropping systems. Remote Sens. Environ. **216**, 245–261 (2018)
13. Denham, T.P., et al.: Origins of agriculture at Kuk Swamp in the highlands of New Guinea. Science **301**(5630), 189–193 (2003)
14. Dobrescu, A., Giuffrida, M.V., Tsaftaris, S.A.: Doing more with less: a multitask deep learning approach in plant phenotyping. Front. Plant Sci. **11**, 141 (2020)
15. Dobrescu, A., Valerio Giuffrida, M., Tsaftaris, S.A.: Understanding deep neural networks for regression in leaf counting. In: Proceedings of the IEEE Conference on Computer Vision and Pattern Recognition Workshops (2019)
16. Etienne, A., Saraswat, D.: Machine learning approaches to automate weed detection by UAV based sensors. In: Autonomous Air and Ground Sensing Systems for Agricultural Optimization and Phenotyping IV, vol. 11008, p. 110080R. International Society for Optics and Photonics (2019)
17. Feng, X., et al.: Hyperspectral imaging combined with machine learning as a tool to obtain high-throughput plant salt-stress phenotyping. Plant J. **101**(6), 1448–1461 (2020)
18. Fuller, D.Q., Madella, M.: Banana cultivation in south Asia and east Asia: a review of the evidence from archaeology and linguistics. Ethnobotany Res. Appl. **7**, 333–351 (2009)
19. Gago, J., et al.: UAVs challenge to assess water stress for sustainable agriculture. Agric. Water Manage. **153**, 9–19 (2015)
20. Genxu, W., Guangsheng, L., Chunjie, L., Yan, Y.: The variability of soil thermal and hydrological dynamics with vegetation cover in a permafrost region. Agric. For. Meteorol. **162**, 44–57 (2012)
21. Gevaert, C.M., Suomalainen, J., Tang, J., Kooistra, L.: Generation of spectral-temporal response surfaces by combining multispectral satellite and hyperspectral UAV imagery for precision agriculture applications. IEEE J. Sel. Top. Appl. Earth Obs. Remote Sens. **8**(6), 3140–3146 (2015)
22. Grbovic, Z., Panic, M., Marko, O., Brdar, S., Crnojevic, V.: Wheat ear detection in RGB and thermal images using deep neural networks. Environments **11**(12), 13 (2019)
23. He, K., Zhang, X., Ren, S., Sun, J.: Deep residual learning for image recognition. In: Proceedings of the IEEE Conference on Computer Vision and Pattern Recognition, pp. 770–778 (2016)
24. Howard, A.G., et al.: MobileNets: efficient convolutional neural networks for mobile vision applications. arXiv preprint arXiv:1704.04861 (2017)
25. Humplík, J.F., Lazár, D., Husičková, A., Spíchal, L.: Automated phenotyping of plant shoots using imaging methods for analysis of plant stress responses–a review. Plant Methods **11**(1), 29 (2015)
26. Ishimwe, R., Abutaleb, K., Ahmed, F., et al.: Applications of thermal imaging in agriculture–a review. Adv. Remote Sens. **3**(03), 128 (2014)
27. Kallarackal, J., Milburn, J., Baker, D.: Water relations of the banana. iii. effects of controlled water stress on water potential, transpiration, photosynthesis and leaf growth. Funct. Plant Biol. **17**(1), 79–90 (1990)
28. Keras: Keras CIFAR-10. https://keras.io/examples/cifar10_cnn/

29. Khanal, S., Fulton, J., Shearer, S.: An overview of current and potential applications of thermal remote sensing in precision agriculture. Comput. Electron. Agric. **139**, 22–32 (2017)

30. Khayat, E., Ortiz, R.: Genetics of important traits in Musa. In: Banana Breeding: Progress and Challenges, pp. 71–83. CRC Press, New York (2011)

31. Kim, J., Kim, S., Ju, C., Son, H.I.: Unmanned aerial vehicles in agriculture: a review of perspective of platform, control, and applications. IEEE Access **7**, 105100–105115 (2019)

32. Korthals, T., Kragh, M., Christiansen, P., Karstoft, H., Jørgensen, R.N., Rückert, U.: Multi-modal detection and mapping of static and dynamic obstacles in agriculture for process evaluation. Front. Robot. AI **5**, 28 (2018)

33. Kragh, M., et al.: Multi-modal obstacle detection and evaluation of occupancy grid mapping in agriculture. In: International Conference on Agricultural Engineering (2016)

34. Kuznichov, D., Zvirin, A., Honen, Y., Kimmel, R.: Data augmentation for leaf segmentation and counting tasks in rosette plants. In: Proceedings of the IEEE Conference on Computer Vision and Pattern Recognition Workshops (2019)

35. Lebot, V.: Biomolecular evidence for plant domestication in Sahul. Genet. Resour. Crop Evol. **46**(6), 619–628 (1999)

36. Li, C., Cheng, H., Hu, S., Liu, X., Tang, J., Lin, L.: Learning collaborative sparse representation for grayscale-thermal tracking. IEEE Trans. Image Process. **25**(12), 5743–5756 (2016)

37. Li, C., Liang, X., Lu, Y., Zhao, N., Tang, J.: RGB-T object tracking: benchmark and baseline. Pattern Recogn. **96**, 106977 (2019)

38. Li, L.F., et al.: Origins and domestication of cultivated banana inferred from chloroplast and nuclear genes. PLoS ONE **8**(11), e80502 (2013)

39. Maes, W.H., Steppe, K.: Perspectives for remote sensing with unmanned aerial vehicles in precision agriculture. Trends Plant Sci. **24**(2), 152–164 (2019)

40. Milella, A., Nielsen, M., Reina, G.: Sensing in the visible spectrum and beyond for terrain estimation in precision agriculture. Adv. Anim. Biosci. **8**(2), 423–429 (2017)

41. Mogili, U.R., Deepak, B.: Review on application of drone systems in precision agriculture. Proc. Comput. Sci. **133**, 502–509 (2018)

42. Mulder, V., De Bruin, S., Schaepman, M.E., Mayr, T.: The use of remote sensing in soil and terrain mapping–a review. Geoderma **162**(1–2), 1–19 (2011)

43. Nouri, M., Gorretta, N., Vaysse, P., Giraud, M., Germain, C., Keresztes, B., Roger, J.M.: Near infrared hyperspectral dataset of healthy and infected apple tree leaves images for the early detection of apple scab disease. Data in Brief **16**, 967–971 (2018)

44. Opgal: Opgal thermal imaging. http://www.opgal.com/products/therm-app/

45. Palmero, C., Clapés, A., Bahnsen, C., Møgelmose, A., Moeslund, T.B., Escalera, S.: Multi-modal RGB-depth-thermal human body segmentation. Int. J. Comput. Vision **118**(2), 217–239 (2016)

46. Perrier, X., et al.: Multidisciplinary perspectives on banana (Musa spp.) domestication. Proc. Natl. Acad. Sci. **108**(28), 11311–11318 (2011)

47. Raeva, P.L., Šedina, J., Dlesk, A.: Monitoring of crop fields using multispectral and thermal imagery from UAV. Eur. J. Remote Sens. **52**(sup1), 192–201 (2019)

48. Rahan: Rahan Meristem (1998) Ltd. http://www.rahan.co.il/

49. Roopaei, M., Rad, P., Choo, K.K.R.: Cloud of things in smart agriculture: intelligent irrigation monitoring by thermal imaging. IEEE Cloud Comput. **4**(1), 10–15 (2017)

50. Sapoukhina, N., Samiei, S., Rasti, P., Rousseau, D.: Data augmentation from RGB to chlorophyll fluorescence imaging application to leaf segmentation of Arabidopsis thaliana from top view images. In: Proceedings of the IEEE Conference on Computer Vision and Pattern Recognition Workshops (2019)
51. Scharr, H., et al.: Leaf segmentation in plant phenotyping: a collation study. Mach. Vis. Appl. **27**(4), 585–606 (2016)
52. Sheperd, K., et al.: Cytogenetics of the genus Musa (1999)
53. Shivakumar, S.S., Rodrigues, N., Zhou, A., Miller, I.D., Kumar, V., Taylor, C.J.: Pst900: RGB-thermal calibration, dataset and segmentation network. arXiv preprint arXiv:1909.10980 (2019)
54. Simic Milas, A., Romanko, M., Reil, P., Abeysinghe, T., Marambe, A.: The importance of leaf area index in mapping chlorophyll content of corn under different agricultural treatments using UAV images. Int. J. Remote Sens. **39**(15–16), 5415–5431 (2018)
55. Simonyan, K., Zisserman, A.: Very deep convolutional networks for large-scale image recognition. arXiv preprint arXiv:1409.1556 (2014)
56. Surendar, K.K., Devi, D.D., Ravi, I., Krishnakumar, S., Kumar, S.R., Velayudham, K.: Water stress in banana–a review. Bull. Env. Pharmacol. Life Sci. **2**(6), 1–18 (2013)
57. Szegedy, C., et al.: Going deeper with convolutions. In: Proceedings of the IEEE Conference on Computer Vision and Pattern Recognition, pp. 1–9 (2015)
58. Tang, J., Fan, D., Wang, X., Tu, Z., Li, C.: RGBT salient object detection: benchmark and a novel cooperative ranking approach. IEEE Trans. Circ. Syst. Video Technol. **30**, 4421–4433 (2019)
59. Tsaftaris, S.A., Minervini, M., Scharr, H.: Machine learning for plant phenotyping needs image processing. Trends Plant Sci. **21**(12), 989–991 (2016)
60. Ubbens, J.R., Stavness, I.: Deep plant phenomics: a deep learning platform for complex plant phenotyping tasks. Front. Plant Sci. **8**, 1190 (2017)
61. Vit, A., Shani, G., Bar-Hillel, A.: Length phenotyping with interest point detection. In: Proceedings of the IEEE Conference on Computer Vision and Pattern Recognition Workshops (2019)
62. Ward, D., Moghadam, P., Hudson, N.: Deep leaf segmentation using synthetic data. arXiv preprint arXiv:1807.10931 (2018)
63. Yang, N., Yuan, M., Wang, P., Zhang, R., Sun, J., Mao, H.: Tea diseases detection based on fast infrared thermal image processing technology. J. Sci. Food Agric. **99**(7), 3459–3466 (2019)
64. Zhu, Y., Aoun, M., Krijn, M., Vanschoren, J., Campus, H.T.: Data augmentation using conditional generative adversarial networks for leaf counting in Arabidopsis plants. In: BMVC, p. 324 (2018)

Sorghum Segmentation by Skeleton Extraction

Mathieu Gaillard[1] (ID), Chenyong Miao[2] (ID), James Schnable[2] (ID),
and Bedrich Benes[1](✉) (ID)

[1] Purdue University, West Lafayette, IN, USA
bbenes@purdue.edu
[2] University of Nebraska-Lincoln, Lincoln, NE, USA
schnable@unl.edu
http://hpcg.purdue.edu, https://schnablelab.org/

Abstract. Recently, several high-throughput phenotyping facilities have been established that allow for an automated collection of multiple view images of a large number of plants over time. One of the key problems in phenotyping is identifying individual plant organs such as leaves, stems, or roots. We introduced a novel algorithm that uses a 3D segmented plant on its input by using a voxel carving algorithm, and separates the plant into leaves and stems. Our algorithm first uses voxel thinning that generates a first approximation of the plant 3D skeleton. The skeleton is transformed into a mathematical tree by comparing and assessing paths from each leaf or stem tip to the plant root and pruned by using biologically inspired features, fed into a machine learning classifier, leading to a skeleton that corresponds to the input plant. The final skeleton is then used to identify the plant organs and segment voxels. We validated our system on 20 different plants, each represented in a voxel array of a resolution 512^3, and the segmentation was executed in under one minute, making our algorithm suitable for the processing of large amounts of plants.

Keywords: 3D plant reconstruction · Phenotyping · Sorghum · Skeleton extraction · Segmentation

1 Introduction

The architecture of plant organs such as leaves, stems, roots, and buds, plays a significant role in determining plant growth, health, and yield. Different individuals of the same species growing in the same environment will exhibit considerable differences in architectural traits due to genetic differences. Mapping and identifying the genes which control variation in plant architecture is a critical

Electronic supplementary material The online version of this chapter (https:// doi.org/10.1007/978-3-030-65414-6_21) contains supplementary material, which is available to authorized users.

© Springer Nature Switzerland AG 2020
A. Bartoli and A. Fusiello (Eds.): ECCV 2020 Workshops, LNCS 12540, pp. 296–311, 2020.
https://doi.org/10.1007/978-3-030-65414-6_21

step in breeding new crop varieties that produce more food, use resources more efficiently, and are more resilient to changing environments.

Mapping genes controlling within-species variation in plant architectural traits requires quantifying these traits, which requires the identification of plant organs. Different approaches have been pioneered and applied in rosette plants (*e.g.,* arabidopsis, brassicas, etc.) using top-down photos and grain crops using side view photos. Skeletonization converts segmented images into graphs that uniquely map onto the plant organs and quantifies architectural traits [3, 4]. The skeletonization of 2D side view images followed by computing on the resulting skeleton is useful in segmenting individual leaves of maize and sorghum plants, but is not robust to leaves that intersect from a single view [4].

Several high throughput phenotyping technology facilities have been built that allow for controlled viewing and photographing of hundreds of plants [8, 10, 13]. The plants grow in a greenhouse, and they are regularly and thoroughly automatically transported by using conveyor belts into imaging chambers where they are photographed from several angles. Various methods for 3D reconstruction of plants from these controlled environments have been developed [11, 17]. We build upon the work of Gaillard et al. [9] that reconstructs several photographs into a voxel grid that approximates the plant's geometry. While the voxel grid is an excellent plant approximation, it does not carry semantic information, such as the number of leaves, curvature, etc.

We present a novel method for extracting high-level semantic features from discrete volumetric arrays obtained by using the voxel carving algorithm into skeletons. The state of the art skeletonization algorithms perform poorly on plants. We claim a contribution in a machine learning-based algorithm that improves the plant skeletonization, allowing us to keep only the essential parts: stem and leaves. These, in turn, are used for efficient segmentation of the input voxel grid. An example in Fig. 1 shows the input voxel grid, the extracted skeleton, and the segmented output (open in Adobe Acrobat for animation).

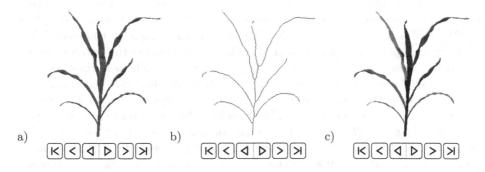

a)   b)   c)

Fig. 1. Please open in Adobe Acrobat to see the animations. a) Input voxel grid, b) extracted skeleton, and c) segmented plant organs. (See Supplementary Material)

2   Related Work

The space carving algorithm [16] retrieves 3D voxel positions that correspond to an input object represented by a set of images. However, this method requires precise calibration to retrieve the 3D positions. Also, an increasing number of images provides better results. This algorithm is highly suitable for controlled environments. Novel algorithms and extensions add reconstruction of small plant parts [12,14,15] or a hierarchical space enumeration by octrees [19]. Our work builds on top of the work of Gaillard et al. [9] that provides 3D voxel reconstruction of Sorghum plants by using an improved voxel carving algorithm. Overall 336 Sorghum plants grown in the UNL phenotyping facility [10] were reconstructed from only six RGB images (five side and one from the top) per plant.

3D reconstruction algorithms working in voxel space are not suitable for direct measurements of organ-level features such as leaf size [12]. Skeletons [7] provide an intuitive and simplified information about the topology of the shape they represent. Skeletons can significantly help shape segmentation by guiding it. However, while 2D skeletonization is well-understood and studied, the skeletonization of 3D shapes is much more complicated. First, 3D shapes can be represented in different ways: for example, as a triangle mesh or a voxel grid. Various data give rise to multiple types of 3D skeletons: surface skeletons have a good correspondence to the input shape but are slower to compute and more challenging to analyze. Curves skeletons do not strictly follow the mathematical properties of skeletons, but provide better shape analysis capabilities as they decompose shapes into a set of 1D curves. They are of particular interest for plants, which often have a tubular shape. For a detailed survey on 3D skeletonization, we refer the reader to [20], in which Figure 21 is of particular interest to understand the difference between surface and curve skeletons. Existing skeletonization algorithms [1], despite their strong mathematical properties, produce noisy results on plants, and many specialized methods for plant skeletonization have been developed. The work of [14] segments the 3D surface of a voxel grid using the eigenvalues of the second-moments tensor. Also, a database of predefined leaves has been fit on the skeletons extracted from the 2D views of the plant in [21]. Golbach et al. [12] use a flood fill algorithm to identify the stem. They measure the spread of voxels added in each iteration during the graph traversal to detect branches. Scharr et al. [19] find clusters of voxels in horizontal slices of the plant from top to bottom. When merging two clusters, special rules are applied to keep track of leaves. The most similar approach to ours [1] uses a skeletonization algorithm and then filter skeleton branches based on their reprojections. Wu et al. [22,23] use a Laplacian contraction, a generic skeletonization algorithm, to shrink a point cloud and then post-process it to output measurements. Xiang et al. [24] also skeletonize and segment a point cloud to measure traits in Sorghum plants. Our algorithm can be thought of as a post-process of a segmentation from voxels. We attempt to improve the skeleton with a particular focus on the precise branching point used in a follow-up segmentation.

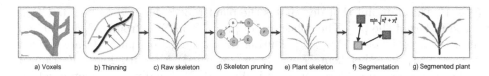

a) Voxels   b) Thinning   c) Raw skeleton   d) Skeleton pruning   e) Plant skeleton   f) Segmentation   g) Segmented plant

Fig. 2. System overview: (round boxes are processes and squared boxes are data): a) the 3D model of the plant represented as a set of voxels is converted into a raw skeleton by using b) the voxel thinning algorithm. c) The raw skeleton is d) filtered to remove noise resulting in e) a smooth plant skeleton. The input voxels are then f) segmented by using the skeleton resulting in g) a segmented set of voxels representing plant organs.

3   Overview

Our method works in three steps shown in Fig. 2: 1) thinning 2) skeleton filtering, and 3) segmentation.

Our algorithm's input is a 3D reconstruction of a sorghum plant in the form of a set of voxels. Our algorithm has two outputs: e) is the plant skeleton, and g) is the segmented plant represented as sets of voxels corresponding to plant organs: the stem and individual leaves. Both the plant and the skeleton are represented in a voxel grid at the same resolution. The segmentation assigns a unique identifier to each voxel.

Thinning: Removes voxels from the input voxel set until only the raw skeleton remains (Sect. 4.1). The input to this step is a plant represented in a voxel grid. Here we use the output of a voxel carving algorithm [9]. Although a smooth, dense, and connected set of voxels provides good input, we designed our algorithm to be robust against noise to successfully process reconstructed plants that do not strictly adhere to these conditions.

Skeleton Pruning: The output of the previous step is a raw skeleton that is equidistant to boundaries. However, the noisy input leads to incorrect results if used directly for leaf counting and plant segmentation. We introduce a novel bio-inspired algorithm in Sect. 4.2 to prune the skeleton removing noisy parts and keeping only the skeleton corresponding to actual leaves.

Segmentation: uses the skeleton calculated from the previous step to generate a segmented plant. We post-process the skeleton to identify the stem and each unique leaf (see Sect. 5), and we assign all voxels from the full plant to its nearest segment and represent them by a unique identifier.

Terminology: Let $\mathcal{V}$ denote the binary voxel grid that contains the reconstructed plant (we use resolution of 512^3 voxels). We refer to each voxel as $v_{i,j,k}$ where $0 \leq i, j, k \leq 511$ denote its discrete coordinates. A voxel $v_{i,j,k} = 1$ belongs to the reconstructed plant, zero voxels identify an empty space.

The raw (generic) skeleton is a set of voxels in $\mathcal{V}$ output by a thinning algorithm, is denoted by $\mathcal{S}_{raw}$. The set of endpoints in $\mathcal{S}_{raw}$ is denoted $\mathcal{E} = \{E_i\}$ with

$E_i \in \mathcal{V}$. The endpoint that designates the plant root is denoted $v_0 \in \mathcal{E}$, and R_{pot} is the radius of the pot. The collection of paths P_i in $\mathcal{S}_{raw}$ from endpoints E_i to the root v_0 is denoted $\mathcal{P} = \{P_i\}$. During the pruning step, paths P_i are labeled in $\mathcal{K} = \{K_i\}$ as either kept ($K_i = 1$) or discarded ($K_i = 0$). The collection of paths that are retained after pruning is denoted $\mathcal{P}^*$. Finally, the skeleton, which is a set of voxels in $\mathcal{V}$ output by our algorithm, is denoted by $\mathcal{S}$.

4   Skeletonization

4.1   Thinning

The input to the first step is the voxel grid $\mathcal{V}$ model of the plant. The voxels are converted to a raw skeleton by using a thinning algorithm; in particular, we use the 3D critical kernel thinning algorithm from [6] implemented in the DGtal library [5]. We chose the listhmus thinning algorithm because it is the fastest algorithm in the library that outputs a curve skeleton. The algorithm is left to default dmax, and the persistence is set to one. In principle, higher persistence values decrease skeleton noise. However, for our dataset, this was not always the case, and our domain-specific pruning algorithm performed better.

Although the critical kernel thinning algorithm is mathematically correct and outputs a skeleton that accurately represents the topology of the plant from the input voxel grid $\mathcal{V}$, it also responds to the inevitable noise in the input leading to a noisy skeleton that cannot be directly used for leaf counting and segmentation. As shown in Fig. 3 column c), the skeleton is noisy, includes small branches and some big lumps, mostly near the stem. Also, it is not guaranteed to have a tree topology; in other words, it may contain loops. Therefore, we further process the raw skeleton by pruning the unnecessary branches and filtering voxels in lumps.

4.2   Skeleton Pruning

A desired output of this algorithm would be one skeleton curve for the stem and one per each leaf. Moreover, the branching pattern for sorghum plants should only have one level of hierarchy $i.e.$, the skeleton should not contain T-junctions unless they are located between the stem and a leaf. Based on these biological observations, our pruning method works in four steps: 1) endpoint identification, 2) root identification, 3) branch finding, and 4) branch pruning.

Endpoints Identification: We identify all endpoints $\mathcal{E} = \{E_i\}$ in the skeleton obtained by thinning and denoted by $\mathcal{S}_{raw}$. Endpoints are voxels having at most one neighbor:

$$\mathcal{E} = \left\{ v_{i,j,k} \in \mathcal{V} \ \Big| \sum_{|(i,j,k)-(x,y,z)|_\infty = 1} v_{x,y,z} \leq 1 \right\} \tag{1}$$

and this includes isolated voxels without neighbors and voxels at the end of skeleton branches. The set of endpoints is likely to be located on the leaf ends, but it also includes some false positives.

Root identification finds the plant's root v_0 that is the starting point of the stem. We define the root as the lowest endpoint that is located within the radius of the pot R_{pot}:

$$v_0 = \underset{\substack{v_{i,j,k} \in \mathcal{E} \\ d_{e_z}(v_{i,j,k}) < R_{pot}}}{\arg \min} \quad (k), \tag{2}$$

where $d_{e_z}(v_{i,j,k})$ is the distance from $v_{i,j,k}$ to e_z, the z-axis. We constrain the root voxel within the pot because we noticed that some leaf tips outside the container could extend below the plant/soil interface level. The lowest endpoint without this restriction could be a low-hanging leaf tip instead of the actual root.

Branch Finding: The raw skeleton $\mathcal{S}_{raw}$ resulting from thinning is not guaranteed to have a tree topology: it can contain loops. To convert the graph into a tree and discard some noisy parts, we run a single source shortest path on $\mathcal{S}_{raw}$ starting from the root voxel v_0. The output is the shortest path from every voxel of $\mathcal{S}_{raw}$ to the root voxel v_0. To accommodate potential disconnections in $\mathcal{S}_{raw}$, we allow the shortest path algorithm to jump between two voxels even if they are not connected. We use the Dijkstra algorithm and add a penalty on the distance if two voxels are not connected.

The distance $d(v_1, v_2)$ between voxels v_1 and v_2 is computed as follow: if v_2 is in the 26-connected neighborhood of v_1, then $d(v_1, v_2) = 1$. If v_2 is within a cube of length 48 around v_1, we still consider them as connected and the penalized distance is computed as a function of their Manhattan distance $i.e.,$ $|v_1 - v_2|$:

$$d(v_1, v_2) = \begin{cases} 1 & if \quad \|v_1 - v_2\|_\infty = 1 \\ \frac{|v_1 - v_2|(1 + |v_1 - v_2|)}{2} & if \quad \|v_1 - v_2\|_\infty \leq 24. \end{cases} \tag{3}$$

By still considering v_1 and v_2 connected but with a distance penalty, we allow the algorithm to connect to the nearest voxel when there is a small discontinuity between two parts of the plant. We empirically chose a maximum distance of 24 in infinity norm for jumping from a voxel to another. It provides a good compromise between the computation time and the distance to which we want to connect two plant parts.

The single-source shortest path algorithm generates a tree of voxels starting from the root v_0. As explained above in the *Endpoints Identification* step and in Eq. (1) we may miss some leaf tips if there is a loop in $\mathcal{S}_{raw}$. Therefore, we update the list of endpoints based on the output of the Dijkstra algorithm. Any voxel that has no predecessor and has at most two neighbors in $\mathcal{S}_{raw}$ is added to the list of endpoints $\mathcal{E}$.

Finally, we output a list of shortest paths $\mathcal{P} = \{P_0, P_1, \dots\}$. The path P_i starts from the endpoint E_i and goes all the way down to the root voxel v_0, and every path potentially includes a plant leaf. We sort paths in $\mathcal{P}$ by descending length (longest leaf first). We also discard paths with a length of only one voxel because they cannot connect to the root voxel.

Branch Pruning: The collection of paths $\mathcal{P}$ represents a plant skeleton that does not include loops nor lumps, and it is less noisy than the raw skeleton $\mathcal{S}_{raw}$.

However, it may still include small spurious branches that do not correspond to any actual leaf. These branches are a likely result of artifacts and irregularities in the 3D reconstruction. They mostly consist of a small spike starting from the middle of a leaf and going to another direction.

This is mitigated by pruning the paths and keep only those that bring the most information. That is achieved by processing each path P_i, decreasing order of length, and considering every other shorter path P_j (therefore $j > i$). When comparing two paths, P_i and P_j, we compute a set of features and use a machine learning classifier to decide whether we keep or discard the branch P_j. We detail the different classifiers and the learning procedure we used in Sect. 6. We repeat the classification for each path in $\mathcal{P}$, making in total $n(n-1)/2$ comparisons, with n being the number of paths in $\mathcal{P}$. The procedure is detailed in Algorithm 1. Paths that are not discarded are kept in a new list $\mathcal{P}^*$

$$\mathcal{P}^* = \{P_i \in \mathcal{P} \mid K_i = 1\}. \tag{4}$$

Finally, the union of voxels from all paths in $\mathcal{P}^*$ forms the plant skeleton $\mathcal{S}$. An example of such a skeleton can be seen in Fig. 3 b).

Data: $\mathcal{P}$ is a list of paths ordered by decreasing length
Result: $\mathcal{K}$ a label for each path stating that a path should be kept or discarded in the
output skeleton.

```
   /* This algorithm discards paths that are not relevant.               */
 1 K ← {1}₁ⁿ; /* By default, we keep all paths.                          */
 2 for i ∈ {1...n} do
 3 │   if Kᵢ = 1 then
 4 │   │   for j ∈ {i + 1...n} do
 5 │   │   │   if Kⱼ = 1 then
   │   │   │   │   /* Run the classifier for the paths: Pᵢ and Pⱼ.       */
 6 │   │   │   │   if runClassifier(Pᵢ, Pⱼ) == false then
   │   │   │   │   │   /* Let's discard Pⱼ.                              */
 7 │   │   │   │   │   Kⱼ ← 0;
 8 │   │   │   end
 9 │   │   end
10 │   end
11 │ end
12 end
```

Algorithm 1: Creation of the training data set $\mathcal{L}$ from an annotated skeleton

5   Segmentation

In this section, we detail how we segment the input set of voxel $\mathcal{V}$ into individual leaves by using the skeleton $\mathcal{S}$. This algorithm proceeds in two steps: we first segment the skeleton and then the full plant.

Skeleton segmentation segments each path in $\mathcal{P}^*$ from the skeleton $\mathcal{S}$ into a stem part and a leaf part. The stem is composed of all voxels that are shared between at least two paths. To find voxels that are shared, we compute the histogram of occurrences for each voxel of the skeleton $\mathcal{S}$. If a voxel appears at least twice in $\mathcal{S}$, it is considered to be a part of the stem. Once the stem has been identified, we remove all stem voxels from all other paths in $\mathcal{P}^*$, and the remaining voxels correspond to the leaves. Although our approach works on well-reconstructed plants, by construction, we do not guarantee that the skeleton has the right topology with poorly reconstructed plants, which have merged leaves. Therefore, it is necessary to check whether the topology, *i.e.,* the stem should not have any T-junctions. We discuss this issue in Sect. 7. Moreover, we also consider the root to be inside the pot.

Plant Segmentation: We segment the full plant according to the skeleton $\mathcal{S}$. Each voxel $v_{i,j,k}$ of the plant is assigned to the nearest segment of the segmented skeleton $\mathcal{S}$. Figure 3 c) shows the final segmented plants.

6   Branch Classification

Below we explain how we designed and trained machine learning classifiers used in Sect. 4.2 to prune a raw plant skeleton. We trained three different classifiers for this task (linear SVM, SVM with RBF kernels, and a Multi-Layer Perceptron) and compared them.

6.1   Data Set

To train our classifiers, we manually annotated 100 plant skeletons by choosing from a pool of 351 sorghum plants. We decided to discard plants that are poorly reconstructed, including merged leaves, missing leaves, or excessive numbers of noise voxels (for example, leaf-like reconstruction artifacts). Since we feed our machine learning classifiers with handcrafted features and are not learning the data representation, we do not use any form of data augmentation.

For each skeleton, we looked at each path $P_i \in \mathcal{P}$ before the pruning step (see Sect. 4.2) and we annotated it in $\mathcal{K}$ by marking whether it should be kept ($K_i = 1$) or discarded ($K_i = 0$) in the final plant skeleton $\mathcal{P}^*$. In our data set, $\mathcal{P}$ contains 108 paths on average before pruning, and $\mathcal{P}^*$ contains only seven paths after pruning. Annotation is a tedious task, and completing annotations for 100 plant skeletons took about a week of effort by a trained expert.

To train the classifiers, we transform the data set into a list $\mathcal{L}$ of triplets $[P_i, P_j, keep]$, with a binary label *keep* that indicates whether P_j should be kept or discarded when compared to P_i. We detail the method to generate triplets in Algorithm 2. The main idea is that a path that is kept should never discard another shorter path that is kept. Moreover, only one longer path is needed to reject a given path that is not supposed to be kept. When a path P_j has to be discarded, we only add the triplet with P_i, the shortest longer path that is the most similar to P_j. By doing so, we guarantee that P_i is discarded at least once. Any other decision made regarding P_i is not essential.

Once the data set $\mathcal{L}$ of triplets is built, we compute a set of features for each pair of paths that are input into the classifier. After trying different combinations, the set of features we retained includes: 1) the number s_{ij} of voxels that P_i

Data: $\mathcal{P}$ is a list of paths, ordered by increasing length and $\mathcal{K}$ is a label for each path stating that a path should be kept or discarded in the output skeleton.

Result: $\mathcal{L}$ is the set of triplets $[P_i, P_j, keep]$ for training.

```
 1  for j ∈ {1 ... n} do
 2      if Kⱼ = 1 then
 3          for i ∈ {j + 1 ... n} do
 4              if Kᵢ = 1 then
 5                  |  L ← L ∪ Pᵢ, Pⱼ, true;
 6              end
 7          end
 8      else
            /* We look for the shortest most similar path longer than Pⱼ
               That is, the first path Pᵢ that maximizes sᵢⱼ.              */
 9          k ← j;
10          mostNbCommonVoxels ← 0;
11          for i ∈ {j + 1 ... n} do
12              if Kᵢ = 1 then
13                  sᵢⱼ ← computeNumberCommonVoxel(Pᵢ, Pⱼ);
14                  if sᵢⱼ > mostNbCommonVoxels then
15                      mostNbCommonVoxels ← sᵢⱼ;
16                      k ← i;
17                  end
18              end
19          end
20          if k > j then
21              |  L ← L ∪ {Pₖ, Pⱼ, true};
22          end
23      end
24  end
```

Algorithm 2: Creating the training data set $\mathcal{L}$ from an annotated skeleton.

and P_j have in common 2) the number $e_j = length(P_j) - s_{ij}$ of voxels that are included only in P_j, and 3) the ratio of the shorter path P_j that is shared with P_i: $p_{ij} = s_{ij}/length(P_j)$.

Our interpretation of the role of the features is that: 1) estimates the position of a branch in the plant (bottom or top), 2) estimates the length of a branch (short or long), and 3) estimates the information brought by a branch compared to a longer branch.

Classes in our data set are unbalanced, and $\mathcal{L}$ contains significantly more pairs of branches that should be discarded than couples to keep. Thus, we balance the data set by applying weights on each sample.

We randomly split the 100 annotated skeletons into 64 for training, 16 for validation, and 20 for testing. As a reference, on the one hand, the 80 skeletons in the training and validation set generate 9,877 triplets. On the other hand, the test set generates 2,422 triplets.

6.2   Classifiers

The input to our classifier is a set of three features computed on a pair of paths P_i and P_j in the raw skeleton $\mathcal{S}_{raw}$. The output is a binary decision: true means that we keep path P_j, false means that we discard it, and we tested three different classifiers. We used machine learning models with handcrafted features because we do not have an extensive data set to allow automatic feature learning. After all, manual labeling is a very time-consuming process. Moreover, it is easier for a human to interpret the predictions.

We tested a **linear SVM** based on the shared proportion feature p_{ij}. This classifier finds the cutoff value of p_{ij} that linearly separates branches that should be kept from those that should be discarded. Although simple, it performs relatively well and can be easily implemented as a single *if* condition in the code. Experimentally, we found the cutoff value to be $p_{ij} = 0.65$, and we used this classifier as a reference for more complicated models.

The second model we tested is a **SVM with RBF kernels** based on the three features. We trained by using grid search and ten-fold cross-validation on the training set.

The third classifier is a **Multi Layer Perceptron** containing an input layer with three units, followed by two eight-units hidden layers, and finally a one-unit output layer. Input features are normalized, and all units use a Sigmoid activation function. For training we use the batch RPROP algorithm [18] implemented in the OpenCV [2] library.

6.3 Evaluation

We run the branch pruning algorithm (Algorithm 1 in Sect. 4.2) on the test set of 20 skeletons, and we used information retrieval evaluation measures to compare the different classifiers. For each skeleton, we compared the set of branches selected by the branch pruning algorithm to the ground truth $\mathcal{K}$. Let's denote by tp (true positive) the number of successfully reconstructed branches *i.e.,* selected branches that are in $\mathcal{K}$; let's further denote fp (false positive) the number of branches that were selected even if they were not in $\mathcal{K}$; we also denote fn (false negative) the number of branches that were not selected even if they were in $\mathcal{K}$. Finally, tn (true negative) is the number of successfully discarded branches. We compute the precision

$$precision = \frac{tp}{tp + fp}, \tag{5}$$

which is the proportion of selected branches that are relevant, recall

$$recall = \frac{tp}{tp + fn}, \tag{6}$$

is the proportion of relevant branches that are selected, and the F-measure

$$Fmeasure = 2 \cdot \frac{precision \cdot recall}{precision + recall}, \tag{7}$$

is the harmonic mean of precision and recall and gives an overall score for the quality of the branch pruning.

Evaluations of our three different classifiers on the training and validation set are given in Table 1 and on the test set are given in Table 2. While the three classifiers performed comparatively, the RBF SVM classifier provided the best results based on the evaluation metrics used in this paper.

Table 1. Evaluation of the classifiers on the Training + Validation set.

	tp	fp	fn	tn	Precision	Recall	F-measure
Linear SVM	535	27	29	8,057	0.952	0.949	0.950
RBF SVM	549	15	15	8,069	0.973	0.973	0.973
MLP	551	24	13	8,060	0.958	0.977	0.968

Table 2. Evaluation of the classifiers on the Test set.

	tp	fp	fn	tn	Precision	Recall	F-measure
Linear SVM	131	8	8	1,990	0.942	0.942	0.942
RBF SVM	135	5	4	1,993	0.964	0.971	0.968
MLP	136	7	3	1,991	0.951	0.978	0.965

7 Results

We have implemented our method in C++ and run it on a workstation equipped
with an Intel Xeon W-2145. We did not use any GPU acceleration, and, on
average, the thinning step takes 43 s, and the pruning and the segmentation
steps take 10 s per plant.

We use the 20 plants in the test set to validate our method. We visually
inspected each plant to make sure that the leaves are at the right location and
the topology of the skeleton is correct. Table 2 from Sect. 6.3 shows that 97.1%
of branches were successfully reconstructed and among reconstructed branches,
96.4% were relevant. Note that the RBF SVM classifier misclassified only 9
branches out of 2,128 (135 true positives and 1,993 true negatives).

We used two measures to automatically get an insight into how well a plant
is skeletonized. 1) We compute the maximum distance from any voxel in V to
the skeleton. If this distance is too high (*e.g.,* more than 5 cm), it indicates that
one leaf is likely missing in the skeleton. If this distance is small, it could suggest
that some spurious branches were not discarded. 2) We look at the skeleton's
topology with the assumption that the stem should not include any T-junctions.
If these two measures fail, the plant needs to be visually inspected.

One plant (1) from the training set and four plants (2, 3, 4, 5) from the test
set are shown in Fig. 3 and additional results are in Fig. 4. Column (b) shows
the skeleton output by our method, and segmented plants are in column (c).

Failure Cases: Some plants have a wooden stick in their pot, and in some cases,
a leaf happens to be reaching under the lowest part of the stem. If such features
are reconstructed within the container radius R_{pot}, the algorithm will erroneously
identify them as the root voxel v_0. Although it does not affect counting and
measuring leaves, we get the wrong length for the stem.

If the input voxel plant is poorly reconstructed and contains leaves that are
merged because they are too close, the algorithm for the shortest path computation can jump between the leaves, because the overall cost is cheaper than

staying on the correct leaf. In this case, the two leaves will have the same starting point but two distinct tips. Depending on the rest of the plant, its topology may even become wrong with a T-junction on the stem. This case is shown in Fig. 3 (2), where the top left leaf is merged to the top leaf. This failure case is not critical when counting leaves, but causes errors in length estimation, by up to a few centimeters.

If the branch classifier makes an erroneous decision during the skeleton pruning step, a leaf can be missed because it has been filtered, as shown in Fig. 3 (3). Conversely, an extra leaf can be left after the pruning step, as can be seen in Fig. 3 (4). It happens especially when voxel carving with a low number of views has been used because it creates leaf-like artifacts.

If a plant is poorly reconstructed and some voxels at the tips are missed, our method, which is only based on input voxels from $\mathcal{V}$, will only output a truncated skeleton because it does not take in account 2D pictures of the plant. This is the reason why skeletons in column (b) of Fig. 3 do not always extend to the end of leaves, because our input voxel plants are not entirely reconstructed. This problem is inherent to voxel carving. To mitigate it, we would need to improve calibration and use multiple cameras instead of a single camera with a turntable.

8   Conclusions and Future Work

We have introduced a novel method for generating skeletons and segmentation of voxel data generated by the voxel carving algorithm that is commonly used by phenotyping facilities to create 3D approximations of measured plants. We tested three different classifiers to prune leaves from raw plant skeletons, and our validations using manually labeled data show that the RBF SVM classifier provided the best results. Furthermore, we used the generated skeleton to segment the input voxel data into the plant's leaves and stem. Each voxel is assigned an identifier that defines the plant organ it belongs.

A limitation of our work, and a possible extension as a future work, is the lack of labeled data. The input data is complicated, and an educated person can use only manually label branches. Having an automatic annotation tool would greatly help to expand the training set. It would be interesting to provide a pixel-based performance metric for the segmentation. We do not guarantee that the constructed skeleton's topology is correct because it can have T-junctions if the input plant is poorly reconstructed. Although it covers a vast majority of cases we encountered, we think it is possible to extend the algorithm with backtracking to ensure that the skeleton has a correct topology, even if it features merged leaves. Top and small leaves are hard to identify, as shown in Sect. 7. We could extend the algorithm and use many stages of the same plant over time to discriminate growing leaves from artifacts using a later stage of the plant where the difference is noticeable. Finally, starting from annotated voxels generated by our tool and corrected by experts, one could develop fully automatic machine-learning classifiers that would allow for fast and reliable segmentation at the voxel level.

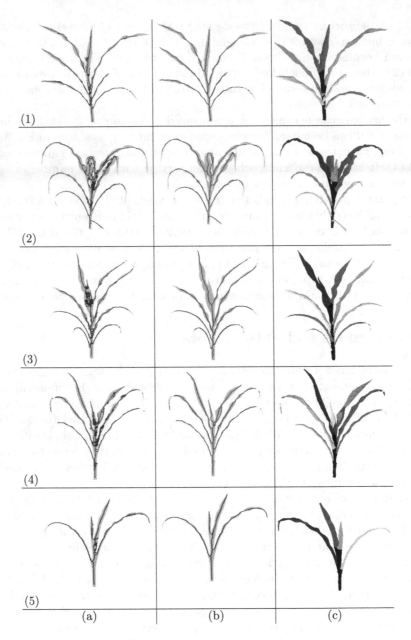

Fig. 3. Columns: (a) Raw skeleton resulting from thinning, (b) skeleton after the pruning step, and (c) segmented leaves and stem. (1) a correctly segmented plant, (2) a successfully segmented complex plant from the test set. The blue leaf (the top left one) is off because in the reconstructed plant, leaves are merged. (3) A plant from the test set with the top leaf missing. (4) Another plant from the test set with a spurious missing (see the dark green area between the orange and red leaves). (5) A correctly reconstructed small plant from the test set. (Color figure online)

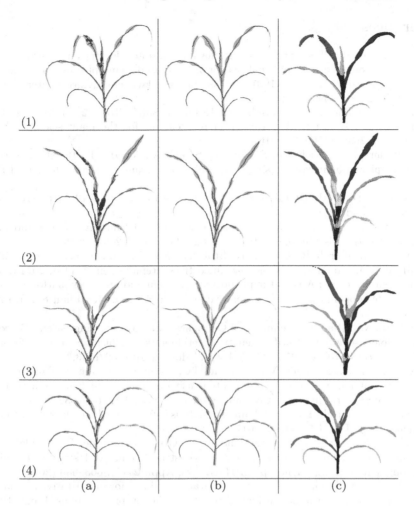

Fig. 4. Additional examples (a) Raw skeleton resulting from thinning, (b) skeleton after the pruning step, and (c) segmented leaves and stem.

Acknowledgements. This research was supported by the Foundation for Food and Agriculture Research Grant ID: 602757 to Benes and Schnable. The content of this publication is solely the responsibility of the authors and does not necessarily represent the official views of the foundation for Food and Agriculture Research. This research was supported by the Office of Science (BER), U.S. Department of Energy, Grant no. DE-SC0020355 to Schnable. This work was supported by a National Science Foundation Award (OIA-1826781) to Schnable.

References

1. Artzet, S., et al.: Phenomenal: an automatic open source library for 3D shoot architecture reconstruction and analysis for image-based plant phenotyping. bioRxiv (2019). https://doi.org/10.1101/805739. https://www.biorxiv.org/content/early/2019/10/21/805739
2. Bradski, G.: The OpenCV library. Dr Dobb's J. Softw. Tools **25**, 120–125 (2000)
3. Bucksch, A.: A practical introduction to skeletons for the plant sciences. Appl. Plant Sci. **2**(8), 1400005 (2014)
4. Choudhury, S.D., Bashyam, S., Qiu, Y., Samal, A., Awada, T.: Holistic and component plant phenotyping using temporal image sequence. Plant Methods **14**(1), 35 (2018)
5. Coeurjolly, D., et al.: DGtal-team/dgtal: Release 1.0, March 2019. https://doi.org/10.5281/zenodo.2611275
6. Couprie, M., Bertrand, G.: Asymmetric parallel 3D thinning scheme and algorithms based on isthmuses. Pattern Recogn. Lett. **76**, 22–31 (2016)
7. Du, S., Lindenbergh, R., Ledoux, H., Stoter, J., Nan, L.: AdTree: accurate, detailed, and automatic modelling of laser-scanned trees. Remote Sen. **11**(18), 2074 (2019)
8. Fahlgren, N., et al.: A versatile phenotyping system and analytics platform reveals diverse temporal responses to water availability in Setaria. Mol. Plant **8**(10), 1520–1535 (2015)
9. Gaillard, M., Miao, C., Schnable, J.C., Benes, B.: Voxel carving-based 3D reconstruction of sorghum identifies genetic determinants of light interception efficiency. Plant Direct **4**(10), e00255 (2020). https://doi.org/10.1002/pld3.255
10. Ge, Y., Bai, G., Stoerger, V., Schnable, J.C.: Temporal dynamics of maize plant growth, water use, and leaf water content using automated high throughput RGB and hyperspectral imaging. Comput. Electron. Agric. **127**, 625–632 (2016)
11. Gehan, M.A., et al.: Plantcv v2: image analysis software for high-throughput plant phenotyping. PeerJ **5**, e4088 (2017)
12. Golbach, F., Kootstra, G., Damjanovic, S., Otten, G., Zedde, R.: Validation of plant part measurements using a 3D reconstruction method suitable for high-throughput seedling phenotyping. Mach. Vis. Appl. **27**(5), 663–680 (2016)
13. Junker, A., et al.: Optimizing experimental procedures for quantitative evaluation of crop plant performance in high throughput phenotyping systems. Front. Plant Sci. **5**, 770 (2015)
14. Klodt, M., Cremers, D.: High-resolution plant shape measurements from multiview stereo reconstruction. In: Agapito, L., Bronstein, M.M., Rother, C. (eds.) ECCV 2014. LNCS, vol. 8928, pp. 174–184. Springer, Cham (2015). https://doi.org/10.1007/978-3-319-16220-1_13
15. Koenderink, N., Wigham, M., Golbach, F., Otten, G., Gerlich, R., van de Zedde, H.: Marvin: high speed 3d imaging for seedling classification. In: van Henten, E., Goense, D., Lokhorst, C. (eds.) Precision Agriculture 2009: Papers Presented at the 7th European Conference on Precision Agriculture, Wageningen, The Netherlands, 6–8 July 2009, pp. 279–286. Wageningen Academic Publishers (2009)
16. Kutulakos, K.N., Seitz, S.M.: A theory of shape by space carving. Int. J. Comput. Vision **38**(3), 199–218 (2000)
17. Lobet, G.: Image analysis in plant sciences: publish then perish. Trends Plant Sci. **22**(7), 559–566 (2017)
18. Riedmiller, M., Braun, H.: A direct adaptive method for faster backpropagation learning: The RPROP algorithm. In: IEEE International Conference on Neural Networks, pp. 586–591. IEEE (1993)

19. Scharr, H., Briese, C., Embgenbroich, P., Fischbach, A., Fiorani, F., Müller-Linow, M.: Fast high resolution volume carving for 3D plant shoot reconstruction. Front. Plant Sci. **8**, 1680 (2017). https://doi.org/10.3389/fpls.2017.01680. https://www.frontiersin.org/article/10.3389/fpls.2017.01680
20. Tagliasacchi, A., Delame, T., Spagnuolo, M., Amenta, N., Telea, A.: 3D skeletons: a state-of-the-art report. In: Computer Graphics Forum, vol. 35, pp. 573–597. Wiley Online Library (2016)
21. Ward, B., et al..: A model-based approach to recovering the structure of a plant from images (2015). http://search.proquest.com/docview/2081688123/
22. Wu, S., et al.: MVS-pheno: a portable and low-cost phenotyping platform for maize shoots using multiview stereo 3D reconstruction. Plant Phenomics 2020 (2020). https://doaj.org/article/bd4ae8082b0c45c2a1d4c6ff935d9ff1
23. Wu, S., et al.: An accurate skeleton extraction approach from 3D point clouds of maize plants. Front. Plant Sci. **10** (2019)
24. Xiang, L., Bao, Y., Tang, L., Ortiz, D., Salas-Fernandez, M.G.: Automated morphological traits extraction for sorghum plants via 3D point cloud data analysis. Comput. Electron. Agric. **162**, 951–961 (2019)

Towards Confirmable Automated Plant Cover Determination

Matthias Körschens[1]([⊠]) [iD], Paul Bodesheim[1] [iD], Christine Römermann[1,2,3] [iD],
Solveig Franziska Bucher[1,3] [iD], Josephine Ulrich[1] [iD], and Joachim Denzler[1,2,3] [iD]

[1] Friedrich Schiller University Jena, Jena, Germany
{matthias.koerschens,paul.bodesheim,christine.roemermann,
solveig.franziska.bucher,josephine.ulrich,joachim.denzler}@uni-jena.de
[2] German Centre for Integrative Biodiversity Research (iDiv) Halle-Jena-Leipzig,
Leipzig, Germany
[3] Michael Stifel Center Jena, Jena, Germany

Abstract. Changes in plant community composition reflect environmental changes like in land-use and climate. While we have the means to record the changes in composition automatically nowadays, we still lack methods to analyze the generated data masses automatically.

We propose a novel approach based on convolutional neural networks for analyzing the plant community composition while making the results explainable for the user. To realize this, our approach generates a semantic segmentation map while predicting the cover percentages of the plants in the community. The segmentation map is learned in a weakly supervised way only based on plant cover data and therefore does not require dedicated segmentation annotations.

Our approach achieves a mean absolute error of 5.3% for plant cover prediction on our introduced dataset with 9 herbaceous plant species in an imbalanced distribution, and generates segmentation maps, where the location of the most prevalent plants in the dataset is correctly indicated in many images.

Keywords: Deep learning · Machine learning · Computer vision · Weakly supervised segmentation · Plants · Abundance · Plant cover

1 Introduction

In current times the effect of anthropogenic activities is affecting ecosystems and biodiversity with regard to plants and animals alike. Whereas poaching and clearing of forests are only some of the smaller impacts of humans, one of the biggest is the anthropogenic effect on climate change.

Plants are strong indicators of climate change, not only in terms of phenological responses [7,9,13,31,32,37], but also in terms of plant community compositions [27,29,37]. However, these compositions do not only reflect changes

Electronic supplementary material The online version of this chapter (https:// doi.org/10.1007/978-3-030-65414-6_22) contains supplementary material, which is available to authorized users.

© Springer Nature Switzerland AG 2020
A. Bartoli and A. Fusiello (Eds.): ECCV 2020 Workshops, LNCS 12540, pp. 312–329, 2020.
https://doi.org/10.1007/978-3-030-65414-6_22

in climate, but also in other aspects, like land use [2,14] and insect abundance [39]. Hence, plant community compositions are a valuable metric for determining environmental changes and are therefore focus of many experiments [6,14,27,39].

In the last years, technology enabled us to develop systems that can automatically collect images of such experiments in high resolution and high frequency, which would be too expensive and time-consuming if done manually. This process also creates large masses of data displaying complex plant compositions, which are also hard to analyse by hand. As we are missing methods to survey the data automatically, this is usually still done manually by biologists directly in the field. However, this process is bound to produce subjective results. Therefore, an automated, objective method would not only enable fast evaluation of the experimental data, but also greatly improve comparability of the results.

Krizhevsky et al. [24] showed in the ILSVRC 2012 challenge [38] that convolutional neural networks (CNNs) can be used to analyze large numbers of images by outperforming alternative approaches by a large margin. Following this, deep learning became a large area of research with many different developments, but only a small number of approaches deal with the analysis of plants and therefore there are only few existing solutions for the very specific problems in this area.

With our approach we propose a system for an important task: the analysis of plant community compositions based on plant cover. The plant cover, i.e., the amount of ground covered by each plant species, is an indicator for the plant community composition. The information on the spatio-temporal distribution of plant communities leads to a better understanding of effects not only related to climate change, but also concerning other environmental drivers of biodiversity [5,6,44]. We present an approach using a custom CNN architecture, which we train on plant cover percentages that are provided as annotations. We treat this as a pixel-wise classification problem known as semantic segmentation and aggregate the individual scores to compute the cover predictions.

CNNs are often treated as black boxes, returning a result without any information to the user what it is based on. To prevent trust issues resulting from this, we also focus on providing a segmentation map, which the network learns by training on the cover percentage labels only. With this map, the user can verify the network detections and whether the output of the network is reasonable. For implausible cases or manual inspections of random samples, the user can look at the segmentations. If detections of the network are deemed incorrect, a manual evaluation of the images can be suggested in contrast to blind trust in the output of the network. To the best of our knowledge, we are first in applying CNNs to plant cover prediction by training on the raw cover labels only and using relative labels (cover percentages) to train a network for generating segmentation maps.

In the next section we will discuss related work, followed by the dataset we used and its characteristics in Sect. 3. In Sect. 4 we will then present our approach, with the results of our experiments following in Sect. 5. We end the paper with a conclusion and a short discussion about future work in Sect. 6.

2   Related Work

An approach also dealing with plant cover is the one by Kattenborn et al. [20], who developed their approach using remote sensing image data, specifically images taken from UAVs. They developed a small convolutional neural network (CNN) architecture with 8 layers to determine the cover of different herb, shrub and woody species. In contrast to our approach, their network was trained on low-resolution image patches with delineations of tree canopies directly in the images. In addition to this, their approach was mostly concerned with the distinction of 2–4 tree species with heterogeneous appearances, which makes the classification easier as compared to our problem.

While, to the best of our knowledge, the aforementioned approach appears to be the only one dealing with plant cover, there are many methods which tackle plant identification in general, e.g. [4,15,25,48]. One example for such a method is the one by Yalcin et al. [48], who applied a pre-trained CNN with 11 layers on fruit-bearing agricultural plants. Another, more prominent project concerned with plant identification is the Flora Incognita project of Wäldchen et al. [45], in which multiple images of a single plant can be used for identification. These approaches, however, are usually applied on one or multiple images of a single plant species in contrast to pictures of plant communities with largely different compositions like in our dataset.

Weakly-supervised segmentation, i.e., the learning of segmentation maps using only weak labels, is an established field in computer vision research. Therefore, we can also find a multitude of different approaches in this area. Some of them use bounding boxes for training the segmentation maps [10,21] while others use merely image-level class annotations [3,18,23,33,46], as these are much easier to acquire than bounding box annotations. However, most of these approaches are only applied on images with mostly large objects like the PASCAL-VOC dataset [12] as opposed to high-resolution images with small fine-grained objects like in our dataset. In addition to this, in our dataset we have a new kind of weak labels: plant cover percentage labels. This type of label enables new approaches for learning segmentation maps, which we try to exploit in this paper.

At first glance the task of predicting the cover percentage appears similar to counting or crowd-counting tasks, which are often solved by training a model on small, randomly drawn image patches and evaluating them on complete images, or also evaluating them on patches and aggregating information afterwards [19, 28,47]. This can be done, because only absolute values have to be determined, which are usually completely independent from the rest of the image. However, in our dataset the target values, i.e., the cover percentages, are not absolute, but relative and therefore depend on the whole image. Because of this, we have to process the complete images during training and cannot rely on image patches.

Fig. 1. A selection of example images from the image series of a single camera in a single EcoUnit. The complete life cycle is captured in the image series, including flowering and senescence.

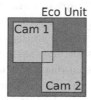

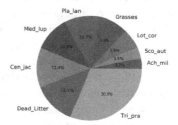

Fig. 2. An EcoUnit from the Ecotron system.

Fig. 3. An example camera setup in an EcoUnit. The two cameras are placed at opposite corners of the EcoUnits and can have an overlapping field of view. In some cases not the complete unit is covered by the cameras.

Fig. 4. The mean cover percentages of the plant species over all annotated images in the dataset in a long-tailed distribution. The abbreviations are explained in Sect. 3.

3   Dataset

For our experiments we used a dataset comprising images from the InsectArmageddon project[1] and therefore we will refer to this dataset as the InsectArmageddon dataset. During this project the effects of invertebrate density on plant composition and growth were investigated. The experiments were conducted using the iDiv Ecotron facilities in Bad Lauchstädt [11,42], which is a

---

[1] https://www.idiv.de/en/research/platforms_and_networks/idiv_ecotron/experimen ts/insect_armageddon.html

system comprising 24 so-called EcoUnits. Each of these EcoUnits has a base area of about 1.5 m × 1.5 m and contains a small, closed ecosystem corresponding to a certain experimental setup. An image of an EcoUnit is shown in Fig. 2.

Over the time span of the project, each of the EcoUnits was equipped with two cameras, observing the experiments from two different angles. One example of such a setup in shown in Fig. 3. It should be noted that the cameras have overlapping fields of view in many cases, resulting in the images from each unit not being independent of each other. Both cameras in each unit took one image per day. As the duration of the project was about half a year, 13,986 images have been collected this way over two project phases. However, as annotating this comparatively large number of images is a very laborious task, only about one image per recorded week in the first phase has been annotated per EcoUnit. This is drastically reducing the number of images available for supervised training.

The plants in the images are all herbaceous, which we separate in nine classes with seven of them being plant species. These seven plants and their short forms, which are used in the remainder of the paper, are: *Trifolium pratense* (tri_pra), *Centaurea jacea* (cen_jac), *Medicago lupulina* (med_lup), *Plantago lanceolata* (pla_lan), *Lotus corniculatus* (lot_cor), *Scorzoneroides autumnalis* (sco_aut) and *Achillea millefolium* (ach_mil). The two remaining classes are grasses and dead litter. These serve as collective classes for all grass-like plants and dead biomass, respectively, mostly due to lack of visual distinguishability in images.

As with many biological datasets, this one is heavily imbalanced. The mean plant cover percentages over the complete dataset are shown in Fig. 4. There, we can see that tri_pra represents almost a third of the dataset and the rarest three classes, ach_mil, sco_aut and lot_cor together constitute only about 12% of the dataset.

3.1 Images

The cameras in the EcoUnits are mounted in a height of about 2 m above the ground level of the EcoUnits and can observe an area of up to roughly 2 m × 2 m, depending on zoom level. Equal processing of the images however is difficult due to them being scaled differently. One reason for this is that many images have different zoom levels due to technical issues. The second reason is that some plants grew rather high and therefore appear much larger in the images.

As mentioned above, the images cover a large time span, i.e., from April to August 2018 in case of the annotated images. Hence, the plants are captured during their complete life cycle, including the different phenological stages they go through, like flowering and senescence.

Occlusion is one of the biggest challenges in the dataset, as it is very dominant in almost every image and makes an accurate prediction of the cover percentages very difficult. The occlusion is caused by the plants overlapping each other and growing in multiple layers. However, as we will mostly focus on the visible parts of the plants, tackling the non-visible parts is beyond the scope of this paper. A small selection from the images of a camera of a single EcoUnit can be seen in Fig. 1. Each of the images has an original resolution of 2688 × 1520 px.

As already discussed in Sect. 2, we are not able to split up the images into patches and train on these subimages, as we only have the cover annotations for the full image. Therefore, during training we always have to process the complete images. This circumstance, in conjunction with the rather high resolution of the images, the similarity of the plants and massive occlusion, makes this a tremendously hard task.

3.2 Annotations

As already mentioned above, the annotations for the images are cover percentages of each plant species, i.e., the percent of ground covered by each species, disregarding occlusion. The cover percentages have been estimated by a botanist using both images of each EcoUnit, if a second image was available. As perfect estimation is impossible, the estimates have been quantized into classes of a modified Schmidt-scale [34] (0, 0.5, 1, 3, 5, 8, 10, 15, 20, 25, 30, 40, 50, 60, 70, 75, 80, 90 and 100%). While such a quantization is very common for cover estimation in botanical research [30,34], it introduces label noise and can, in conjunction with possible estimation errors, potentially impair the training and evaluation process of machine learning models. In addition to the cover percentages, we also estimated vegetation percentages, specifying the percentage of ground covered by plants in general, which we use as auxiliary target value.

While both images of each EcoUnit have been used for estimating a single value, the distribution of plants should approximately be the same for both images. Therefore, we increase the size of our dataset by using one annotation for both images, which leads us to 682 image-annotation pairs.

4   Approach

Due to the necessity of using the complete image for the training process, we require a setting, in which it is feasible to process the complete image efficiently without introducing too strong limitations on hyperparameters like the batch size. The most important part of such a setting is the image resolution. As it is hard to train models on very high resolutions due to GPU memory limitations, we chose to process the images at a resolution of 672×336 px, which is several times larger than other common input image resolutions for neural network architectures, like e.g. ResNet [17] training on the ImageNet dataset [38] with a resolution of 224×224 px. To make the results confirmable, we aim to create a segmentation map during prediction that designates, which plant is located at each position in the image. This segmentation map has to be learned implicitly by predicting the cover percentages. Due to the plants being only very small in comparison to the full image, this segmentation map also has to have a high resolution to show the predicted plants as exactly as possible.

The usage of standard classification networks, like ResNet [17] or Inception [40,41], is not possible in this case, as the resolution of the output feature maps is too coarse for an accurate segmentation map. Additionally, these networks

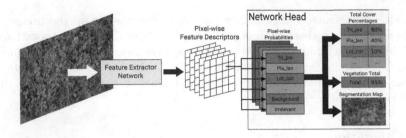

Fig. 5. The basic structure of the network. It consist of a feature extractor network as backbone, which aggregates information from the input image in a high resolution, and a network head, which performs the cover percentage calculation and generates the segmentation map.

Table 1. A detailed view of the network architecture. We use the following abbreviations: k - kernel size, s - stride, d - dilation rate

Layers				Output shape
Conv k:5x5, s:2x2				336x168x128
Conv k:5x5, s:2x2				168x84x256
9xResidual Bottleneck				168x84x256
Conv k:1x1	Conv k:3x3	Conv k:3x3 Conv k:3x3 Conv k:3x3	Conv k:3x3 Conv k:3x3, d:3,3 Conv k:3x3	168x84x512
Conv k:1x1				168x84x128

and most segmentation networks with a higher output resolution, like Dilated ResNet [49], have large receptive fields. Thus, they produce feature maps that include information from large parts of the image, most of which is irrelevant to the class at a specific point. This leads to largely inaccurate segmentation maps.

We thus require a network, which can process the images at a high resolution, while only aggregating information from a relatively small, local area without compressing the features spatially to preserve as much local information as possible. Our proposed network is described in the following.

4.1   General Network Structure

The basic structure of our network is shown in Fig. 5. We do a logical separation of the network into two parts: backbone and network head, similar to Mask R-CNN [16]. The backbone, a feature extractor network, extracts the local information from the image approximately pixel-wise and thus generates a high-resolution feature map, which can then be used by the network head for the cover calculation and generation of the segmentation map. In the network head the pixel-wise probabilities for each plant are calculated, which are then aggregated

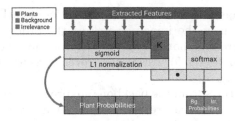

Fig. 6. The calculation in the network head. We use a sigmoid function to determine the plant probabilities and a softmax function for the background and irrelevance probabilities. To bring these into a relationship with each other, we use a hyperparameter κ, L1-normalization and a multiplication, denoted with $\cdot$.

to calculate the total cover percentage of the complete image. The maxima of the intermediate probabilities are used for generating the segmentation map.

4.2 Feature Extractor Network

Feature extractor network initially applies two downscaling operations with 2-strided convolutions, bringing the feature maps to a resolution of 25% of the original image, which is kept until the end of the network. The downscaling layers are followed by nine residual bottleneck blocks as defined in the original ResNet paper [17]. To aggregate information quickly over multiple scales, an inception block, similar to the ones introduced in the papers by Szegedy et al. [40, 41] is used. The inception block consists of four branches with different combinations of convolutions, resulting in four different receptive field sizes: 1×1, 3×3, 7×7 and 11×11. In Table 1 the network architecture is shown in detail.

4.3 Network Head and Calculation Model

In the network head we try to calculate the cover percentages as exact as possible. For this, we first introduce two additional classes to the ones already described in Sect. 3: the background and the irrelevance class. While very similar at the first glance, these two classes differ significantly in meaning. The background class represents every part of the image that is not a plant, but still relevant to cover percentage calculation. The most obvious example for this is the bare soil visible in the images. This class will be abbreviated with *bg* in the following. The irrelevance class, denoted with *irr* in the following, represents all image parts that are not a plant but also not relevant for the cover calculation. Here, the most obvious example are the walls of the EcoUnits, which are visible in many images. The aim of differentiating between these two classes is to separate unwanted objects from the actual plantable area of the EcoUnits and therefore enable the network to work on images without manual removal of such objects, which can be very laborious. If not handled in any way, such objects like the walls of the EcoUnits in our dataset can strongly distort the calculation of cover percentages.

For the latter, we require the pixel-wise probabilities of each plant being at the corresponding location in the image as well as the probabilities for both the location being background that is still relevant for the cover percentages, and the location being irrelevant for estimating cover percentages. The calculation scheme is shown in Fig. 6.

The extracted features from the backbone are processed by a 1×1 convolution to create the classification features for each plant as well as background and irrelevance. As due the occlusion multiple plants can be detected at the same location, we do not consider their probabilities to be mutually exclusive. Hence, we use a sigmoid function to calculate the probability for each plant appearing at this location or not. However, a softmax activation is applied to the classification features for background and irrelevance, as they are mutually exclusive. We also introduce a hyperparameter κ, which we use within the L1-normalization of the probabilities for each plant, and an additional multiplication for the normalized κ to relate the appearance probabilities to those for background and irrelevance, as they depend on each other. The detailed equations for the complete calculation process are explained in the following.

While the plants already have separate classes, for our formalization we introduce the abstract biomass class, abbreviated with bio, which simply represents the areas containing plants. For the introduced classes the following holds:

$$A_{total} = A_{bio} + A_{bg} + A_{irr}, \tag{1}$$

where A represents the area covered by a certain class. For improved readability we also define the area relevant for cover calculation as

$$A_{rel} = A_{bio} + A_{bg} = A_{total} - A_{irr} \tag{2}$$

As mentioned above we consider the classes of the plants, denoted with C^{plants}, to be not mutually exclusive due to occlusion enabling the possibility of multiple plants at the same location. However, the classes bio, bg and irr are mutually exclusive. We will refer to these as area classes and denote them with C^{area}.

Based on this formulation we describe our approach with the following equations. Here, we select a probabilistic approach, as we can only estimate the probabilities of a pixel containing a certain plant. With this, the following equation can be used to calculate the cover percentages for each plant:

$$cover_p = \frac{A_p}{A_{rel}} = \frac{\sum_{\forall x} \sum_{\forall y} P(C^{plants}_{x,y} = p)}{\sum_{\forall x} \sum_{\forall y} 1 - P(C^{area}_{x,y} = irr)}, \tag{3}$$

with p being the class of a plant. whereas x and y determine a certain location in the image. $C^{\cdot}_{x,y}$ is the predicted class at location (x, y) and $P(C^{\cdot}_{x,y} = c)$ is the probability of class c being located at the indicated position.

As mentioned before, we also use the vegetation percentages for training to create an auxiliary output. The vegetation percentage represents how much of the relevant area is covered with plants. This additional output helps for determining the area actually relevant for calculation. It can be calculated as follows:

$$vegetation = \frac{A_{bio}}{A_{rel}} = \frac{\sum_{\forall x}\sum_{\forall y} 1 - P(C_{x,y}^{area} = bg) - P(C_{x,y}^{area} = irr)}{\sum_{\forall x}\sum_{\forall y} 1 - P(C_{x,y}^{area} = irr)}. \tag{4}$$

The notation is analogous to Eq. 3.

While the probabilities for each plant as well as for background and irrelevance can be predicted, we are still missing a last piece for the construction of the network head: the calculation of the biomass class bio. We mentioned above that this class is abstract. This means it cannot be predicted independently, as it is mostly dependent on the prediction of plants in an area. We solve this by introducing the hyperparameter κ as mentioned above, which represents a threshold at which we consider a location to contain a plant (in contrast to background and irrelevance). We concatenate this value with the plant probabilities $P(C_{x,y}^{plants})$ to form a vector $v_{x,y}$. We normalize this vector using L1-normalization, which can then be interpreted as the dominance of each plant with the most dominant plant having the highest value. As the values of this normalized vector sum up to 1, they can also be treated as probabilities. The value at the original position of κ, which basically represents the probability for the absence of all plants, is higher, if no plant is dominant. Hence, we can define:

$$P(C_{x,y}^{area} = bio) = 1 - \left(\frac{v_{x,y}}{\|v_{x,y}\|_1}\right)_\kappa \tag{5}$$

where $(\cdot)_\kappa$ designates the original position of the value κ in the vector. The value $1 - P(C_{x,y}^{area} = bio)$ can then be multiplied with the background and irrelevance probabilities to generate the correct probabilities for these values. This results in the probabilities of the area classes summing up to one:

$$1 = P(C_{x,y}^{area} = bio) + P(C_{x,y}^{area} = bg) + P(C_{x,y}^{area} = irr). \tag{6}$$

Based on these equations we can construct our network head, which is able to accurately represent the calculation of plant cover in our images.

To generate the segmentation map, we use the maximum values of sigmoidal probabilities of the plant classes together with the ones for background and irrelevance. As these values only have 25% of the original resolution, they are upsampled using bicubic interpolation, resulting in a segmentation map that has the original image resolution.

5   Experiments

In the following, we will show our experimental setup, then explain the error measures we used and afterwards will go over the numerical results followed by evaluation of the segmentation maps.

Table 2. The mean cover percentages used for scaling in Eq. 8 during evaluation.

Tri_pra	Pla_lan	Med_lup	Cen_jac	Ach_mil	Lot_cor	Sco_aut	Grasses	Dead Litter
33.3%	11.5%	11.7%	13.4%	3.4%	6.4%	3.8%	10.3%	14.1%

Table 3. The mean values and standard deviations of the absolute errors and scaled absolute errors.

Plants	Tri_pra	Pla_lan	Med_lup	Cen_jac	Ach_mil
MAE	9.88 (± 10.41)	5.81 (± 5.19)	7.36 (± 6.13)	5.53 (± 5.04)	2.15 (± 2.62)
MSAE	0.30 (± 0.31)	0.50 (± 0.45)	0.63 (± 0.52)	0.41 (± 0.38)	0.63 (± 0.77)
Plants	Lot_cor	Sco_aut	Grasses	Dead_Litter	
MAE	3.20 (± 3.75)	2.31 (± 3.44)	4.49 (± 6.51)	7.23 (± 9.01)	
MSAE	0.50 (± 0.59)	0.61 (± 0.91)	0.44 (± 0.63)	0.51 (± 0.64)	

5.1 Setup

During our experiments we used an image resolution of 672×336 px and a batch size of 16. We trained the network for 300 epochs using the Adam [22] optimizer with a learning rate of 0.01, decreasing by a factor of 0.1 at epoch 100, 200 and 250. As loss we used the MAE both for the cover percentage prediction as well as for the vegetation prediction weighted equally. Furthermore, we used L2 regularization with factor of 0.0001. The activation functions in the backbone were ReLU functions and we used reflective padding instead of zero padding, as this produces fewer artifacts at the border of the image. During training the introduced hyperparameter κ was set to 0.001. For data augmentation we used horizontal flipping, small rotations in the range of $-20°$ to $20°$, coarse dropout, and positional translations in the range of -20 to 20 pixels. We trained the model using the Tensorflow framework [1] with Keras [8] using mixed precision. For a fair evaluation, we divided the images into training and validation parts based on the EcoUnits. We use 12-fold cross validation, such that each cross validation split consists of 22 EcoUnits for training and 2 for testing. While the cover percentages are not equally distributed over the EcoUnits, this should only have little effect on the results of the cross validation.

5.2 Error Measures

To evaluate the numerical results of our approach, we will take a look at two different error measures. The first one is the mean absolute error (MAE), which is defined as follows:

$$MAE(t, p) = \frac{1}{n} \sum_{i=1}^{n} |t_i - p_i|, \tag{7}$$

where t and p are the true and predicted cover values, respectively. As the mean absolute error can be misguiding when comparing the goodness of the predictions

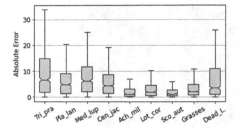 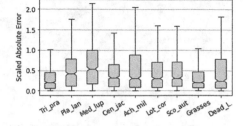

Fig. 7. An overview over the MAE of the plant cover prediction in the dataset.

Fig. 8. An overview over the MSAE of the plant cover prediction in the dataset.

for imbalanced classes, we also propose a scaled version of the MAE: the mean scaled absolute error (MSAE), which is defined as follows:

$$MSAE(t,p) = \frac{1}{n}\sum_{i=1}^{n} \frac{|t_i - p_i|}{m_i}. \tag{8}$$

The absolute error values for each class are scaled by a value m_i, which is the mean cover percentage averaged over the different annotations within the respective class in the dataset. This error will provide a better opportunity for comparing the predictions between the classes. The values that have been used for scaling can be found in Table 2.

5.3   Experimental Results

Cover Predictions. Our model achieves an overall MAE of 5.3% and an MSAE of 0.50. The detailed results for each species are shown in Table 3 as well as in Fig. 7 and Fig. 8. With respect to the MAE, we can see that the error of tri_pra appears to be the highest, while the error of the less abundant plants (ach_mil, lot_cor, sco_aut) appears to be much lower. However, as mentioned above, the distribution of the MAE mostly reflects the distribution of the plants in the whole dataset, as the errors for the more abundant plants are expected to be higher. Therefore, to compare the goodness of the results between plants, we take a look at the MSAE depicted in Fig. 8, where we can see that tri_pra actually has the lowest relative error compared to the other plants, partially caused by the comparably large amounts of training data for this class. The most problematic plants appear to be ach_mil, sco_aut and med_lup, with MSAE error values of 0.63, 0.61 and 0.63, respectively. For ach_mil the rather high error rate might result from multiple circumstances. The plant is very rare in the dataset, small in comparison to many of the other plants in the dataset and also has a complex leaf structure, most of which might get lost using smaller resolutions. The large error for med_lup might be caused by its similarity to tri_pra, which is very dominant in the dataset. Therefore, the network possibly predicts *Trifolium* instead of

Fig. 9. Segmentation results for an image with a high zoom level from the validation set. We can see that grasses, *P. lanceolata*, *T. pratense* and background area are segmented correctly in many cases.

Medicago on many occasions, causing larger errors. The same might be the case for sco_aut and pla_lan or cen_jac, especially since sco_aut is one of the least abundant plants in the dataset making a correct recognition difficult.

To put these results into perspective, we also provide the results using a constant predictor, which always predicts the mean of the cover percentages of the training dataset, and the results using a standard U-Net [36] as feature extractor. These achieved an MAE of 9.88% and MSAE of 0.84, and an MAE of 5.54% and MSAE of 0.52 for the constant predictor and the U-Net respectively. We can see that our proposed network outperforms the constant predictor by a large margin and also slightly improves the accuracy of a U-Net, despite having less than 10% of the number of parameters compared to the U-Net (3 million vs. 34 million). More details can be found in the supplementary material.

Segmentations. To evaluate the result of our network, we also take a look at the results of the segmentation. The first image, shown in Fig. 9, is one with a comparably high zoom level. There we can see that tri_pra is detected correctly in the areas on the left and right sides of the image, while the segmentations are not perfect. pla_lan has been segmented well in many cases, especially on the right side of the image. On the left we can see that it is also segmented correctly, even though it is partially covered by grass. Therefore, the approach appears to be robust to minor occlusions to some extent. Despite these results, the segmentation is still mostly incorrect in the top center of the image. Grasses are also detected correctly in most regions of the image, whereas above the aforementioned instances of pla_lan they are not segmented at all, which is mostly caused by the low resolution of the segmentation map. This low resolution also

Fig. 10. Segmentation results for a zoomed out image from the validation set. While the network captures the signals of many plants correctly, the segmentations are rather inaccurate leading to a large number of wrongly segmented plant species.

appears to impair the segmentation results in many other occasions and we would like to tackle this problem in the future.

The second segmentation image is shown in Fig. 10. Here, the zoom level is lower than in the image before, which results in the segmentations getting increasingly inaccurate. We can see that the network correctly captured the presence of most plant species. Notably, the approximate regions of med_lup and tri_pra are marked correctly. However, the detailed segmentation results are not very accurate. It also appears that some parts of the wall are wrongly recognized as tri_pra, while other parts are correctly marked as irrelevant for cover calculation. The segmentations with a U-Net feature extractor can be found in the supplementary materials. All in all, the segmentations appear to be correct for the more prominent plants in the dataset shown in the images with high zoom level and at least partially correct in the images without zoom. Therefore, the segmentation maps can be used to explain and confirm the plant cover predictions for some plants from the dataset.

6   Conclusions and Future Work

We have shown that our approach is capable of predicting cover percentages of different plant species while generating a high-resolution segmentation map. Learning is done without any additional information other than the original cover annotations. Although not perfect, the segmentation map can already be used to explain the results of the cover prediction for the more prevalent plants in the dataset. Many original images have a very high resolution and are currently downscaled due to computational constraints. Making our approach applicable to images of higher resolution could be one improvement. This would also increase the resolution of the segmentation map, resulting in much finer segmentations. The recognition of the less abundant plants, but also of very similar plants like *T. pratense* and *M. lupulina*, might be improved by applying transfer learning techniques. For example, we could pretrain the network on the iNaturalist datasets [43], since they contain a large number of plant species. Heavy occlusions are still a big challenge in our dataset, making predictions of correct plants and their abundances very hard. While there are already some approaches for segmenting occluded regions in a supervised setting [26,35,50], it is a completely unexplored topic for weakly-supervised semantic segmentation.

Acknowledgements. Matthias Körschens thanks the Carl Zeiss Foundation for the financial support. In addition, we would like to thank Mirco Migliavacca for additional comments on the manuscript.

References

1. Abadi, M., et al.: TensorFlow: a system for large-scale machine learning. In: 12th USENIX Symposium on Operating Systems Design and Implementation (OSDI 2016), pp. 265–283 (2016)
2. Aggemyr, E., Cousins, S.A.: Landscape structure and land use history influence changes in island plant composition after 100 years. J. Biogeogr. **39**(9), 1645–1656 (2012)
3. Ahn, J., Cho, S., Kwak, S.: Weakly supervised learning of instance segmentation with inter-pixel relations. In: Proceedings of the IEEE Conference on Computer Vision and Pattern Recognition, pp. 2209–2218. IEEE (2019)
4. Barré, P., Stöver, B.C., Müller, K.F., Steinhage, V.: LeafNet: a computer vision system for automatic plant species identification. Ecol. Inform. **40**, 50–56 (2017)
5. Bernhardt-Römermann, M., et al.: Drivers of temporal changes in temperate forest plant diversity vary across spatial scales. Glob. Change Biol. **21**(10), 3726–3737 (2015)
6. Bruelheide, H., et al.: Global trait-environment relationships of plant communities. Nat. Ecol. Evol. **2**(12), 1906–1917 (2018)
7. Bucher, S.F., König, P., Menzel, A., Migliavacca, M., Ewald, J., Römermann, C.: Traits and climate are associated with first flowering day in herbaceous species along elevational gradients. Ecol. Evol. **8**(2), 1147–1158 (2018)
8. Chollet, F., et al.: Keras (2015). https://keras.io
9. Cleland, E.E., et al.: Phenological tracking enables positive species responses to climate change. Ecology **93**(8), 1765–1771 (2012)
10. Dai, J., He, K., Sun, J.: BoxSup: exploiting bounding boxes to supervise convolutional networks for semantic segmentation. In: Proceedings of the IEEE International Conference on Computer Vision, pp. 1635–1643. IEEE (2015)
11. Eisenhauer, N., Türke, M.: From climate chambers to biodiversity chambers. Front. Ecol. Environ. **16**(3), 136–137 (2018)
12. Everingham, M., Van Gool, L., Williams, C.K., Winn, J., Zisserman, A.: The pascal visual object classes (VOC) challenge. Int. J. Comput. Vis. **88**(2), 303–338 (2010)
13. Fitter, A., Fitter, R.: Rapid changes in flowering time in British plants. Science **296**(5573), 1689–1691 (2002)

14. Gerstner, K., Dormann, C.F., Stein, A., Manceur, A.M., Seppelt, R.: Editor's choice: review: effects of land use on plant diversity-a global meta-analysis. J. Appl. Ecol. **51**(6), 1690–1700 (2014)
15. Ghazi, M.M., Yanikoglu, B., Aptoula, E.: Plant identification using deep neural networks via optimization of transfer learning parameters. Neurocomputing **235**, 228–235 (2017)
16. He, K., Gkioxari, G., Dollár, P., Girshick, R.: Mask R-CNN. In: Proceedings of the IEEE International Conference on Computer Vision, pp. 2961–2969. IEEE (2017)
17. He, K., Zhang, X., Ren, S., Sun, J.: Deep residual learning for image recognition. In: Proceedings of the IEEE Conference on Computer Vision and Pattern Recognition, pp. 770–778. IEEE (2016)
18. Huang, Z., Wang, X., Wang, J., Liu, W., Wang, J.: Weakly-supervised semantic segmentation network with deep seeded region growing. In: Proceedings of the IEEE Conference on Computer Vision and Pattern Recognition, pp. 7014–7023. IEEE (2018)
19. Idrees, H., et al.: Composition loss for counting, density map estimation and localization in dense crowds. In: Ferrari, V., Hebert, M., Sminchisescu, C., Weiss, Y. (eds.) ECCV 2018. LNCS, vol. 11206, pp. 544–559. Springer, Cham (2018). https://doi.org/10.1007/978-3-030-01216-8_33
20. Kattenborn, T., Eichel, J., Wiser, S., Burrows, L., Fassnacht, F.E., Schmidtlein, S.: Convolutional neural networks accurately predict cover fractions of plant species and communities in unmanned aerial vehicle imagery. Remote Sen. Ecol. Conserv. (2020)
21. Khoreva, A., Benenson, R., Hosang, J., Hein, M., Schiele, B.: Simple does it: weakly supervised instance and semantic segmentation. In: Proceedings of the IEEE Conference on Computer Vision and Pattern Recognition, pp. 876–885. IEEE (2017)
22. Kingma, D.P., Ba, J.: Adam: a method for stochastic optimization. CoRR abs/1412.6980 (2015)
23. Kolesnikov, A., Lampert, C.H.: Seed, expand and constrain: three principles for weakly-supervised image segmentation. In: Leibe, B., Matas, J., Sebe, N., Welling, M. (eds.) ECCV 2016. LNCS, vol. 9908, pp. 695–711. Springer, Cham (2016). https://doi.org/10.1007/978-3-319-46493-0_42
24. Krizhevsky, A., Sutskever, I., Hinton, G.E.: ImageNet classification with deep convolutional neural networks. In: Pereira, F., Burges, C.J.C., Bottou, L., Weinberger, K.Q. (eds.) Advances in Neural Information Processing Systems 25, pp. 1097–1105. Curran Associates, Inc. (2012). http://papers.nips.cc/paper/4824-imagenet-classification-with-deep-convolutional-neural-networks.pdf
25. Lee, S.H., Chan, C.S., Wilkin, P., Remagnino, P.: Deep-plant: Plant identification with convolutional neural networks. In: 2015 IEEE International Conference on Image Processing (ICIP), pp. 452–456. IEEE (2015)
26. Li, K., Malik, J.: Amodal instance segmentation. In: Leibe, B., Matas, J., Sebe, N., Welling, M. (eds.) ECCV 2016. LNCS, vol. 9906, pp. 677–693. Springer, Cham (2016). https://doi.org/10.1007/978-3-319-46475-6_42
27. Liu, H., et al.: Shifting plant species composition in response to climate change stabilizes grassland primary production. Proc. Nat. Acad. Sci. **115**(16), 4051–4056 (2018)
28. Liu, L., Qiu, Z., Li, G., Liu, S., Ouyang, W., Lin, L.: Crowd counting with deep structured scale integration network. In: Proceedings of the IEEE International Conference on Computer Vision, pp. 1774–1783. IEEE (2019)

29. Lloret, F., Peñuelas, J., Prieto, P., Llorens, L., Estiarte, M.: Plant community changes induced by experimental climate change: seedling and adult species composition. Perspect. Plant Ecol. Evol. Systemat. **11**(1), 53–63 (2009)
30. Van der Maarel, E., Franklin, J.: Vegetation Ecology. Wiley, Hoboken (2012)
31. Menzel, A., et al.: European phenological response to climate change matches the warming pattern. Glob. Change Biol. **12**(10), 1969–1976 (2006)
32. Miller-Rushing, A.J., Primack, R.B.: Global warming and flowering times in Thoreau's concord: a community perspective. Ecology **89**(2), 332–341 (2008)
33. Pathak, D., Krahenbuhl, P., Darrell, T.: Constrained convolutional neural networks for weakly supervised segmentation. In: Proceedings of the IEEE International Conference on Computer Vision, pp. 1796–1804. IEEE (2015)
34. Pfadenhauer, J.: Vegetationsökologie - ein Skriptum. IHW-Verlag, Eching, 2. verbesserte und erweiterte auflage edn. (1997)
35. Purkait, P., Zach, C., Reid, I.: Seeing behind things: extending semantic segmentation to occluded regions. In: 2019 IEEE/RSJ International Conference on Intelligent Robots and Systems (IROS), pp. 1998–2005. IEEE (2019)
36. Ronneberger, O., Fischer, P., Brox, T.: U-Net: convolutional networks for biomedical image segmentation. In: Navab, N., Hornegger, J., Wells, W.M., Frangi, A.F. (eds.) MICCAI 2015. LNCS, vol. 9351, pp. 234–241. Springer, Cham (2015). https://doi.org/10.1007/978-3-319-24574-4_28
37. Rosenzweig, C., et al.: Assessment of observed changes and responses in natural and managed systems. In: Climate Change 2007: Impacts, Adaptation and Vulnerability. Contribution of Working Group II to the Fourth Assessment Report of the Intergovernmental Panel on Climate Change, pp. 79–131 (2007)
38. Russakovsky, O., Deng, J., Su, H., Krause, J., Satheesh, S., Ma, S., Huang, Z., Karpathy, A., Khosla, A., Bernstein, M., et al.: Imagenet large scale visual recognition challenge. Int. J. Comput. Vision **115**(3), 211–252 (2015)
39. Souza, L., Zelikova, T.J., Sanders, N.J.: Bottom-up and top-down effects on plant communities: nutrients limit productivity, but insects determine diversity and composition. Oikos **125**(4), 566–575 (2016)
40. Szegedy, C., Ioffe, S., Vanhoucke, V., Alemi, A.A.: Inception-v4, inception-resnet and the impact of residual connections on learning. In: Thirty-First AAAI Conference on Artificial Intelligence (2017)
41. Szegedy, C., Vanhoucke, V., Ioffe, S., Shlens, J., Wojna, Z.: Rethinking the inception architecture for computer vision. In: Proceedings of the IEEE Conference on Computer Vision and Pattern Recognition, pp. 2818–2826. IEEE (2016)
42. Türke, M., et al.: Multitrophische biodiversitätsmanipulation unter kontrollierten umweltbedingungen im idiv ecotron. In: Lysimetertagung, pp. 107–114 (2017)
43. Van Horn, G., et al.: The inaturalist species classification and detection dataset. In: Proceedings of the IEEE Conference on Computer Vision and Pattern Recognition, pp. 8769–8778. IEEE (2018)
44. Verheyen, K., et al.: Combining biodiversity resurveys across regions to advance global change research. Bioscience **67**(1), 73–83 (2017)
45. Wäldchen, J., Mäder, P.: Flora incognita-wie künstliche intelligenz die pflanzenbestimmung revolutioniert: Botanik. Biologie unserer Zeit **49**(2), 99–101 (2019)
46. Wang, Y., Zhang, J., Kan, M., Shan, S., Chen, X.: Self-supervised equivariant attention mechanism for weakly supervised semantic segmentation. In: Proceedings of the IEEE/CVF Conference on Computer Vision and Pattern Recognition, pp. 12275–12284. IEEE (2020)

47. Xiong, H., Lu, H., Liu, C., Liu, L., Cao, Z., Shen, C.: From open set to closed set: counting objects by spatial divide-and-conquer. In: Proceedings of the IEEE International Conference on Computer Vision, pp. 8362–8371. IEEE (2019)
48. Yalcin, H., Razavi, S.: Plant classification using convolutional neural networks. In: 2016 Fifth International Conference on Agro-Geoinformatics (Agro-Geoinformatics), pp. 1–5. IEEE (2016)
49. Yu, F., Koltun, V., Funkhouser, T.: Dilated residual networks. In: Proceedings of the IEEE Conference on Computer Vision and Pattern Recognition, pp. 472–480. IEEE (2017)
50. Zhan, X., Pan, X., Dai, B., Liu, Z., Lin, D., Loy, C.C.: Self-supervised scene de-occlusion. In: Proceedings of the IEEE/CVF Conference on Computer Vision and Pattern Recognition, pp. 3784–3792. IEEE (2020)

Unsupervised Domain Adaptation for Plant Organ Counting

Tewodros W. Ayalew[✉], Jordan R. Ubbens, and Ian Stavness

University of Saskatchewan, Saskatoon, SK S7N 5A8, Canada
{tewodros.ayalew,jordan.ubbens,ian.stavness}@usask.ca

Abstract. Supervised learning is often used to count objects in images, but for counting small, densely located objects, the required image annotations are burdensome to collect. Counting plant organs for image-based plant phenotyping falls within this category. Object counting in plant images is further challenged by having plant image datasets with significant domain shift due to different experimental conditions, e.g. applying an annotated dataset of indoor plant images for use on outdoor images, or on a different plant species. In this paper, we propose a domain-adversarial learning approach for domain adaptation of density map estimation for the purposes of object counting. The approach does not assume perfectly aligned distributions between the source and target datasets, which makes it more broadly applicable within general object counting and plant organ counting tasks. Evaluation on two diverse object counting tasks (wheat spikelets, leaves) demonstrates consistent performance on the target datasets across different classes of domain shift: from indoor-to-outdoor images and from species-to-species adaptation.

1 Introduction

Object counting is an important task in computer vision with a wide range of applications, including counting the number of people in a crowd [21,34,42], the number of cars on a street [37,46], and the number of cells in a microscopy image [32,44,49]. Object counting is a relevant task in image-based plant phenotyping, notably for counting plants in the field to estimate the rate of seedling emergence, and counting plant organs to estimate traits relevant for selection in crop breeding programs. For example, counting spikes or heads in cereal crops is a relevant trait for estimating grain yield [2,27,33], counting flowers for estimating the start and duration of flowering is relevant for demarcating plant growth stages [3,23,45], and counting leaves and tillers is a relevant trait to assess plant health [1,16]. Leaf counting, in particular, has been a seminal plant phenotyping task, thanks to the CVPPP leaf counting competition dataset [29,30].

Convolutional neural networks (CNN) provide state-of-the-art performance for object counting tasks. Supervised learning is most common in previous work, but object detection [20,22] and density estimation [21,27,31] approaches both require fairly tedious annotation of training images with either bounding boxes

© Springer Nature Switzerland AG 2020
A. Bartoli and A. Fusiello (Eds.): ECCV 2020 Workshops, LNCS 12540, pp. 330–346, 2020.
https://doi.org/10.1007/978-3-030-65414-6_23

or dots centered on each object instance. In the context of plant phenotyping, plant organ objects are often small and densely packed making the annotation process even more laborious. In addition, unlike general objects which can be annotated reliably by any individual, identifying and annotating plant organs in images often requires specialized training and experience in plant science [14,48]. This makes it difficult to obtain large annotated plant image datasets.

Another challenge for computer vision tasks in plant phenotyping is that, unlike large image datasets of general objects, plant image datasets usually include highly self-similar images with a small amount of variation among images within the dataset. An individual plant image dataset is often acquired under similar conditions (single crop type, same field) and therefore trying to directly use a CNN trained on a single dataset to new images from a different crop, field, or growing season will likely fail. This is because a model trained to count objects on one dataset (source dataset) will not perform well on a different dataset (target dataset) when these datasets have different prior distributions. This challenge is generally called *domain-shift*. An extreme case of domain shift in plant phenotyping is using a source dataset of plant images collected in an indoor controlled environment, which can be more easily annotated because plants are separated with controlled lighting on a blank background, and attempting to apply the model to a target dataset of outdoor field images with multiple, overlapping plants with variable lighting and backgrounds, and blur due to wind motion. Domain shift is often handled by fine-tuning a model, initially trained on a source dataset, with samples taken from a target dataset. Fine-tuning, however, still requires some annotated images in the target dataset. Another approach used to solve this problem is domain adaptation. Domain adaptation techniques typically try to align the source and the target data distributions [15].

In this paper, we propose a method that applies an unsupervised domain adaptation technique, first proposed for image classification [9], to jointly train a CNN to count objects from images with dot annotations (i.e., source domain) and adapt this knowledge to related sets of images (i.e., target domain) where labels are absent. We modeled the object counting problem as a density map estimation problem. We evaluated our proposed domain adaptation method on two object counting tasks (wheat spikelet counting and rosette leaf counting) each with a different challenge. The wheat spikelet counting task adapts from indoor images to outdoor images and presents challenges due to self-similarity, self-occlusion, and appearance variation [2,33]. The leaf counting task adapts from one plant species to a different plant species and presents variability in leaf shape and size, as well as overlapping and occluded leaves [19]. Results show consistent improvements over the baseline models and comparable results to previous domain adaptation work in leaf counting. The contributions of our paper include: 1) the extension of an adversarial domain adaptation method for density map estimation that learns and adapts in parallel, 2) the evaluation of our method on two diverse counting tasks, and 3) a new public dataset of annotated wheat spikelets imaged in outdoor field conditions.

2   Related Work

Object Counting. A number of approaches have been proposed to train CNNs for object counting tasks. Among these, the most prominent methods are counting by regression [1,8,16,40], counting by object detection [12,13,28], and counting by density estimation [11,21,33,36]. Here, we focus on density estimation methods, which estimate a count by generating a continuous density map and then summing over the pixel values of the density map. Density estimation uses weak labels (dot annotations on object centers), which are less tedious to annotate than bounding boxes or instance segmentation masks. Density estimation based counting was first proposed by Lempitsky & Zisserman [21] who demonstrated higher counting accuracy with smaller amounts of data.

Spikelet Counting. Pound et al. [33] presented a method for counting and localizing spikes and spikelets using stacked hourglass networks. They trained and evaluated their model on spring wheat plant images taken in a controlled environment. Their approach achieved a 95.91% accuracy for spike counting and a 99.66% accuracy for spikelet counting. Even though their results show high accuracy, counting spikelets from real field images requires further fine-tuning of their model. Alkhudaydi et al. [2] proposed a method to count spikelets from infield images using a fully convolutional network named SpikeletFCN. They compared the performance of training SpikeletFCN from scratch on infield images and fine-tuning the model, which was initially trained using images taken in a controlled setting [33]. They also showed that manually segmenting the background increased performance, but manually annotating spikelets and manually removing backgrounds for field images is a time-consuming process.

Leaf Counting. Leaf counting in rosette plants has been performed with CNN-based regression approaches. Dobrescu et al. [8] used a pre-trained ResNet architecture fine-tuned on rosette images to demonstrate state-of-the-art leaf counting performance in the CVPPP 2017 leaf counting competition. Aich & Stavness [1] presented a two-step leaf counting method, first using an encoder-decoder network to segment the plant region from the background, and then using a VGG-like CNN to estimate the count. Giuffrida et al. [16] extended the CNN regression method to show that using multiple image modalities can improve counting accuracy. They also demonstrated the generality of the approach by evaluating with different plant species and image modalities. Itzhaky et al. [18] proposed two deep learning based approaches that utilize a Feature Pyramid Network (FPN) for counting leaves, namely using direct regression and density map estimation. Their density map based approach achieved 95% average precision.

Domain Adaptation. In recent years, a number of deep domain adaptation approaches have been proposed to minimize the domain shift between a source dataset and a target dataset [5,17,25,26,38,39]. Tzeng et al. [39] proposed an adversarial approach for an unsupervised domain adaptation named Adversarial Discriminative Domain Adaptation (ADDA). ADDA minimizes the representation shift between the two domains using a Generative Adversarial Network

(GAN) loss function where the domains are allowed to have independent mappings. As an alternative to a GAN loss, Ganin & Lempitsky [9] presented an unsupervised domain adaptation technique by creating a network that has a shared feature extraction layers and two classifiers. The adaptation process works by minimizing the class label prediction loss and maximizing the domain confusion. These previous works have only been evaluated on image classification problems. Domain adaptation has garnered less attention for other computer vision tasks, such as density estimation or regression problems.

Domain Adaptation in Plant Phenotyping. Giuffrida et al. [15] proposed a method to adapt a model trained to count leaves in a particular dataset to an unseen dataset in an unsupervised manner. Under the assumption that the images in the source domain are private, their method uses the representations of images from the source domain (from a pretrained network) to adapt to the target domain. They employed the Adversarial Discriminative Domain Adaptation approach to solving the domain shift problem in the leaf counting task across different datasets. Their method aligns the predicted leaf count distribution with the source domain's prior distribution, which limits the adapted model to learning a leaf count distribution similar to the source domain. They have shown their method achieved a mean square error of 2.36 and 1.84 for intra-species and inter-species domain adaptation experiments, respectively. For fruit counting in orchard images, Bellocchio et al. [4] proposed a Cycle-GAN based domain adaptation method combined with weak presence/absence labels. They demonstrated state-of-the-art counting results for many combinations of adapting between plant species (almond, apple, olive).

These promising results on leaf and fruit counting motivate a broader investigation of different domain adaptation approaches, which could be used for counting different plant organs and different categories of domain shift, e.g. from indoor image to field images. To the best of our knowledge, this is the first work that has applied domain adaptation to wheat spikelet counting and to a counting task with an indoor-to-outdoor domain shift.

3   Method

In this section, we discuss the problem setting for our proposed method, followed by a description of the proposed architecture and cost function. Finally we describe the training procedures used for all experiments.

3.1   Problem Setting

We model the task of domain adaptation in object counting as an unsupervised domain adaptation problem. We assume that the data is sampled from two domains: a source domain ($\mathcal{D}^s$) and a target domain ($\mathcal{D}^t$). We also assume that labels only exist for the images sampled from the source domain. Therefore, the source domain is composed of a set of images ($\mathcal{X}^s$) and their corresponding labels ($\mathcal{Y}^s$). Whereas, the target domain only has images ($\mathcal{X}^t$) without labels.

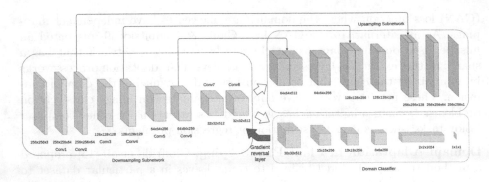

Fig. 1. The proposed Domain-Adversarial Neural Network composed of two networks that share weights between Conv1 and Conv8. The downsampling subnetwork (G_d), the upsampling subnetwork (G_u), and the domain classifier (G_c) are denoted by the blue, red, and green boxes respectively. The red arrow shows the Gradient Reversal Layer used to reverse the gradient in the backpropagation step as proposed by [9]. (Color figure online)

3.2   Architecture

Our proposed model is a Domain-Adversarial Neural Network (DANN), a class of models designed for unsupervised domain adaptation tasks [10]. These networks are commonly characterized by two parallel classification networks that share weights in the first n layers and a custom layer called Gradient Reversal Layer (GRL) [43]. One of the parallel networks is designed for the main classification task, while the second network is a classifier designed to discriminate whether an input is sampled from the source domain or from the target domain. We customize this general architecture by replacing the main classification network with a U-Net network [35] that is used for density map estimation.

Our proposed architecture is a fully convolutional network composed of two parallel networks, as can be seen in Fig. 1. These parallel networks share weights from *Conv1* to *Conv8*. Given these shared weights, the architecture can be seen as one network containing three subnetworks: the downsampling subnetwork (G_d), the upsampling subnetwork (G_u), and the domain classifier subnetwork (G_c). In Fig. 1 these subnetworks are denoted by different color boxes. The model takes an image x_j and predicts a density map as $\hat{y}_{density} = G_u(G_d(x_j))$ and a class probability prediction representing whether x_j is sampled from the source domain or from the target domain given by $\hat{y}_{domain} = G_c(G_d(x_j))$.

A downsampling subnetwork followed by an upsampling subnetwork (sometimes referred to as an Hourglass network or encoder-decoder network) is a commonly used architecture for density estimation tasks[24,44]. The general structure for the Hourglass network we used is adapted from U-Net [35] with slight modifications. The downsampling subnetwork is composed of blocks of two 3×3 padded convolutions followed by a 2×2 max pooling layer with a stride of 2. The upsampling network is composed of blocks of 2×2 transpose convolutions followed by a pair of 3×3 padded convolutions. The feature maps generated from each convolution preceded by a transpose convolution are concatenated with their corresponding feature maps from the upsampling subnetwork (shown in Fig. 1 as dashed boxes). The domain classifier subnetwork is formed from 3×3 unpadded convolutions, a 2×2 maxpooling layers, and a 1×1 convolution (which is used to reduce the dimension of the output to a single value). All of the convolutions in our network, with the exception of the ones in the output layers, are followed by batch normalization and a ReLU activation. The output layer is followed by a sigmoid activation function.

The output of the downsampling subnetwork passes through a Gradient Reversal Layer (GRL) before passing through to the domain classifier subnetwork. The GRL, a custom layer proposed by [9], is designed to reverse the gradient in the backpropagation step by multiplying the gradient by a negative constant $(-\lambda)$. The purpose of reversing the gradient is to force the network into finding domain-invariant features common to both the source- and the target-domain. This aligns the distributions of the feature representations of images sampled from the two domains.

3.3   Cost Function

The parameters in our proposed model are optimized by minimizing a two-part cost function, as follows:

$$\mathcal{L}_{density} = \log(\frac{1}{N} \sum_{i-0}^{N} (G_u(G_d(x_i^s)) - y_i^s)^2) \tag{1}$$

$$\mathcal{L}_{domain} = -\mathbb{E}_{x^s \sim \mathcal{X}^s}[\log(G_c(G_d(x_i^s)))] - \mathbb{E}_{x^t \sim \mathcal{X}^t}[1 - \log(G_c(G_d(x_i^t)))] \tag{2}$$

$$\mathcal{L} = \mathcal{L}_{density} + \mathcal{L}_{domain} \tag{3}$$

where (x_i^s, y_i^s) represents the ith image and density map label sampled from the source domain $(\mathcal{D}^s)$. The first part $(\mathcal{L}_{density})$ accounts for errors in the density map estimation for samples taken from the source domain. This is implemented as the logarithm of mean square error between the ground truth sampled from the source dataset and the predicted density map from the model. The second part $(\mathcal{L}_{domain})$ is a Binary Cross-Entropy loss for domain classification.

3.4   Training

Conceptually, the general training scheme we use to train our model can be decomposed into two sub-tasks: supervised density estimation on the source dataset, and domain classification on the merged source and target datasets. The goal of this training scheme is to extract relevant features for the density estimation task and, in parallel, condition the downsampling network to extract domain-invariant features.

In all experiments, we randomly partitioned the source domain dataset into training and validation sets. Since we only have ground truth density maps for the source domain, in-training validations are carried out using the validation set taken from the source domain.

The model was trained with a batch size of 8. The input images in the training set were resized to 256×256 pixels. We used the Adam optimizer with tuned learning rates for each subnetwork. We used a learning rate of 10^{-3} for the downsampling and upsampling subnetworks, and 10^{-4} for the domain classifier.

The constant parameter λ—which multiplies the gradient in the GRL layer— is updated on each iteration during training. Following [10], the value is provided to the network as a scheduler in an incremental manner ranging from 0 to 1.

4   Experiments

In this section, we present two object counting tasks to evaluate the proposed method: wheat spikelet counting and leaf counting.

Our proposed method's performance in all of the experiments is evaluated using target domain data as our test set. Our experiments are mainly assessed using Root Mean Square Error (RMSE) and Mean Absolute Error (MAE):

$$RMSE = \sqrt{\frac{1}{N} \sum_{i=0}^{N} (\hat{y}_i - y_i)^2}$$

$$MAE = \frac{1}{N} \sum_{i=0}^{N} |\hat{y}_i - y_i|$$

where N represents the number of test images, $\hat{y}_i$ and y_i represent the predicted value and the ground truth for the ith sample, respectively.

In addition to these metrics, we include task-specific metrics that are commonly used in the literature for each task. For the wheat spikelet task, we provide coefficient of determination (R^2) between the predicted and ground truth counts. For the leaf counting task, we report metrics that are provided for the CVPPP 2017 leaf counting competition. These metrics are Difference in Count (DiC), absolute Difference in Count $(|DiC|)$, mean squared error (MSE) and percentage agreement (%).

To demonstrate the counting performance improvement provided by domain adaptation, we use a vanilla U-Net as a baseline model by removing the domain classifier sub-network from our proposed architecture. We train this baseline model exclusively on the source domain and evaluate performance on the target domain data. All experiments were performed on a GeForce RTX 2070 GPU with 8GB memory using the Pytorch framework. The implementation is available at: https://github.com/p2irc/UDA4POC.

4.1   Wheat Spikelet Counting

For the wheat spikelet counting task, we used the general problem definition of wheat spikelet counting as stated in [2]. Under this setting, our second experiment tries to adapt the wheat spikelet counting task from images taken in a

Fig. 2. Example images for wheat spikelet counting experiments. Top: Source dataset, ACID [33]. Middle: Target dataset 1, Global Wheat Head Dataset [7]. Bottom: Target dataset 2, CropQuant dataset [47].

greenhouse to images captured in an outdoor field. The datasets used for this task are as follows:

1. *ACID dataset* [33]: This dataset is composed of wheat plant images taken in a greenhouse setting. Each image has dot annotations representing the position of each spikelet. The ground truth density maps were generated by filtering the dot annotations with a 2D gaussian filter with $\sigma = 1$.
2. *Global Wheat Head Dataset* (GWHD) [7]: A dataset presented in the Kaggle wheat head detection competition. The dataset comes with bounding box annotations for each wheat head, but these are not used in the present study.
3. *CropQuant dataset* [47]: This dataset is composed of images collected by the Norwich Research Park (NRP). As a ground truth, we used the dot annotations created by [2], where they annotated 15 of the images from the CropQuant dataset. The ground truth has a total of 63,006 spikelets.

We chose the ACID dataset as the source domain and the GWHD dataset as the target domain. Sample images from these datasets are displayed in Fig. 2. We randomly sampled 80% of the images in the source dataset for training. These images were mixed with 200 images taken from the target dataset, making up the training set. Using this training set, the network is trained on the ACID dataset for spikelet counting and, in parallel, adapted to the GWHD. To evaluate our method, we created dot annotations for 67 images from the GWHD which are used as ground truth. These annotations are made publicly available at https://doi.org/10.6084/m9.figshare.12652973.v2.

In the second experiment, we used the CropQuant dataset as the target domain. To be consistent with the experiments presented in [2], we randomly extracted 512×512 sized patches from each image in the CropQuant dataset. We generated 300 patches as the target domain. We partitioned the source domain with 80:20 training-validation split. We merged the training split with the generated patches and trained the model.

4.2   Leaf Counting

For the leaf counting task, we follow [15] in order to compare to their baseline results. Therefore, we used the following datasets:

1. *CVPPP 2017 LCC Dataset* [29,30]: A dataset compiled for the CVPPP leaf counting competition, containing top-view images of *Arabidopsis thaliana* and tobacco plants. To be consistent with [15], we use the A1, A2 and A4 *Arabidopsis* image subsets. The dataset includes annotations for leaf segmentation masks, leaf bounding boxes, and leaf center dot annotations. We use the leaf center annotations and generate the groudtruth density map by filtering the center points with a 2D gaussian filter with $\sigma = 3$.
2. *KOMATSUNA Dataset* [41]: Top-view images of Komatsuna plants developed for 3D plant phenotyping tasks. We use the RGB images for the experiments presented here.
3. *MSU-PID* [6]: A multi-modal imagery dataset which contains images of *Arabidopsis* and bean plants. Among these, we used the RGB images of the *Arabidopsis* subset.

Figure 3 shows sample images from these datasets. Using these datasets, we performed three leaf-counting experiments in total. In all experiments, the CVPPP dataset was used as the source domain.

The first experiment involved adapting leaf counting from the CVPPP dataset to the KOMATSUNA dataset. For this experiment, we randomly selected 80% of the images from the CVPPP dataset and merged them with the images from the KOMATSUNA dataset. The remaining 20% of the source domain is used for validation. With this setting, we trained our model for 150 epochs. Finally, we evaluated performance on the images from the KOMATSUNA dataset.

In the second experiment, we used the MSU-PID dataset as the target domain. The data from the source domain was partitioned into training and validation sets in a similar procedure as experiment 1. The model was evaluated using the MSU-PID dataset.

In the final experiment, we aimed to verify that our domain adaptation scheme is independent of the leaf count distributions in the source domain. To

Fig. 3. Example images for leaf counting experiments. Top: Source dataset, *Arabidopsis*, CVPPP Dataset [29,30]. Middle: Target dataset 1, KOMATSUNA Dataset [41]. Bottom: Target dataset 2, *Arabidopsis* MSU-PID Dataset [6].

test this, we took the model trained in the first experiment and evaluated it on composite images created from KOMATSUNA containing multiple plants. The composite images are generated by randomly selecting four images from the KOMATSUNA dataset and stitching them together form a 2 × 2 grid.

5   Results

In this section, we present the performance of the proposed model on two counting tasks. We also compare these results with the baseline model and with other existing methods, where possible.

5.1   Wheat Spikelet Counting

Table 1 summarizes the results from the wheat spikelet counting adaptation experiment. Figure 4 provides a qualitative comparison of example density maps output by the baseline model and by the proposed model.

Table 1. Domain adaptation results for the wheat spikelet counting tasks.

	MAE	$RMSE$	R^2
ACID to GWHD			
Baseline (No Adaptation)	91.65	114.4	0.005
Our method	29.48	35.80	0.66
ACID to CropQuant			
Baseline (No Adaptation)	443.61	547.98	−1.70
SpikeletFCN [2] (Scratch)	498.0	543.5	–
SpikeletFCN [2] (Fine-tuned)	77.12	107.1	–
Our method	180.69	230.12	0.43

ACID to GWHD: On the GWHD, our method reduced the MAE by 67.8% and the RMSE by 68.7% as compared to the baseline model without adaptation. The proposed method also achieved an R^2 value of 0.66 on the target domain.

ACID to CropQuant: On patches extracted from the *CropQuant* dataset, our method reduced the MAE by 59.3% and the RMSE by 58.0% as compared to the baseline model. The proposed method was also compared to two supervised methods from [2]. One was trained solely on the target domain (from scratch), and the other was pre-trained on the ACID dataset and fine-tuned on the target domain (fine-tuned). Our model outperformed the SpikeletFCN trained from scratch by 63.72%, while the fine-tuned SpikeletFCN model provides better performance than the proposed method. However, both of the reference methods require annotated datasets in the target domain while ours does not.

5.2 Leaf Counting

Figure 5 provides sample images taken from the KOMOTSUNA and MSU-PID datasets and shows that the proposed method accurately produces hot spots in the density map at the center of each leaf.

CVPPP to KOMATSUNA: We report results from the baseline model trained without adaptation, a previously proposed domain adaptation method [15], and the proposed method (Table 2). The adapted model resulted in a 71.6% drop in MSE, while the percentage agreement increased from 0% to 29.33%. Additionally, we compared the proposed approach with a previous domain adaptation technique proposed by [15]. Our method outperformed the previous result when using the least-squares adversarial loss (LS), and achieves similar performance when using the cross-entropy loss (XE).

CVPPP to MSU-PID: On the MSU-PID dataset, our method outperformed the no-adaptation baseline model in all metrics (Table 2). The proposed model provided a 23.9% increase in percentage agreement and a 10.93% decrease in $|DiC|$. The proposed method outperformed the previous domain adaptation

(a) Input   (b) GT   (c) Baseline   (d) Ours

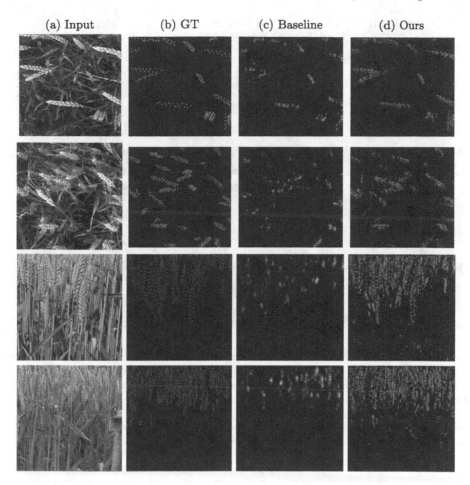

Fig. 4. Qualitative results for density map estimation. The first two rows are sampled from the Global Wheat Head Dataset and the last two are 512×512 patches extracted from CropQuant dataset (a) Input images. (b) Ground truth. (c) Predicted density map from the baseline model. (d) Predicted density map from the proposed model.

technique in this task when using the least-squares (LS) loss function, but under-performed it when using the cross-entropy (XE) loss.

Composite-KOMATSUNA: Applying the model from *CVPPP to KOMAT-SUNA* to composite images of multiple plants without retraining resulted in a mean DiC of 9.89 and mean $|DiC|$ of 9.90. Figure 6 shows a sample output from our model on the composite dataset. These results show that our model does not rely on assumptions about the distribution of object counts in the source domain, whereas previous work [15] used a KL divergence loss between the distribution of leaf counts in the source domain and the predictions of their model to avoid posterior collapse.

(a) Input   (b) Baseline   (c) Ours

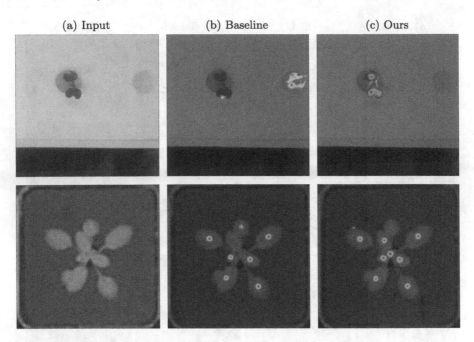

Fig. 5. Sample input images (left) and density map estimations generated from the baseline model (middle) and from the adapted model (right) in the leaf counting task. Top row: **CVPPP to KOMATSUNA** experiment. Bottom row: **CVPPP to MSU-PID** experiment.

Table 2. Domain adaptation results for the leaf counting tasks. XE is cross-entropy loss, LS is least-squares loss, ↓ denotes lower is better, ↑ denotes higher is better.

	DiC ↓	$|DiC|$ ↓	% ↑	MSE ↓
CVPPP to KOMATSUNA				
Baseline (No Adaptation)	4.09 (1.32)	4.09 (1.32)	0	18.49
Giuffrida, et al. [15] (XE)	−0.78 (1.12)	1.04 (0.87)	26	1.84
Giuffrida, et al. [15] (LS)	−3.72 (1.93)	3.72 (1.93)	2	17.5
Our method	−0.95 (2.09)	1.56 (1.67)	29.33	5.26
CVPPP to MSU-PID				
Baseline (No Adaptation)	1.21 (2.04)	1.83 (1.52)	0	5.65
Giuffrida, et al. [15] (XE)	−0.81 (2.03)	1.68 (1.39)	20	4.78
Giuffrida, et al. [15] (LS)	−0.39 (1.49)	1.18 (0.98)	26	2.36
Our method	1.17 (1.85)	1.63 (1.62)	23.96	4.25

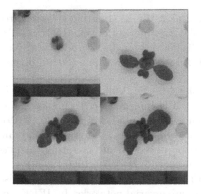 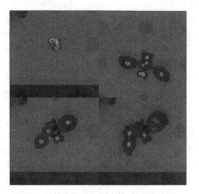

Fig. 6. Sample composite image with multiple plants (left) and density map estimation (right) generated from the model trained to adapt leaf counting using domain and target datasets with one plant per image.

6 Summary

In this paper, we attempted to address performance problems that arise in image-based plant phenotyping due to domain shift. Our study focused specifically on density estimation based object counting, which is a common method for plant organ counting. We proposed a custom Domain-Adversarial Neural Network architecture and trained it using the unsupervised domain adaptation technique for density estimation based organ counting. Our evaluation showed performance improvements compared to baseline models trained without domain adaptation for both wheat spikelet and leaf counting. For leaf counting, our results show similar performance to a previously proposed domain adaptation approach without the need to manually select a loss function for each dataset. This study is also the first to investigate the use of domain adaptation from an indoor source dataset to an outdoor target dataset. This may be a viable method for many plant phenotyping contexts, such as plant breeding experiments with both controlled environment and field trials. We plan to confirm this indoor-to-outdoor adaptation approach through additional testing with different plant species, different organ counting tasks, and additional field conditions, as future work.

Acknowledgement. This research was undertaken thanks in part to funding from the Canada First Research Excellence Fund.

References

1. Aich, S., Stavness, I.: Leaf counting with deep convolutional and deconvolutional networks. In: IEEE International Conference on Computer Vision Workshops (ICCVW), pp. 2080–2089 (2017)

2. Alkhudaydi, T., Zhou, J., De La lglesia, B.: SpikeletFCN: counting spikelets from infield wheat crop images using fully convolutional networks. In: Rutkowski, L., Scherer, R., Korytkowski, M., Pedrycz, W., Tadeusiewicz, R., Zurada, J.M. (eds.) ICAISC 2019. LNCS (LNAI), vol. 11508, pp. 3–13. Springer, Cham (2019). https://doi.org/10.1007/978-3-030-20912-4_1

3. Aslahishahri, M., Paul, T., Stanley, K.G., Shirtliffe, S., Vail, S., Stavness, I.: KI-divergence as a proxy for plant growth. In: IEEE Information Technology, Electronics and Mobile Communication Conference (IEMCON), pp. 120–126 (2019)

4. Bellocchio, E., Costante, G., Cascianelli, S., Fravolini, M.L., Valigi, P.: Combining domain adaptation and spatial consistency for unseen fruits counting: a quasi-unsupervised approach. IEEE Robot. Autom. Lett. 5(2), 1079–1086 (2020)

5. Bousmalis, K., Silberman, N., Dohan, D., Erhan, D., Krishnan, D.: Unsupervised pixel-level domain adaptation with generative adversarial networks. In: IEEE Conference on Computer Vision and Pattern Recognition (CVPR), pp. 3722–3731 (2017)

6. Cruz, J.A., et al.: Multi-modality imagery database for plant phenotyping. Mach. Vis. Appl. 27(5), 735–749 (2015). https://doi.org/10.1007/s00138-015-0734-6

7. David, E., et al.: Global wheat head detection (GWHD) dataset: a large and diverse dataset of high resolution RGB labelled images to develop and benchmark wheat head detection methods. Plant Phenomics (2020, in press)

8. Dobrescu, A., Valerio Giuffrida, M., Tsaftaris, S.A.: Leveraging multiple datasets for deep leaf counting. In: IEEE International Conference on Computer Vision Workshops (ICCVW), pp. 2072–2079 (2017)

9. Ganin, Y., Lempitsky, V.: Unsupervised domain adaptation by backpropagation. In: International Conference on Machine Learning (ICML), pp. 1180–1189 (2015)

10. Ganin, Y., et al.: Domain-adversarial training of neural networks. J. Mach. Learn. Res. 17(1), 2096–2030 (2016)

11. Gao, G., Gao, J., Liu, Q., Wang, Q., Wang, Y.: CNN-based density estimation and crowd counting: a survey. arXiv preprint arXiv:2003.12783 (2020)

12. Ghosal, S., et al.: A weakly supervised deep learning framework for sorghum head detection and counting. Plant Phenomics 2019, 1525874 (2019)

13. Gibbs, J.A., Burgess, A.J., Pound, M.P., Pridmore, T.P., Murchie, E.H.: Recovering wind-induced plant motion in dense field environments via deep learning and multiple object tracking. Plant Physiol. 181(1), 28–42 (2019)

14. Giuffrida, M.V., Chen, F., Scharr, H., Tsaftaris, S.A.: Citizen crowds and experts: observer variability in plant phenotyping. Plant Methods 14, 12 (2018)

15. Giuffrida, M.V., Dobrescu, A., Doerner, P., Tsaftaris, S.A.: Leaf counting without annotations using adversarial unsupervised domain adaptation. In: IEEE Conference on Computer Vision and Pattern Recognition Workshops, pp. 1–8 (2019)

16. Giuffrida, M.V., Doerner, P., Tsaftaris, S.A.: Pheno-deep counter: a unified and versatile deep learning architecture for leaf counting. Plant J. 96(4), 880–890 (2018)

17. Hu, J., Lu, J., Tan, Y.P.: Deep transfer metric learning. In: IEEE Conference on Computer Vision and Pattern Recognition, pp. 325–333 (2015)

18. Itzhaky, Y., Farjon, G., Khoroshevsky, F., Shpigler, A., Bar-Hillel, A.: Leaf counting: multiple scale regression and detection using deep CNNs. In: British Machine Vision Conference Workshops (BMVCW), p. 328 (2018)

19. Kuznichov, D., Zvirin, A., Honen, Y., Kimmel, R.: Data augmentation for leaf segmentation and counting tasks in rosette plants. In: IEEE Conference on Computer Vision and Pattern Recognition Workshops (2019)

20. Leibe, B., Seemann, E., Schiele, B.: Pedestrian detection in crowded scenes. In: IEEE Conference on Computer Vision and Pattern Recognition (CVPR), vol. 1, pp. 878–885 (2005)
21. Lempitsky, V., Zisserman, A.: Learning to count objects in images. In: Advances in Neural Information Processing Systems, pp. 1324–1332 (2010)
22. Li, M., Zhang, Z., Huang, K., Tan, T.: Estimating the number of people in crowded scenes by mid based foreground segmentation and head-shoulder detection. In: IEEE International Conference on Pattern Recognition (CVPR), pp. 1–4 (2008)
23. Lin, P., Chen, Y.: Detection of strawberry flowers in outdoor field by deep neural network. In: IEEE International Conference on Image, Vision and Computing (ICIVC), pp. 482–486 (2018)
24. Liu, M., Jiang, J., Guo, Z., Wang, Z., Liu, Y.: Crowd counting with fully convolutional neural network. In: IEEE International Conference on Image Processing (ICIP), pp. 953–957. IEEE (2018)
25. Liu, M.Y., Tuzel, O.: Coupled generative adversarial networks. In: Advances in Neural Information Processing Systems, pp. 469–477 (2016)
26. Long, M., Zhu, H., Wang, J., Jordan, M.I.: Deep transfer learning with joint adaptation networks. In: International Conference on Machine Learning (ICML), pp. 2208–2217 (2017)
27. Lu, H., Cao, Z., Xiao, Y., Zhuang, B., Shen, C.: TasselNet: counting maize tassels in the wild via local counts regression network. Plant Methods **13**, 79 (2017)
28. Madec, S., et al.: Ear density estimation from high resolution RGB imagery using deep learning technique. Agric. For. Meteorol. **264**, 225–234 (2019)
29. Minervini, M., Fischbach, A., Scharr, H., Tsaftaris, S.: Plant phenotyping datasets (2015). http://www.plant-phenotyping.org/datasets
30. Minervini, M., Fischbach, A., Scharr, H., Tsaftaris, S.A.: Finely-grained annotated datasets for image-based plant phenotyping. Pattern Recogn. Lett. (2015). https://doi.org/10.1016/j.patrec.2015.10.013. http://www.sciencedirect.com/science/article/pii/S0167865515003645
31. Olmschenk, G., Tang, H., Zhu, Z.: Crowd counting with minimal data using generative adversarial networks for multiple target regression. In: IEEE Winter Conference on Applications of Computer Vision (WACV), pp. 1151–1159 (2018)
32. Paul Cohen, J., Boucher, G., Glastonbury, C.A., Lo, H.Z., Bengio, Y.: Countception: counting by fully convolutional redundant counting. In: IEEE International Conference on Computer Vision Workshops, pp. 18–26 (2017)
33. Pound, M.P., Atkinson, J.A., Wells, D.M., Pridmore, T.P., French, A.P.: Deep learning for multi-task plant phenotyping. In: IEEE International Conference on Computer Vision Workshops (ICCVW), pp. 2055–2063 (2017)
34. Ranjan, V., Le, H., Hoai, M.: Iterative crowd counting. In: European Conference on Computer Vision (ECCV), pp. 270–285 (2018)
35. Ronneberger, O., Fischer, P., Brox, T.: U-Net: convolutional networks for biomedical image segmentation. In: Navab, N., Hornegger, J., Wells, W.M., Frangi, A.F. (eds.) MICCAI 2015. LNCS, vol. 9351, pp. 234–241. Springer, Cham (2015). https://doi.org/10.1007/978-3-319-24574-4_28
36. Sindagi, V.A., Patel, V.M.: A survey of recent advances in CNN-based single image crowd counting and density estimation. Pattern Recogn. Lett. **107**, 3–16 (2018)
37. Tayara, H., Soo, K.G., Chong, K.T.: Vehicle detection and counting in high-resolution aerial images using convolutional regression neural network. IEEE Access **6**, 2220–2230 (2017)

38. Tzeng, E., Hoffman, J., Darrell, T., Saenko, K.: Simultaneous deep transfer across domains and tasks. In: IEEE International Conference on Computer Vision, pp. 4068–4076 (2015)
39. Tzeng, E., Hoffman, J., Saenko, K., Darrell, T.: Adversarial discriminative domain adaptation. In: IEEE Conference on Computer Vision and Pattern Recognition (CVPR), pp. 7167–7176 (2017)
40. Ubbens, J.R., Stavness, I.: Deep plant phenomics: a deep learning platform for complex plant phenotyping tasks. Front. Plant Sci. **8**, 1190 (2017)
41. Uchiyama, H., et al.: An easy-to-setup 3D phenotyping platform for Komatsuna dataset. In: IEEE International Conference on Computer Vision Workshops, pp. 2038–2045 (2017)
42. Wang, C., Zhang, H., Yang, L., Liu, S., Cao, X.: Deep people counting in extremely dense crowds. In: ACM International Conference on Multimedia, pp. 1299–1302 (2015)
43. Wang, M., Deng, W.: Deep visual domain adaptation: a survey. Neurocomputing **312**, 135–153 (2018)
44. Xie, W., Noble, J.A., Zisserman, A.: Microscopy cell counting and detection with fully convolutional regression networks. Comput. Methods Biomech. Biomed. Eng. Imaging Vis. **6**(3), 283–292 (2018)
45. Zhang, C., et al.: Image-based phenotyping of flowering intensity in cool-season crops. Sensors **20**(5), 1450 (2020)
46. Zhang, S., Wu, G., Costeira, J.P., Moura, J.M.: FCN-RLSTM: deep spatio-temporal neural networks for vehicle counting in city cameras. In: IEEE International Conference on Computer Vision (ICCV), pp. 3667–3676 (2017)
47. Zhou, J., et al.: CropQuant: an automated and scalable field phenotyping platform for crop monitoring and trait measurements to facilitate breeding and digital agriculture. bioRxiv (2017). https://doi.org/10.1101/161547
48. Zhou, N., et al.: Crowdsourcing image analysis for plant phenomics to generate ground truth data for machine learning. PLoS Comput. Biol. **14**(7), e1006337 (2018)
49. Zhu, R., Sui, D., Qin, H., Hao, A.: An extended type cell detection and counting method based on FCN. In: IEEE International Conference on Bioinformatics and Bioengineering (BIBE), pp. 51–56 (2017)

Automatic Differentiation of Damaged and Unharmed Grapes Using RGB Images and Convolutional Neural Networks

Jonas Bömer[1], Laura Zabawa[2(✉)], Philipp Sieren[1], Anna Kicherer[4],
Lasse Klingbeil[2], Uwe Rascher[3], Onno Muller[3], Heiner Kuhlmann[2],
and Ribana Roscher[2]

[1] Bonn University, Bonn, Germany
[2] Institute of Geodesy and Geoinformation, Bonn University, Bonn, Germany
`zabawa@igg.uni-bonn.de`
[3] Forschungszentrum Jülich, IBG-2, Jülich, Germany
[4] Julius Kühn-Institut, Institute for Grapevine Breeding Geilweilerhof,
Siebeldingen, Germany

Abstract. Knowledge about the damage of grapevine berries in the vineyard is important for breeders and farmers. Damage to berries can be caused for example by mechanical machines during vineyard management, various diseases, parasites or abiotic stress like sun damage. The manual detection of damaged berries in the field is a subjective and labour-intensive task, and automatic detection by machine learning methods is challenging if all variants of damage should be modelled. Our proposed method detects regions of damaged berries in images in an efficient and objective manner using a shallow neural network, where the severeness of the damage is visualized with a heatmap.

We compare the results of the shallow, fully trained network structure with an ImageNet-pretrained deep network and show that a simple network is sufficient to tackle our challenge. Our approach works on different grape varieties with different berry colours and is able to detect several cases of damaged berries like cracked berry skin, dried regions or colour variations.

Keywords: Deep learning · Classification · Grapevine · Precision viticulture · Plant phenotyping

1 Introduction

In viticulture, fungal pathogens such as *Oidium tuckeri*, *Plasmopara viticola* or *Botrytis cinerea*, play a major role in yield and quality loss. However, the use of pesticides in the production of wine is cost intensive and has a negative impact

---

J. Bömer and L. Zabawa—Contributed equally to this work.

© Springer Nature Switzerland AG 2020
A. Bartoli and A. Fusiello (Eds.): ECCV 2020 Workshops, LNCS 12540, pp. 347–359, 2020.
https://doi.org/10.1007/978-3-030-65414-6_24

on the ecosystem [8]. Identifying damaged or infected areas of grapevine berries could be beneficial for the evaluation of breeding material and it could improve the harvest planing in commercial vineyards. Because the manual identification and screening of infected areas is very time consuming, there is a need for automated procedures to detect such defects [2].

Neural networks have proven to be a promising approach for object detection as well as classification of objects in images. Some work has already been done in the field of disease detection for plants, showing the potential of neural networks. Pound et al. [10] trained a network for root and shoot feature identification and localization. They extended their classification pipeline with a sliding window approach to localize the features in images. Förster et al. [4], for example, use generative adversarial networks to forecast the spread of a disease on leaves. Particularly convolutional neural networks (CNNs) have proven to be useful because they are able to capture the relevant spatial characteristics of data, especially images. Amara et al. [1], for example, use CNNs for the detection and classification of diseases in banana leaf images. Furthermore, Zacharias et al. [13] use CNNs for the UAV supported detection of poisonous Senecio jacobaea in grassland areas while Halstead et al. [6] use CNNs for the detection of sweet pepper, followed by a ripeness classification. In the context of grapevine diseases, Strothmann et al. [11] show first attempts to use an unsupervised convolutional auto-encoder to detect damaged berries.

In this paper, we present a framework to automatically detect damaged grapes in viticulture using RBG images and CNNs. Hereby, we do not distinguish between damage caused by diseases or mechanical machines, for example injuries induced by the harvester during thinning procedures. We train a CNN model from scratch which classifies image patches showing different varieties of grapevine berries into the classes *unharmed* and *damaged*. We apply the CNN with a patch-based sliding window approach, similar to [10], operating in a region of interest showing bunches of grapes. This method reduces the labeling complexity in contrast to semantic segmentation approaches where pixels or regions have to be annotated in detail. The patch-based classification results are combined into a heatmap which indicates infected and damaged regions. These heatmaps can be of use to farmers since they indicate damaged or infected areas which call for action. Furthermore, they can be used by breeders to evaluate breeding material.

2   Materials and Methods

2.1   Data

The data set consists of 129 RGB images with a size of 5472×3648 pixels taken by a digital single-lens reflex camera (DSLR). The data were collected in September 2017 under different daylight conditions with flash (see Fig. 1 for example patches showing different illumination situations). The distance to the canopy is roughly 1 meter, since the space between vine rows is limited. This results in a spatial resolution 0.3 mm per pixel. Images of 39 different varieties, including green, red and blue varieties, were collected and used to achieve a good generalization of the learned network. The composition of the data set consists of 34 images showing green (example shown in Fig. 6), 42 images showing red (example shown in Fig. 7) and 53 images showing blue berries (example shown in Fig. 8). Each image shows a different plant and each of the 39 varieties is covered by between 1 to 6 images.

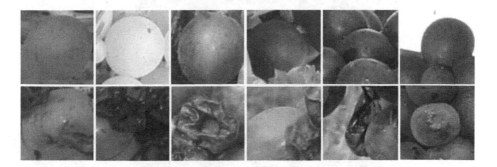

Fig. 1. Examples of image patches used for training and testing of our network. The upper row shows patches of the class *unharmed* while the lower row shows examples of damaged berries. The examples represent different varieties.

Twenty of the 129 images, including representative images of each color variety, were divided into two equal parts. One part of each image was used to generate patches for the test set, which are spatially independent from the train set, as well as the generation of the heatmaps, the other part was used together with the remaining 109 images to extract patches for the training set. The generation of the heat maps is performed on the independent test data set. In the images, unharmed and damaged grapes are manually annotated with dots in the middle of each berry. Around damaged berries no unharmed berries are annotated to prevent mixed annotations. This means that no damaged berries are present in patches that are annotated as *unharmed*.

For training of the network, a patch size of 120 × 120 pixels is used (see Fig. 1). This ensures that each patch contains an entire grape berry, since one pixel has an approximate spatial resolution of 0.3 mm.

2.2 Patch Generation and Data Augmentation

Since the training of neural networks generally requires a lot of training data, data-augmentation is applied to extend the previously designed training set. Due to the annotation strategy, damaged berries always appear in the center of patches. To reduce a negative influence of this systematic effect, we cut eight additional patches around each damaged berry with a random offset of maximum 60 pixels to generate patches with damaged spots at edges and in corners. An illustration of the procedure is shown in Fig. 2, where additional image patches used for training are exemplarily indicated by pink boxes. Unharmed berries are represented by patches where the annotation lies in the middle (see light blue boxes in Fig. 2). Other unharmed berries are visible in the surrounding area and no systematic effects due to dominant positions are present.

Fig. 2. Annotations and patch generation. Unharmed berries are annotated with a light blue dot, damaged berries with a pink one. Around each unharmed berry a patch is extracted (light blue rectangle). Since each annotated unharmed berry is surrounded by other unharmed berries, there is no systematic in the berry positions. Around damaged berries, multiple patches are extracted randomly to prevent patches where damaged berries are always positioned in the center of a patch (pink rectangles). (Color figure online)

Since we already generated nine patches per damaged berry by the previously described procedure, no data augmentation is applied for the patches showing damaged berries in order to prevent class imbalance. Overall, the training set contains 51064 patches and the test set contains 7007 patches belonging to the class *damaged*. For the class *unharmed*, the training set consists of 18275 patches and the test set consists of 2438 patches. To get a better balance of the two classes and to extend the data set, data-augmentation is applied. Therefore, patches of the class *unharmed* are rotated two times by 120° and 240° and saved additional to the original patch, resulting in 54825 patches for the training set and 7314 patches for the test set. Because rotating a 120×120 pixel patch and saving it in the same size will produce empty corners, patches of the "unharmed" class are first cut with 170 × 170 pixel size and are cut to 120 × 120 pixels during the data-augmentation process.

2.3   Structure of the Convolutional Neural Network

We chose a shallow network structure, since our classification problem is comparably simply with only 2 classes. To show the advantages of a fully trained, shallow network, we will compare our results with a deeper, pre-trained network.

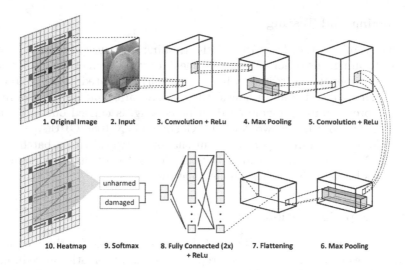

1. Original Image   2. Input   3. Convolution + ReLu   4. Max Pooling   5. Convolution + ReLu

10. Heatmap   9. Softmax   8. Fully Connected (2x)   7. Flattening   6. Max Pooling
                                + ReLu

Fig. 3. Schematic representation of the generation of the heatmaps including the structure of the CNN.

Shallow Network

The basis of the CNN used in this study is the LeNet architecture, developed by LeCun et al. [9]. The network consists of two layer sets, each with a convolutional layer, an activation function and a max pooling layer. These layer sets are followed by a fully connected layer, an activation function and a last fully connected layer as well as a softmax classifier (see Fig. 3). In our first layer set, the convolutional layer consists of 20 convolutional filters (50 in the second layer set) with a size of 5×5 pixels. The activation function was the Rectified-Linear-Unit function (ReLu) [5] which sets any negative input to zero. In the 2×2 max pooling layer, superfluous information is discarded, thus increasing the speed of the CNNs computation and the receptive field. In the fully connected layers, which are completed by the softmax classifier, the decision for a class is realized. Therefore, the number of neurons in the last fully connected layer is equal to the number of categories for the classification, in our case two. The output of the CNN is a number between 1 and 0, representing the probability of the patch belonging to one of the two classes.

Deep Network

We compare our results with a deeper, ImageNet [3] pre-trained network, namely ResNet50. This network is widely used for classification tasks and is used as a baseline. For details regarding the network structure we refer the reader to He et al. [7].

2.4　Training and Testing

During the training procedure, 75% of the training set is used for actual model parameter training, the other 25% are withheld for validation of the CNN during training. The independent test set which is spatially separated from the training set is used for qualitative and quantitative evaluation by means of a confusion matrix. All programming, including data processing, is carried out in Python. We train the network using two NVIDIA GeForce GTX 1080 GPU's.

We perform 25 training epochs for our shallow network with a batch size of 32. The ResNet50 was trained for 60 epochs, considering that the number of parameters is higher than for the shallow network.

2.5　Heatmap Generation

For our experiments, we manually mask bunches of grapes for the creation of heatmaps. However, this step can be replaced by an automatic procedure, for example a detector, which can determine a region of interest where berries are located. Heatmaps are created utilizing a sliding window approach with different strides in the region of interest, where the stride determines the level of detail of the heatmap. Since the classification result per patch is assumed for all contained pixels and thus multiple results are obtained for pixels in overlapping areas, an averaging process is performed that leads to the final heatmap for each pixel.

3　Results

3.1　Classification Performance of the Our CNN

The network obtains a training accuracy of 96.4% and validation accuracy of 97% during the training of the model. The evaluation by means of a confusion matrix (see Fig. 4) turns out a bit inferior, but basically supports these observations.

N = 14 321		prediction		
		unharmed	damaged	
actual	unharmed	6827	487	7314
	damaged	265	6742	7007
		7092	7229	Total error: 5.25 %

Fig. 4. Evaluation of the test set with our Network

N = 14 3210		prediction		
		unharmed	damaged	
actual	unharmed	6019	1295	7314
	damaged	242	6765	7007
		6261	8060	Total error: 10.73 %

N = 14 3210		prediction		
		unharmed	damaged	
actual	unharmed	6192	1121	7314
	damaged	262	6745	7007
		6455	7866	Total error: 9.66 %

(a) Evaluation of the test set with a pre-trained ResNet50.

(b) Evaluation of the test set with a ResNet50 which was trained from sratch.

Fig. 5. Comparison of network performances between a pre-trained deep network and one trained from scratch.

We evaluate the confusion matrix with the classical measures precision and recall. Precision depicts the number of correctly identified unharmed patches (true positives) in relation to all patches which were identified as unharmed (prediction positive). The recall describes the relation between the correctly identified unharmed patches in relation to all unharmed patches (condition positive). Our network yields a precision of 96.26% and a recall of 93.34%. In contrast to that the type one error is the relation between patches showing damaged berries which are identified as unharmed (false positives) and all patches which are identified as unharmed (condition positive). Type two error relate the number of patches showing unharmed berries which are identified as damaged (false positive) with the number of all unharmed patches (condition positive). The type one error with 3.74% is lower than the type two error with 6.66%. This means that the network tends to wrongly classify more unharmed patches as damaged than the other way around. But nonetheless the amount of misclassifications is small.

3.2 Classification Performance of ResNet50

We compare our network with a ResNet50 and train it twice, once with ImageNet pre-trained weights and a second time completely from scratch. Both networks are trained for more than 60 epochs.

The evaluation of the pre-trained ResNet50 yields a precision of 96.13% and a recall of 82.29%. The type one error is 3.87% while the type two error is noticebly larger with 17.71%. The ResNet50 which was trained from scratch yields comparable results. The precision is 95.94%, which slightly lower than for the pre-trained model. The recall on the other hand is slightly higher with 84.67%. The type one error (4.06%) and the type two error (15.33%) show the same characteristics.

Figure 5 shows that both versions of the ResNet50 tend to misclassify unharmed patches as damaged (false negatives). This leads to the high type two error. The number of patches which show damaged berries but are classified as unharmed (false positives) is comparable to our network. The total error of both networks is nearly twice as high as for our shallow network structure.

3.3 Heatmaps

For our experiments, coarse heatmaps (shown in Fig. 6a, 7b and 8c) and a fine heatmap (shown in Fig. 6a, 7b and 8c) are created, where the coarse heatmaps are created with a stride of 60 pixels and the fine heatmap with a stride of 10 pixels. For illustration, all pixels located in the region of interest are overlayed in colour levels of 10% according to their probability of damage.

The heatmaps show a good classification performance that differs in accuracy among the various varieties. Due to our annotation strategy, a patch is classified as damaged as soon as a damaged berry or parts of it appear in the patch. Therefore the areas which are detected as damaged are larger than the real damaged berries.

Specific classification problems that occur are mostly related to shadows caused by leaves or grapes themselves inside the annotated area (Fig. 7). Bark, stipes or wire cause similar problems and are mostly seen on the edge zones of the grape bunches (Fig. 9).

4 Discussion

4.1 Network Structure

We compared a shallow network structure with a ResNet50, a deep network which is known to perform well for classification applications. The same network was trained twice, one time with ImageNet pre-trained weights and one time from scratch. Our shallow network was trained from scratch as well. We trained both versions of ResNet50 and our shallow network with the same dataset and compared the results on the same test set.

We show that the shallow network performs better than both versions of the ResNet50. Especially the number of unharmed patches which are classified as damaged are higher for both ResNet50. This leads to a high type two error and increases the total error rate.

The ImageNet pre-trained network performs slightly worse than the not pre-trained one. One possible explanation is that ImageNet inspired features do not generalize well for plant problems. Although if one trains long enough this influence might be reduced. The fact that our shallow network performs better than the deep ResNet50 structure encourages our network choice. A simple and shallow network is sufficient to tackle the comparably simple two class problem.

4.2 Heatmaps

It can be seen that there is a difference in the accuracy of the heatmaps regarding the different varieties (Fig. 6, 7 and 8). Figure 6 shows the heatmap of a green variety. The damaged regions are well identified. Figure 7 shows a red variety.

(a) Original Image   (b) Fine Heatmap   (c) Coarse Heatmap

100 to >90%   90 to >80%   80 bis >70%   70 to >60%   60 to >50%   50 to >40%   40 to >30%   30 to >20%   20 to >10%   10 to 0%

damaged ———————————————————————————————→ unharmed

Fig. 6. Examples for the Heat Map Generation for a green variety. The left columns shows the original image. The middle column shows an example for a fine heatmaps while the right one shows a coarse heatmaps. The damaged part on the upper grape bunch are correctly identified. Only small artifacts can be seen at the edges of the lower grape bunch. (Color figure online)

The damaged berries in the upper and lower part of the grape bunch are correctly identified, but in the middle we have incorrect classification due to an opening in the middle of the bunch. Background can be seen as well as non typical small green berries and stems. This caused the network to detect an abnormality.

The network seems to have issues with correctly classifying dark blue grapes. An example can be seen in Fig. 8. Damaged areas can be clearly seen on the left side of the grape bunch. The right side, where more areas are detected as severely damaged, the berries feature a high variety of colour shades. Most berries have an inhomogeneous colour with many dark spots. As mentioned before, the procedure for the heatmap generation promotes large damaged areas. As Fig. 8 shows, patches which are classified as damaged result in large infected areas in the heat map, regardless of the accuracy of the prediction. Other classification problems could likely be caused by sporadic lower light conditions evoked through shadows of leaves or grapes themselves, stems and bark included in the patch. Furthermore it must be noted that a decision whether a berry is damaged is oftentimes also difficult for a human annotator. The inhomogeneous colour with darker or lighter areas on each berry pose a difficult challenge for our network. In addition, the network has partial issues dealing with patches containing structures like wire or bark (see Fig. 9), which leads to classification errors mainly in the edge zone of the bunch of grapes. This problem could be overcome by using more than two classification categories, for example introducing a background class.

Furthermore, the outcome of the heatmaps is influenced by the way they are created. By cutting and classifying a patch only every sixtieth pixel in x and y direction, as was done for the coarse heatmaps, we are able to reduce the

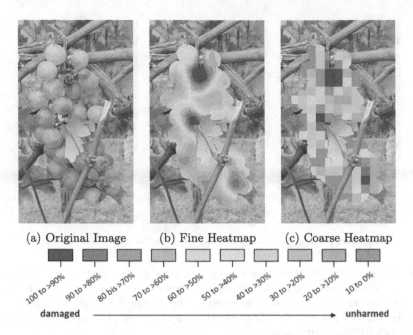

(a) Original Image   (b) Fine Heatmap   (c) Coarse Heatmap

100 to >90% 90 to >80% 80 bis >70% 70 to >60% 60 to >50% 50 to >40% 40 to >30% 30 to >20% 20 to >10% 10 to 0%

damaged ⟶ unharmed

Fig. 7. Examples for the Heat Map Generation of a red variety. The left columns shows the original image. The middle column shows an example for a fine heatmaps while the right one shows a coarse heatmaps. The damaged berries in the lower and upper part of the bunch are correctly identified. In the middle a wrong classification can be seen, where background is visible in the middle of the bunch as well as some uncharacteristic small green berries. The network recognizes this region as an anomaly and decides to label it as damaged. (Color figure online)

calculation time by a factor of 360 in comparison to the fine heatmap. However, this method results in a more inaccurate heatmap with far less classifications per pixel. In contrast to the coarse heatmap, the fine one shows a more differentiated result with a larger transition zone between certainly *damaged* or *unharmed* spots. This is an approach to reduce single incorrect classifications, as can be seen in Fig. 7 at the left edge of the bunch of grapes, but can also cause the opposite effect of enlarging transition zones with no explicit assignment of a pixel to belong to one or the other class, as can be seen in Fig. 8. Despite the mentioned issues, Fig. 6 and 7 show a good representation of the damaged spots on the grape bunches. The CNN in combination with the heatmap generation is capable of detecting and visualising even single damaged grapes so that a very precise mapping of the health condition of the future crop is possible. In combination with an algorithm able to detect grapes in images, similar to the one in Zabawa et al. [12], this method could be a useful tool in viticulture.

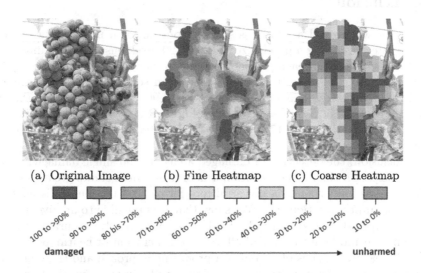

(a) Original Image   (b) Fine Heatmap   (c) Coarse Heatmap

100 to >90%   90 to >80%   80 bis >70%   70 to >60%   60 to >50%   50 to >40%   40 to >30%   30 to >20%   20 to >10%   10 to 0%

damaged ⟶ unharmed

Fig. 8. Examples for the Heat Map Generation of a blue variety. The left columns shows the original image. The middle column shows an example for a fine heatmaps while the right one shows a coarse heatmaps. On the left upper side of the bunch, multiple damaged berries can be seen. They are correctly identified in the heatmaps as damaged. In the lower right part of the bunch, damaged berries are detected as well. The berries feature dark spots, where berry was is removed, which the network detects as damaged. (Color figure online)

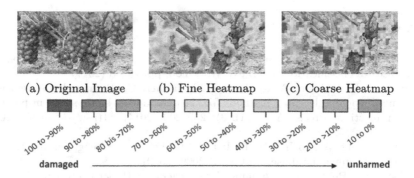

(a) Original Image   (b) Fine Heatmap   (c) Coarse Heatmap

100 to >90%   90 to >80%   80 bis >70%   70 to >60%   60 to >50%   50 to >40%   40 to >30%   30 to >20%   20 to >10%   10 to 0%

damaged ⟶ unharmed

Fig. 9. Examples for Problems with the Heat Map Generation of a red variety. The left columns shows the original image. The middle column shows an example for a fine heatmaps while the right one shows a coarse heatmaps. In the middle of the left picture we see damaged berries which are correctly identified in the heat maps. But we see additional regions, which are classified as damaged, where a lot of rachis are present (left and right part of the image). (Color figure online)

5 Conclusion

In summary, we trained a two-class shallow convolutional neural network with LeNet architecture using a patch size of 120 × 120 pixel, ensuring a training with whole berries covered in each patch. We compare our results with an ImageNet pre-trained ResNet50 as well as one trained from scratch. In our experiments with different varieties, we achieved a precision of 96.26% with our network with a total error rate of 5.25%. The error rate of both ResNet50 is twice as high due to a high type two error (15–17%). The ResNet50s tend to misclassify unharmed patches as damaged.

In conclusion, the detection of damaged areas or even single damaged grapes with the help of a shallow convolutional neural networks shows great potential and has the capability to be a powerful tool for early detection of diseases and damages of grapes in viticulture. We show that it is possible to archive a classification between damaged and unharmed grapes with a total accuracy of about 95% even with moderate technical effort. The generation of heatmaps showing damaged areas in bunches of grapes is a promising approach and there are multiple possibilities to improve the results of this study. In the context of precision viticulture, the combination of such a detection tool with autonomous robotic systems could save manual labour and improve the management of vineyards and breeding material.

Acknowledgment. This work was partially funded by German Federal Ministry of Education and Research (BMBF, Bonn, Germany) in the framework of the project novisys (FKZ 031A349) and partially funded by the Deutsche Forschungsgemeinschaft (DFG, German Research Foundation) under Germanys Excellence Strategy - EXC 2070-390732324.

References

1. Amara, J., Bouaziz, B., Algergawy, A.: A deep learning-based approach for banana leaf diseases classification. In: BTW Workshop, pp. 79–88 (2017)
2. Behmann, J., Mahlein, A.-K., Rumpf, T., Römer, C., Plümer, L.: A review of advanced machine learning methods for the detection of biotic stress in precision crop protection. Precis. Agric. **16**(3), 239–260 (2014). https://doi.org/10.1007/s11119-014-9372-7
3. Deng, J., Dong, W., Socher, R., Li, L.J., Li, K., Fei-Fei, L.: ImageNet: a large-scale hierarchical image database. In: CVPR 2009 (2009)
4. Foerster, A., Behley, J., Behmann, J., Roscher, R.: Hyperspectral plant disease forecasting using generative adversarial networks. In: International Geoscience and Remote Sensing Symposium (2019)
5. Hahnloser, R.H., Sarpeshkar, R., Mahowald, M.A., Douglas, R.J., Seung, H.S.: Digital selection and analogue amplification coexist in a cortex-inspired silicon circuit. Nature **405**(6789), 947–951 (2000)
6. Halstead, M., McCool, C., Denman, S., Perez, T., Fookes, C.: Fruit quantity and ripeness estimation using a robotic vision system. IEEE Robot. Autom. Lett. **3**(4), 2995–3002 (2018)

7. He, K., Zhang, X., Ren, S., Sun, J.: Deep residual learning for image recognition. In: IEEE Conference on Computer Vision and Pattern Recognition (CVPR), pp. 770–778 (2016)
8. Hildebrandt, A., Guillamón, M., Lacorte, S., Tauler, R., Barceló, D.: Impact of pesticides used in agriculture and vineyards to surface and groundwater quality (North Spain). Water Res. **42**(13), 3315–3326 (2008). https://doi.org/10.1016/j.watres.2008.04.009, http://www.sciencedirect.com/science/article/pii/S0043135408001516
9. LeCun, Y., Bottou, L., Bengio, Y., Haffner, P.: Gradient-based learning applied to document recognition. IEEE Proc. **86**(22), 2278–2324 (1998)
10. Pound, M.P., et al.: Deep machine learning provides state-of-the-art performance in image-based plant phenotyping (2017)
11. Strothmann, L., Rascher, U., Roscher, R.: Detection of anomalous grapevine berries using all-convolutional autoencoders. In: International Geoscience and Remote Sensing Symposium (2019)
12. Zabawa, L., Kicherer, A., Klingbeil, L., Reinhard, T.: Counting of grapevine berries in images via semantic segmentation using convolutional neural networks. ISPRS J. Photogrammetry Remote Sens. **164**, 73–83 (2020)
13. Zacharias, P.: Uav-basiertes grünland-monitoring und schadpflanzenkartierung mit offenen geodaten. GeoForum MV 2019 - Geoinformation in allen Lebenslagen, At Warnemünde (2019)

Germination Detection of Seedlings in Soil: A System, Dataset and Challenge

Hanno Scharr[1]([⊠])[iD], Benjamin Bruns[1][iD], Andreas Fischbach[1][iD],
Johanna Roussel[1,2][iD], Lukas Scholtes[1][iD], and Jonas vom Stein[1][iD]

[1] Institute of Bio- and Geosciences (IBG), IBG-2: Plant Sciences,
Forschungszentrum Jülich, Jülich, Germany
`h.scharr@fz-juelich.de`
[2] Medical Engineering and Technomathematics, University of Applied Sciences
Aachen, Jülich, Germany

Abstract. In phenotyping experiments plants are often germinated in high numbers, and in a manual transplantation step selected and moved to single pots. Selection is based on visually derived germination date, visual size, or health inspection. Such values are often inaccurate, as evaluating thousands of tiny seedlings is tiring. We address these issues by quantifying germination detection with an automated, imaging-based device, and by a visual support system for inspection and transplantation. While this is a great help and reduces the need for visual inspection, accuracy of seedling detection is not yet sufficient to allow skipping the inspection step. We therefore present a new dataset and challenge containing 19.5k images taken by our germination detection system and manually verified labels. We describe in detail the involved automated system and handling setup. As baseline we report the performances of the currently applied color-segmentation based algorithm and of five transfer-learned deep neural networks.

Keywords: Transplantation guidance · High-throughput plant phenotyping · Automation · Arabidopsis · Two leaf stadium

1 Introduction

We present a new dataset and challenge containing 19.5k images taken by an automated germination detection system and manually verified labels.

A plant's phenotype is constituted by its traits as reaction to diverse environmental conditions. Plant phenotyping has been identified to be key for progress in plant breeding and basic plant science [26]. To increase throughput of plant experiments and overcome the phenotyping bottleneck [10] many automated technologies have been and are developed. Such systems are often image-based and image analysis builds a performance bottleneck [17].

Electronic supplementary material The online version of this chapter (https://doi.org/10.1007/978-3-030-65414-6_25) contains supplementary material, which is available to authorized users.

A. Bartoli and A. Fusiello (Eds.): ECCV 2020 Workshops, LNCS 12540, pp. 360–374, 2020.
https://doi.org/10.1007/978-3-030-65414-6_25

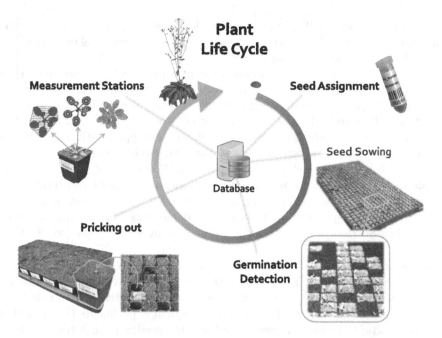

Fig. 1. Life cycle of a plant. In a typical plant phenotyping experiment seeds are selected and assigned to the experiment, they are sown, plants germinated, seedlings pricked out and transplanted to bigger pots, and then measured at different stations with a selection of measurement systems. Some plants are then brought to flowering to produce seeds, closing the circle.

When running larger-scale experiments with hundreds of plants in soil, due to space limitations, it is reasonable to sow seeds in smaller pots until germination and transplant them later into larger pots (see Fig. 1). We seed into special multi-well-plates with 576 wells of size 15 mm × 15 mm. They are too small for the final experiment, but allow for undisturbed germination in soil and, importantly, undisturbed transplantation of a seedling to its final typically 7 cm × 7 cm pot. The number of seeds to sow strongly depends on germination rates which can differ from genotype to genotype. Often the three- to fourfold amount of required seedlings are sown to ensure enough plants germinate within a given time-frame needed for synchronisation of treatments in an experiment. For typical medium to high-throughput experiments several thousand seeds need to be sown.

For larger seedlings, like e.g. Tobacco plants, visual germination detection and manual selection of seedlings for subsequent experiments is routinely and reliably done. For tiny seedlings like Arabidopsis, with a typical diameter of 1 mm to 2 mm, it is tiring and frequent errors are unavoidable. However, this inspection is crucial for subsequent experiments, as seedlings are selected according to the germination date, size or health and thus it should to be automated to increase reliability and throughput [7,14,19,20]. Further, without an automated system, size and health are just available by visual inspection and after transplantation no mapping between seedlings' and plants' identities is available. However, when data-handling is intended to cover the whole plant life cycle, plants' identities need to be tracked through this process.

We designed and built two systems, allowing for image-based automated germination detection, support of visual seedling inspection and guided transplantation of seedlings. Imaging for germination detection, or more exactly, detection of the two leaf stadium, is done by a variant of GrowScreen [13,27]. Further, we designed a special workplace, a *Handling Station*, equipped with a camera and video projector for visual seedling inspection and guided transplantation. The systems are coupled via a database (see Sect. 2.1), storing e.g. layout of multi-well-plates or trays, seed properties and identification number (seed ID) for each well, germination date and seedling size when detected etc. Germination can be verified and transplantation automatically documented in the database using the Handling Station. Both systems are described in detail below (see Sect. 2.2 and 2.3).

In the workflow individual seeds/seedlings are identified by their location on the soil-filled tray. The germination detection system automatically measures relevant seedling data and stores it in a database. The system allows rule-based selection of seedlings for pricking out and subsequent experiments based on their measured traits. For visual inspection all selected seedlings on a tray are highlighted by the video projector. After validation, the user is guided through transplantation by highlighting only the next single seedling to process. The system generates an adhesive label indicating among other things the ID and (randomized) target position of the newly potted plant. By this a one-to-one mapping between seedling and newly potted plant is generated in the database. Using this setup for germination detection and pricking out reduced time and labor, as well as increased reliability of seedling-to-plant assignment. In addition, the tedious work of selecting the right plant among hundreds of others is now done by the system, allowing for faster and less tiring working.

While the two systems significantly increased manual throughput and enabled reliable seed to plant tracking, it not yet allows for fully automatic seedling selection. Especially reliability of two-leaf stage detection is still too low. We aim at well above 99.8% reliability (i.e. human rater performance), but as we are not yet there, visual inspection of germination detection results is still needed. Therefore, image data of the germination detection system together with binary labels 'positive', i.e. germinated seedling visible versus 'negative', i.e. no germinated seedling visible, are provided together with this paper [21]. As a baseline for further algorithm development on the challenge of predicting binary labels from images, we provide the performance of a simple color-segmentation based algorithm (see Sect. 2.6), as well as five transfer learned deep neural networks (DNNs). The dataset is described in detail in Sect. 2.5 and the experiments for the baseline in Sect. 3.

2   Materials and Methods

2.1   Database

Automated plant phenotyping commonly is performed in an environment of specialized, rarely interconnected systems with highly diverse datasets and custom analysis tools [8,12,13,18,22]. In lab or greenhouse situations, where potted plants are monitored over a longer time span, in different situations, and with different measurement systems, a wealth of often weakly structured data is collected. Keeping track of the individual plants in an experiment can then be cumbersome. Database systems handling individual plants data, e.g. by using unique IDs for each plant, allow to join such data in a user-friendly and reliable fashion [3,22].

We use the information system PhenOMIS [3,22], a distributed information system for phenotyping experiments, and integrated the systems presented here. PhenOMIS allows to track plant histories by acquiring (and guiding) manual treatments during activity, covering plants' whole life cycle (cmp. Fig. 1). The information system allows central access to distributed, heterogeneous phenotyping data and integration of spatial, temporal and sensor data. To this end it loosely couples different components, not integrating all data in one database, but rather enabling and supporting data co-existence. Web services for encapsulation allow to further extend functionality.

Here, seeds are sown into multi-pot trays (Fig. 2B) using a robotic system [12] assigning each seed a unique ID. Seed ID and pot position in the tray are stored in the database for identification. Subsequent measurements from germination detection are then assigned to this ID.

2.2   Automated Germination Detection System

Imaging. A variant of the *Growscreen Fluoro* [13] system is used for automated germination detection (Fig. 2A), fit into a 19" rack. It consists of an x-y-z-moving stage (blue components in Fig. 2A), where we mounted a 5 MP RGB color camera (red in Fig. 2A) with a 25 mm optical lens, instead of the fluorescence camera. This offers a spatial resolution of 29.0 pixel/mm. A white LED ring is used for illumination.

Different plant species need different pot sizes for germination. The current tray is identified by a bar code, the tray's layouts read from the database, allowing the system to calculate camera positions for suitable image tiling. The camera is automatically moved to the calculated positions by x-y-moving stages. Optimal working distance and thus focus is ensured by the z-moving stage according to the known pot height. Scanning a tray takes several seconds to a few minutes, depending on tray layout. Some sample images are shown in Fig. 2C. They are automatically cropped into images showing one pot each from the known tray layout. RGB images in the dataset (see Sect. 2.5) are collected from this step. Figure 2D top shows a time series of images for one pot.

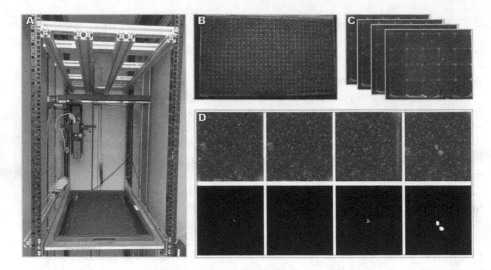

Fig. 2. Germination detection. A: Imaging setup. B: A soil-filled tray used for germination, together with germinated seedlings. C: Acquired images. D: Temporal image sequence of one pot, showing the first time point when green tissue is visible (left) and two leaf stage (right). (Color figure online)

Processing. In the system, single pot images are segmented into a binary foreground-background mask using thresholds in HSV-space [27]. Thresholds are manually optimized once per plant species and stored in the database for this species, in order to adapt to seedling color and illumination conditions. Figure 2D bottom shows some segmentation results.

Two points in time need to be derived from the images:

- germination date, defined by the first occurrence of green pixels in the image. To avoid artefacts a size threshold of 10 pixels ($0.0119\,\text{mm}^2$) is used for Arabidopsis.
- emergence of separate cotyledons, i.e. two leaf stage. This is where the data challenge is aiming at.

In the current system, the two leaf stage is detected using the algorithm described in Sect. 2.6. In our plant experiments, we require seedlings to reach this stage to be suitable for transplantation and subsequent experiments. In addition, for *Arabidopsis thaliana* the seedlings are required to have a minimum leaf diameter of four pixels ($0.138\,\text{mm}$) and a projected leaf area larger than 100 pixels ($0.119\,\text{mm}^2$).

2.3   A Handling Station for Pricking Out and Randomization

For visual inspection selected seedlings are highlighted using a video projector. The same method guides the user through the transplantation process by highlighting the next single seedling to pique.

Fig. 3. Visualization of germination results. A: The setup for visual inspection and transplantation guidance. Inset: Video projector for highlighting. B: Germinated seedlings highlighted in their pots. C: print out showing the germination dates and rates for the same tray.

Materials. The handling station consists of a video projector (ViewSonic Pro 8600) and an RGB camera (Point Grey Grashopper GS2-GE-20S4C-C) mounted above a table and aiming towards the table workspace (see Fig. 3A). A cutout in the table allows to keep standard trays in a fixed position (Fig. 3B) and enables single pot access from below for the pricking out process (Fig. 4D).

The camera is used to calibrate the projector to the tray. This is done in two steps. In the first step an image of the table with a tray is acquired. On this image a user marks the four corners of the tray clockwise by mouse-click. From these positions and the known tray dimension a homography $H_{C,T}$ is calculated mapping tray coordinates x_T to camera image coordinates x_C. In the second step five red square markers are projected on the table and imaged by the camera. Four squares are used as calibration markers, the fifth marks the projector origin. For the known positions of the markers in the camera as well as projected image homography $H_{P,C}$ from camera coordinates x_C to projector coordinates x_P is derived. Combining the two homographies the pixel position x_P in the projector's plane can be calculated for a given location x_T on the tray by $x_P = H_{P,C} \cdot H_{C,T} \cdot x_T$. With the tray layout information available from the database, size and location of a pot or well in the projector image can

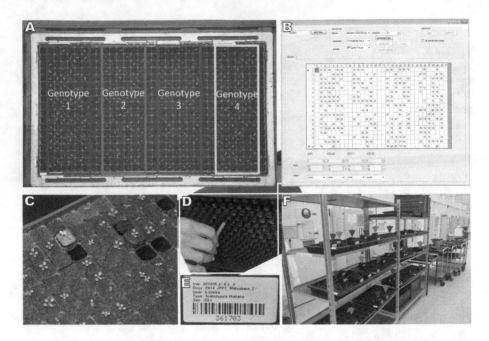

Fig. 4. Transplantation and Randomization. A: Tray with four genotypes. B: Graphical user interface for germination. C: Next seedling to pick. D: Pricking out from below the table without moving the tray. E: Label with destination. F: 22 target trays for 880 plants with randomized destinations, two genotypes symbolically indicated by arrows.

be calculated. For higher accuracy more markers could be used, however the described procedure enables sufficiently accurate highlighting of single pots.

Visual Inspection. For Visual inspection a tray is set into the cutout and identified by its ID, triggering highlighting of all seedlings with two or more leaves (see Fig. 3B). Germination detection can be toggled by mouse click on the respective pot. The toggling event is stored in the database. In case a non-detected seedling is now marked as being in two-leaves stage the time point when the mouse click was performed is used as detection date. Overviews and statistics of the germination dates and rates can be printed and used for further experiment design (Fig. 3C).

2.4   Transplantation Guidance

Seedlings are pricked out and transplanted into bigger pots for subsequent experiments. Which seedlings to pick is decided by selection rules, e.g. medium sized or the largest seedlings, germinated at a certain day, with the highest growth rate, or random selection. This improves repeatability of experiments compared to visual selection of seedlings. Thus, in a first step for transplantation, a graphical user interface allows selection of trays to calculate statistics like seedling size distributions (see Fig. 4B) and to define selection rules.

We use a randomized plant order on trays to mitigate border and microclimate effects. To enable randomization, the desired number of seedlings per genotype need to be provided together with the layout of destination trays. The system automatically randomizes seedlings' target positions over all destination trays and generates labels for each destination tray, which are printed directly.

Transplantation is done source tray after source tray and plant after plant. For each plant (i) the system highlights the seedling (Fig. 4C) using the projector, (ii) the user pushes the pot content from below (Fig. 4D), and (iii) transplants the seedling into a bigger pot, prefilled with soil. Simultaneously the system generates a new plant object in the database, links it to the seed ID, and prints out the label for the pot in which the seedling is implanted. To simplify randomization, the printed plant label also contains the destination tray ID and position on that tray (Fig. 4E). The user sticks the label to the pot, waters the seedling and brings the pot to its destination position (Fig. 4F). The destination position printed on the label speeded up the randomization process considerably and reduced mistakes. A foot switch (Fig. 4A, yellow object on the floor) is used to step to the next seedling, avoiding dirt on mouse or keyboard, keeping the users hands free for transplantation.

Besides considerable speedup, improved reliability, and less tiring work due to user guidance, the system automatically joins results of seedling- and plant measurements in the database. This is enabled by unique plant and seed identities and appropriate position tracking of pots during the pricking out process.

2.5   Dataset

By using this system, we collected 19,486 images and initially labelled them with 'plant', when one or more healthy seedlings were visible having at least two fully developed and unfolded leaves, i.e. their cotyledons, or 'no plant' else. However, a finer-grained classification may be of interest, when fully automated systems need to assess germination of seedlings. We therefore relabelled the images by visual inspection in four 'no plant' subclasses and four 'plant' subclasses. The table in Fig. 6 shows the number of images available per class and subclass and Fig. 5 shows example images of the eight subclasses.

The negative 'no plant' subclasses are:

- class 0 'no plants': no plant material visible at all
- class 1 'plants with less than two leaves visible': there is a plant visible, but it clearly does not have two leaves fully unfolded yet; or larger parts of the plant are covered by soil.
- class 2 'multiple plants with less than two leaves visible': same as class 1, but multiple plants are visible.
- class 3 'plants with almost two leaves visible': This class contains the hard to decide cases, where seedlings are very close to have two fully developed leaves, but are just not yet there.

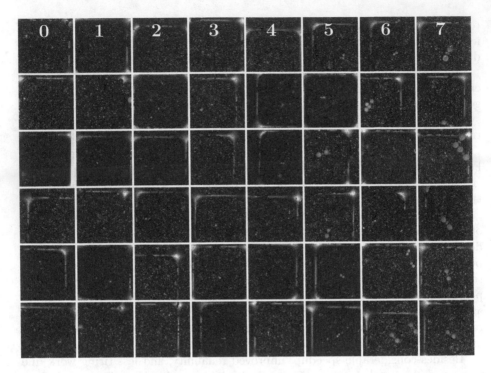

Fig. 5. Example images from each subclass, ordered from left to right. Each column shows examples from the same subclass.

The positive 'plant' subclasses are:

- class 4 'plants with just two leaves': This class contains the hard to decide cases, where seedlings have just unfolded their cotyledons. They may be tiny or by are fully developed.
- class 5 'plants with two leaves': clear cases of plants in two-leaves stage.
- class 6 'multiple plants with two leaves': like class 5, but multiple plants visible.
- class 7 'plants with four leaves': at least one of the visible seedlings has more than two leaves.

The cases in subclasses 3 and 4 are the borderline cases of the two main classes. They are hard to distinguish even for human raters, as the opening of cotyledons is a gradual process. However, the more accurate the transition between still closed and already opened cotyledons can be pinpointed, the higher the temporal resolution possible in an automated system. Currently, imaging takes place once a day and visual inspection only just before transplantation.

Labelling was done by five human raters. Per image assignments to the different subclasses are represented as normalized probabilities. Their values can be found in the 'ImageDescription' tag of each image, stored in tiff format. A

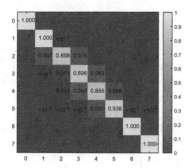

class	#images	subclass	#images
no plant	14862	0	13161
		1	1476
		2	24
		3	201
plant	4624	4	274
		5	2609
		6	287
		7	1454

Fig. 6. Confusion matrix for subclasses as labelled by human raters and number of images per subclass.

confusion matrix is shown in Fig. 6. For binary 'plant' (positive) or 'no plant' (negative) assignment we summed up probabilities of the four respective subclasses. An image then was assigned to the 'plant' or 'no plant' subclass with the highest probability. In doubt, the class with lower index was selected. Number of images per class are also given in Fig. 6.

True positive and true negative rates are 99.44% and 99.87%, respectively, and the overall accuracy for the whole dataset being above 99.77%. The target performance of an automated solution should therefore be similarly high.

In Fig. 5 we observe that cropping of captured images to images of single pots according to the tray layout does not always yield well centered pots. A more precise mechanical setup or image-based pot detection and cropping may help. Please note, that non-perfect pot alignment does not significantly influence the presented detection task and challenge.

2.6 Algorithm for Detection of Two-Leaves Stadium

Input RGB images are segmented into a binary foreground-background mask using thresholds in HSV-space [27] (cmp. Sect. 2.2). On the binary mask a Euclidean distance map is computed (using *openCV* [2]), containing distances between the object pixel and the nearest background pixel. Local maxima in the distance map are potential leaf center points x_i. A maximum value gives the radius r_i of a leaf candidate i, more specifically, the radius r_i of the biggest inscribed circle. Candidates i with a radius r_i smaller than a predefined, species-specific threshold (minimal leaf radius) are deleted. The threshold is needed to delete maxima detected on the petiole. Images with two or more detected leaves are labelled as 'germinated'. Leaf diameters $d_i = 2r_i$, center points x_i, detected projected leaf area (i.e number of foreground pixels) etc. are then assigned to the seed ID and written to the database.

3   Experiments

Experiments involve transfer learning [28] of five well established deep neural networks, namely VGG16 [24], VGG19 [24], DenseNet121 [11], and InceptionV3 [25], where we distinguish two different optimizer configurations for InceptionV3. The networks have been pre-trained for image classification on ImageNet [6], i.e. for classification of 1000 different classes of single dominant foreground objects. Seedling detection can be considered as classification of just two (or eight) other classes. The selected networks are known to perform well on ImageNet and their pretrained versions are easily available e.g. in Keras [4].

For all tested architectures, we keep the pretrained feature extraction layers fixed. The original classifier layers are removed and the simple classifier – dense layer 256, dropout, dense layer 1 – introduced in [23] is used instead. As optimizer we either use RMSProp (setup 'A', with VGG's and InceptionV3) or Adam with $\beta_1 = 0.9$ and $\beta_2 = 0.999$ (setup 'B', with DenseNet121 and InceptionV3). Best batch size, step size and number of epochs are derived by grid search. Training is done with class weighted cross entropy loss, as usual.

For the challenge, we provide a fixed split of the data into training, validation and test set in an $80/10/10$ split. In order to ensure that each of the sets represent the same distribution, we perform a 5-fold cross-validation within the grid search. To this end, we split the data into 5 sets $S_i, i \in \{1, ..., 5\}$ of approximately same size. For the ith validation run, we leave set S_i out as validation set and keep the other 4 sets for training of our baseline transfer learned neural nets. In addition, we split the 20% validation set into two subsets being validation and test set candidates for the final split. The grid search results are shown in supplemental Figures S1–S3. In Fig. 7 we show boxplots of the accuracies[1] and losses of our five different network configurations, trained using the best values from the grid search. Shown are results for training as well as 20% validation set and its split into 10% test and 10% validation subsets. We observe that for all tested networks all validation values lie close together, respectively, indicating that the different data splits are equivalent. We provide the split with least deviation from the average performance. With the resulting $80/10/10$ split, we transfer learned the neural nets, again, where we reinitialized the trainable parameters. Results are shown in Fig. 8 as 'confusion' wrt. the binary labels, i.e. per subclass accuracy. The nets achieve accuracies between 96.1% (VGG16) and 98.3% (DenseNet121). We observe, that wrong classification mostly happens in the hard to decide subclasses 3 and 4, but even for the clear cases in subclasses 0 and 6, 7, accuracies are mostly not 100%.

For comparison we also show the results of our classic detection algorithm from Sect. 2.6 and results of human raters. We see that results of the classic algorithm are excellent in terms of recall (99.8%) at the cost of a lower precision (71.7%). High recall is preferred by human raters manually cleaning the data using the handling stations. Overall accuracy is 90.6%. The deep networks

---

[1] For definitions of accuracy, precision and recall, please see Section B in the supplemental material.

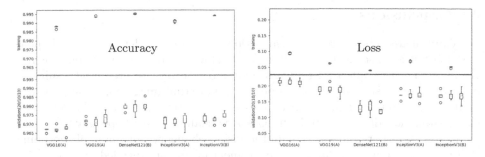

Fig. 7. Results of cross-validation experiment. Plots show results per net. Top part shows training results, bottom the results for the 20% validation set and its splits into two 10% sets. Left: Accuracy. Right: Loss

Per Class Accuracys

	0	1	2	3	4	5	6	7	total accuracy
VGG16(A)	0.9592	0.973	1.0	0.95	0.3333	0.8199	0.931	0.9726	0.9605
VGG19(A)	1.0	0.9787	1.0	0.9	0.4444	0.8621	0.931	0.9726	0.9682
DenseNet121(B)	0.9977	0.9662	1.0	0.75	0.7407	0.9617	1.0	0.9726	0.9826
InceptionV3(A)	0.9932	0.9595	1.0	0.8	0.7037	0.8812	0.931	0.9726	0.9672
InceptionV3(B)	0.9932	0.9595	1.0	0.75	0.7407	0.931	0.931	0.9361	0.9743
Human Raters	1.0	1.0	1.0	0.9075	0.9117	0.9898	1.0	1.0	0.9977
Algorithm	0.9407	0.4262	0.2083	0.1642	0.9745	1.0	1.0	1.0	0.9065

classes

Fig. 8. Confusion in the subclasses for the transfer learned nets using the 80/10/10 split. Human raters and the algorithm (see Sect. 2.6) are evaluated on the whole dataset.

perform better in terms of accuracy, and may gain from fine-tuning and more sophisticated classifiers, which we will test in future work. However, available solutions so far are not yet close to human raters with well above 99.7% accuracy. Even the best achieved training accuracy of 99.5% (DenseNet121) is not yet there. For fully automated germination detection there is still a considerable gap to fill, which we hope to achieve by challenge results.

4   Conclusions

Fully automated above ground germination detection, i.e. detection of full opening of cotyledons, seems to be well in reach. However, detection reliability is not yet high enough to allow for fully automated operation. We presented a suitable imaging-based system, where we hope and believe that image processing can still be improved. Hence we make our image data and labels freely available and pose the germination detection problem as a new challenge. We hope that this data complements other available computer vision challenges in plant phenotyping [1,5,9,15,16], allowing a further improvement of urgently needed methods [17].

In conjunction with the germination detection system, we presented a handling system supporting manual greenhouse work in plant phenotyping experiments. It not only facilitates visual inspection and transplantation of seedlings, making such work less tiring, less error-prone, but at the same time increases throughput even at a reduced human workload. We observed that transplantation can be done at 15s per plant over hours without slowing down when teams of two people cooperate. The system was designed along well established greenhouse procedures which increased usability and user acceptance. Especially the avoidance of data to be keyed in during transplantation was very well received.

The system allows for automated, measurement-based seedling selection, making this process quantifiable and less subjective. Heterogeneous data describing the plants status is automatically kept consistent with the database during the transplantation process. This enables reliable tracking of plants from seed to plant, an experimental option we want to explore in future plant phenotyping experiments.

Authors' Contributions

HS initiated the research leading to this manuscript and drafted the manuscript. HS, BB, AF, and JR have made substantial contributions to conception and design of the overall system described here and contributed to algorithmic solutions and interfaces. BB designed, implemented, and tested the database system. AF designed, implemented, and tested the germination detection system. JR and AF designed, implemented, and tested the pricking out station jointly. JvS and LS performed the deep learning studies. All authors contributed to writing the manuscript and proofreading. All authors read and approved the final manuscript.

Acknowledgements. Part of this work has been supported by the network for phenotyping science: CROP.SENSe.net, funded by German BMBF (0315531C). The authors thank Silvia Braun, Thorsten Brehm, and Birgit Bleise for testing the system in their practical work and giving feedback for improvements.

References

1. Bell, J., Dee, H.: Aberystwyth Leaf Evaluation Dataset (2016). https://doi.org/10.5281/zenodo.168158

2. Bradski, G.: The OpenCV Library. Dr. Dobb's J. Softw. Tools (2000)
3. Bruns, B., Scharr, H., Schmidt, F.: Entwicklung einer Multi-Plattform-Benutzerschicht zur tätigkeitsbegleitenden Verwaltung von Phänotypisierungsexperimenten und Pflanzenbestandsdaten. In Komplexität versus Bedienbarkeit Mensch-Maschine-Schnittstellen, Referate der 35. GIL-Jahrestagung, 23–24 February 2015, Geisenheim, Germany, pp. 1–4 (2015)
4. Chollet, F., et al.: Keras (2015). https://keras.io
5. Cruz, J.A., Yin, X., Liu, X., Imran, S.M., Morris, D.D., Kramer, D.M., Chen, J.: Multi-modality imagery database for plant phenotyping. Mach. Vis. Appl. **27**(5), 735–749 (2015). https://doi.org/10.1007/s00138-015-0734-6
6. Deng, J., Dong, W., Socher, R., Li, L.-J., Li, K., Fei-Fei, L.: ImageNet: a large-scale hierarchical image database. In: IEEE Conference on Computer Vision and Pattern Recognition (CVPR) (2009)
7. Fiorani, F., Schurr, U.: Future scenarios for plant phenotyping. Ann. Rev. Plant Biol. **64**, 267–291 (2013)
8. Granier, C., et al.: PHENOPSIS, an automated platform for reproducible phenotyping of plant responses to soil water deficit in Arabidopsis thaliana permitted the identification of an accession with low sensitivity to soil water deficit. New Phytologist **169**(3), 623–635 (2006)
9. Guo, W., et al.: Global WHEAT dataset (2020). http://www.global-wheat.com/
10. Houle, D., Govindaraju, D.R., Omholt, S.: Phenomics: the next challenge. Nat. Rev. Genet. **11**(12), 855–866 (2010)
11. Huang, G., Liu, Z., Van Der Maaten, L., Weinberger, K.Q.: Densely connected convolutional networks. In: 2017 IEEE Conference on Computer Vision and Pattern Recognition (CVPR), pp. 2261–2269 (2017)
12. Jahnke, S., et al.: Phenoseeder-a robot system for automated handling and phenotyping of individual seeds. Plant Physiol. **172**(3), 1358–1370 (2016)
13. Jansen, M., et al.: Simultaneous phenotyping of leaf growth and chlorophyll fluorescence via GROWSCREEN FLUORO allows detection of stress tolerance in Arabidopsis thaliana and other rosette plants. Func. Pant Biol. Spec. Issue: Plant Phenomics **36**(10/11), 902–914 (2009)
14. MacLeod, N., Benfield, M., Culverhouse, P.: Time to automate identification. Nature **467**(7312), 154–155 (2010)
15. Minervini, M., Fischbach, A., Scharr, H., Tsaftaris, S.A.: Plant phenotyping datasets (2015). http://www.plant-phenotyping.org/datasets
16. Minervini, M., Fischbach, A., Scharr, H., Tsaftaris, S.A.: Finely-grained annotated datasets for image-based plant phenotyping. Pattern Recogn. Lett. **81**, 80–89 (2016)
17. Minervini, M., Scharr, H., Tsaftaris, S.A.: Image analysis: the new bottleneck in plant phenotyping. IEEE Sig. Process. Mag. **32**(4), 126–131 (2015)
18. Nagel, K.A., et al.: GROWSCREEN-Rhizo is a novel phenotyping robot enabling simultaneous measurements of root and shoot growth for plants grown in soil-filled rhizotrons. Func. Plant Biol. **39**, 891–904 (2012)
19. Poland, J.A., Nelson, R.J.: In the eye of the beholder: the effect of rater variability and different rating scales on QTL mapping. Phytopathology **101**(2), 290–298 (2010)
20. Rousseau, D., Dee, H., Pridmore, T.: Imaging methods for phenotyping of plant traits. In: Kumar, J., Pratap, A., Kumar, S. (eds.) Phenomics in Crop Plants: Trends, Options and Limitations, pp. 61–74. Springer, New Delhi (2015). https://doi.org/10.1007/978-81-322-2226-2_5

21. Scharr, H., Bruns, B., Fischbach, A., Roussel, J., Scholtes, L., vom Stein, J.: Juelich dataset for germination detection of soil-grown plants (2020). https://doi.org/10.25622/FZJ/2020/1

22. Schmidt, F., Bruns, B., Bode, T., Scharr, H., Cremers, A.B.: A distributed information system for managing phenotyping mass data. In: Massendatenmanagement in der Agrar- und Ernährungswirtschaft, Erhebung - Verarbeitung - Nutzung, Referate der 33. GIL-Jahrestagung, 20–21 February 2013, Potsdam, Germany, pp. 303–306 (2013)

23. Shen, X., Zhang, J., Yan, C., Zhou, H.: An automatic diagnosis method of facial acne vulgaris based on convolutional neural network. Sci. Rep. 8(1), 5839 (2018)

24. Simonyan, K., Zisserman, A.: Very deep convolutional networks for large-scale image recognition. In: International Conference on Learning Representations (2015)

25. Szegedy, C., Vanhoucke, V., Ioffe, S., Shlens, J., Wojna, Z.: Rethinking the inception architecture for computer vision. In: 2016 IEEE Conference on Computer Vision and Pattern Recognition (CVPR), pp. 2818–2826 (2016)

26. Tardieu, F., Schurr, U.: White paper on plant phenotyping. In: EPSO Workshop on Plant Phenotyping, Jülich, November 2009. http://www.plantphenomics.com/phenotyping2009

27. Walter, A., et al.: Dynamics of seedling growth acclimation towards altered light conditions can be quantified via GROWSCREEN: a setup and procedure designed for rapid optical phenotyping of different plant species. New Phytologist 174(2), 447–455 (2007)

28. Yosinski, J., Clune, J., Bengio, Y., Lipson, H.: How transferable are features in deep neural networks? In: Ghahramani, Z., Welling, M., Cortes, C., Lawrence, N.D., Weinberger, K.Q. (eds.) Advances in Neural Information Processing Systems 27, pp. 3320–3328. Curran Associates Inc. (2014)

Detection in Agricultural Contexts: Are We Close to Human Level?

Omer Wosner[1]([⊠]), Guy Farjon[1], Faina Khoroshevsky[1], Lena Karol[2], Oshry Markovich[3], Daniel A. Koster[2], and Aharon Bar-Hillel[1]

[1] Ben Gurion University of the Negev, Be'er Sheva, Israel
{wosnero,guyfar,brodezki}@post.bgu.ac.il,
barhille@bgu.ac.il
[2] Hazera Seeds Ltd., Berurim M.P Shikmim, Be'er Sheva 7983700, Israel
{Lena.Karol,daniel.koster}@hazera.com
[3] Rahan Meristem (1998) Ltd., Kibutz Rosh-Hanikra, 2282500 Western Galilee, Israel
oshrym@rahan.co.il

Abstract. We consider detection accuracy in agricultural contexts. Five challenging datasets were collected and benchmarked, with three recent networks tested. Based on an initial analysis showing the importance of image resolution, models were trained and tested with a multiple-resolution procedure. Detection results were compared to human performance, judged based on the consistency of multiple annotators. A quantitative analysis was made highlighting the role of object scale and occlusion as detection failure causes. Finally, novel detection accuracy metrics were suggested based on the needs of agriculture tasks, and used in detector performance evaluation.

Keywords: Detection · Precision agriculture · Human performance

1 Introduction

Object detection in the agriculture environment is important for a variety of agricultural tasks and applications such as robotic manipulation, counting, and fine phenotyping. Robotic manipulation tasks as fruit [22] and vegetable [27] harvesting were recognized as an important task to automate more than 50 years ago [23]. Other robotic tasks requiring a detection module include plant spraying [3] and detection and handling of pests and diseases [9]. Counting tasks are common for the purpose of yield estimation [18,28], or blooming intensity estimation [8], and at least in some approaches require explicit object detection. Fine phenotyping tasks involve examining an object's traits and features to evaluate a plant's growth, resistance, physiology condition, or any other observable parameter [5]. For example, in [1,26] various length or height parameters of

Electronic supplementary material The online version of this chapter (https://doi.org/10.1007/978-3-030-65414-6_26) contains supplementary material, which is available to authorized users.

© Springer Nature Switzerland AG 2020
A. Bartoli and A. Fusiello (Eds.): ECCV 2020 Workshops, LNCS 12540, pp. 375–390, 2020.
https://doi.org/10.1007/978-3-030-65414-6_26

plant parts were estimated. A successful object detector is crucial for achieving practical performance in each of the above tasks.

Detecting objects in field or orchard conditions is not an easy task. In 2001 it was recognized by Li et al. [14] that improvements in detection and localization of objects are the main obstacles preventing harvesting robots from reaching human capabilities. In recent years, Convolutional Neural Networks (CNNs) based detectors dramatically improved, bridging some of the gap between human and machine performance. CNNs based detectors can be divided into two natural groups - single stage and two stage detectors. Single stage detectors, such as YOLO [19], RetinaNet [16], and EfficientDet [25], consider hundreds of thousands of possible object locations in the image, and classify them in a single unified network. Two stage detectors, such as Faster-RCNN [20] or Mask R-CNN [11], start by generating a smaller set of object candidates (a few hundreds or thousands), then classify and refine them in a second network.

Detection in the field context is different in some characteristics from traditional detection benchmarks [6,17] and in some respects more challenging. First, field images may contain dozens of objects, with high scale variance. Naturally both near and far objects are captured, and objects in many octaves often exist in a single image. Second, in many cases, such as apple flowers or tomatoes, the objects of interest grow in clusters. Hence occlusion is very common, with many objects suffering from high occlusion degrees. Third, the objects of interest often have a challenging shape with similarity to background structures. Tomatoes and avocados for example, are simple and round without discriminative details, and can often be confused with round leafs in the foliage. Cucumbers are green and stick-like, with high similarity to some branches and stalks. Tomato whole plants, on the other hand, are non-convex and skeletal. Finally, there are challenges introduced by the outdoors illumination conditions, including coping with severe cast shadows and required invariance to capturing hour. Some of these difficulties can be observed in Fig. 1.

With the rapid advance of detection networks, several questions arise with respect to the agricultural context: What are the better network architectures and training procedures for the agricultural domains? With the best models, is fruit and plant detection now approaching human level? If there is still a gap, can we characterize the main reasons for detection errors and quantify them? finally, assuming human level has not been reached yet: is the accuracy of current detectors satisfactory for practical applications? Which measurements may help us in answering this question? In this paper we try to make some progress regarding these issues.

In order to characterize detection performance we collected images from five different agricultural contexts, in which the tasks are detection of tomatoes, cucumbers, avocados, banana bunches, and whole tomato plants. These represent a diverse set of challenges related to non-discriminative shape (cucumbers), non-convex shape (whole plants), natural clustering and occlusion (tomatoes). Datasets were annotated with strict annotation, aiming to include all objects visible by a human (including small and highly occluded ones). To obtain good

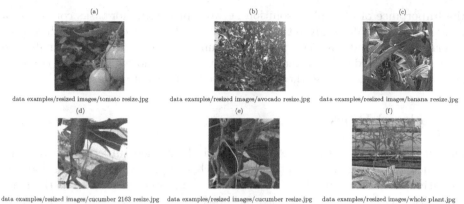

(a)   (b)   (c)

data examples/resized images/tomato resize.jpg   data examples/resized images/avocado resize.jpg   data examples/resized images/banana resize.jpg

(d)   (e)   (f)

data examples/resized images/cucumber 2163 resize.jpg   data examples/resized images/cucumber resize.jpg   data examples/resized images/whole plant.jpg

Fig. 1. Challenges for visual detection in the agricultural context. (a) severe occlusion and scale variations (b) dozens of avocado objects in a single image (c) severe occlusion (d) scale variation (e) poor illumination (f) challenging discrimination. Image (b) shows a full image, and the others are sub-images showing the difficulties

detectors, we have experimented with two main dimensions. First, we tested three leading network architectures: the two staged Mask R-CNN [11], the single-stage RetinaNet [16], and the recent EfficientDet [25]. More importantly, we conducted an analysis showing that accuracy is highly affected by object scales and processing resolutions. In light of that realization we have experimented with training and inference procedures involving multiple image resolutions.

The best detectors obtained were analysed in two informative respects: first, accuracy was compared to estimated human accuracy (on 4 of the 5 datasets where the required annotation existed). To estimate human accuracy, annotations of the same dataset made by 2 or 3 different annotators were used. A second analysis was pursued to quantify the role of object scale and occlusion in detection difficulty. To this end, additional occlusion annotation was added to the most challenging dataset in this respect, the tomato dataset. Detection accuracy was then measured for subsets of objects characterized by their size and occlusion level, and compared to the corresponding human performance.

The suitability of a certain detector to a certain application is clearly an application-dependent question, which cannot be answered here. However, we claim that in order to answer such questions, detector accuracy should be characterized with a richer set of measurements than the commonly used Average Precision (AP) [7] or F_1 statistic [4]. Phenotyping, for example, usually does not have to detect all the objects [26], but measurements should not be made on non-objects. Hence recall at high precision levels is the most relevant. We suggest a set of accuracy measurements tailored to agriculture applications and measure the best detectors trained.

Our contribution in this work is four-fold. We report a benchmark of several recent detector architectures on a diverse dataset of five agricultural detection tasks. We analyze the effect of object scale on detection accuracy and show the importance of multiple resolution network processing for tasks containing a wide scale range. We analyze the performance of the best detectors with respect to error source and with comparison to human level performance. Finally, we suggest a set of detection accuracy measurements more tailored to common agricultural tasks, and use these to measure the best detectors trained.

2   Related Work

Object detection in agriculture is an extensively studied subject, with the outdoor field environment presenting unique challenges. In [10] variable lighting condition, occlusions, and fruit or flower clustering were mentioned as the main difficulties. Other works [2,8] have acknowledged and faced the challenges of having many objects with small scale. Most early published work was based on explicit formation of color, texture or geometric features enabling detection of the target objects. The review [10] published in 2015, provides a good overview of these techniques. With the advance of CNN models in recent years, they were found a good fit to cope with the challenges, and avoid the manual feature construction. Given enough data, deep networks learn a good representation including discriminating features, which enable detection of target objects with accuracy superior to previous methods. A review of deep learning techniques in agriculture, including some examples of successful detection applications is presented in [13].

While there are numerous studies that use a deep learning based detector in agricultural tasks, we focus here on the benchmarks [2,21,29], which are the most relevant to our work. Sa and his collaborators [21] use Faster R-CNN [20] to detect sweet peppers and rock-melons by combining RGB and Near-IR information. Specifically, the two modalities were combined by adding the NIR map to the network input, and this addition is shown to contribute to accuracy (rising the F_1 score from 0.813 to 0.838). The work shows the generality of the approach by considering 7 different fruit kinds, and the F_1 results obtained are very good. However, the datasets used are significantly easier than in our work. Images are mostly taken in plantation conditions, with small distance between the camera and the relevant fruits. Hence typically an image includes less than 5 fruits (rarely more than 10), and these are usually big and clearly visible. In contrast, the images used in our experiments often contain many dozens of objects, and with significant scale variation including many small and far objects. In [29] a large dataset, containing $49,000$ annotated objects from 31 classes was collected and benchmarked for detection and classification with deep networks. However, this data is even more extreme than the datasets of [21], with most images containing a single large object of interest.

Bargoti and Underwood [2] used Faster R-CNN to detect mangoes, apples, and almonds. Their dataset is similar to ours with respect to the image size and the number of objects in each frame. To resolve scale and number of objects issues they also use a tiling system - image was divided into tiles of 500×500 pixels with 50 pixels overlap. While this work is the most similar to ours, our benchmark

Table 1. Data set sizes and partitions shown as (# images, #objects), image sizes, and object size statistics. Object size is in pixels, defined by $log_2(\sqrt{width \times height})$. The table shows mean object size and $std(log_2(size))$ in parenthesis

Crop	Train set	Validation	Test set	Image size	object size
Banana	(133, 642)	(28, 128)	(28, 134)	3024 × 4032	8.36(1.24)
Cucumber	(21, 457)	(3, 75)	(4, 118)	6000 × 4000	7.76(1.07)
Avocado	(17, 613)	(2, 110)	(5, 143)	3024 × 4032	6.87(0.59)
Tomato	(22, 572)	(3, 107)	(6, 173)	5184 × 3456	7.54(0.94)
Tomato whole plant	(30, 223)	(7, 82)	(10, 198)	6000 × 4000	9.62(0.89)

work is wider in scope. Specifically it includes comparison of several (newer) networks, a comparison to human performance, detailed analysis of the main error causes: occlusion and scale, and consideration of performance measurement beyond the general F_1 or AP statistics.

3 Method

The datasets used in this work are briefly presented in Sect. 3.1, followed by a short description of the networks used in Sect. 3.2. In Sect. 3.3 image resolution and detection with multiple scales are described. Section 3.4 describes human performance estimation and Sect. 3.5 discusses agriculture-related performance measurements.

3.1 Datasets

Datasets were collected and annotated for five different crops: banana bunches, cucumbers, avocados, tomato fruits, and tomato plants. Images sizes differ between datsets in the range of 12–24 mega pixel. The number of objects per frame varies between 4 up to 72 objects. For detailed information on the number of objects and image sizes see Table 1. As seen in Fig. 1, the images include large scale and occlusion variation, with some of the dataset (tomato, cucumber, avocado), dominated by the large amount of far and small objects. The challenge in shape and color varies between dataset: cucumbers are often small and hard to differentiate from branches. Tomatoes change color before harvesting, but the majority of the tomatoes in our data are unripe and therefore green and blending in with the green background of the foliage. Tomato whole plants have an irregular non-convex shape making it harder to demarcate one from the other.

3.2 Detection Models

We tested three state-of-the-art detection algorithms: Mask R-CNN [11], RetinaNet [16], and EfficientDet [25]. Each model was tested on binary problem for each dataset separately. While there are significant differences, all these networks share a common general structure. First, a pre-trained classification network, termed the 'backbone' network, is applied in a fully-convolutional manner to produce a dense representation for the entire image. At a second stage a variant of Feature Pyramid Network (FPN) [15] is applied. It creates tensors of similar representation but different resolutions, representing the image in multiple octaves to enable multiple scale detection. The model then tests for object existence in a pre-defined set of (position, scale) candidate rectangles termed 'anchors', which typically contains hundreds-of-thousands of candidates. The candidates, or a filtered subset of them, are then passed to processing by several parallel 'head' modules. A classification head is trained to classify candidates among non-objects and classes of interest. A second 'bounding-box refining' head is trained to refine the proposed rectangle, in case it contains an object, to a tighter rectangle with better fit to the object extent.

Mask R-CNN [11], has evolved as an improved variant of Faster R-CNN [20] with optional object segmentation capabilities. This is a two-stage model: the first stage, termed a Region Proposal Network (RPN) [20], filters from the possible anchors a few hundreds/thousands for further processing. It uses a ResNet-50 [12] backbone network to produce the initial representation, and the FPN, to create the multi-scale pyramid representation. An object/non-object initial classification is made for each anchor for the filtering. While positive object proposal are carefully chosen, negative 'no object' candidates (required for classification training at the second stage), are chosen heuristically to balance the number of positives. The object candidate regions are sampled from the representation tensors using a sampling layer (RoI-Align) and sent to the second stage, which includes the classification and bounding-box refining heads. There is no gradient flow between the stages in training, and they are essentially trained separately.

RetinaNet [16] is similar to Mask R-CNN in its usage of ResNet-50 network as backbone, and the FPN [15] for multiple scale representation. However, unlike Mask R-CNN, this is a single stage network trained end-to-end. Instead of filtering object candidates, all the hundreds-of-thousands anchors are considered as candidates and go through the object classification and bounding box regression heads. While enabling end-to-end training, this creates a problem of class imbalance, as classification is trained with hundreds of thousands of negative examples (non object candidates) versus a few positive examples in each image. The problem is addressed using a modification to the standard cross entropy loss termed 'Focal Loss', in which 'easy examples', including most of the negatives, are down-weighted in training. This mechanism channels the network learning effort efficiently to the hard examples, both positive and negative.

EfficientDet [25] is a one-stage network similar to the RetinaNet, and like it uses the focal-loss in training. However, it includes several improvements, and was recently (2019) reported to achieve state-of-the-art results on the MS-COCO detection challenge. The backbone used in this model is the B4-EfficientNet [24], reported to have higher ImageNet accuracy than ResNet-50 while using only one fifth of the parameters and running 10× faster. A second module in which significant changes were made is the FPN. EfficientDet uses a modified version termed

Bi-FPN, which includes top-down connections between consecutive resolutions, and a weighting mechanism for fusion of information in these connections.

3.3   Image Resolution

The datasets' images are typically very large (see Table 1). The networks cannot accept such resolutions due to GPU memory limitations, and are limited to 1024×1024 input size. In a simple treatment, each image is resized such that its larger dimension is resized to 1024 pixels, keeping the aspect ratio, and padded with zeros in the shorter dimension. This down-scaling clearly diminish the object's size by a significant factor. For example in the cucumber dataset the scale factor is $max(\frac{6000}{1024}, \frac{4000}{1024}) = 5.85$, so each object's size is $5.85X$ smaller and their areas is $34X$ smaller than in the raw data.

We overcame this issue by working with images in two resolutions. Instead of using only the down-sampled original image, a set of 1024×1024 sub-images covering the original image were cropped with a fixed overlap. Both the down sampled original image and the cropped sub-images are used in network training and inference. As will be discussed below, both resolutions are required, and this two-resolution policy provides the detectors more opportunities to detect an object either in the sub-images or in the resized original image. The detected bounding boxes set of both resolutions are unified before Non Maxima Supression (NMS). The overlap parameter was chosen using initial empirical tests with the tomato datset and was set to 581 pixels. Note that with such overlap, which is close to half the sub-image size, each object is typically seen in 4 sub-images and one time in the single full-image, so the system has 5 opportunities to detect it.

An analysis of the effect of object scale on detection performance is presented in Fig. 2. The detection bounding boxes of the Mask R-CNN model for the tomato dataset were used to compute several statistics of interest. It turns out object scale has a profound impact on detection performance. Larger objects are more likely to be detected, have higher IoU (Intersection over Union) with the correspond ground truth rectangles, and higher confidence scores. Specifically detection probability arises linearly with object scale (measured logarithmically) in a significant scale range. A particular statistic of interest in our system is the 'second chance' probability: the probability to find an object in its larger scale (the sub-images) given that its detection has failed in the down-scaled image. Surprisingly, this probability is high not only for the small objects, but more for medium size objects, where it gets to values in $[0.4 - 0.5]$.

Since there are many sub-images and only few full downscaled images, the former dominate the dataset statistics. This sometimes create a problem in training, since large objects usually appear in the sub-images as partial objects. A very big object is typically seen in the data once as a full object in the full image, and many times as a partial object in sub-images. Upon training, this creates a tendency of the models to detect large objects with multiple bounding box corresponding to their parts, which is detrimental to performance. To avoid this tendency, we use two means. First, an object is annotated in a sub-image only

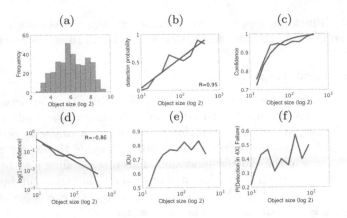

Fig. 2. The relation between object scale and detection quality. The graphs show empirical analysis conducted on detected bounding boxes of a tomato Mask R-CNN model. Object scale is measured as $\sqrt{height \cdot weight}$ in logarithmic scale with base 2. (a) Detection scale histogram. (b) Detection probability as a function of object scale. A close-to-linear relation holds for a significant scale range. The linear approximation is shown in red. (c) network detections' confidence as a function of scale. The red line shows the model based on (d). (d) $log(1 - confidence)$ as a function of scale. The red line shows a linear model fitted. (e) IoU with object rectangle as a function of scale. It rises, then saturates for large objects. (f) The probability for an object to be detected in the sub-images given that it was not detected in the original down-scaled image (Color figure online)

if at least 60% of it is visible. Second, the full downscaled images are assigned a higher weight (8× higher) in training, to enhance the importance of detecting whole objects.

3.4   Analysis and Human Performance Estimation

In a difficult detection task as presented here humans do not usually reach perfect performance. Specifically severe occlusion cases and small far objects can be easily missed, and foliage's texture creates false alarms. In addition, annotators are different in their skills and capabilities. As a fact, different annotators produce very different annotations when annotating the same dataset, as can be seen in Table 2. While we do not know which annotator is better, human performance can be measured by checking the degree of agreement between annotators. Specifically, we can define the task as predicting the annotations of a specific human, and compare a network to other humans in this task.

A comparison between algorithms and humans is hence made by temporarily setting one annotator as the Ground Truth (GT) annotator. The other human annotators are considered 'detectors', and are measured just like a detection algorithm would. However, human annotators do not provide confidence score for their annotations, so a recall-precision curve cannot be plotted for them and an AP score cannot be computed. Instead they provide a single (recall, precision) working point. While AP cannot be computed, an F_1 score for the (recall, precision) point can be computed and compared to the best F_1 score obtainable by a competing algorithm. The competing algorithms are trained on mixed annotation by sampling randomly a single annotator per image at training. All the non-GT annotators and the algorithms are evaluated and compared on the same test set. Since no human annotator is a-priory preferable to the others, the process is repeated for every human annotator at the GT annotator role.

Table 2. The number of annotations for each annotator and dataset. Tomato and cucumber datasets were annotated by three annotators, banana and avocado by two

Data-set	First annotator	Second annotator	Third annotator
Banana	118	139	—
Cucumber	119	108	127
Avocado	143	141	—
Tomato	175	175	174

A different analysis direction is to evaluate detector performance on object subsets of interest. Specifically, we suggest to compute the recall precision curve in six sub-categories of objects: all, small, big, occluded, non-occluded, big and non-occluded. Such category breakdown can helps a lot in understanding the model's strengths and weaknesses. Moreover, it enables understanding of the model capability to perform certain applications. For example, detector usage as the first stage for phenotyping only required success for fully visible and big objects where the phenotype can be measured. Such sub-category analysis required additional annotation, and it was performed for a single dataset - the tomato set. For size, a threshold between small and big objects was chosen at $100.5Kpixels$ based on manual inspection, keeping a portion of 24.55% as big objects. Object occlusion was determined by manual annotation.

The recall precision curve for such a subset of interest cannot be measure with standard recall and precision definitions. The reason is that the classifier of interest was trained to detect all objects, not just the subset. If the set of ground truth object is trimmed to a subset, for example of small objects only, than detector hits on the complementary set (big) are considered false alarms, leading to low and irrelevant precision rates. To avoid this, one has to keep using the full object set in the false alarm definition. Formally, denote the full set of ground truth rectangles by GT, and the subset of interest by S. A detection rectangle D is now defined to be True Positive (TP) or False Negative (FN) by

$$D \in TP \ \text{ iff } \ \exists R \in S \ \text{ s.t } \ IoU(R, D) > 0.5 \tag{1}$$
$$D \in FP \ \text{ iff } \ \sim \exists R \in GT \ \text{ s.t } \ IoU(R, D) > 0.5$$

The suggested definition enables measuring recall precision curves for subsets of interest, but it is not well suited when a single (True Positive Rate (TPR), False Positive Rate (FPR)) working point exist, as is the case with humans. Since the FPR is fixed across subsets, small subsets (hence with low TPR) obtain low precision $P = TPR/(TPR + FPR)$ and hence low F_1 scores. For comparison with human on subsets, we hence look at recall rates (of the algorithm and the human) at the same FPR, determined by the human working point.

3.5   Performance Estimation for Agriculture Tasks

Detectors are commonly evaluated using the Average Precision (AP) score, which measured the area under the recall-precision curve and provides a robust and threshold-independent performance measure. However, for the task families common in agricultural applications this measure is too general, and more specific additional measurements can be more informative.

Robotic applications typically require high localization accuracy, to enable robotic interaction with the plant. Localization quality is not measured at all in the standard AP measure. For a single successful detection, localization accuracy can be measured by considering the center pixel deviation, i.e the distance between centers of detection and GT rectangles. This deviation is in pixel units, and can be divided by the GT object scale to get the relative deviation, which is a unit-less, more intuitive fraction. The relative deviation can be averaged over all successful detections and provide the mean relative deviation.

Another requirement in some robotic applications is finding all the objects, i.e. high recall, in order to perform the task for all of them (like harvesting or overcoming noxious entities). While high recall have the cost of low precision, most false alarms can be corrected by moving the robot closer to the object. Practically, we can measure the recall at 0.1 precision as an estimate of this 'total recall', measuring the fraction of relevant objects found by the detector.

In counting applications, the detector is typically applied with a certain threshold and its output rectangles are counted. To measure performance, count deviation from the true count is measured, and divided by the true count to get the relative count error - the deviation as a fraction of the true count. The natural threshold to use is the one without bias: the one for which the expected number of false positives (non objects identified as objects) is equal to the amount of false negatives (objects not identified). In that case, the counting error expectation is zero. Hence we propose to use the average relative count error, of the count estimates at the non biased threshold.

In Detection-based phenotyping, detectors are used as a first stage to enable phenotype measurements (e.g. [1,26]). The detector finds the objects, then another model measures the desired feature. Typically the breeder is interested in statistics of the feature across a field or plot, like average and std of cucumber lengths [26] or spikelet count in a wheat spike. For estimation of such statistics, the detector does not have to consider all objects: a small sample of 'measurable objects' is enough. However, false positives are harmful, as each FP detection produces a 'noise' measurement contaminating the statistics. Therefore an appropriate detector measure will be the recall at 0.99 or 0.9 precision, where a minimal number of FP occur.

Table 3. Average Precision (AP) on the test set for each crop and network model

Data set	Mask R-CNN	RetinaNet	EfficientDet
Banana	0.741	0.604	0.455
Cucumber	0.507	0.516	0.453
Avocado	0.801	0.774	0.714
Tomato	0.580	0.646	0.522
Tomato whole plant	0.718	0.703	0.443

Fig. 3. Detection results examples: Red bounding-boxes are the detection results and blue bounding-boxes represents the ground truth annotations. Additional examples can be found in the supplementary material (Color figure online)

4 Experimental Results

We start by comparing the results of the three tested models on the five datasets. The best models are compared to human performance using F_1 scores. We continue with analysis of the obtained AP using the break-down of performance into object sub-categories. Finally, the additional agriculture-related performance measurements are reported and discussed.

Networks Results: Table 3 shows the results of the three tested networks, trained with the dual-resolution approach. As can be seen, the comparison didn't yield a superior model, but Mask R-CNN and RetinaNet performed better than EfficientDet over all crops. Mask R-CNN works better on the 'easier' (as indicated by the obtained accuracy) datasets avocado, banana, and tomato whole plant, and RetinaNet performed better for the more difficult cucumber and tomato datasets. The results show similarity between the two leading models indicate that accuracy is primarily a function of dataset difficulty, not of chosen network. Surprisingly, scale variance, and not only mean scale, is a prominent source of difficulty. The avocado data, despite having the smallest mean object size was successfully handled, probably because is has low object size variance (see Table 1) and a rather fixed view point across images. Banana and tomato whole plant, which are larger and do not suffer from high occlusion rates are of medium difficulty. The most difficult are tomato and cucumber, which are small, have high scale variance, and severe occlusion problems.

Comparison to Human Performance: Table 4 shows comparisons between the best trained models and a human detector, in terms of obtained F_1 scores. For the networks, the (recall, precision) working point with the highest F_1 score was chosen for comparison. The results show that for most tasks, a significant gap still exist between human and network detection. For Banana bunches, the network practically achieves human level detection: its agreement with human annotators is similar to the agreement between themselves. For Avocado, the network achieves high detection rates, yet humans are approximately 10% better. For the difficult tomato and cucumber data, significant gaps of 30–40% exist. The reasons for this bug gaps and analyzed next.

Table 4. F_1 scores obtained for (data set, annotator) tasks, with the annotator index defining the ground truth. The best model chosen based on the AP scores from Table 3 is compared to the human annotators. The model type (Mask-R-CNN or RetinaNet) is indicated by (R) or (M) in parenthesis. Duplicate figures for human F_1 score are due to the symmetry of using one annotator to estimate the other and vice-versa

Crop, annotator	Model F_1	Human F_1	Human F_1
Banana, 1st	0.826 (M)	0.794	—
Banana, 2nd	0.748 (M)	0.794	—
Cucumber, 1st	0.589 (R)	0.789	0.802
Cucumber, 2nd	0.543 (R)	0.789	0.791
Cucumber, 3rd	0.548 (R)	0.802	0.791
Avocado, 1st	0.801 (M)	0.894	—
Avocado, 2nd	0.824 (M)	0.894	—
Tomato, 1st	0.64 (R)	0.903	0.923
Tomato, 2nd	0.648 (R)	0.903	0.905
Tomato, 3rd	0.636 (R)	0.923	0.905

Error Cause Analysis: Recall-Precision graphs for object subsets of the tomato dataset are shown in Fig. 4. Graphs are plotted for 6 subsets based on occlusion (occluded/non occluded) and scale (small/big) binary variables

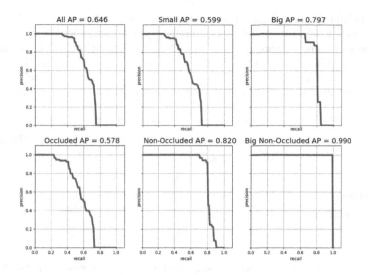

Fig. 4. Recall-Precision graphs of the RetinaNet tomato model on object subsets of interest. Average Precision (AP) scores are stated in the titles. Similar graphs for the other datasets, but only for big/small division, may be found in the supplementary material

defined in Sect. 3.4. The results clearly reveal scale and occlusion as the dominant causes of detection errors, and quantify their impact. Specifically, Moving from big to small objects causes 25% degradation in accuracy (from 0.797 AP to 0.6), and introducing occlusion degrades accuracy by 30% (from 0.82 to 0.578). The detection accuracy is practically perfect for big non-occluded objects, where these two causes of difficulty are gone. The latter result encourages the usage of detectors to extract objects for secondary phenotyping measurements, where only measurable un-occluded objects are of interest.

In Table 5 accuracy of networks and humans is compared according to scale and occlusion sub-categories. Occlusion and scale are the error causes of both human and the network, but the degradation form is different and more severe for the network. It can be seen that humans keep close to perfect performance as long as the object is either big or non-occluded, so their errors arise when objects are both small and occluded. The network significantly degrades and looses half of the recall rate due to size and occlusion independently.

Agriculture-Related Performance Measurements: The performance indices from Sect. 3.5 were measured for the best models. The results, presented in Table 6, give rise to several observations. For counting, relative deviation depends not only on detector accuracy, but also on the typical number of objects per image. With more objects per image, relative count deviation gets smaller due to the law of large numbers. Hence better accuracy is obtained mainly for datasets with high number of objects per image (see Table 1), like cucumber and avocado. For detection-based phenotyping, were a sample is required with minimal number

Table 5. Recall rates at specific False-Positive Rates (FPR) for network and humans on tomato dataset. Each row considers a different object subset. Columns present results obtained when different annotators are providing the ground truth. The average FPR of the Non-GT annotators is stated in the columns title in parentheses. Recall rates of the RetineNet model at the same FPR are reported in the 'Network' columns. Rates reported in 'Human' columns are the average recall of the two non-GT annotators

Category	Annotator 1 (16)		Annotator 2 (17)		Annotator 3 (15)	
	Network	Human	Network	Human	Network	Human
Small	0.46	0.90	0.48	0.89	0.43	0.90
Big	0.79	0.98	0.77	0.98	0.85	1
Occluded	0.44	0.89	0.47	0.90	0.41	0.90
Non occluded	0.79	0.98	0.79	0.98	0.79	1
Big non-occluded	1	1	1	1	0.89	1

Table 6. Performance measurements suggested in Sect. 3.5, measured for each crop by the best model

Measurement	Banana	Cucumber	Avocado	Tomato	Tomato plant
Count deviation	0.452	0.117	0.152	0.328	0.354
Recall@0.99	0.430	0.22	0.13	0.34	0.11
Recall@0.9	0.680	0.31	0.68	0.51	0.52
Recall@0.1	0.780	0.62	0.89	0.75	0.85
Localization dev.	0.341	0.221	0.133	0.139	0.184

of false alarms, the results indicate significant maturity of current detectors. With 1% of false positives, the detectors can sample $13 - 43\%$ of the objects, and if a noise of 10% false alarms can be tolerated most detectors can retrieve more than half of the objects. For localization the results are encouraging, with relative deviation lower than 14% obtained for 3 of the 5 datasets. However, since robotic applications require accuracy mainly for near (hence big) objects, characterization of localization error as a function of scale is required.

5   Concluding Remarks

The benchmark reveals that current detection networks are able to achieve human level accuracy for banana bunch detection, and get close to this level for avocado. However in more difficult tasks significant gaps exist. The two dominant causes of error were identified to be small object scale and occlusion. Each of these variables causes performance degradation of 25–30%, and when these are removed detection is nearly perfect. The results were obtained with relatively small samples and may clearly improve with data size, but they nevertheless suggest clear directions for focusing work on detectors improvement.

The task-related performance measurement show high potential for counting and detection-based phenotyping. For counting, 14% error were obtained for several datasets, even without usage of further mechanisms common in counting networks. For detection-based phenotyping current detectors were shown to be mature enough, as they are able to provide representative samples with high precision, and detect almost flawlessly big and un-occluded objects.

Acknowledgement. This research was supported by the Phenomics consortium, the Israel innovation authority, and the Ministry of Science & Technology, Israel.

References

1. Baharav, T., Bariya, M., Zakhor, A.: In situ height and width estimation of sorghum plants from 2.5 d infrared images. Electron. Imag. **2017**(17), 122–135 (2017)
2. Bargoti, S., Underwood, J.: Deep fruit detection in orchards. In: 2017 IEEE International Conference on Robotics and Automation (ICRA), pp. 3626–3633. IEEE (2017)
3. Berenstein, R., Shahar, O.B., Shapiro, A., Edan, Y.: Grape clusters and foliage detection algorithms for autonomous selective vineyard sprayer. Intel. Serv. Robot. **3**(4), 233–243 (2010)
4. Chinchor, N., Sundheim, B.M.: Muc-5 evaluation metrics. In: Fifth Message Understanding Conference (MUC-5): Proceedings of a Conference, Baltimore, Maryland, 25–27 August 1993 (1993)
5. Costa, C., Schurr, U., Loreto, F., Menesatti, P., Carpentier, S.: Plant phenotyping research trends, a science mapping approach. Front. Plant Sci. **9**, 1933 (2019)
6. Everingham, M., Van Gool, L., Williams, C.K.I., Winn, J., Zisserman, A.: The pascal visual object classes (voc) challenge. Int. J. Comput. Vis. **88**(2), 303–338 (2010)
7. Everingham, M., Van Gool, L., Williams, C.K., Winn, J., Zisserman, A.: The pascal visual object classes (voc) challenge. Int. J. Comput. Vis. **88**(2), 303–338 (2010)
8. Farjon, G., Krikeb, O., Hillel, A.B., Alchanatis, V.: Detection and counting of flowers on apple trees for better chemical thinning decisions. Precis. Agric. **21**(3), 503–521 (2019). https://doi.org/10.1007/s11119-019-09679-1
9. Fuentes, A., Yoon, S., Kim, S.C., Park, D.S.: A robust deep-learning-based detector for real-time tomato plant diseases and pests recognition. Sensors **17**(9), 2022 (2017)
10. Gongal, A., Amatya, S., Karkee, M., Zhang, Q., Lewis, K.: Sensors and systems for fruit detection and localization: a review. Comput. Electron. Agric. **116**, 8–19 (2015)
11. He, K., Gkioxari, G., Dollár, P., Girshick, R.: Mask R-CNN. In: Proceedings of the IEEE International Conference on Computer Vision, pp. 2961–2969 (2017)
12. He, K., Zhang, X., Ren, S., Sun, J.: Deep residual learning for image recognition. In: Proceedings of the IEEE Conference on Computer Vision and Pattern Recognition, pp. 770–778 (2016)
13. Kamilaris, A., Prenafeta-Boldú, F.X.: Deep learning in agriculture: a survey. Comput. Electron. Agric. **147**, 70–90 (2018)
14. Li, P., Lee, S.H., Hsu, H.Y.: Review on fruit harvesting method for potential use of automatic fruit harvesting systems. Procedia Eng. **23**, 351–366 (2011)

15. Lin, T.Y., Dollár, P., Girshick, R., He, K., Hariharan, B., Belongie, S.: Feature pyramid networks for object detection. In: Proceedings of the IEEE Conference on Computer Vision and Pattern Recognition, pp. 2117–2125 (2017)
16. Lin, T.Y., Goyal, P., Girshick, R., He, K., Dollár, P.: Focal loss for dense object detection. In: Proceedings of the IEEE International Conference on Computer Vision, pp. 2980–2988 (2017)
17. Lin, T.-Y., et al.: Microsoft COCO: common objects in context. In: Fleet, D., Pajdla, T., Schiele, B., Tuytelaars, T. (eds.) ECCV 2014. LNCS, vol. 8693, pp. 740–755. Springer, Cham (2014). https://doi.org/10.1007/978-3-319-10602-1_48
18. Linker, R.: A procedure for estimating the number of green mature apples in night-time orchard images using light distribution and its application to yield estimation. Precis. Agric. **18**(1), 59–75 (2016). https://doi.org/10.1007/s11119-016-9467-4
19. Redmon, J., Divvala, S., Girshick, R., Farhadi, A.: You only look once: unified, real-time object detection. In: Proceedings of the IEEE Conference on Computer Vision and Pattern Recognition, pp. 779–788 (2016)
20. Ren, S., He, K., Girshick, R., Sun, J.: Faster R-CNN: towards real-time object detection with region proposal networks. In: Advances in Neural Information Processing Systems, pp. 91–99 (2015)
21. Sa, I., Ge, Z., Dayoub, F., Upcroft, B., Perez, T., McCool, C.: Deepfruits: a fruit detection system using deep neural networks. Sensors **16**(8), 1222 (2016)
22. Santos, T.T., de Souza, L.L., dos Santos, A.A., Avila, S.: Grape detection, segmentation, and tracking using deep neural networks and three-dimensional association. Comput. Electron. Agric. **170**, 105247 (2020)
23. Schertz, C., Brown, G.: Basic considerations in mechanizing citrus harvest. Trans. ASAE **11**(3), 343–0346 (1968)
24. Tan, M., Le, Q.V.: EfficientNet: rethinking model scaling for convolutional neural networks. arXiv preprint arXiv:1905.11946 (2019)
25. Tan, M., Pang, R., Le, Q.V.: EfficientDet: scalable and efficient object detection. arXiv preprint arXiv:1911.09070 (2019)
26. Vit, A., Shani, G., Bar-Hillel, A.: Length phenotyping with interest point detection. In: Proceedings of the IEEE Conference on Computer Vision and Pattern Recognition Workshops (2019)
27. Vitzrabin, E., Edan, Y.: Adaptive thresholding with fusion using a RGBD sensor for red sweet-pepper detection. Biosyst. Eng. **146**, 45–56 (2016)
28. Xiong, H., Cao, Z., Lu, H., Madec, S., Liu, L., Shen, C.: TasselNetv2: in-field counting of wheat spikes with context-augmented local regression networks. Plant Methods **15**(1), 150 (2019)
29. Zheng, Y.Y., Kong, J.L., Jin, X.B., Wang, X.Y., Su, T.L., Zuo, M.: CropDeep: the crop vision dataset for deep-learning-based classification and detection in precision agriculture. Sensors **19**(5), 1058 (2019)

AutoCount: Unsupervised Segmentation and Counting of Organs in Field Images

Jordan R. Ubbens(✉), Tewodros W. Ayalew, Steve Shirtliffe, Anique Josuttes, Curtis Pozniak, and Ian Stavness

University of Saskatchewan, Saskatoon, SK S7N 5A8, Canada
{jordan.ubbens,tewodros.ayalew,steve.shirtliffe,acj450,
curtis.pozniak,ian.stavness}@usask.ca

Abstract. Counting plant organs such as heads or tassels from outdoor imagery is a popular benchmark computer vision task in plant phenotyping, which has been previously investigated in the literature using state-of-the-art supervised deep learning techniques. However, the annotation of organs in field images is time-consuming and prone to errors. In this paper, we propose a fully unsupervised technique for counting dense objects such as plant organs. We use a convolutional network-based unsupervised segmentation method followed by two post-hoc optimization steps. The proposed technique is shown to provide competitive counting performance on a range of organ counting tasks in sorghum ($S.$ $bicolor$) and wheat ($T.$ $aestivum$) with no dataset-dependent tuning or modifications.

Keywords: Computer vision · Organ counting · Unsupervised segmentation

1 Introduction

Object counting is a common real-world use case for deep learning and computer vision. In plant science and agriculture in particular, the counting of plant organs such as heads, tassels, flowers, pods, or fruiting bodies in field environments is relevant, as the number of organs correlates with agronomically important factors such as grain yield. Organ counting is a labor-intensive manual task which is practical to replace with automated methods. There is a substantial body of existing literature describing the counting of organs via supervised learning methods [2,9–11,14]. Among the most common supervised techniques are counting by density estimation or local counts regression, object detection, and counting by segmentation. The majority of these supervised methods were originally designed for much more difficult multi-class object detection and object counting problems, such as the challenging COCO benchmark [8]. In contrast, plant organs have low intra-class variability, but are dense and overlapping. Supervised methods also require data annotation in the form of either point annotations, bounding boxes, or segmentation masks. Collecting such annotations is difficult

© Springer Nature Switzerland AG 2020
A. Bartoli and A. Fusiello (Eds.): ECCV 2020 Workshops, LNCS 12540, pp. 391–399, 2020.
https://doi.org/10.1007/978-3-030-65414-6_27

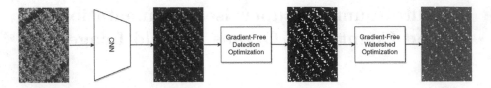

Fig. 1. The proposed process from image to instance segmentation.

and time-consuming, given the typically high density of organs in field images. In addition, poor quality labels can contribute to non-convergence and overfitting as the label noise overpowers the signal in the annotations. Label quality has been shown to be degraded when annotations are provided by domain non-experts, who may be unfamiliar with the plant structures in question [3]. Principally, these supervised methods need to be retrained on annotated data for each unique dataset, because of the issue of domain shift [15,17].

In this paper, we propose a novel unsupervised counting technique for counting plant organs in aerial images of fields. We show that a combination of unsupervised segmentation using a CNN, a gradient-free optimization of threshold parameters, and a search over watershed segmentation parameters is able to perform well in these tasks. A complete diagram of the process is shown in Fig. 1. We call this proposed object counting system *AutoCount*. We report unsupervised object counting results for two public datasets of sorghum heads (botanical term panicles), as well as a new dataset of wheat heads (botanical term spikes). These results represent new baselines for performing plant organ counting in a fully unsupervised setting, making organ counting easy to apply routinely on a wide variety of aerial image datasets.

1.1   Related Work

Organ counting is a common task in image-based plant phenotyping. Early organ counting efforts focused on leaf counting in *A. thaliana* rosettes grown in controlled, indoor conditions such as growth chambers and greenhouses [1, 4,16]. Other work with indoor imaging data has localized wheat spikes and spikelets [11,12]. While performing organ counting with controlled imaging is relevant for many experiments, organ counting is also more relevant in outdoor field-based experiments and is frequently assessed using less-controlled imaging techniques such as with unmanned aerial vehicles (UAVs). Vision tasks in outdoor contexts are substantially different than those in indoor contexts, as factors such as lighting and plant motion cannot be controlled.

Previous work in organ counting in outdoor field environments has explored both image processing as well as deep learning methods. [5] used a decision tree-based segmentation with an image processing feature set, followed by the classification of regions using Bag-of-Visual-Words features for head detection in sorghum. Deep learning has provided some of the strongest performance for plant organ counting tasks in field imagery. Notably, [9] used local counts regression on a density map estimated by a CNN. The network showed strong performance for counting the tassels of maize plants from field images. A follow-up work extended the technique to the counting of wheat heads [18]. Organ counting has also been explored using object detection methods, such as R-CNN [6], Faster R-CNN [10], as well as the state-of-the-art RetinaNet meta-architecture [2]. Others have demonstrated counting by segmentation. [14] proposed a superpixel segmentation followed by further segmentation using a fully convolutional network. A similar pipeline has been used for rice panicles [19].

2   Unsupervised Instance Segmentation

In contrast to existing organ counting methods, we propose a strictly unsupervised technique which does not require any annotated data. The technique is agnostic to any particular dataset, and does not require any hand-tuning to apply. An open source implementation is available[1]. In the initial step, we make use of an existing unsupervised segmentation technique to train a segmentation network on the entire dataset [7]. This method utilizes a Simple Linear Iterative Clustering (SLIC) superpixel segmentation to first roughly segment the image. Next, a fully convolutional network consisting of three blocks of only convolution and batch normalization, is used to assign class labels to pixels. This network uses a stride of one on each convolutional layer to maintain the full size of the input image through the network. Pixel-level labels are assigned to superpixels based on the maximum class membership within each region, and the segmentation network is trained on this target using cross-entropy loss. Because of the trivial clustering solution where every region is assigned the same label, intra-axis normalization is used to encourage a larger number of unique class labels. This process is repeated until either a maximum number of iterations or a minimum number of unique class labels has been attained.

Next, the trained segmentation network is used to perform inference on each image in the dataset and the softmax function is applied to the raw activations of the final layer. The activations for the organ class in the output are extracted. The index of this label can either be determined by selecting the label which attains the lowest value of the objective function in Eq. 2, or it can be indicated by the user. The result is a probability density map for the organ class for each image in the dataset.

In order to obtain a binary segmentation, a gradient-free optimization is performed on the activations using the portfolio-discrete One Plus One optimizer implemented in the Nevergrad library [13]. We find that this optimization procedure consistently provides more accurate segmentations of object regions than the class predictions from the segmentation CNN. Since the pixel values in this map correspond to probabilities, a log-likelihood loss function is optimized over individual activation maps $\mathbf{k}_i \in \{\mathbf{k}_0 \dots \mathbf{k}_N\}$ as

---

[1] https://github.com/p2irc/autocount.

$$\operatorname*{argmin}_{\theta} \;\; -\log\left(\frac{\sum m_\theta(\mathbf{k}_i)}{\sum \mathbf{k}_i}\right) - \log\left(\frac{\sum m_\theta(\mathbf{k}_i)}{|m_\theta(\mathbf{k}_i)|}\right) \tag{1}$$

where m_θ is a binary masking function which applies a Gaussian filter, a pixel threshold, and then a morphological dilation and erosion to the selected areas. This function returns its input masked by the selected pixels. $|m_\theta|$ indicates the number of selected pixels in that masking, and $\theta = (\tau, \lambda, \phi_{erosion}, \phi_{dilation})$ are the standard deviation of the Gaussian filter, the pixel threshold value, and the diameter of a circular kernel for morphological erosion and dilation, respectively. The first term of the loss function encourages the optimization to include as much of the total probability density as possible, while the second term encourages the average probability of selected pixels to be high.

Finally, objects smaller than a pixels are removed, and watershed segmentation is performed using the distance transform of the binary mask with local maxima no closer than b pixels apart used as watershed markers. These parameters are optimized over binary masks as

$$\operatorname*{argmin}_{a,b} \;\; \frac{1}{N}\sum_{i=0}^{N}[W(\mathbf{s}_i,\hat{\mathbf{s}}_i) + W(\hat{\mathbf{s}}_i,\mathcal{N}_{\hat{\mathbf{s}}_i}) + W(\Gamma_{\hat{\mathbf{s}}_i},\mathcal{N}_{\hat{\mathbf{s}}_i})]^2 \tag{2}$$

where $\mathbf{s}$ and $\hat{\mathbf{s}}$ are the empirical distributions of object sizes before and following the procedure, W is the Wasserstein distance, $\mathcal{N}_{\hat{\mathbf{s}}} = \mathcal{N}\left(\mu(\hat{\mathbf{s}}), \sigma(\hat{\mathbf{s}})^2\right)$ is the normal distribution parameterized by $\hat{\mathbf{s}}$, and $\Gamma_{\hat{\mathbf{s}}} = \Gamma\left(\alpha(\hat{\mathbf{s}}), \beta(\hat{\mathbf{s}})\right)$ is the gamma distribution parameterized by $\hat{\mathbf{s}}$. This optimization is performed via an exhaustive search, because of the low dimensionality of the search space. In contrast to Eq. 1 which is optimized on a per-sample basis, Eq. 2 is optimized over the entire dataset. We find that this mitigates outlier samples which would otherwise find suboptimal solutions. Following optimization, the number of objects is reported as the number of object regions in the watershed segmentation.

Intuitively, for objects such as plant organs from top-down aerial images, we assume that the object sizes should be roughly normally distributed. The tension between the first term, which acts as a prior, and the second two terms, which shape the distribution by affecting its normality and skewness, finds a balance between over-segmentation and under-segmentation of object regions. Note that, if $\mathbf{s}$ is already normally distributed, then the optimal solution to Eq. 2 is to leave the distribution unchanged. Figure 2 shows the distributions of object sizes for the sorghum and wheat experiments before and after this step.

Although the training of the CNN in the initial step includes several tuneable hyperparameters such as the learning rate and the number of convolutional blocks, we use the default settings provided by the authors' implementation for all experiments [7]. We adjust the number of output channels to 32 and the final number of labels (used as a stopping criteria) to eight, based on the low level of object diversity present in field datasets.

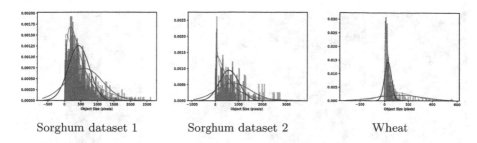

| Sorghum dataset 1 | Sorghum dataset 2 | Wheat |

Fig. 2. Distributions of object sizes for a single example image before (red) and after (blue) the segmentation step. A normal distribution fit to the samples is shown as a solid line and a gamma distribution is shown as a dashed line. (Color figure online)

3   Results

We evaluate the counting performance of the proposed method using the Mean Absolute Error (MAE), the Root Mean Square Error (RMSE), and the coefficient of determination between the annotated and predicted counts (R^2). Table 1 shows a comparison against two previously reported results from the literature on two publicly available datasets of sorghum heads [5]. The resolution for these datasets is 1154 by 1731 pixels for dataset 1 and 1394 by 357 pixels for dataset 2. Although counting performance fell short of the previous standard set by the powerful RetinaNet object detection meta-architecture applied in [2], AutoCount shows that a high degree of performance can be attained without using labels. Examples from the public sorghum datasets are shown in Fig. 3.

Table 1. Counting results for the sorghum and wheat head datasets.

		MAE	RMSE	R^2
Sorghum 1	Segmentation + Classification [5]	–	–	0.84
	RetinaNet [2]	–	–	0.82
	AutoCount (unsupervised)	29.92	36.39	0.79
Sorghum 2	Segmentation + Classification [5]	–	–	0.56
	RetinaNet [2]	–	–	0.76
	AutoCount (unsupervised)	4.95	6.22	0.48
Wheat	AutoCount (unsupervised)	297.22	344.89	0.17

The proposed method was also tested on a dataset of wheat field images (Table 1). The wheat head dataset consisted of 81 images from a *T. aestivum* breeding trial captured with a camera extended from an all terrain vehicle pointed downward at a height of approximately 6 m. The images were captured in RAW format with a 18 MP camera with an ultra-wide angle rectilinear lens. Wheat head centers were annotated by domain experts using a custom

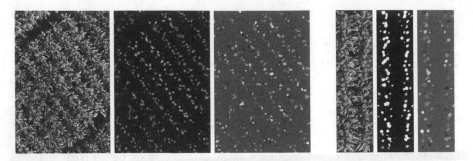

Fig. 3. Example images and intermediate outputs from the sorghum datasets. Left to right: original image, softmax activations, watershed output.

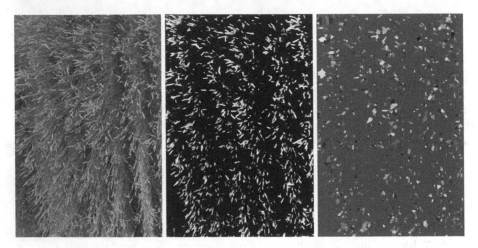

Fig. 4. Example image and intermediate outputs from the wheat dataset. Left to right: original image, softmax activations, watershed output.

tool and the total number of boxes was used for the annotated count value. The images were resized to 400 by 600 pixels for expediency because of the extremely high resolution, although this resizing is not required and the original resolution can be used as was the case for the sorghum datasets. Compared to the sorghum dataset, the wheat images included substantially more visual variation among images, including different head densities, different levels of maturity, and both awned and unawned heads. Figure 4 shows an example from the wheat dataset. Despite this variation, AutoCount was applied to the wheat dataset with no modifications. Although the detection of heads is positive, the arrangement of the heads in dense clumps proved difficult to segment accurately using the watershed method, resulting in a lower R^2 value than in the sorghum experiments. This issue is discussed further in Sect. 4.

4   Discussion

AutoCount is best suited to datasets where organs are represented as contiguous regions, making aerial imagery a good candidate. In contrast, high-resolution proximal imaging could create a scenario where the organ region is over-segmented, for example, by segmenting a wheat head into individual spikelets. The counting performance of the method is expected to suffer under these conditions. The method also is subject to the weaknesses of watershed segmentation. One such drawback is the potential over-segmentation of objects which are not compact. For example, the performance of the method suffered significantly in the second dataset of sorghum due to a small subset of images containing a high density of ring-shaped panicles. An example of this mode of failure is shown in Fig. 5. Another example of a failure case for the watershed method can be seen in the dense clumps of wheat heads shown in Fig. 4. Many of these regions do not admit a shape which can be accurately segmented by the watershed method. Without the ability to accurately separate these large regions, the optimization of segmentation parameters ultimately settles on an over-segmentation of the wheat heads as can be seen in Fig. 2c.

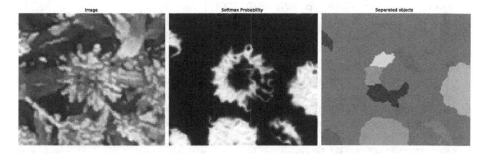

Fig. 5. A failure case from the second sorghum dataset. Ring-shaped organs are over-segmented by the watershed method.

Although we focus on the organ count in the present work as it is the most agronomically relevant trait, the method also provides an instance segmentation of organs by way of the final watershed segmentation following optimization. This mask could potentially be used for other applications such as measuring organ size. Future work could potentially address replacing the watershed method with a more robust segmentation technique, such as one that is able to act on the activation map directly as opposed to requiring a thresholded input. We expect that a more robust segmentation technique would help boost performance for wheat head images and for other challenging object instances which tend to become over-segmented with our current approach, and could be dropped into the overall system without the need to change the segmentation loss function.

5   Conclusion

Here we have introduced the first unsupervised object counting algorithm specifically tailored for counting plant organs in field environments, called AutoCount. Although it does not rely on labelled data, the counting performance of the method is comparable to that of previously reported supervised methods on two datasets of sorghum. Additional positive results are presented on a new dataset of wheat heads. Some limitations of the method and avenues for future work are identified. The results presented here represent the first fully unsupervised baselines for performing plant organ counting of arbitrary plant organs in diverse field imagery.

Acknowledgments. This research was funded by a Canada First Research Excellence Fund grant from the Natural Sciences and Engineering Research Council of Canada.

References

1. Dobrescu, A., Giuffrida, M.V., Tsaftaris, S.A.: Leveraging multiple datasets for deep leaf counting. In: Proceedings of the IEEE International Conference on Computer Vision Workshops, pp. 2072–2079 (2017)
2. Ghosal, S., et al.: A weakly supervised deep learning framework for sorghum head detection and counting. Plant Phenomics **2019**, 1–14 (2019). https://doi.org/10.34133/2019/1525874
3. Giuffrida, M.V., Chen, F., Scharr, H., Tsaftaris, S.A.: Citizen crowds and experts: observer variability in image-based plant phenotyping. Plant Methods **14**(1), 12 (2018)
4. Giuffrida, M.V., Minervini, M., Tsaftaris, S.A.: Learning to count leaves in rosette plants. In: Proceedings of the Computer Vision Problems in Plant Phenotyping (CVPPP) 2015. BMVA Press (2016)
5. Guo, W., et al.: Aerial imagery analysis-quantifying appearance and number of sorghum heads for applications in breeding and agronomy. Front. Plant Sci. **9**, 1544 (2018)
6. Hasan, M.M., Chopin, J.P., Laga, H., Miklavcic, S.J.: Detection and analysis of wheat spikes using Convolutional Neural Networks. Plant Methods **14**(1), 1–13 (2018). https://doi.org/10.1186/s13007-018-0366-8
7. Kanezaki, A.: Unsupervised image segmentation by backpropagation. In: 2018 IEEE International Conference on Acoustics, Speech and Signal Processing (ICASSP), pp. 1543–1547. IEEE (2018)
8. Lin, T.-Y., et al.: Microsoft COCO: common objects in context. In: Fleet, D., Pajdla, T., Schiele, B., Tuytelaars, T. (eds.) ECCV 2014. LNCS, vol. 8693, pp. 740–755. Springer, Cham (2014). https://doi.org/10.1007/978-3-319-10602-1_48
9. Lu, H., Cao, Z., Xiao, Y., Zhuang, B., Shen, C.: TasselNet: counting maize tassels in the wild via local counts regression network. Plant Methods 1–14 (2017). https://doi.org/10.1186/s13007-017-0224-0
10. Madec, S., et al.: Ear density estimation from high resolution RGB imagery using deep learning technique. Agric. Forest Meteorol. **264**, 225–234 (2019). https://doi.org/10.1016/j.agrformet.2018.10.013

11. Pound, M.P., Atkinson, J.A., Wells, D.M., Pridmore, T.P., French, A.P.: Deep learning for multi-task plant phenotyping. In: Proceedings of the IEEE International Conference on Computer Vision Workshops, pp. 2055–2063 (2017)
12. Qiongyan, L., Cai, J., Berger, B., Okamoto, M., Miklavcic, S.J.: Detecting spikes of wheat plants using neural networks with laws texture energy. Plant Methods **13**(1), 83 (2017)
13. Rapin, J., Teytaud, O.: Nevergrad - A gradient-free optimization platform (2018). https://GitHub.com/FacebookResearch/Nevergrad
14. Sadeghi-Tehran, P., Virlet, N., Ampe, E.M., Reyns, P., Hawkesford, M.J.: Deep-Count: in-field automatic quantification of wheat spikes using simple linear iterative clustering and deep convolutional neural networks. Front. Plant Sci. **10**, 1–16 (2019). https://doi.org/10.3389/fpls.2019.01176
15. Ubbens, J., Cieslak, M., Prusinkiewicz, P., Stavness, I.: The use of plant models in deep learning: an application to leaf counting in rosette plants. Plant Methods **14**(1), 6 (2018). https://doi.org/10.1186/s13007-018-0273-z. https://plantmethods.biomedcentral.com/articles/10.1186/s13007-018-0273-z
16. Ubbens, J.R., Stavness, I.: Deep plant phenomics: a deep learning platform for complex plant phenotyping tasks. Front. Plant Sci. **8**, 1190 (2017)
17. Valerio Giuffrida, M., Dobrescu, A., Doerner, P., Tsaftaris, S.A.: Leaf counting without annotations using adversarial unsupervised domain adaptation. In: Proceedings of the IEEE Conference on Computer Vision and Pattern Recognition Workshops (2019)
18. Xiong, H., Cao, Z., Lu, H., Madec, S., Liu, L., Shen, C.: TasselNetv2: in - field counting of wheat spikes with context - augmented local regression networks. Plant Methods (2019). https://doi.org/10.1186/s13007-019-0537-2
19. Xiong, X., et al.: Panicle-SEG: a robust image segmentation method for rice panicles in the field based on deep learning and superpixel optimization. Plant Methods **13**(1), 104 (2017)

CorNet: Unsupervised Deep Homography Estimation for Agricultural Aerial Imagery

Dewi Endah Kharismawati[1,2,3,4], Hadi Ali Akbarpour[1,2],
Rumana Aktar[1,2], Filiz Bunyak[1,2], Kannappan Palaniappan[1,2],
and Toni Kazic[1,2,3,4(✉)]

[1] Department of Electrical Engineering and Computer Science,
University of Missouri, Columbia, USA
{aliakbarpourh,bunyak,pal,kazict}@missouri.edu
[2] MU Plant Science Foundry, University of Missouri, Columbia, USA
[3] Missouri Maize Center, University of Missouri, Columbia, USA
[4] Interdisciplinary Plant Group, University of Missouri, Columbia, USA
{dek8v5,rayy7}@mail.missouri.edu

Abstract. Efficient and accurate estimation of homographies among images is the first step in mosaicking crop fields for phenotyping. The current strategy uses sophisticated vehicles that have excellent telemetry to hover over a grid of waypoints, imaging each one. This approach simplifies homography estimation, but precludes more flexible, adaptive protocols that can collect richer information. It also makes aerial phenotyping impractical for many researchers and farmers. We are developing an alternative strategy that uses consumer-grade vehicles, freely flown over a variety of trajectories, to collect video. We have developed an unsupervised deep learning network that estimates the sequence of planar homography matrices of our corn fields from imagery, without using any metadata to correct estimation errors. The vehicle was freely flown using a variety of trajectories and camera views. Our system, *CorNet*, performed faster than and with comparable accuracy to the gold standard ASIFT algorithm in many challenging cases.

Keywords: Unsupervised convolutional neural network · Homography estimation · UAV · Freely flown · Maize · Mosaicking

1 Introduction

The small size, high image resolution, and low cost of unmanned aerial vehicles (UAV) offer researchers and farmers a powerful tool for capturing the raw data of plant phenotypes. Drones are emerging as phenotyping and agricultural tools because of their low cost, convenient use, and high image resolution [19,22,32,36]. So far, the prevailing strategy for transforming the UAV sensor data into quantitative biological information has relied on canopy reflectances at different frequencies that are collected from nadir views at precomputed, fixed

© Springer Nature Switzerland AG 2020
A. Bartoli and A. Fusiello (Eds.): ECCV 2020 Workshops, LNCS 12540, pp. 400–417, 2020.
https://doi.org/10.1007/978-3-030-65414-6_28

waypoints. Though information collected this way has been successfully used to phenotype crop plants [14, 16, 20, 24, 43, 46], this strategy has three significant limitations. The first is that canopy observations require extensive, often line-specific, validation before they can be used as surrogates for morphological and developmental phenotypes [10]. The second is the heavy reliance on nadir views at fixed waypoints. These simplify image registration by constraining the possible camera poses and trajectories and ensuring high overlap between images, and allow one to pre-compute the matrices that map the spatial transformations between frames. However, nadir views inherently lose significant morphological information that is visible in oblique views and fixing many waypoints requires significant effort and longer missions. Finally, grid-based trajectories lose information for other reconstructions, including three-dimensional ones [6, 42]. In principle, researchers and farmers should not need to "engineer up" to exploit UAV technologies.

An alternative strategy is to use cheap equipment to freely fly any trajectory with any camera pose, then compute the orthomosaics and three-dimensional reconstructions directly from video imagery. This strategy has several advantages, especially in vehicle cost and complexity, ease of deployment, speed of data acquisition, the inherent sequencing of video, and information richness. It discards vehicle telemetry and georegistration, which are usually only marginally accurate for consumer-grade vehicles, and avoids the labor-intensive placement and maintenance of ground control points. Perhaps most importantly, this strategy is tailor-made for developing adaptive protocols that would dynamically survey large areas at low resolution, identify phenotypic outliers, and reimage those for more and different details. However, this alternative entails significant computational costs, beginning with registering successive pairs of images that vary much more in their spatial relationships than those from fixed, constant altitude trajectories. Without prior knowledge of these relationships or metadata from which to derive them, the matrices that transform the local coordinate system of one image into that of its successor, and these into the global coordinate system of the mosaic, must be computed *de novo*.

Image registration has benefitted from many years of research in computer vision [3, 4, 9, 11, 13, 29, 30, 39, 45, 51]. Its essential problem is to compute the homography matrix that optimally maps the local coordinate system of one image and its population of recognizable features onto another's, accounting for the relative translations, rotations, and scalings between the two images' features. Not all members of the first image's feature population will exhibit identical motions relative to the second image's. This is obvious for rotations: the greater a feature's displacement from the axis of rotation, the greater its Cartesian motions. But it can also occur in translations and scalings, notably when a vehicle experiences minor air currents that push it horizontally or vertically during flight. Camera lenses invariably distort the image around the lenses' peripheries, and minor errors in gimbal pointing foreshorten one side of the image relative to another.

Algorithmic approaches to registration rely on identifying *key points*—highly localized, statistically distinct features in one image—and finding their correspondents in the next [41]. Sparse features such as lines, corners, and edges are used to increase the spatial separation among them and to exploit standard feature recognition algorithms and feature descriptor functions [8,33,35,40]. The features are then matched between images under the assumption that the field of interest is sufficiently planar and the matrix estimated using, for example, direct linear transform, singular value decomposition, or RANSAC [17,21,28,34]. Mosaics built this way significantly benefit from metadata to correct the drift that accumulates as the homography matrices are multiplied together during mosaicking [31,34,50]. This two-dimensional approach can perform well on built environments, but it is problematic for imagery with fuzzy features that repeat frequently, such as plants in crop fields. Significant improvements in algorithmically mosaicking maize fields imaged by free flight from imagery alone depend on the statistical feature descriptor, on selecting the strongest features, and on minimizing the chain of matrix multiplications that build the mosaic by first mosaicking subsequences of frames [5]. Nonetheless, Aktar *et al.*'s VMZ pipeline can still geometrically distort the mosaic and mismatch repetitive features. Structure from Motion (SfM) finds the third dimension for unconstrained trajectories without metadata at higher accuracy, so it would seem perfect for our alternative strategy [2,18]. However, its computational cost grows exponentially with the number of images unless the imagery can be pre-processed to partition it or to select key frames. This makes SfM prohibitive for routine use on the large data sets needed for agricultural landscapes [2,49]. More modern incremental methods, such as iSAM2, may produce better geometric models more efficiently for agricultural imagery without so much intervention [25].

In contrast, convolutional neural networks (CNN) estimate homography by learning the regression of the displacement of patches between two frames. DeTone *et al.*'s supervised HomographyNet estimated the homography of 128×128 pixel patches bounded by four points using a VGG8-like architecture [44]. They augmented the original images by synthesizing new ones using random projective transformations to the images. They minimized L_2 loss between the four-point patch locations in the synthesized training data and those predicted by the network [15,44]. The average corner error was 9.2 pixels compared to the 11.2 pixels average corner error computed by ORB and RANSAC [17,40]. Kang *et al.* proposed a hybrid framework that incorporates deep learning and energy minimization of photometric refinement for an accurate and efficient homography estimation [26]. It simplifies HomographyNet to reduce the computational time significantly and performs well on images of room ceilings interspersed with large fluorescent light fixtures using 32×32 pixel patches. The light fixtures have very sharp edges, and it is unclear how the patches are chosen. Without available code to test, it seems unlikely this approach would be suitable for agricultural imagery. Nguyen *et al.* extended HomographyNet to an unsupervised CNN they called Unsupervised Deep Homography (UDH). These investigators used the estimated homography matrix to warp one image and compared the warped image to the successor image, minimizing the L_1 photometric loss function [37]. They trained two models using UDH: one with synthetic data and the other with aerial imagery. Throughout this paper, we will refer to the aerial model as UDH. While UDH did as well or better on built landscapes than HomographyNet, our experiments showed it does not generalize well to crop field imagery, especially fields with a high vegetation index (see Sect. 3.1).

Rapid, accurate image registration for freely flown trajectories is the first step in making it possible for researchers and farmers to routinely phenotype their fields with minimal effort and cost. Ideally, a flexible, efficient framework that could accommodate any trajectory and camera pose could serve as the back-end for a public computational resource. Here we present our experiments extending the unsupervised CNN of Nguyen *et al.* to video of maize nurseries imaged with a freely flown consumer-grade vehicle. Our trained network, *CorNet*, ran faster and produced results comparable to ASIFT, the gold standard feature descriptor for algorithmic homography estimation. No vehicle or camera telemetry or ground control points are used in matrix estimation, but accurate mosaicking requires training data that cover a range of vehicle movements and camera poses. Once trained, *CorNet* performs well on a wide variety of landscapes.

2 Materials and Methods

2.1 Maize Nurseries

Imagery was collected from four different fields during the summer, 2019 field season in Columbia, MO. Three were planted with 36 in. row spacing. Two genetic nurseries, fields 33 and 22, were planted in 20 ft rows with 4 ft alleys; the central file of rows was left unplanted to help orient the pilots. Field 33 was hand-planted with inbred lines and disease lesion mimic mutants using an Almaco jab planter at 12 in. spacing and machine-planted with an elite line using a Jang TD-1 push planter at 8 in. spacing [48]. Field 22 was planted late in the season with inbred lines (hand-planted) and the elite line (Jang-planted). Field 30 was machine-planted with sweet corn at $\approx$ 6 in. spacing without alleys. A field of maize varieties for drought trialing planted at 30 in. row spacing with $\approx$ 6 in. spacing between plants was imaged after fertilization and before sensecence.

2.2 *CorNet*'s Architecture

DeTone *et al.* used a supervised regression to predict the four point homography and used the L_2 loss as a cost function relative to the ground truth. Nguyen *et al.* began with a set of unsupervised convolution layers, passed the result to DeTone *et al.*'s regression layer, and converted the four point homography to a 3 × 3 matrix. This was used to warp the frames and produce a prediction. The L_1 photometric loss, used as the cost function, was back-propagated to the regression model. *CorNet* uses UDH's basic architecture, but adds four layers to the input, as shown in Fig. 1. These layers allow us to use a larger patch size

$(512 \times 512 \times 2)$ compared to Nguyen *et al.* $(128 \times 128 \times 2)$. The larger patch size prevents many estimation errors arising from the repetitive nature of the imagery and information loss. Using half the number of filters used by UDH prevented loss of the fuzzier plant features (data not shown). Figure 2 shows *CorNet*'s workflow, which ranged from field operations to mosaicking.

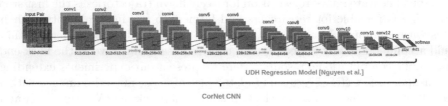

Fig. 1. Architecture of *CorNet*'s CNN.

2.3   Data Collection

We used a DJI Phantom 4 manufactured by Da-Jiang Innovations, Shenzhen, China, which carries an RGB camera to image our fields, forests, and buildings (https://www.dji.com/phantom-4). No structural alterations to the vehicle or its camera were made; by 2019, the Phantom 4 had accumulated errors in its gimbal position. All flights were flown manually using the Autopilot mobile app (https://autoflight.hangar.com/autopilot/flightschool). Flight trajectories included all four translations (forward, backward, right slides, and left slides); scaling (changing altitude); rotations (clockwise and counter-clockwise); and combinations of two or more movements. In forward and backward movements, the camera is pointed parallel to the direction of vehicle motion; in "slides", the camera is pointed perpendicular to the vehicle's motion. In rotations, the position of the camera is fixed and the vehicle rotates, sometimes with other translations and scalings. Six different types of trajectories were used: straight to target; serpentine, parallel or perpendicular to the rows, with slides or rotations on each pass; and orbits around the field. Camera poses included both nadir

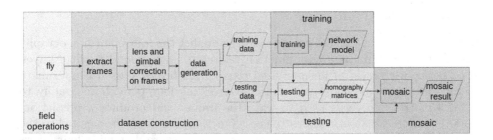

Fig. 2. *CorNet* workflow.

and a variety of oblique orientations. All flights were at relatively low altitudes, *ca.* 25–100 ft above ground level (AGL). The vehicle was flown in relatively light winds, but occasionally experienced horizontal and vertical displacements due to air currents.

Video was collected at 24 frames/sec in 24-bit color depth on Secure Digital High Capacity (SDHC) cards inserted in the vehicle and manually copied to the computer. The Phantom 4 can record a maximum of about 19 min 42 s of video, providing about 30,000 frames of 4069 × 2160 each for a total of about 850 GB. All code was run on a Lambda Labs Intel Core i9-9920X, 2 NVIDIA RTX 2080Ti, and 128 GB memory.

2.4 Training and Testing

We used Nguyen *et al.*'s published code and model to retune their UDH aerial model. We used a similar approach to training *CorNet*, with the necessary modifications for network architecture and emphasizing a large data set of diverse motions, rather than cross-validating a smaller sample. All new code was written in the TensorFlow framework using stochastic gradient descent, a batch size of 128 for rUDH and 32 for *CorNet* [1]. The Adam Optimizer was used with $\beta_1 = 0.9$, $\beta_2 = 0.999$, $\varepsilon = 10^{-8}$, and a learning rate of 0.0001 [27]. The mean intensity of all pixels was used to normalize images during training. We fine-tuned rUDH for 250,000 iterations in approximately 36 h. *CorNet* was trained from scratch using the lower half of Table 1 for 150,000 iterations in 47 h. Training *CorNet* was essentially complete by $\approx 30,000$ iterations. For testing, the pairwise sucessive frames are fed to the network, which returns the four points' homographic prediction and the L_1 photometric loss. We then compute the 3×3 homography.

2.5 Mosaicking

The output of homography estimation is a 3×3 matrix that projects the current frame to the next frame ($H_{n \rightarrow n+1}$). We used the first frame as a base frame, projecting the inverse of $H_{n \rightarrow n+1}$ for each successive frame back to the first, $H_{1 \leftarrow n}$:

$$H_{1 \leftarrow n} = H_{1 \leftarrow 1} \times H_{1 \leftarrow 2} \times H_{2 \leftarrow 3} \times \ldots \times H_{n-1 \leftarrow n} \qquad (1)$$

We estimated the global canvas by transforming the four corners of each frame to the accumulated $H_{1 \leftarrow n}$, shifting any projections with negative values to $(0,0)$ with a translation offset matrix, M_o. Thus, for every n^{th} frame, the total projection is

$$I_n^w = (M_o \times H_{1 \leftarrow n}) \times I_n, \qquad (2)$$

where the original image I_n is projected and translated to produce the warped image I_n^w. We used a simple pixel replacement blending to produce the final mosaic [23].

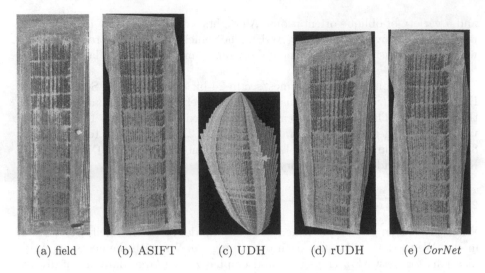

(a) field   (b) ASIFT   (c) UDH   (d) rUDH   (e) *CorNet*

Fig. 3. Homography accuracy by the four methods, visualized as mosaics, compared to an image of the whole field. Panel (a), the whole field. Panels (b)—(e), mosaics by the four methods. The blue arrow in panel (c) marks the single irrigation stand pipe that is repeated five times. The pink arrows in panels (d) and (e) mark the same field location in the two mosaics. (Color figure online)

2.6   Model and Mosaic Evaluation

To evaluate the quality of the estimated homography matrices, we picked the four corners of a random rectangular patch in the first frame and constructed the corners of a ground truth patch in the second frame with ASIFT [35], then compared these constructed corners to the ones predicted by CorNet. The root mean squared error (RMSE) between the n corners of the estimated, *Pred*, and ASIFT ground truth *Gt*, patches was computed using

$$RMSE = \sqrt{\frac{\sum_{i=1}^{n}(Pred_i - Gt_i)^2}{n}}. \tag{3}$$

We estimated geometric distortion of the mosaics relative to a whole field image as $Area(Field)/Area(OrientedBoundingBox)$ [47]. We also visually evaluated the homography by comparing the outlines of the *CorNet* and ASIFT patches in the second frame. Finally, we evaluated the quality of the mosaics by visual inspection, since the different errors are quite characteristic. Run time performance was also benchmarked.

3   Results

3.1   Motivation

The acid test of homography estimation is how well it maps between images when mosaicked. Errors that seem insignificant when comparing between two frames accumulate during mosaicking, since the matrices are multiplied together (Eq. 1). Figure 3 shows how errors in homography estimation can accumulate. For these frames, the trajectory was backwards over the center of a nearly rectangular field, with the camera pointing downwards in nadir view. Lens and gimbal errors were corrected.

ASIFT (panel (b)) produces the best results, with only minor geometric distortion of the field, seen as bending of the two irrigation pipes on either side. UDH's difficulties in computing the homographies are apparent from the extreme geometric distortion; the area around the single irrigation stand pipe marked by the blue arrow is repeated five times (panel (b)). While the field is not perfectly rectangular (0.9612), the geometric distortion of each mosaic relative to an image of the field, measured as deviation from a perfect rectangle, are 0.9613, 0.7034, 0.8428, and 0.9169, for ASIFT, UDH, rUDH, and *CorNet*, respectively. *CorNet*'s performance is superior to those of UDH and rUDH, though not quite as good as ASIFT.

If the problem of UDH was simply a breakdown in transfer learning, then retraining it with our own imagery should improve matters (panel (d)). It does, but geometric distortion is visible moving from the top to the bottom of the mosaic. The left edges of the mosaics are shorter than the right edges and the overall length is too small. Registration errors between frames are also apparent when looking along the pipes and at the maize itself (pink arrow). These results led to the development of *CorNet* (Sect. 2.2). Panel (e) shows *CorNet* reduced the false joins and produces a better overall field geometry than rUDH, though errors are still visible at the extreme ends of the field. The pink arrow in panel (e) marks the same position as in panel (d) so that the relative magnitude of the errors between the two models can be compared. Compared to ASIFT, *CorNet* compresses the field slightly from top to bottom; for example, the left border rows wiggle slightly and the overall row lengths are slightly shorter in panel (e).

3.2 Training and Testing Data

Table 1 summarizes the data sets used to retrain UDH, which we denote rUDH, and *CorNet*. Frames grouped under translation include five different movements (forward, backward, left slide, right slide, and both types of serpentines) and both orientations relative to the rows. We retrained UDH for maize with 13,614 pairs of frames covering both seedling and adult maize. No lens or gimbal corrections were applied (Table 1, upper portion). The training data for *CorNet* only slightly overlapped those for UDH (Table 1, lower portion). Seedling imagery was excluded, more movements were added, and lens and gimbal errors were corrected using the procedure of [5]. Both data sets were augmented for movements by frame rotation and swapping, doubling the total frames; and each frame had some random illumination changes [37].

To evaluate the accuracy of homography estimation and mosaicking, run time, and generalizability, we used a wide variety of imagery that was corrected for lens and gimbal errors (Table 2). Maize testing data were drawn from multiple videos shot over the growing season on seedlings, before flowering, and mature maize, again with multiple trajectories and camera poses. The same vehicle imaged farm buildings, patches of forest in south-central Missouri in summer and fall, and the surface of a pond. We used the telemetrized versions of the soil, urban, and countryside imagery of UMDC, but without using the telemetry [7].

3.3 Homography Estimation for Complex Movements

Table 3 compares the run time and errors of ASIFT, UDH, our rUDH, and *CorNet* for eight different movements. UDH consistently has the best run time performance, with the exception of an orbital trajectory. *CorNet*'s run time lies between that of ASIFT and either version of UDH, running between ≈ 0.1 to ≈ 0.5 of ASIFT. *CorNet* is the most accurate for all movements except the right slide, consistently outperforming UDH and rUDH. Examination of the predicted homographies supports these quantitative comparisons. Figure 4 compares homographies for a pair of frames for UDH, rUDH, and *CorNet*, using ASIFT as the ground truth. UDH and rUDH both produce more errors in homography estimation compared to *CorNet*: the RMSEs for UDH, rUDH, and *CorNet* are 7.656, 3.190, and 2.096 for these pairs, respectively.

Table 1. Training data for retraining UDH (upper half) and *CorNet* (lower half). The translation data include forward, backward, left slide, right slide, and serpentine movements.

Movement	Camera pose	Growth stage, sampling	Raw frame pairs
Translation	Nadir	Seedling, 3 fps	2817
		Adult, 1/4 – 3 fps	5333
	Oblique	Seedling, 3 fps	2038
		Adult, 1/2 fps	185
Rotation	Nadir	Adult, 1/4– 3 fps	653
	Oblique	Seedling, 3 fps	931
Scale	Nadir	Adult, 1/4 – 3 fps	561
Orbit	Oblique	Adult, 1–2 fps	1096
Total rUDH			13,614
Translation	Nadir	Adult, 3 fps	2130
	Oblique	Adult, 1/3 – 3 fps	2132
Rotation	Nadir	Adult, 1/3 – 5 fps	2152
Scale	Nadir	Adult, 1/5 – 5 fps	1367
Orbit	Oblique	Adult, 1/3 – 3 fps	1430
Total *CorNet*			9211

Table 2. Test data. Target is maize unless otherwise indicated.

Video no., data set	Sampling rate (fps)	No. pairs
174, forward	4	164
167, backward	3	83
201, right slide	3	101
201, left slide	3	140
362, scaling	3	180
413, dense maize	3	179
466, forest, summer	2	120
517, forest, fall	2	180
202, rotation	4	120
360, oblique parallel to row	3	96
274, orbit	3	180
179, maize seedlings	3	180
urban, UMDC	3	141
soil, UMDC	2	150
countryside, UMDC	3	119
368, farm buildings	1	166
520, water	3	90

Table 3. Comparison of the methods' performance over different aerial movements. Run time is in seconds, RMSE is in pixels; the best values for each are shown in **bold** font.

Movement	ASIFT Run time	UDH Run time	UDH RMSE	retrained UDH Run time	retrained UDH RMSE	*CorNet* Run time	*CorNet* RMSE
Forward	2101.5102	**73.7880**	38.3918	80.0166	3.2817	181.9988	**1.7370**
Backward	750.5475	**67.8707**	132.6135	74.9763	2.7282	130.0924	**2.3456**
Right slide	941.6449	**69.7518**	120.3709	76.8579	**4.5235**	140.5475	8.1995
Left slide	1609.9070	**74.6321**	108.6612	78.8836	4.0709	166.6295	**2.3374**
Scale	347.4869	**89.2801**	91.8952	91.8022	2.3649	161.6219	**0.6875**
Rotation	1954.2377	**90.6798**	114.5680	93.4896	3.4074	164.3279	**2.0111**
Oblique	653.4463	**85.7609**	47.6286	89.0532	11.4312	143.9081	9.5610
Orbit	2106.3459	103.8676	110.0210	**98.7048**	4.3955	220.8464	**2.9083**

Combinations of simple movements can produce even greater mosaicking errors. Figure 5 shows a very stringent test: the UAV was flown forward and backward without rotating the camera. Minor errors in homography estimation accumulate over twice as many frames, and are readily apparent as mismatches between the beginning and end of the sequence. In this type of trajectory, the perspective changes as the UAV moves. ASIFT (panel (a)) is the least sensitive

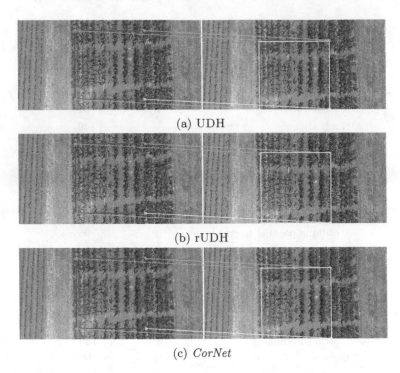

(a) UDH

(b) rUDH

(c) *CorNet*

Fig. 4. Sample four point homography estimation by ASIFT (ground truth), UDH, rUDH, and *CorNet* for a frame and the adjacent sampled frame. The red squares in the left frames are the randomly chosen patches; in the right frames, these are the matching patches predicted by ASIFT. The yellow frames are predicted by each of the CNNs. (Color figure online)

to this change: the tilt of the top edge and the imperfect registration at the bottom of the image illustrate this error accumulation. None of the CNNs perform as well as ASIFT, but *CorNet* (panel (d)) registers each pass fairly well individually. The worse case is visible in the figure; the forward pass lying underneath is better, similar to Fig. 3, panel (d). Misalignment of the two passes increases moving down the field than in ASIFT (panel (a)). Consistent with this, starting the mosaic in the center of the sequence produces a better overlay of the two passes because there is less error to accumulate (data not shown). The perspective change combines with these errors to compress the mosaic. UDH and rUDH illustrate more extreme versions of these errors (panels (b) and (c)). UDH cannot register the backward pass at all, and rUDH exaggerates the backward pass errors so much they form a small tail at the top of the mosaic.

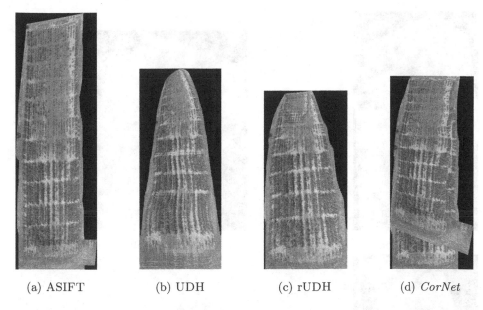

(a) ASIFT   (b) UDH   (c) rUDH   (d) *CorNet*

Fig. 5. Mosaics by the four methods for forward and backward motion, with a slightly oblique camera pose. (Color figure online)

Table 4. Run time and RMSE for the four methods on challenging data.

Movement	ASIFT Run time	UDH Run time	RMSE	Retrained UDH Run time	RMSE	CorNet Run time	RMSE
Seedling maize	1899.9222	**77.1553**	41.8223	80.9572	**4.0044**	192.3239	16.0899
Dense maize	2393.3591	**77.0296**	92.7713	81.3696	4.7668	194.0659	**1.7360**
Farm buidings	595.6757	**78.0988**	100.4067	80.5361	3.2728	183.6212	**3.2293**
Summer forest	1114.0191	**74.3189**	54.9672	77.2994	3.2922	151.8185	**1.6663**
Fall forest	1457.5384	**79.2406**	66.6672	81.7336	5.6429	195.0878	**3.3883**
UMDC, urban	632.2847	**73.8767**	58.7524	77.9679	4.9085	166.0596	**1.6666**
UMDC, dirt	555.7996	**78.6077**	39.9764	83.8195	5.2164	166.3824	**3.2201**
UMDC, countryside	590.4839	**75.0135**	69.2046	77.0026	6.2166	150.5176	**1.5583**
Irrigation pond	200.3115	**83.6731**	555.6703	86.0839	535.4424	136.5369	**529.0653**

3.4 *CorNet*'s Generalizability to Different Landscapes

CorNet performs well on a variety of landscapes. Table 4 summarizes the performance of the four methods. Apart from seedling maize, *CorNet* has the best RMSE; as before, UDH has the best run time. Figure 6 gives some examples of *CorNet*'s mosaics. The major difficulty it experiences is when the vehicle hovers for some time. This is visible in panels (a) and (d). In hovering, the vehicle's motion becomes more comparable to the error in homography estimation. This can produce very large errors, even in built landscapes (data not shown).

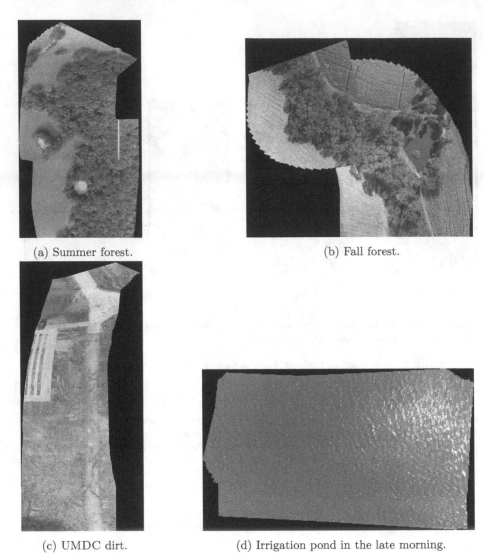

(a) Summer forest.

(b) Fall forest.

(c) UMDC dirt.

(d) Irrigation pond in the late morning.

Fig. 6. *CorNet* generalizes to other landscapes. Panel (a): the white object is the landing gear when the UAV was tipped slightly by an updraft. Just below that is an area of poor mosaicking due to the vehicle hovering. Panel (b): a different forest. Panel (c): the blue arrow marks a misalignment in the road's edge. Panel (d): minor misalignments of frames are visible in the highlighted ripples (pink arrow). (Color figure online)

4   Discussion

CorNet is successful for several reasons. First, to a homography estimator, all corn looks alike. A simple thought experiment illustrates why: in a relatively uniform stand of maize, a patch the size of a corn plant would lack the adjacent features that help to distinguish one plant from another. Increasing the patch size allows more of these features in the patch and the combinations of those features helps distinguish patches. Second, *CorNet* was trained with data from many different movements. Our initial model used only forward movement in training and failed miserably on backwards movements. As we added more movements, more mosaics of different trajectories were computed successfully. Since the training data used for rUDH and *CorNet* overlapped only slightly, it is possible the former was trained with more right slides than the latter. Diversifying the training imagery also helps: the only time another model was more accurate than *CorNet* was rUDH on seedlings, imagery of which was included in that training set. Third, the trajectories used to train rUDH and *CorNet* varied considerably. The majority of UDH's predictions lie on the same location regardless of camera movements. Though no details of camera movements are given in [37], we surmise their data were collected from a vehicle flown automatically, resulting in approximately uniform overlap. This partially explains why their model did not transfer to maize imagery.

Several areas need improvement. Four classes of movement need additional training data: forwards and backwards over long serpentine trajectories; turns, especially by rotation; scaling due to altitude changes; and hovers. We also need to find substitutes for checkerboard-based lens correction and empirical gimbal correction, since few researchers and farmers will have calibration checkerboards. We are examining ways to exploit imagery of farm buildings and other structures with perpendicular corners to automatically determine gimbal error.

Several algorithmic changes to homography estimation and mosaicking could also improve our results. Adaptive frame sampling to assure sufficient overlap between frames without bloating the data would resolve some of *CorNet*'s issues with turns and scale changes. Comparison of *CorNet*'s mosaics to those produced by AutoStitch and LF-Net show the strengths and weaknesses of each approach [12,38]. Though outside the scope of the present paper, some preliminary conclusions suggest several improvements. All the methods perform best with nadir, as opposed to oblique, camera poses, and all have difficulty mosaicking serpentine trajectories. AutoStitch and LF-Net's use of more features over the entire frame, as opposed to the random image patches of *CorNet* and UDH, sometimes produce superior mosaics. However, *CorNet* often generalizes better, especially when the features are more uniform, as in the pond imagery. Finally, our mosaics would benefit from a more sophisticated mosaicking algorithm, such as the minimosaicking algorithm in VMZ [5]. This is particularly true for complex motions, such as those in Sect. 3.3.

5   Conclusions

In this paper we have presented a new unsupervised homography estimation network, sets of videos captured by a consumer-grade UAV, and a set of training strategies for generating mosaics of crop fields that suffer from repetitive structure and fuzzy features. *CorNet* homography estimation outperforms both

ASIFT and other CNNs over a variety of UAV movements and landscapes. While UDH and our retrained version of it are much faster, their results are far less accurate than *CorNet* in nearly all cases. Finally, *CorNet* generalizes well to non-maize landscapes.

Acknowledgments. We are grateful for the support of the NSF Midwest Big Data Hub Digital Agriculture Spoke, the Missouri Maize Center, the National Corn Growers Association, the U.S. Army Research Laboratory (cooperative agreement W911NF1820285), the Army Research Office (DURIP W911NF1910181), an Executive Women's Forum doctoral fellowship through the University of Missouri College of Engineering (to R.A.), and an anonymous gift. We thank Shizeng Zhou, Behirah Hartranft, Steven Suddarth, and Vinny for helpful discussions. We are especially grateful to the referees for their very helpful comments. Any opinions, findings, and conclusions or recommendations expressed in this publication are those of the authors' and do not necessarily reflect the views of the U. S. Government or any agency thereof. The authors declare no conflict of interest.

References

1. Abadi, M., et al.: TensorFlow: Large-scale machine learning on heterogeneous distributed systems (2016). https://arxiv.org/abs/1805.09662
2. Agarwal, S., et al.: Building Rome in a day. Commun. ACM **54**(10), 105–112 (2011)
3. Aktar, R., Aliakbarpour, H., Bunyak, F., Seetharaman, G., Palaniappan, K.: Performance evaluation of feature descriptors for aerial imagery mosaicking. In: Applied Imagery Pattern Recognition Workshop (AIPR), pp. 1–7. IEEE (2018)
4. Aktar, R., Prasath, V.S., Aliakbarpour, H., Sampathkumar, U., Seetharaman, G., Palaniappan, K.: Video haze removal and Poisson blending based mini-mosaics for wide area motion imagery. In: Applied Imagery Pattern Recognition Workshop (AIPR), pp. 1–7. IEEE (2016)
5. Aktar, R., et al.: Robust mosaicking of maize fields from aerial imagery. Appl. Plant Sci. **8**, e11387 (2020). https://doi.org/10.1002/aps3.11387, https://bsapubs.onlinelibrary.wiley.com/doi/full/10.1002/aps3.11387
6. Aliakbarpour, H., Palaniappan, K., Seetharaman, G.: Stabilization of airborne video using sensor exterior orientation with analytical homography modeling. In: Sergiyenko, O., Flores-Fuentes, W., Mercorelli, P. (eds.) Machine Vision and Navigation, pp. 579–595. Springer, Cham (2020). https://doi.org/10.1007/978-3-030-22587-2_17
7. Avola, D., Cinque, L., Foresti, G.L., Martinel, N., Pannone, D., Piciarelli, C.: A UAV video dataset for mosaicking and change detection from low-altitude flights. IEEE Trans. Syst. Man Cybern. Syst. **PP**, 1–11 (2018)
8. Bay, H., Tuytelaars, T., Van Gool, L.: SURF: speeded up robust features. In: Leonardis, A., Bischof, H., Pinz, A. (eds.) ECCV 2006. LNCS, vol. 3951, pp. 404–417. Springer, Heidelberg (2006). https://doi.org/10.1007/11744023_32
9. Bentoutou, Y., Taleb, N., Kpalma, K., Ronsin, J.: An automatic image registration for applications in remote sensing. IEEE Trans. Geosci. Remote Sens. **43**(9), 2127–2137 (2005)
10. Blancon, J., et al.: A high-throughput model-assisted method for phenotyping maize green leaf area index dynamics using unmanned aerial vehicle imagery. Fron. Pl. Sci. **10**, 685 (2019). https://doi.org/10.3389/fpls.2019.00685

11. Brown, L.G.: A survey of image registration techniques. ACM Comput. Surv. (CSUR) **24**(4), 325–376 (1992)
12. Brown, M., Lowe, D.G.: Automatic panoramic image stitching using invariant features. Int. J. Comput. Vis. **74**(1), 59–73 (2006). https://doi.org/10.1007/s11263-006-0002-3
13. Chen, H.M., Arora, M.K., Varshney, P.K.: Mutual information-based image registration for remote sensing data. Int. J. Remote Sens. **24**(18), 3701–3706 (2003)
14. Condorelli, G.E., et al.: Comparative aerial and ground based high throughput phenotyping for the genetic dissection of NDVI as a proxy for drought adaptive traits in durum wheat. Frontiers Plant Sci. **9** (2018). https://doi.org/10.3389/fpls.2018.00893
15. DeTone, D., Malisiewicz, T., Rabinovich, A.: Deep image homography estimation. arXiv preprint arXiv:1606.03798 (2016)
16. Enciso, J., et al.: Validation of agronomic UAV and field measurements for tomato varieties. Comput. Electron. Agric. **158**, 278–283 (2019). https://doi.org/10.1016/j.compag.2019.02.011
17. Fischler, M.A., Bolles, R.C.: Random sample consensus: a paradigm for model fitting with applications to image analysis and automated cartography. In: Readings in Computer Vision, pp. 726–740. Elsevier (1987)
18. Gao, K., AliAkbarpour, H., Palaniappan, K., Seetharaman, G.: Evaluation of feature matching in aerial imagery for structure-from motion and bundle adjustment. In: Geospatial Informatics, Motion Imagery, and Network Analytics VIII, vol. 10645, p. 106450J. International Society for Optics and Photonics (2018)
19. Gnädinger, F., Schmidhalter, U.: Digital counts of maize plants by unmanned aerial vehicles (UAVs). Remote Sens. **9**(6), 544 (2017)
20. Gracia-Romero, A., Kefauver, S.C., Fernandez-Gallego, J.A., Vergara-Díaz, O., Teresa Nieto-Taladriz, M., Araus, J.L.: UAV and ground image-based phenotyping: a proof of concept with durum wheat. Remote Sens. **11**(10), 1244 (2019). https://doi.org/10.3390/rs11101244
21. Hartley, R., Zisserman, A.: Multiple View Geometry in Computer Vision. Cambridge University Press, Cambridge (2004)
22. Huang, Y., Thomson, S.J., Hoffmann, W.C., Lan, Y., Fritz, B.K.: Development and prospect of unmanned aerial vehicle technologies for agricultural production management. Int. J. Agric. Biol. Eng. **6**(3), 1–10 (2013)
23. Jain, P.M., Shandliya, V.: A review paper on various approaches for image mosaicing. Int. J. Comput. Eng. Res. **3**(4), 106–109 (2013)
24. Johansen, K., Morton, M.J.L., Malbeteau, Y.M., Aragon, B., Al-Mashharawi, S.K., Ziliani, M.G., Angel, Y., Fiene, G.M., Negrão, S.S.C., Mousa, M.A.A., Tester, M.A., McCabe, M.F.: Unmanned Aerial Vehicle-based phenotyping using morphometric and spectral analysis can quantify responses of wild tomato plants to salinity stress. Frontiers Plant Sci. **10**, 370 (2019). https://doi.org/10.3389/fpls.2019.00370
25. Kaess, M., Johannsson, H., Roberts, R., Ila, V., Leonard, J.J., Dellaert, F.: iSAM2: incremental smoothing and mapping using the Bayes tree (2011). https://doi.org/10.1177/0278364911430419, https://arxiv.org/abs/1908.02002v1
26. Kang, L., Wei, Y., Xie, Y., Jiang, J., Guo, Y.: Combining convolutional neural network and photometric refinement for accurate homography estimation. IEEE Access, pp. 109460–109473 (2019)
27. Kingma, D.P., Ba, J.: Adam: A method for stochastic optimization. arXiv preprint arXiv:1412.6980 (2014)
28. Kriegman, D.: Homography estimation. Lecture Computer Vision I, CSE a 252 (2007)

29. Le Moigne, J., Netanyahu, N.S., Eastman, R.D.: Image Registration for Remote Sensing. Cambridge University Press, Cambridge (2011)
30. Li, Q., Wang, G., Liu, J., Chen, S.: Robust scale-invariant feature matching for remote sensing image registration. IEEE Geosci. Remote Sens. Lett. **6**(2), 287–291 (2009)
31. Lin, Y., Medioni, G.: Map-enhanced UAV image sequence registration and synchronization of multiple image sequences. In: IEEE Conference on Computer Vision and Pattern Recognition, 2007. CVPR 2007, pp. 1–7. IEEE (2007)
32. López-Granados, F., Torres-Sánchez, J., De Castro, A.I., Serrano-Pérez, A., Mesas-Carrascosa, F.J., Peña, J.M.: Object-based early monitoring of a grass weed in a grass crop using high resolution UAV imagery. Agron. Sustain. Dev. **36**(4), 67 (2016)
33. Lowe, D.G.: Distinctive image features from scale-invariant keypoints. Int. J. Comput. Vision **60**(2), 91–110 (2004)
34. Molina, E., Zhu, Z.: Persistent aerial video registration and fast multi-view mosaicing. IEEE Trans. Image Process. **23**(5), 2184–2192 (2014)
35. Morel, J.M., Yu, G.: ASIFT: a new framework for fully affine invariant image comparison. SIAM J. Imaging Sci. **2**(2), 438–469 (2009)
36. Nasir, A.K., Tharani, M.: Use of Greendrone UAS system for maize crop monitoring. In: ISPRS-International Archives of the Photogrammetry, Remote Sensing and Spatial Information Sciences, pp. 263–268 (2017)
37. Nguyen, T., Chen, S.W., Shivakumar, S.S., Taylor, C.J., Kumar, V.: Unsupervised deep homography: a fast and robust homography estimation model. IEEE Rob. Autom. Lett. **3**(3), 2346–2353 (2018)
38. Ono, Y., Trulls, E., Fua, P., Yi, K.M.: LF-Net: learning local features from images. arXiv preprint arXiv:1805.09662v2 (2018)
39. Pohl, C., Van Genderen, J.L.: Review article multisensor image fusion in remote sensing: concepts, methods and applications. Int. J. Remote Sens. **19**(5), 823–854 (1998)
40. Rublee, E., Rabaud, V., Konolige, K., Bradski, G.: ORB: an efficient alternative to sift or surf. In: 2011 International Conference on Computer Vision, pp. 2564–2571. IEEE (2011). https://doi.org/10.1109/ICCV.2011.6126544
41. Schonberger, J.L., Frahm, J.M.: Structure-from-motion revisited. In: Proceedings of the IEEE Conference on Computer Vision and Pattern Recognition, pp. 4104–4113 (2016)
42. Seetharaman, G., Palaniappan, K., Akbarpour, H.A.: Method for fast camera pose refinement for wide area motion imagery (2019). U.S. Patent 9,959,625
43. Shi, Y., et al.: Unmanned Aerial Vehicles for high-throughput phenotyping and sgronomic research. PLoS ONE **11**(7), e0159781 (2016). https://doi.org/10.1371/journal.pone.0159781
44. Simonyan, K., Zisserman, A.: Very deep convolutional networks for large-scale image recognition. arXiv preprint arXiv:1409.1556 (2014)
45. Teters, E., AliAkbarpour, H., Palaniappan, K., Seetharaman, G.: Real-time geoprojection and stabilization on airborne GPU-enabled embedded systems. In: Geospatial Informatics, Motion Imagery, and Network Analytics VIII, vol. 10645, p. 106450H. International Society for Optics and Photonics (2018)
46. Wang, X., et al.: Dynamic plant height QTL revealed in maize through remote sensing phenotyping using a high-throughput unmanned aerial vehicle (UAV). Sci. Rep. **9**(1) (2019). https://doi.org/10.1038/s41598-019-39448-z
47. Wirth, M.A.: Shape Analysis & Measurement. University of Guelph (2004). http://www.cyto.purdue.edu/cdroms/micro2/content/education/wirth10.pdf

48. Woodward Crossings: Jang TD-1 Push Planter. Woodward Crossings (2019)
49. Wu, C.: Towards linear-time incremental structure from motion. In: 2013 International Conference on 3D Vision, pp. 127–134 (2013)
50. Zhu, Z., Riseman, E.M., Hanson, A.R., Schultz, H.: An efficient method for georeferenced video mosaicing for environmental monitoring. Mach. Vis. Appl. **16**(4), 203–216 (2005)
51. Zitova, B., Flusser, J.: Image registration methods: a survey. Image Vis. Comput. **21**(11), 977–1000 (2003)

Expanding CNN-Based Plant Phenotyping Systems to Larger Environments

Jonas Krause[✉], Kyungim Baek, and Lipyeow Lim

Department of Information and Computer Sciences, University of Hawai'i at Mānoa,
1680 East-West Road, Honolulu, HI, USA
{krausej,kyungim,lipyeow}@hawaii.edu

Abstract. Plant phenotyping systems strive to maintain high categorization accuracy when expanding their scopes to larger environments. In this paper, we discuss problems associated with expanding the plant categorization scope. These problems are particularly complicated due to the increase in the number of species and the inter-species similarity. In our approach, we modify previously trained Convolutional Neural Networks (CNNs) and integrate domain-specific knowledge in the fine-tuning process of these models to maintain high accuracy while expanding the scope. This process is the key idea behind our CNN-based expanding approach resulting in plant-expert models. Experiments described in this paper compare the accuracy of an expanded phenotyping system using different plant-related datasets during the training of the CNN categorization models. Although it takes much longer to train these models, our approach achieves better performance compared to models trained without the integration of domain-specific knowledge, especially when the number of species increases significantly.

Keywords: Plant phenotyping · Convolutional Neural Networks · Integration of domain-specific knowledge

1 Introduction

The categorization of species and plant phenotyping are challenging problems relevant to both disciplines of Botany and Computer Science. Classifying plant images at the species level, considering specific characteristics of their phenotype, is called fine-grained categorization. Despite the availability of various applications, categorizing plants in an environment with a large number of species remains an unsolved problem. Furthermore, an automated system capable of addressing the complexity of this problem and handle larger environments has important implications, not only in preserving ecosystem biodiversity but also in numerous agricultural activities.

In this paper, we extend CNN-based phenotyping systems [5–7] and present a new approach to maintain their high accuracy when applied to a larger environment. The proposed approach implements a novel scale-up process that adapts

© Springer Nature Switzerland AG 2020
A. Bartoli and A. Fusiello (Eds.): ECCV 2020 Workshops, LNCS 12540, pp. 418–432, 2020.
https://doi.org/10.1007/978-3-030-65414-6_29

the CNNs so that it can handle environments with a broader range of plant species. In this process, we replace the top classification layers to accommodate a more significant number of plant species. However, due to the lack of training data, pre-trained weights have to be used to achieve satisfactory performance. Recently, Cui et al. [3], Xiangxi et al. [11], and Ngiam et al. [12] addressed this problem by introducing domain-specific models, fine-tuning their CNNs to other fine-grained categorization problems. Following their ideas, we implement the integration of domain-specific knowledge for plant phenotyping problems. Also called targeted fine-tuning, the pre-training process generates plant-related knowledge from multiple datasets to create expert CNN models. These fine-tuned CNNs take much longer to train but better performance is achieved when the number of species is significantly increased.

The contributions of our research are the process to expand the scope (i.e., the number of species to be categorized) of CNN-based phenotyping systems, the publication of a new plant species dataset, and the creation of plant-expert CNNs used to fine-tune our categorization models. More specifically, this paper describes modifications made to successfully deploy a plant categorization system (previously designed to categorize 100 species) in a larger environment, expanding its scope to 300 species without significant loss of performance.

We organized this paper as follows: Sect. 2 presents the related work by describing how most of the plant categorizing systems operate. In Sect. 3, we describe the adaptations made to expand the scope of the previously trained CNN models. Section 4 details the integration of domain-specific knowledge, a systematic pre-training process developed to integrate previously learned datasets. And Sects. 5 and 6 present the experimental results and observations showing the prediction accuracy of our expanded approach and compares them with other commonly used training methods. Finally, Sect. 7 provides a conclusion and describes future work.

2 Related Work

Nowadays, applications created for categorizing plants implement different CNN models. Most of them are well-known CNN models adapted to work with plant images. However, few of them present new approaches designed to address specific aspects of plant phenotyping, such as the challenges of expanding the categorization scope.

Implementing a simple CNN model, Barré et al. [1] present the application called *LeafNet*. They used three datasets (*Flavia*, *Foliage*, and *LeafSnap*) to train and test their CNN-based plant identification system. By comparing CNN models with hand-designed feature methods such as the *LeafSnap* [9] application, Barré et al. show that learning features by using a CNN (even with a simple architecture) provides a better representation of leaf images and consequently better discrimination. However, their application does not inherit plant-related knowledge from other datasets, which limits the learning process to the extraction of discriminative features out of only one plant dataset at the time.

Adapting the CNN model called *AlexNet* [8] and using a dataset of 44 plant species, Lee *et al.* [10] also focus on the classification of preprocessed leaf images. They detect leaves in the images by using HSV (Hue, Saturation, and Lightness) color space information to extract the foreground pixels. For the initial experiments, they pre-trained a CNN using the ImageNet dataset and fine-tuned it using the segmented samples. Initial results are not as expected, so they decided to create a deconvolutional structure to visually verify what features the CNN has learned. During this process, Lee *et al.* noticed that the trained model is focusing almost exclusively on the contour and shape of leaves. Based on the low performance, they conclude that leaf shape is not a good choice to identify plants, which is not necessarily true. Morphological features of leaves have been heavily and successfully used for plant species categorization. In this case, non-satisfactory results may be a consequence of a poor pre-training process of the classification model, making the CNN focus almost exclusively on the counter of the leaves. Integrating domain-specific knowledge may assist with this problem.

Sun *et al.* [15] implement plant categorization models customized to 100 plant species. They use images collected from the Beijing Forestry University campus, available online in the BJFU100 dataset[1]. These images present a variety of backgrounds, different illumination conditions, shadows, and it is not always possible to identify a leaf of the plant. Knowing these challenges, Sun *et al.* implement a modified version of the Residual Network (*ResNet*) [4] to classify these images. In their model, a pre-trained *ResNet* works as a bottleneck structure between an initial convolutional block and the last layers of the network. Like this, they adapted the *ResNet* architecture to their needs, customizing this successful CNN model and fine-tuning it with their dataset. Despite implementing a successful customized CNN model, Sun *et al.* limit their work to 100 plant species. And its scope expansion may be harder to perform due to the customized model.

Using the BJFU100 dataset, the work of Krause *et al.* [5] explores multi-scale methods to improve the categorization process of plant species. They present better results when compared with the work of Sun *et al.* by implementing a CNN-based system called *WTPlant*. This system implements a scene parsing method to locate different plant organs and delimit the most representative areas in the images for the categorization of plant species. Additionally, Krause *et al.* present experiments using another dataset with 100 plant species called UHManoa100[2]. As a result, the plant phenotyping system designed by Krause *et al.* has a limited categorization scope. Because of that, our approach focuses on expanding CNN-based systems like the *WTPlant* to deploy it in larger environments. Furthermore, the proposed approach has to face the challenge of maintaining the high accuracy results that other systems reported when trained over small environments (with 100 species or less).

---

[1] https://pan.baidu.com/s/1jILsypS.
[2] https://github.com/jonaskrause/UHManoa100.

3   Expanding the Plant Categorization Scope

The accurate categorization of 100 plant species from multiple datasets suggests the work of Krause *et al.* [5] as an effective method for classifying natural images of plants. However, expanding its scope brings new problems: The first problem is to correctly identify which species exist in the target environment and collect representative images of the listed plants. This process requires the assistance of a botany specialist to define the number of species and annotate the training images. With the correctly annotated species, we collect images of plants in the wild to create a new dataset and define the expanded scope. Considering that this new dataset represents the flora biodiversity of a specific region of the globe, classification models trained over these images compose a plant categorization system with that particular scope.

The second problem is to include new species in the plant categorization scope while maintaining the knowledge previously learned and, consequently, minimizing the loss in performance. A simple solution to this problem would be training new CNN models from scratch using the target dataset. However, experiments performed by Krause *et al.* [5] using the ImageNet [13] pre-trained weights showed how valuable the knowledge integration is for the fine-tuning of these plant categorization models. Therefore, a solution that expands the plant categorization scope for more than 100 species must take into account the pre-training process of the classification models. To integrate previously learned knowledge, we implement a modification of the classification models by replacing the top layers of the CNNs with new extended ones. These new layers accommodate a more significant number of plant species but do not guarantee a high categorization accuracy. To address it, we create expert models by training the modified CNNs over domain-specific datasets and use their pre-trained weights to fine-tune the classification models over the new dataset. The pre-trained weights of these plant-expert CNN models are available online[3] and can be used by other researches to fine-tune their models. In this way, knowledge extracted from CNNs pre-trained over plant-related datasets assist in the fine-tuning of the models over new target datasets. And we expand the plant categorization scope by adapting the classification models to inherit powerful discriminative features previously learned during the training over multiple datasets.

Experiments performed using the proposed solution compare the accuracy of the CNN classification models when pre-trained with different datasets like the ImageNet, the UHManoa100 published by Krause *et al.*, and the iNat682 (a plant dataset from *iNaturalist*[4] with 682 species). We use them to pre-train the models before the fine-tuning process over the new target dataset. Although it takes much longer to train these CNNs, the resulting models with integrated domain-specific knowledge categorize plants more accurately throughout all experiments. For much larger scopes that encompass more than one environment (e.g., different regions of the globe, continents, or countries), multiple systems can operate

---

[3] https://github.com/jonaskrause/Plant_Flower-Expert_CNN_Models.

[4] https://www.inaturalist.org/.

in parallel using other guidance methods (such as geolocation of the testing image) to indicate which version to use. In this way, we can deploy multiple CNN categorization models to cover an even larger environment and categorize the entire flora of that specific region.

3.1   Creating a Dataset of Broader Range of Species

The first step to increasing the number of species analyzed by CNN-based phenotyping systems is to gather a representative dataset of the listed plants. So we invited a botanist specialist to perform a sanity check on all of our images and ensure that each one of them contains visible traits of the selected species. We also eliminate the incorrectly labeled images as well as the ones with poor quality and low-resolution (smaller than 400 × 400 pixels). This process is necessary due to the lack of datasets with annotated species available for the experiments conducted in this study.

Following this initial process, we organize the new dataset as a collection of 300 plant species with 50 natural images per species, totalizing 15,000 correctly annotated images. This new dataset, called UHManoa300[5], comprises diverse images with different sizes from 400 × 400 to 6000 × 4000 pixels, plants appearing in varying angles, scales, and stages of life. Consequently, it becomes more difficult to categorize this dataset as the appearance of plants changes considerably within the same species. For example, the 50 images of the species *Acacia koa* in Fig. 1 illustrate how diverse the within-species plant appearance can be. In addition, similar to the UHManoa100, different plants may appear in the background or even in front of the dominant plant. The annotation of the plant in the images indicates the dominant species (*Acacia koa*) covering the most substantial areas of the images. Also, as shown in the Figure, images in this new dataset contain plants at various scales ranging from leaves and flowers to the entire tree.

3.2   Modifying CNN Models to Accommodate Expanded Scope

After constructing a new dataset containing the plants in the target environment, we adapt the classification models to work with a larger number of species. In this process, we remove the top classification layers of the CNNs initially trained to categorize 100 plant species, saving the weights of each CNN model without the top layers. For each model, pre-trained weights create a basic knowledge of what the model learned in previous training processes (also called base models). A new and larger classification layer added at the top of a base model creates a new CNN with similar architecture but adapted to work with an extended scope. Figure 2 shows this process by expanding the plant categorization scope from 100 to 300 species. Thus, we can load knowledge learned from previous experiments into the same models but with a larger classification layer at the top. Retraining the modified models over the new images, we fine-tune the CNNs

---

[5] https://github.com/jonaskrause/UHManoa300.

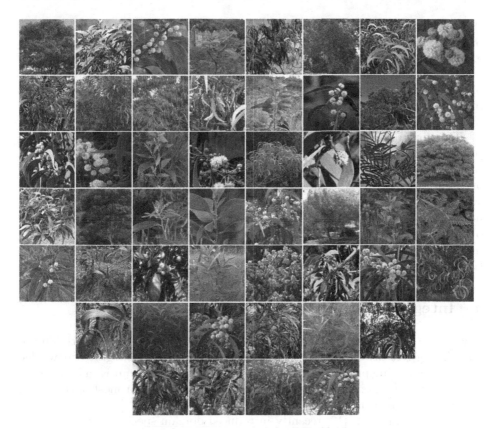

Fig. 1. All 50 images of the *Acacia koa* in the UHManoa300 dataset.

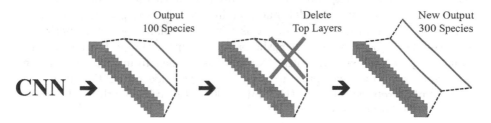

Fig. 2. Process of replacing the top classification layers of CNN models to expand the plant categorization scope.

to learn discriminative features between a more significant number of species using the pre-trained weights as a starting point for this process.

The number of new species to be integrated into the scope determines the size of the new classification layer at the top of the model. In this paper, we expand the plant categorization scope from 100 to 300 species. For that, we exclude the two dense layers at the top of the CNN models and replace them with new ones that accommodate the expanded scope. The first layer has the same size as the previously excluded one, but the second layer (the last one) is customized to the exact number of classes in the new dataset (300 plant species). These two top layers work together and are responsible for producing the output predictions of the model. Thereby, modified CNN models have the same architecture as the previously trained ones but are loaded with pre-trained weights and ready to work with more classes (plant species).

The fine-tuning of modified models over the target dataset updates the parameter values for the entire CNN based on the pre-trained weights. Consequently, well-trained base models lead to a better fine-tuning process of CNNs over the target dataset. Implementing this adaptation of the classification models, we present a solution to expand the scope of CNN-based systems using multiple pre-trained models. The integration of knowledge from the continuous pre-training processes of the CNNs suggests the creation of domain-specific models. These classification models require an intensive computational effort for training but yield more accurate results.

4   Integration of Domain-Specific Knowledge

Due to the lack of training data for most of the fine-grained categorization problems, ImageNet pre-trained weights are frequently used to integrate knowledge during the training process of CNN models. These weights comprise a base model trained over a million images, and this knowledge is useful for most of the visual classification problems. As previously described, Cui *et al.* [3], Xiangxi *et al.* [11], and Ngiam *et al.* [12] recently introduced domain-specific models for fine-tuning their CNNs to different fine-grained categorization problems. Exploring these approaches, we expand the CNN-based plant categorization systems by adapting the classification models and searching for the best pre-trained weights to fine-tune the models over a new dataset with 300 plant species. Experiments using this new target dataset (UHManoa300) compare the performance of the plant categorization system with CNN models trained from scratch, with the ImageNet pre-trained weights, and with the knowledge integration using two different plant-related datasets (UHManoa100 and iNat682) with the ImageNet pre-trained weights as a starting point.

4.1   Training with Random Initialization of CNN Parameters

Initial experiments start by extracting samples from the training images of the UHManoa300 dataset and randomly dividing them into 80% for training and 20% for validation. We begin by training three CNN models (*Inception-v3* [17], *Inc-ResNet-v2* [16], and *Xception* [2]) using the random initialization of the model's weights. In this process, we initialize the training of each CNN model with random parameter values (weights and biases) and, consequently, no pre-trained knowledge. The training process runs for 100 epochs with hyperparameters (such as the number of convolutional filters, their sizes, padding, and stride) of each CNN set as suggested in the papers that presented the models. We use the backpropagation algorithm to propagate the error backward throughout the CNN and update its parameter values (weights). This process is performed for each training sample while the validation set is used to calculate the model's accuracy at each epoch. During the training process, CNN models are not over-fitting after 100 epochs. We observe this behavior by monitoring performance on the validation set and stop the training at 100 epochs because we are limited by computation. The final trained model is the one with the smallest validation error after completing the training process.

4.2   Training with ImageNet Pre-trained Weights

We extend previous experiments by using the ImageNet [13] pre-trained weights to initialize the training process of the CNN models. We use these pre-trained weights as base models for the next experiments over the UHManoa300 dataset. As suggested by previous results on datasets with 100 plant species (Krause *et al.* [5]), CNN models implementing inception modules (such as the *Inception-v3*, *Inc-ResNet-v2*, and *Xception*) take the most advantage of the ImageNet pre-trained weights. These CNN models have millions of parameters and are better fine-tuned over small datasets when using pre-trained weights as initial parameter values.

4.3   Training with ImageNet and UHManoa100 Pre-trained Weights

We use ImageNet pre-trained weights as a starting point for the training of the CNN models over different datasets to integrate domain-specific knowledge from plant images. The first one is the UHManoa100 dataset, a collection of plant samples from 100 species previously presented. The best performing CNN model after 100 epochs of training creates a new base model with domain-specific knowledge, enabling the extraction of the ImageNet+UHManoa100 pre-trained weights. Consequently, these base models are an integration of previously learned knowledge built over the ImageNet initial weights and what is learned during the training process of these models over the UHManoa100 dataset.

As described previously, we remove the top layers of these models and save their weights and biases (parameter values). In this way, knowledge learned during previous experiments can be used during the fine-tuning process of models adapted for the UHManoa300 dataset. It should be noted that the knowledge integration of the ImageNet pre-trained weights trained over the UHManoa100 dataset creates a domain-specific model that may provide better initial weights for fine-tuning models with new target datasets.

4.4   Training with ImageNet and iNat682 Pre-trained Weights

The iNat682 dataset is a collection of plant images from *iNaturalist* community with 682 species. We downloaded this dataset from the *iNaturalist* challenge website[6] for the classification of animals and plants. However, our training process selects only those images from the *Plantae* category, excluding other groups such as animals, insects, and fungus. The resulting training set is a highly unbalanced collection of 158,463 images over 682 plant species (and these images are not part of the UHManoa100 or UHManoa300 datasets). Ranging from 19 to 503, the number of images per class varies according to the endemic nature of each species worldwide. Furthermore, these images vary in resolution, orientation, and focus, making this dataset a very diverse collection of plant images.

In an attempt to create more robust domain-specific CNN models, experiments using the ImageNet+iNat682 pre-trained weights integrate the general knowledge from the ImageNet dataset and the knowledge from the iNat682 dataset to create plant-expert models. For this process, we initiate the training of the CNN over the iNat682 dataset with the ImageNet pre-trained weights for 100 epochs. In such a way, the iNat682 dataset contributes to the intermediate training process undertaken to create a powerful domain-specific CNN model that learns useful plant-related features for the fine-grained categorization of other plant datasets. We extract the learned knowledge by saving pre-trained weights of the CNN models with the best validation performance, creating the ImageNet+iNat682 pre-trained weights.

5   Experimental Results

Focusing on the UHManoa300 dataset, we implement a CNN-based approach to expand the scope of plant categorization systems to a larger environment of 300 species. Due to the clean-up previously performed in this dataset (Sect. 3.1), preprocessing these plant images creates highly representative species samples for training and testing. For this balanced dataset, the testing set comprises 10% of the data and the rest is the training set. We perform the fine-tuning process of the CNNs over the UHManoa300 dataset for 100 epochs. During this process, we divide the training set of extracted samples into training (80%) and validation (20%). It is important to reinforce that images selected for the validation set have all their samples for validation only. In this way, we use training and validation samples exclusively in their respective sets for fine-tuning the models. We evaluate the trained CNNs using the testing set of images unseen by the trained models, containing five different images of each plant species. As a metric for performance evaluation, we use the prediction accuracy, e.g., the percentage of images correctly categorized in the testing set. We consider an image is correctly categorized when the Top-1 prediction matches the annotated species of the plant.

Table 1 presents the Top-1 classification accuracy of CNN models pre-trained on different integrated datasets and fine-tuned over the UHManoa300 dataset. As shown in the Table, the CNNs' performance is improved by pre-training them multiple times to integrate domain-specific knowledge. This integration starts with the commonly used ImageNet weights and trains models initially loaded

---

[6] https://www.kaggle.com/c/inaturalist-challenge-at-fgvc-2017.

with these parameter values on different plant datasets. It includes pre-trained weights from 100 plant species (ImageNet+UHManoa100) and expert models trained with a larger domain-specific dataset (ImageNet+iNat682). The use of pre-trained weights allows the CNN-based systems to improve their accuracy for the categorization of 300 plant species. More specifically, this approach correctly categorized 84% of the testing images when the *Xception* is pre-trained as a plant-expert model and fine-tuned over the target dataset.

Table 1. Accuracy results with CNNs pre-trained on different dataset for classifying plant species in the UHManoa300 dataset.

CNN model	Random Initialization	ImageNet Weights	ImageNet+ UHManoa100	ImageNet+ iNat682
Inception-v3	51.93%	75.67%	76.07%	78.80%
Inc-ResNet-v2	52.00%	76.73%	77.07%	82.33%
Xception	52.40%	81.20%	81.40%	**84.00%**

As shown in Table 1, CNNs fine-tuned for the UHManoa300 achieved more accurate results when pre-trained as domain expert models. Initially, ImageNet pre-trained weights bring a general knowledge with models trained to classify 1,000 common objects. The ImageNet pre-trained models are commonly employed in numerous computer vision problems, but they are limited to the lack of domain-specific knowledge required for fine-grained categorization problems. In the process of creating plant-expert models, we use domain-specific datasets to train the CNNs before fine-tuning them over the target dataset. We collect the ImageNet+UHManoa100 pre-trained weights from CNN models that yielded the best performance over the UHManoa100 dataset. However, CNNs fine-tuned with these pre-trained weights resulted in just slightly more accurate models than those with no domain-specific knowledge integration (ImageNet). Although being a well-organized dataset with multi-scale samples of the original images, the UHManoa100 lacks diversity and is limited to 100 plant species. Hence, knowledge integration is not that evident since the CNN models inherited just a few discriminating features from the UHManoa100 dataset.

In the process of creating plant-expert models, we use a dataset covering much more variety of plant species to train the CNNs before fine-tuning them for the target environment. The training process over this domain-specific dataset helps the models to learn more discriminative features and better generalize for objects from that domain. Thus, we use natural images of different species in the iNat682 dataset to train plant-expert CNN models. These models produce plant-related pre-trained weights that serve as initial parameter values for the fine-tuning process over the UHManoa300 dataset. Consequently, the knowledge integration from an extensive dataset such as ImageNet and a domain-specific dataset such as the iNat682 resulted in better pre-trained CNN models for the plant categorization.

6   Observations and Discussions

The proposed approach described in this paper focuses on expanding the plant categorization scope of CNN-based systems to 300 plant species. As the first step, a botanist helped with the creation and organization of a new dataset (UHManoa300), ensuring a collection of good quality natural images of correctly annotated plants. Subsequently, classification CNN models need to be trained over this new dataset, and common methods such as retraining them from scratch resulted in poor performance. These state-of-the-art CNN models trained from scratch with randomly initialized parameters tend to overfit on training data with few images per class, resulting in incorrect predictions for test images. However, an adaptation on the top layers of the classification models accommodates a larger number of species and allows the use of pre-trained weights, improving the models' accuracy.

State-of-the-art CNN architectures implementing inception modules have shown improved results when using the ImageNet pre-trained weights (Krause et al. [5]). These base models trained to categorize the UHManoa100 dataset achieved highly accurate results, serving as plant-related pre-trained weights for the experiments with the UHManoa300 dataset. Furthermore, we integrate domain-specific knowledge using a much larger dataset (iNat682) into the categorization CNNs to create plant-expert models. The integration of domain-specific knowledge also helps to avoid the overfitting problem often encountered when training large CNN models over small datasets. Presented in Table 1, experimental results show the improvement of the CNN-based system when the classification models are pre-trained with domain-specific datasets. This accuracy gain is more evident when a large dataset, such as the iNat682, is used to generate the base models. Training over a big plant-related dataset, base models become plant-expert CNNs and assist on the fine-tuning of the classification models over the UHManoa300 dataset. As a result, the adaptation of previously trained CNNs accommodates an extended categorization scope and allows the CNN-based system to upgrade its models for target datasets with a larger number of classes.

Even with the ease of downloading ImageNet pre-trained weights[7], the creation of plant-expert models demands the learning of thousands of plant images. For instance, integrate domain-specific knowledge from the iNat682 dataset requires a lot of additional computational effort. The ImageNet+iNat682 plant-expert models required the full training of each CNN over 158,463 images before the final fine-tuning process. As an example, for the experiments presented in this paper, we used three GPUs (two *GeForce RTX 2070* and one *GeForce GTX 1080*), and it took almost two months to completely train all CNN models. Among those, the best performing model (*Xception*) took three weeks to be fully trained and fine-tuned.

---

[7] https://keras.io/applications/.

Experimental results presented in Table 1 show that CNN models improved their predictions when fine-tuned using domain-specific pre-trained weights. The results also show that the *Xception* [2] is the most effective CNN model for classifying 300 plant species. Focusing on the predictions of this model, this fine-tuned CNN can correctly categorize an average of 2.7% (81 testing images) more when using the ImageNet+iNat682 pre-trained weights (compared to other pre-trained weights). In particular, the *Xception* model using the ImageNet+iNat682 pre-trained weights correctly categorize ten images that are not even listed in the Top-5 predictions of models using other pre-trained weights. Figure 3 presents some of these images showing possible discriminative features that this CNN (*Xception*) has inherited from its respective plant-expert model. Visually reviewing these images, they all resembled the shape of small trees with voluptuous treetops. The iNat682 dataset has multiple annotated images of trees (not including any images from the UHManoa300 dataset) that create transferable knowledge related to these types of plants. Representative images of entire trees are not common for some species in the target dataset, causing categorization errors. As suggested by experiments performed in this paper, these errors can be remediated by integrating plant domain-specific knowledge.

Expanding the plant categorization scope brings the challenges of gathering a new target dataset and adapt the classification models to work with the new scope. This paper addresses both of these challenges, but one problem is noticeable in all plant species categorization system: the categorization of different plant components such as leaves, flowers, fruits, barks, etc. As shown in Fig. 4, some incorrectly categorized plants do not have their leaves completely visible, while the flowers of that species are evident in the images. In these cases, CNNs trained specifically to classify flowers may successfully categorize the plant species. Therefore, we designed our expandable CNN-based approach to be capable of adapting multiple categorization models, even if they focus on different objectives – i.e., new expert CNN models can be created focusing on each one of the plant's components. In the literature, previously implemented solutions using multiple CNN models to analyze different plant components [5,7,14] suggest an improvement in accuracy when categorizing plant images. Therefore, the addition of multiple expert CNN models may be a suitable alternative to better handle natural images of plants. Future experiments will consider expanding the flower categorization scope as well as integrating domain-specific knowledge from flower-related datasets.

Brugmansia x candida   *Catalpa longissima*

Citharexylum spinosum   *Erythrina crista-galli*

Fig. 3. Images correctly categorized by the expanded CNN-based approach using the *Xception* model with ImageNet+iNat682 pre-trained weights.

Hibiscus rosa-sinensis   *Thunbergia grandiflora*

Fig. 4. Images incorrectly categorized by the expanded CNN-based approach.

7   Conclusion

In this paper, we study the problem of expanding the categorization scope of CNN-based plant phenotyping systems. Amongst the many challenges of this problem, we address the particular challenges of gathering a new representative dataset for a larger environment, adapting the CNN classification layers to accommodate the larger scope, and integrating domain-specific knowledge to maintain a high categorization accuracy. The contributions of this paper include the publication of a new dataset (UHManoa300) with 50 correctly annotated images for each of the 300 plant species, the adaptation process implemented in the top layers of the CNN classification models to accommodate an expanded scope, and the integration of domain-specific knowledge to create plant-expert models (also available online). The pre-trained weights of these expert models can be used by other researchers as a starting point for the fine-tuning process over their new datasets.

Among the challenges, the creation of plant-expert models by integrating domain-specific knowledge is the most demanding one. We implement this integration process by repeatedly training the classification models over plant-related datasets to maintain the high categorization accuracy over a more significant number of species. By dedicating an enormous computational effort to train and fine-tune the modified models, we successfully expand the plant categorization scope to a broader environment while maintaining high accuracy.

In summary, the proposed approach provides a scalable solution for the problem of species categorization and expanding an existing plant phenotyping system to a larger environment. The future work of this research includes developing a unified approach to consolidate various plant phenotyping systems and integrating the knowledge previously learned by them to create better trained CNN expert models.

References

1. Barre, P., Stover, B., Muller, K., Steinhage, V.: LeafNet: a computer vision system for automatic plant species identification. Ecol. Inf. **40**, 50–56 (2017)
2. Chollet, F.: Xception: deep learning with depthwise separable convolutions. CoRR abs/1610.02357 (2016)
3. Cui, Y., Song, Y., Sun, C., Howard, A., Belongie, S.J.: Large scale fine-grained categorization and domain-specific transfer learning. CoRR abs/1806.06193 (2018)
4. He, K., Zhang, X., Ren, S., Sun, J.: Deep residual learning for image recognition. CoRR abs/1512.03385 (2015)
5. Krause, J., Baek, K., Lim, L.: A guided multi-scale categorization of plant species in natural images. In: CVPR Workshop on Computer Vision Problems in Plant Phenotyping (CVPPP 2019). IEEE Press (2019)
6. Krause, J., Sugita, G., Baek, K., Lim, L.: What's that plant? WTPlant is a deep learning system to identify plants in natural images. In: BMVC Workshop on Computer Vision Problems in Plant Phenotyping (CVPPP 2018). BMVA Press (2018)

7. Krause, J., Sugita, G., Baek, K., Lim, L.: WTPlant (what's that plant?): a deep learning system for identifying plants in natural images. In: Proceedings of the International Conference on Multimedia Retrieval (ICMR 2018). ACM Press (2018)
8. Krizhevsky, A., Sutskever, I., Hinton, G.E.: ImageNet classification with deep convolutional neural networks. In: Pereira, F., Burges, C.J.C., Bottou, L., Weinberger, K.Q. (eds.) Advances in Neural Information Processing Systems, vol. 25, pp. 1097–1105. Curran Associates, Inc. (2012)
9. Kumar, N., et al.: Leafsnap: a computer vision system for automatic plant species identification. In: Fitzgibbon, A., Lazebnik, S., Perona, P., Sato, Y., Schmid, C. (eds.) ECCV 2012. LNCS, vol. 7573, pp. 502–516. Springer, Heidelberg (2012). https://doi.org/10.1007/978-3-642-33709-3_36
10. Lee, S.H., Chan, C.S., Wilkin, P., Remagnino, P.: Deep-plant: plant identification with convolutional neural networks. In: 2015 IEEE International Conference on Image Processing (ICIP), pp. 452–456 (2015)
11. Mo, X., Cheng, R., Fang, T.: Pay attention to convolution filters: towards fast and accurate fine-grained transfer learning. CoRR abs/1906.04950 (2019)
12. Ngiam, J., Peng, D., Vasudevan, V., Kornblith, S., Le, Q.V., Pang, R.: Domain adaptive transfer learning with specialist models. CoRR abs/1811.07056 (2018)
13. Russakovsky, O., et al.: ImageNet large scale visual recognition challenge. Int. J. Comput. Vis. 115(3), 211–252 (2015). https://doi.org/10.1007/s11263-015-0816-y
14. Rzanny, M., Mäder, P., Deggelmann, A., Chen, M., Wäldchen, J.: Flowers, leaves or both? How to obtain suitable images for automated plant identification. Plant Methods 15(77), 1746–4811 (2019)
15. Sun, Y., Liu, Y., Guan, W., Zhang, H.: Deep learning for plant identification in natural environment. Comput. Intell. Neurosci. 2017(7361042), 6 (2017)
16. Szegedy, C., Ioffe, S., Vanhoucke, V.: Inception-v4, inception-ResNet and the impact of residual connections on learning. CoRR abs/1602.07261 (2016)
17. Szegedy, C., Vanhoucke, V., Ioffe, S., Shlens, J., Wojna, Z.: Rethinking the inception architecture for computer vision. CoRR abs/1512.00567 (2015)

Weakly Supervised Minirhizotron Image Segmentation with MIL-CAM

Guohao Yu[1] , Alina Zare[1(✉)] , Weihuang Xu[1] , Roser Matamala[2] ,
Joel Reyes-Cabrera[3] , Felix B. Fritschi[3] , and Thomas E. Juenger[4]

[1] Department of Electrical and Computer Engineering, University of Florida,
32611 Gainesville, FL, USA
azare@ece.ufl.edu
[2] Argonne National Laboratory, 60439 Lemont, IL, USA
[3] Division of Plant Sciences, University of Missouri, 65211 Columbia, MO, USA
[4] Department of Integrative Biology, University of Texas at Austin,
78712 Austin, TX, USA

Abstract. We present a multiple instance learning class activation map
(MIL-CAM) approach for pixel-level minirhizotron image segmentation
given weak image-level labels. Minirhizotrons are used to image plant
roots *in situ*. Minirhizotron imagery is often composed of soil containing
a few long and thin root objects of small diameter. The roots prove
to be challenging for existing semantic image segmentation methods to
discriminate. In addition to learning from weak labels, our proposed MIL-
CAM approach re-weights the root versus soil pixels during analysis for
improved performance due to the heavy imbalance between soil and root
pixels. The proposed approach outperforms other attention map and
multiple instance learning methods for localization of root objects in
minirhizotron imagery.

1 Introduction

Minirhizotron (MR) imaging plays an important role in plant root studies. It
is a widely-used non-destructive root sampling method used to monitor root
systems over extended periods of time without repeatedly altering critical soil
conditions or root processes [4,11,24,31]. Yet, a significant bottleneck which
impacts the value of MR systems is the analysis time needed to process collected
imagery. Standard analysis approaches require manual root tracing and labeling
of root characteristics. Manually tracing roots collected with MR systems is
very tedious, slow, and error prone. Thus, MR image analysis would greatly
benefit from automated methods to segment and trace roots. There have been
advancements made in this area [33,36,38]. However, the effective automated
methods still require a manually-labeled training set. Although these approaches
provide a reduction in effort needed over hand-tracing an entire collection of data,
the generation of these training sets is still time consuming and labor intensive.
In this paper, we propose a weakly supervised MR image segmentation method
that relies only on image-level labels.

© Springer Nature Switzerland AG 2020
A. Bartoli and A. Fusiello (Eds.): ECCV 2020 Workshops, LNCS 12540, pp. 433–449, 2020.
https://doi.org/10.1007/978-3-030-65414-6_30

G. Yu et al.

By relying only on weak image level labels (e.g., this image does/does not contain roots), the time and labor needed to generate a training set is drastically reduced [21,22,26,30,35,42]. It is also much easier and less error prone to identify when an image does or does not contain roots as opposed to correctly labeling every pixel in an image. However, current weakly-supervised methods used to infer pixel-levels labels do not perform as well as semantic image segmentation methods leveraging full annotation [3,6,16,17,25].

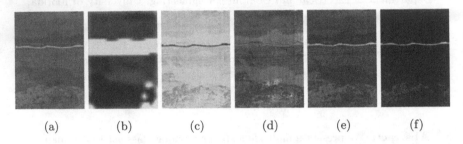

| | | | | | |
| (a) | (b) | (c) | (d) | (e) | (f) |

Fig. 1. Example attention maps from different methods of semantic segmentation of an MR image. (a) Original MR Image. (b) CAM Result (c) Grad-CAM Result (d) Grad-CAM++ Result (e) SMOOTHGRAD Result (f) Result of proposed method, MIL-CAM.

Attention or class activation maps have been widely used to infer pixel-level labels from training data with weak image-level labels [5,27,28,42]. However, existing attention map approaches have been found to be inaccurate in identifying and delineating roots in MR imagery. For example, CAM (the class activation maps) approach [42] overestimates the size of the roots as shown in Fig. 1b. The Gradient-weighted Class Activation Mapping (Grad-CAM) [27] approach incorrectly identifies the background soil as the root target as shown in Fig. 1c. Grad-CAM++ [5] and SMOOTHGRAD [28] shown in Fig. 1d and Fig. 1e, respectively, result in maps with poor contrast between roots and soil and, thus, many false alarms.

In this paper, we propose the multiple instance learning CAM (MIL-CAM) approach to address root segmentation in MR imagery given weak image-level labels. MIL-CAM is outlined in Sect. 3. In Sect. 4, we compare MIL-CAM approach results to existing approaches with both weak- and full-annotation on an MR dataset collected from switchgrass.

2 Related Work

2.1 Attention Maps for Semantic Segmentation

CAM [42] is one of the earliest methods showing that attention maps can localize the discriminative image components for classification. CAM uses a network structure consisting of a block of fully convolutional layers followed by a global

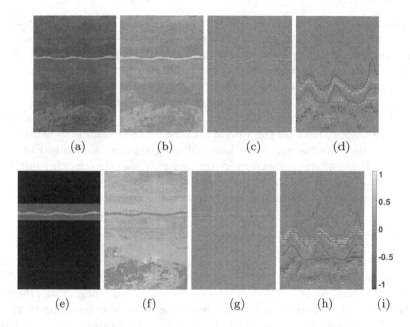

Fig. 2. Examples of the gradients with respect to the image classification score of the target root class using various individual feature maps. (a) Original MR image. (b) Feature map of Figure (a) which highlights the root area. (c) Gradients of feature map in Figure (b) with respect to the root class score. (d) Gradients in Figure (c) around the masked root regions shown in Figure (e). The red lines indicate the boundary of root object. (e) Cropped area from full image as mask in Figure (d) and Figure (h). (f) Feature map of Figure (a) which highlights soil. (g) Gradients of feature map in Figure (g). (h) Gradients in Figure (g) around root object in Figure (e). The red lines indicate the boundary of the root object. Figure (d) and Figure (h) are rescaled to match the size of other subfigure. (i) Colorbar used in for images in Figure.

average pooling layer and a single fully connected layer. The block of fully convolutional layers extract image features. These extracted features are combined linearly using the weights of the final fully connected layer to define the CAM. However, the CAM from this approach often has a low resolution, making it challenging to infer pixel-level labels precisely for semantic segmentation. Grad-CAM [27] is an extension of CAM which can estimate higher resolution attention maps by using features from any convolutional layer in the network. Specifically, Grad-CAM estimates the weights used for combining features as the average of gradients of the image classification score with respect to each value in the corresponding feature maps. Following the introduction of Grad-CAM, many methods were proposed to attempt to improve the quality of attention maps generated. Grad-CAM++ [5] takes a weighted average of only the positive gradients of the image classification score with respect to each feature map. SMOOTH-GRAD [28] averages over several Grad-CAM estimated attention maps with added zero-mean Gaussian noise with the aim of reducing sensitivity to feature map noise. Smooth Grad-CAM++ [18] mimics SMOOTHGRAD but applies the approach to Grad-CAM++ estimated attention maps. Score-CAM [32] attempts to improve Grad-CAM by weighting feature maps based on a metric which measures the increase of confidence for a class associated with the inclusion of each feature map. In application, all of these methods have been found to be either imprecise or sensitive to imbalanced data sets. Specifically in our application, soil pixels having complex gradients (i.e., both positive and negative gradients) which has a huge impact on the weights. Consider the example shown in Fig. 2. Gradients across the feature maps have differing signs as shown in Fig. 2d and Fig. 2h and, thus, when averaged over the map may cancel each other out. Given this, the standard Grad-CAM approach is ineffective since the average of the gradients over the feature map is used to compute the attention map.

2.2   Weakly Supervised Learning

Weakly supervised and multiple instance learning (MIL) algorithms for image segmentation do not require precise pixel-level labels. Under MIL, a set of samples (e.g., an image) is labeled as either "positive" or "negative." Positively labeled images are assumed to have at least one pixel corresponding to the target class (i.e., in our case, roots). The number of target pixels in positively labeled images are unknown. Negatively labeled images are composed of only non-target class (i.e., soil) pixels. Often, MIL approaches iteratively estimate the likelihood each pixel is a target and, using these values, update classifier parameters (and, then, subsequently update likelihood values again) [7,19]. Similarly, the pixels with the lowest target likelihood in each positively labeled image is also commonly assumed to be from the non-target class [7,8]. In contrast to methods that select likely target and non-target pixels, some methods have been proposed which consider all pixels in an image as equally contributing to the image-level label [42]. The Log-Sum-Exp (LSE) algorithm uses a hyper-parameter which trades off between selecting a single pixel as the target representative and considering all pixels in an image as target with equal contribution [22]. Global weighted rank pooling (GWRP) is another way to generalize number of pixels identified as targets [12]. In all of these approaches, it is difficult to select a fixed number of pixels to identify as targets representatives.

One reason that identifying the number of target pixels is challenging is that the size of the target class objects vary across images (e.g., some images contain only very few thin, fine roots whereas others are filled with roots of varying diameter). To alleviate this challenge, some approaches identify target pixels by adapting a threshold [1,10,14,35]. In [35], pixels with target class scores larger than a predefined threshold are labeled as targets and pixels with low saliency values are considered background. However, in this approach pixels are often unassigned to either target or background classes, pixels may be assigned to multiple target labels, or pixels may be assigned to background despite being surrounded in their neighborhood by target pixels. In [34], pixels in all of these cases are ignored. The approach outlined in [10] attempts to deal with these ignored pixels using deep seeded region growing (DSRG). DSRG proposes to propagate labels from labeled pixels to unlabeled pixels. The method presented in [14] extends the DSRG approach [10] by thresholding aggregated localization maps to improve delineation of target regions and adapts their algorithm to accommodate semi-supervised segmentation. Following an initial segmentation, some approaches apply post-processing steps to smooth and improve segmentation labels. These include conditional random fields (CRF) and the GWRP approach [1,12,13].

2.3 MR Image Segmentation

Several methods have been developed for automated minirhizotron image segmentation [9,23,40,41]. Currently, supervised deep learning approaches are the methods that are achieving the state-of-art results in MR image segmentation [29,33,36,37]. Yet, deep learning methods require a large collection of precisely traced root images for training the networks. A small number of approaches have been investigated for weakly supervised MR image segmentation [38]. Yu et al. [38] studied the application of three MIL algorithms: multiple instance adaptive cosine coherence estimator (MI-ACE) [39], multiple instance support vector machine (miSVM) [2], and multiple instance learning with randomized trees (MIForests) [15] for application to MR imagery. These methods, however, did not do feature learning and, so, the authors manually identified color features to be used during segmentation.

3   MIL-CAM Methodology

Semantic segmentation from weak labels using MIL-CAM is achieved in two training stages. The first stage, outlined in Algorithm 1, estimates the set of parameters needed to compute an attention map $\mathbf{S}^c \in \mathbb{R}^{M \times N}$ for a class c where M and N are the numbers of rows and columns of the input image, respectively. The attention map is estimated using the softmax output of a weighted linear combination feature maps extracted from the various layers of a trained CNN as described in Eq. 1,

$$\mathbf{S}^c = \frac{exp(\sum_j w_j^c y(\mathbf{F}_j) + b^c)}{\sum_q exp(\sum_j w_j^q y(\mathbf{F}_j) + b^q)}. \tag{1}$$

where q is an index over all output classes, $w_j^q \in \mathbb{R}$ is the weight estimated for class q and feature map $\mathbf{F}_j \in \mathbb{R}^{A_j \times B_j}$, A_j and B_j are the number of rows and columns in the jth feature map, $b^q \in \mathbb{R}$ is an estimated bias term, and $y(\cdot)$ is an interpolation function to scale an input to the size of $M \times N$.

Once attention maps are obtained using Algorithm 1, a segmentation network is then trained as outlined in Algorithm 2. After training, the segmentation network maps input test imagery to get pixel level segmentation outputs.

3.1   Attention Map Estimation

MIL-CAM estimates the set of parameters needed to obtain attention maps and compute Eq. 1 using the combination of three key components: (a) a pixel-level feature extraction component; (b) a pixel sampling component used to form a bag for each image for MIL analysis; and (c) a linear model that performs the MIL-based segmentation. The sampled pixels with features extracted from the image classification network are used to train the linear model. The approach is illustrated in Fig. 3 and outlined in the following sub-sections.

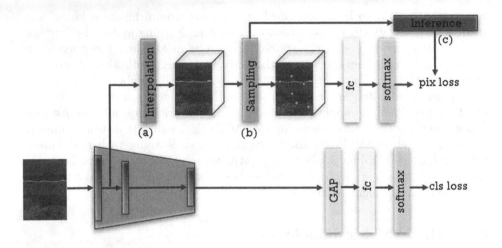

Fig. 3. Architecture of MIL-CAM. GAP represents a global average pooling layer and fc represents a fully connected layer. cls loss represents the loss for image classification into positive (i.e., containing roots) or negative (i.e., does not contain roots). pix loss represents the loss for pixel level classification into root vs. soil.

Feature Extraction and Interpolation. An image-level CNN classification network is first trained to extract coarse feature maps for each image. The training data set, $\{(\mathbf{I}_1, l_1), ...(\mathbf{I}_k, l_k), ...(\mathbf{I}_K, l_K)\}$, consists of K images where each image $\mathbf{I}_k \in \mathbb{R}^{3 \times M \times N}$ is paired with image label $l_k \in \{0, 1\}$ where 0 represents a negative image (i.e., does not contain roots) and 1 represents a positive image (i.e., contains roots). Using this training data an image-level classification network is trained by optimizing the cross-entropy loss as shown in Eq. 2,

$$\min_{\boldsymbol{\theta}_0} \sum_{k=1}^{K} L_{cls-loss}(\mathbf{I}_k; \boldsymbol{\theta}_0, \mathbf{l}_k) = \frac{-1}{K} \sum_{k=1}^{K} \sum_{q=1}^{2} l_{kq} \log f_q(\mathbf{I}_k; \boldsymbol{\theta}_0) \qquad (2)$$

where $\mathbf{l}_k = [l_{k1}, l_{k2}]$ is the one-hot encoded label of the image $\mathbf{I}_k$ and $f_q(\mathbf{I}_k; \boldsymbol{\theta}_0)$ is the qth element of the softmax output layer of the the image classification network defined by parameters $\boldsymbol{\theta}_0$. The assumption is, provided an effective image-level classification network can be trained, that the network is extracting features that are useful for the semantic segmentation problem and these useful features are encoded in the CNN feature maps. Once the classification network is trained, then the coarse CNN feature maps are upsampled using bilinear interpolation to match the size of the input image. Each pixel is represented by the corresponding feature vector obtained from the collection of upsampled feature maps.

Algorithm 1: Estimating Weights and Biases for Attention Maps

---

Data: $(\mathbf{I}, \mathbf{l}) = \{(\mathbf{I}_1, l_1), ...(\mathbf{I}_k, l_k), ...(\mathbf{I}_K, l_K)\}$

Train the image classification network with $(\mathbf{I}, \mathbf{l})$;

Extract feature maps $\mathbf{F}$ from image classification network for $(\mathbf{I}, \mathbf{l})$;

Interpolate the CNN feature maps $y(\mathbf{F})$ for $(\mathbf{I}, \mathbf{l})$;

Sample instances and construct bags, $\{(\mathbf{B}_1, l_1), ...(\mathbf{B}_k, l_k), ...(\mathbf{B}_K, l_K)\}$;

Initialize each instance label with the label of its corresponding bag;

repeat

    Update $\mathbf{w}, \mathbf{b}$ by optimizing Eq. 3 with stochastic gradient descent for one epoch using the instances and updated labels $(\mathbf{x}_k^n, \mathbf{l}_k^n)$;

    Compute $p_k^n = g(\mathbf{x}_k^n; \mathbf{w}, \mathbf{b})$ for each instance;

    for Every positive bag $(\mathbf{B}_k, l_k = 1)$ **do**

        $p_t = \text{Otsu's}(\{p_k^1, ...p_k^{N_k}\})$;

        If $p_k^n \geq p_t$, then set $\mathbf{l}_k^n$ as target, else set $\mathbf{l}_k^n$ as non-target;

    end

    for Every negative bag $(\mathbf{B}_k, l_k = 0)$ **do**

        set $\mathbf{l}_k^n$ as non-target;

    end

until Fixed number of epochs completed;

return $\boldsymbol{\theta}_0, \mathbf{w}, \mathbf{b}$ from epoch with smallest loss

---

Instance Sampling. In order to address some of the imbalance in the data set (i.e., there are many more soil pixels than root pixels), a sampling approach is used to identify representative pixels from each image. The green band of the RGB minirhizotron image is used for instance sampling. The approach draws a single pixel to represent the set of pixels from each possible 8-bit value from the green band in the image. In other words, a 256 bin histogram is built using the values of the green band of the MR imagery. For each non-empty bin, a uniform random draw is used to identify a representative pixel for that green-level. In our application, we found this to be an effective approach to re-balance root-vs-nonroot pixels in positively labeled imagery (given that pixel level labels are unavailable). The sampled pixels are organized into a set of bags, $\{(\mathbf{B}_1, l_1), ...(\mathbf{B}_k, l_k), ...(\mathbf{B}_K, l_K)\}$. Each bag, $\mathbf{B}_k = \left\{\mathbf{x}_k^1, \mathbf{x}_k^2, \ldots, \mathbf{x}_k^{N_k}\right\}$, corresponds to one image $\mathbf{I}_k$ with image label l_K and is composed of N_k instances.

The instance $\mathbf{x}_k^n \in \mathbb{R}^J$ is the feature vector for the nth instance in the kth bag where J is the number of feature maps used to construct the feature vectors.

Algorithm 2: Weakly Supervised Image Segmentation

Data: $\{(\mathbf{I}_1, l_1), ... (\mathbf{I}_k, l_k), ... (\mathbf{I}_K, l_K)\}$

Parameter: s_t

for Every positive image **do**

    | Compute the score-map $\mathbf{S}_k^c$ from MIL-CAM using $(\boldsymbol{\theta}_0, \mathbf{w}, \mathbf{b})$ in Eq. 1 ;

    | Estimate a threshold, $o_t = $ Otsu's $(\mathbf{S}_k^c)$;

    | If $\mathbf{S}_k^c(m, n) \geq o_t$, then set $l_k^0(m, n)$ as target, else set $l_k^0(m, n)$ as non-target;

end

for Every negative image $(\mathbf{I}_k, l_k = 0)$ **do**

    | set $l_k^0(m, n)$ as non-target;

end

Update parameters $\boldsymbol{\theta}_1$ for data set with pixel labels $l_k^0(m, n)$ for a fixed number of epochs;

repeat

    | Compute score-map for each image using the U-Net with updated parameters;

    | **if** A positive image $(\mathbf{I}_k, l_k = 1)$ **then**

        | If $\mathbf{P}_k(m, n) \geq s_t$, then set $l_k(m, n)$ as target, else $l_k(m, n)$ as non-target;

        | If every pixel $\mathbf{P}_k(m, n) < s_t$, then $l_k(m, n) = l_k^0(m, n)$;

    | **else if** A negative image $(\mathbf{I}_k, l_k = 0)$ **then**

        | set $l_k(m, n)$ as non-target;

    | Update parameters $\boldsymbol{\theta}_1$ for dataset with pixel labels $l_k(m, n)$;

until Fixed number of epochs completed;

return Segmentation network parameters $\boldsymbol{\theta}_1$

Estimated Weights and Biases. After instance sampling, the weights and biases used to compute the attention maps as defined in Eq. 1 are estimated by optimizing the cross-entropy loss shown in Eq. 3 given the MIL constraints that for each positive bag, at least one instance must be labeled as root and all instances in every negative bag are labeled as non-root,

$$\min_{l_k^n} \min_{\mathbf{w}, \mathbf{b}} \sum_{\mathbf{x}_k^n} L_{pix-loss}(\mathbf{x}_k^n; \mathbf{w}, \mathbf{b}, l_k^n) = \frac{-1}{\sum_k N_k} \sum_{\mathbf{x}_k^n} \sum_q l_{kq}^n \log g_q(\mathbf{x}_k^n; \mathbf{w}, \mathbf{b}) \quad (3)$$

where l_k^n is the one-hot encoded label of the instance $\mathbf{x}_k^n$ and $g_q(\mathbf{x}_k^n; \mathbf{w}, \mathbf{b})$ is the qth element of the softmax output of the MIL-CAM with parameters $(\mathbf{w}, \mathbf{b})$. The loss is updated iteratively as outlined in Algorithm 1. During the initial epoch, each instance is labeled the same label as its bag. In all subsequent epochs, the probability that an instance belongs to the target class, $p_k^n = g(\mathbf{x}_k^n; \mathbf{w}, \mathbf{b})$, is predicted by the linear model trained from the previous epoch. Then for each positive bag, a threshold p_t is computed using Otsu's threshold [20] and all instances greater than the threshold are labeled as target whereas all others are labeled as non-target. For negative bags, all instances are labeled as non-target.

3.2 Training the Image Segmentation Network

Once MIL-CAM attention maps can be estimated, an image segmentation network is trained as outlined in Algorithm 2. First, target class attention maps for positively labeled images are estimated and thresholded using Otsu's threshold to obtain a label for each pixel. All pixels in negatively labeled images are given a non-target label. These labels are used to estimate the parameters for the U-Net [25] architecture illustrated in lower branch of Fig. 4. After initially training the U-Net with labels obtained from the attetnion maps, the U-Net is iteratively fine-tuned. A score-map, $\mathbf{P}_k \in \mathbb{R}^{M \times N}$, is computed using the soft-max output of the U-Net. The score-map of positively-labeled is thresholded using a fixed (large) threshold parameter, s_t, to obtain updated pixel level labels which highlight more likely positive samples. The updated labels are iteratively used to fine-tune the parameters of the U-Net.

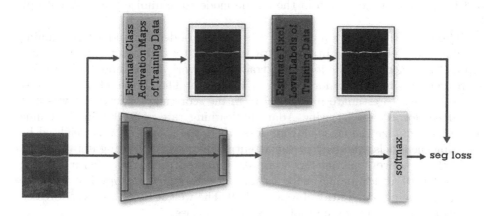

Fig. 4. Architecture of segmentation U-Net with MIL training branch. The bottom branch is the U-Net. The top branch is used to infer label of training data.

4 Experiments

4.1 Data Description

For our experiments, we used a switchgrass (*Panicum virgatum* L.) MR imagery dataset consisting of 561 training images with image-level labels and 30 test and validation images with pixel-level labels. Each image was 2160×2550 in size and was divided into sub-images of size 720×510. 500 sub-images containing roots and 500 sub-images containing only soil were randomly selected as training data for estimating attention map parameters. 1500 root sub-images and 1500 soil sub-images were randomly selected as training data for the U-Net segmentation network. The 30 images with pixel-level labels were randomly divided into 10 validation images and 20 test images.

4.2   Architecture

Our experiments use U-Net [25] with layer depth of 5 as backbone for MR image segmentation. The feature extraction network used to estimate attention map parameters was a 2-class convolutional neural network with the encoder of the U-Net, followed by a global average pooling layer and a fully connected layer. We extract $1024 \times 46 \times 33$ feature maps and vectorize the feature maps to classify each image into 2 classes with a fully connected layer. The feature extraction net is trained using SGD at a learning rate of 0.0001 and momentum of 0.8 in the online mode to minimize the cross entropy loss. The MIL-CAM attention map module extracts a 64-dimensional feature for each sampled instance from the fourth layer of the encoder of the feature extraction network. Then, classifies each sampled instance into one of two classes using a fully connected layer. The MIL-CAM attention map module is trained using SGD at a learning rate of 0.001 and momentum of 0.5 in the online mode to minimize the cross entropy loss.

The image segmentation network was a U-Net of depth 5 and a MIL training branch. The MIL training branch extracts $64 \times 720 \times 510$ features from the first layer of the encoder of the feature extraction network and compute a 720×510 score-map of target class for each training image. The threshold parameter s_t was set to 0.9 to estimate pixel label from the score-map. The U-Net was first initialized for 10 epochs using Adam at learning rate of 0.0001 in the online mode to minimize the cross entropy loss where the root class was weighted by 50 using the labels produced by the attention maps. Then, during iterative fine-tuning, the network parameters were also updated using Adam with learning rate of 0.0001 in the online mode to minimize the cross entropy loss with the root class having an additional weight of 50. The weight on root class addressed the imbalance issue between root class and soil class.

4.3   Experiments: MIL-CAM Attention Maps

The attention maps of MIL-CAM were first qualitatively compared with attention maps of other methods as shown in Fig. 5. As can be seen, MIL-CAM results shown in Fig. 5f more accurately indicate root locations as compared to the attention maps produced by CAM in Fig. 5b. This difference in performance is largely due to the fact that CAM requires interpolating a low resolution attention map to the size of the input image resulting in blurred, oversized detection regions.

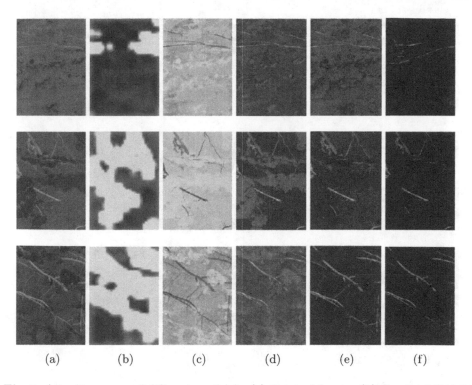

(a)   (b)   (c)   (d)   (e)   (f)

Fig. 5. Attention maps of different methods. (a) Original Image. (b) Result of CAM. (c) Result of Grad-CAM. (d) Result of Grad-CAM++. (e) Result of SMOOTHGRAD. (f) Result of MIL-CAM.

Grad-CAM in Fig. 5c. fails to correctly identify roots and, instead, highlights soil. Furthermore, MIL-CAM produced attention maps with higher contrast between root pixels and background than those Grad-CAM ++ in Fig. 5d and SMOOTH-GRAD in Fig. 5e.

Figure 6 compares attention maps from a selection of approaches after thresholding with Otsu's threshold. Table 1 lists the average and standard devation for precision, recall and F1 score of three training runs of the various approaches to compare the quality these thresholded results. The proposed MIL-CAM method has a significantly higher F1 score among all those compared. The precision of MIL-CAM is an order of magnitude better than the comparison methods without a significant loss in recall as compared with the gains of precision. Although other methods except Grad-CAM have a better recall, the low precision scores of these methods indicate a large amount of background pixels are mislabeled as root pixels. This can be visualized in Fig. 6.

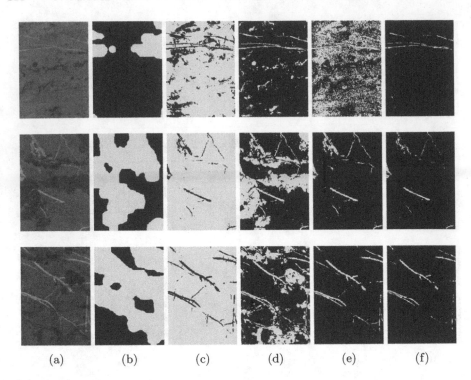

(a)   (b)   (c)   (d)   (e)   (f)

Fig. 6. Thresholded attention maps. (a) Original Image. (b) Result of CAM. (c) Result of Grad-CAM. (d) Result of Grad-CAM++. (e) Result of SMOOTHGRAD. (f) Result of MIL-CAM.

4.4   Experiments: Semantic Segmentation

We also compared the performance of our final MIL segmentation network (i.e., MIL-CAM Th in the table) against other MIL methods (MI-ACE [38], miSVM [38], and MIForest [38]). The average and standard deviation of three runs of the precision, recall, F1 score and mIoU were compared at false positive rate (FPR) is 0.03 in Table 2. Our proposed approach outperformed all other MIL methods. The proposed MIL-CAM Th method (i.e., the thresholded MIL-CAM result) achieved recall = 0.878. The recall of MIL-CAM Th was 10% better than miSVM which was the second best. MIL-CAM Th also had the best precision of all MIL methods.

The segmentation results of the proposed MIL-CAM approach when taking the argmax of the softmax outputs (i.e., argmax MIL-CAM in the table) are shown in the third column in Fig. 7c. The long roots are a challenging problem. Although our proposed method detects most of the root pixels, it expands the boundary of some roots. This expansion results in high recall (0.859) but

Table 1. Compare results of thresholded attention maps

Method	Precision	Recall	F1 score	mIoU
CAM	0.045 ± 0.0053	**0.931 ± 0.0459**	0.085 ± 0.0098	0.045 ± 0.0053
Grad-CAM	0.003 ± 0.0012	0.229 ± 0.0939	0.006 ± 0.0024	0.003 ± 0.0012
Grad-CAM++	0.015 ± 0.0084	0.550 ± 0.1951	0.030 ± 0.0159	0.015 ± 0.0083
SMOOTHGRAD	0.033 ± 0.0028	0.782 ± 0.0191	0.064 ± 0.0052	0.033 ± 0.0028
MIL-CAM	**0.248 ± 0.1870**	0.536 ± 0.1450	**0.289 ± 0.1814**	**0.177 ± 0.1190**

Table 2. Comparison of image segmentation results. All comparison methods use weak image level labels except the U-Net approach from [36]. MIL-CAM Th is the result found after thresholding the U-Net softmax outputs corresponding to the target class at FPR = 0.03; argmax MIL-CAM is the result when taking the argmax of U-Net softmax outputs; and MIL-CAM + CRF method is the result when the argmax MIL-CAM result is postprocessed with a CRF.

Method	Label	Precision	Recall	F1 score	mIoU
U-Net [36]	Pixel	0.307	**0.913**	0.459	0.298
MI-ACE [38]	Image	0.130 ± 0.0010	0.775 ± 0.0067	0.223 ± 0.0017	0.125 ± 0.0011
miSVM [38]	Image	0.134 ± 0.0015	0.798 ± 0.0104	0.229 ± 0.0026	0.129 ± 0.0017
MIForests [38]	Image	0.101 ± 0.0104	0.582 ± 0.0664	0.172 ± 0.0180	0.094 ± 0.0108
MIL-CAM Th	Image	0.145 ± 0.0050	0.878 ± 0.0341	0.249 ± 0.0088	0.142 ± 0.0057
argmax MIL-CAM	Image	0.186 ± 0.0278	0.859 ± 0.0423	0.304 + 0.0364	0.180 ± 0.0251
MIL-CAM + CRF	Image	**0.667 ± 0.0257**	0.692 ± 0.0267	**0.678 ± 0.0058**	**0.513 ± 0.0066**

low precision (0.186) as shown in Table 2. To mitigate this, we also applied a conditional random field (CRF) [13] postprocessing to the segmentation results of our approach. The default parameters of the CRF were used as 0.7 for the certainty of the label, 3 for the parameter of the smoothness kernel, 80 for the spatial parameter of the appearance kernel, 13 for the color parameter of the appearance kernel and 2 inference steps were run. Segmentation results after CRF postprocessing are shown in the fourth column in Fig. 7c. Postprocessing improved the precision of results from 0.186 to 0.667, and the mean Intersection-Over-Union (mIoU) from 0.180 to 0.513 as shown in Table 2. The only approach with that outperformed the proposed MIL-CAM with CRF postprocessing on any metric was the U-Net method outlined in [36]. However, this U-Net was pretrained using a large dataset consisting of 17567 MR images with full pixel-level annotation and, thus, did not have to overcome the weak label challenge.

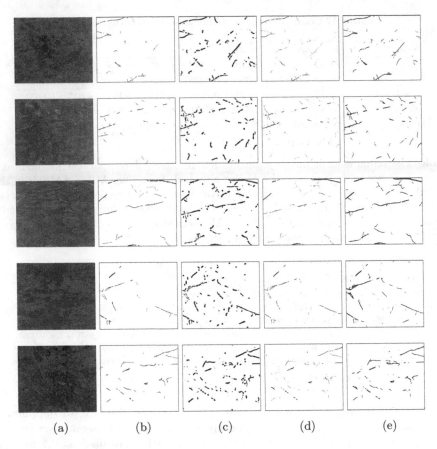

     (a)   (b)   (c)   (d)   (e)

Fig. 7. Qualitative examples of root segmentation results with different method. (a) Original image. (b) groundtruth (GT). (c) Result of argmax MIL-CAM (d) Result of argmax MIL-CAM + CRF. (e) Result of U-Net.

5   Conclusion

In this work, we proposed MIL-CAM for weakly supervised MR image segmentation. The proposed MIL-CAM approach outperformed a variety of comparison attention map approaches as well as a variety of MIL segmentation methods, particularly when incorporating a CRF post-processing.

Acknowledgement. This work was supported by the U.S. Department of Energy, Office of Science, Office of Biological and Environmental Research award number DE-SC0014156 and by the Advanced Research Projects Agency - Energy award number DE-AR0000820.

References

1. Ahn, J., Cho, S., Kwak, S.: Weakly supervised learning of instance segmentation with inter-pixel relations. In: Proceedings of the IEEE Conference on Computer Vision and Pattern Recognition, pp. 2209–2218 (2019)
2. Andrews, S., Tsochantaridis, I., Hofmann, T.: Support vector machines for multiple-instance learning. In: Advances in Neural Information Processing Systems, pp. 577–584 (2003)
3. Badrinarayanan, V., Kendall, A., Cipolla, R.: SegNet: a deep convolutional encoder-decoder architecture for image segmentation. IEEE Trans. Pattern Anal. Mach. Intell. **39**(12), 2481–2495 (2017)
4. Bates, G.: A device for the observation of root growth in the soil. Nature **139**(3527), 966–967 (1937)
5. Chattopadhay, A., Sarkar, A., Howlader, P., Balasubramanian, V.N.: Grad-CAM++: generalized gradient-based visual explanations for deep convolutional networks. In: 2018 IEEE Winter Conference on Applications of Computer Vision (WACV), pp. 839–847. IEEE (2018)
6. Chen, L.C., Papandreou, G., Kokkinos, I., Murphy, K., Yuille, A.L.: DeepLab: semantic image segmentation with deep convolutional nets, atrous convolution, and fully connected CRFs. IEEE Trans. Pattern Anal. Mach. Intell. **40**(4), 834–848 (2017)
7. Durand, T., Mordan, T., Thome, N., Cord, M.: WILDCAT: weakly supervised learning of deep convnets for image classification, pointwise localization and segmentation. In: Proceedings of the IEEE Conference on Computer Vision and Pattern Recognition, pp. 642–651 (2017)
8. Durand, T., Thome, N., Cord, M.: WELDON: weakly supervised learning of deep convolutional neural networks. In: Proceedings of the IEEE Conference on Computer Vision and Pattern Recognition, pp. 4743–4752 (2016)
9. Heidari, M., et al.: A new method for root detection in minirhizotron images: hypothesis testing based on entropy-based geometric level set decision. Int. J. Eng. **27**(1), 91–100 (2014)
10. Huang, Z., Wang, X., Wang, J., Liu, W., Wang, J.: Weakly-supervised semantic segmentation network with deep seeded region growing. In: Proceedings of the IEEE Conference on Computer Vision and Pattern Recognition, pp. 7014–7023 (2018)
11. Johnson, M., Tingey, D., Phillips, D., Storm, M.: Advancing fine root research with minirhizotrons. Environ. Exp. Bot. **45**(3), 263–289 (2001)
12. Kolesnikov, A., Lampert, C.H.: Seed, expand and constrain: three principles for weakly-supervised image segmentation. In: Leibe, B., Matas, J., Sebe, N., Welling, M. (eds.) ECCV 2016. LNCS, vol. 9908, pp. 695–711. Springer, Cham (2016). https://doi.org/10.1007/978-3-319-46493-0_42
13. Krähenbühl, P., Koltun, V.: Efficient inference in fully connected CRFs with Gaussian edge potentials. In: Advances in Neural Information Processing Systems, pp. 109–117 (2011)
14. Lee, J., Kim, E., Lee, S., Lee, J., Yoon, S.: FickleNet: weakly and semi-supervised semantic image segmentation using stochastic inference. In: Proceedings of the IEEE Conference on Computer Vision and Pattern Recognition, pp. 5267–5276 (2019)

15. Leistner, C., Saffari, A., Bischof, H.: MIForests: multiple-instance learning with randomized trees. In: Daniilidis, K., Maragos, P., Paragios, N. (eds.) ECCV 2010. LNCS, vol. 6316, pp. 29–42. Springer, Heidelberg (2010). https://doi.org/10.1007/978-3-642-15567-3_3

16. Lin, G., Shen, C., Van Den Hengel, A., Reid, I.: Efficient piecewise training of deep structured models for semantic segmentation. In: Proceedings of the IEEE Conference on Computer Vision and Pattern Recognition, pp. 3194–3203 (2016)

17. Long, J., Shelhamer, E., Darrell, T.: Fully convolutional networks for semantic segmentation. In: Proceedings of the IEEE Conference on Computer Vision and Pattern Recognition, pp. 3431–3440 (2015)

18. Omeiza, D., Speakman, S., Cintas, C., Weldermariam, K.: Smooth grad-CAM++: an enhanced inference level visualization technique for deep convolutional neural network models. arXiv preprint arXiv:1908.01224 (2019)

19. Oquab, M., Bottou, L., Laptev, I., Sivic, J.: Is object localization for free?-weakly-supervised learning with convolutional neural networks. In: Proceedings of the IEEE Conference on Computer Vision and Pattern Recognition, pp. 685–694 (2015)

20. Otsu, N.: A threshold selection method from gray-level histograms. IEEE Trans. Syst. Man Cybern. 9(1), 62–66 (1979)

21. Papandreou, G., Chen, L.C., Murphy, K.P., Yuille, A.L.: Weakly-and semi-supervised learning of a deep convolutional network for semantic image segmentation. In: Proceedings of the IEEE International Conference on Computer Vision, pp. 1742–1750 (2015)

22. Pinheiro, P.O., Collobert, R.: From image-level to pixel-level labeling with convolutional networks. In: Proceedings of the IEEE Conference on Computer Vision and Pattern Recognition, pp. 1713–1721 (2015)

23. Rahmanzadeh, B.H., Shojaedini, S.: Novel automated method for minirhizotron image analysis: Root detection using curvelet transform. Int. J. Eng. 29, 337–346 (2016)

24. Rewald, B., Ephrath, J.E.: Minirhizotron techniques. In: Plant Roots: The Hidden Half, pp. 1–15 (2013)

25. Ronneberger, O., Fischer, P., Brox, T.: U-net: convolutional networks for biomedical image segmentation. In: Navab, N., Hornegger, J., Wells, W.M., Frangi, A.F. (eds.) MICCAI 2015. LNCS, vol. 9351, pp. 234–241. Springer, Cham (2015). https://doi.org/10.1007/978-3-319-24574-4_28

26. Roy, A., Todorovic, S.: Combining bottom-up, top-down, and smoothness cues for weakly supervised image segmentation. In: Proceedings of the IEEE Conference on Computer Vision and Pattern Recognition, pp. 3529–3538 (2017)

27. Selvaraju, R.R., Cogswell, M., Das, A., Vedantam, R., Parikh, D., Batra, D.: Grad-CAM: visual explanations from deep networks via gradient-based localization. In: Proceedings of the IEEE International Conference on Computer Vision, pp. 618–626 (2017)

28. Smilkov, D., Thorat, N., Kim, B., Viégas, F., Wattenberg, M.: SmoothGrad: removing noise by adding noise. arXiv preprint arXiv:1706.03825 (2017)

29. Smith, A.G., Petersen, J., Selvan, R., Rasmussen, C.R.: Segmentation of roots in soil with U-net. Plant Methods 16(1), 1–15 (2020)

30. Tang, M., Djelouah, A., Perazzi, F., Boykov, Y., Schroers, C.: Normalized cut loss for weakly-supervised cnn segmentation. In: Proceedings of the IEEE Conference on Computer Vision and Pattern Recognition, pp. 1818–1827 (2018)

31. Waddington, J.: Observation of plant roots in situ. Can. J. Bot. 49(10), 1850–1852 (1971)

32. Wang, H., Du, M., Yang, F., Zhang, Z.: Score-CAM: improved visual explanations via score-weighted class activation mapping. arXiv preprint arXiv:1910.01279 (2019)
33. Wang, T., et al.: SegRoot: a high throughput segmentation method for root image analysis. Comput. Electron. Agric. **162**, 845–854 (2019)
34. Wei, Y., Feng, J., Liang, X., Cheng, M.M., Zhao, Y., Yan, S.: Object region mining with adversarial erasing: a simple classification to semantic segmentation approach. In: Proceedings of the IEEE Conference on Computer Vision and Pattern Recognition, pp. 1568–1576 (2017)
35. Wei, Y., Xiao, H., Shi, H., Jie, Z., Feng, J., Huang, T.S.: Revisiting dilated convolution: a simple approach for weakly-and semi-supervised semantic segmentation. In: Proceedings of the IEEE Conference on Computer Vision and Pattern Recognition, pp. 7268–7277 (2018)
36. Xu, W., et al.: Overcoming small minirhizotron datasets using transfer learning. Comput. Electronics in Agriculture **175** (2020). https://doi.org/10.1016/j.compag.2020.105466
37. Yasrab, R., Atkinson, J.A., Wells, D.M., French, A.P., Pridmore, T.P., Pound, M.P.: Rootnav 2.0: deep learning for automatic navigation of complex plant root architectures. GigaScience **8**(11), giz123 (2019)
38. Yu, G., et al.: Root identification in minirhizotron imagery with multiple instance learning. Mach. Visi. Appl. **31** (2020). https://doi.org/10.1007/s00138-020-01088-z
39. Zare, A., Jiao, C., Glenn, T.: Discriminative multiple instance hyperspectral target characterization. IEEE Trans. Pattern Anal. Mach. Intell. **40**(10), 2342–2354 (2017)
40. Zeng, G., Birchfield, S.T., Wells, C.E.: Detecting and measuring fine roots in minirhizotron images using matched filtering and local entropy thresholding. Mach. Vis. Appl. **17**(4), 265–278 (2006)
41. Zeng, G., Birchfield, S.T., Wells, C.E.: Rapid automated detection of roots in minirhizotron images. Mach. Vis. Appl. **21**(3), 309–317 (2010)
42. Zhou, B., Khosla, A., Lapedriza, A., Oliva, A., Torralba, A.: Learning deep features for discriminative localization. In: Proceedings of the IEEE Conference on Computer Vision and Pattern Recognition, pp. 2921–2929 (2016)

BTWD: Bag of Tricks for Wheat Detection

Yifan Wu[1], Yahan Hu[2], and Lei Li[3(✉)]

[1] Beihang University, Beijing 100191, China
yifanwu@buaa.edu.cn
[2] Northwestern Polytechnical University, Xi'an 710072, China
yahanHu@mail.nwpu.edu.cn
[3] Beijing Institute of Technology, Beijing 100081, China
3220180022@bit.edu.cn

Abstract. Accurate detection of wheat heads outdoors is a great challenge. Wheat color and shape distinctions, as well as overlaps and wind blurring in wheat photos, make it difficult to detect wheat heads. We propose a Bag of Tricks for Wheat Detection (BTWD), finding that a reasonable combination of some tricks will bring great improvement to the wheat detection results, and apply it on different networks such as YOLO v5x, YOLO v3, EfficientDet-D5, Faster R-CNN, etc. BTWD has greatly enhanced comparison with the original network without tricks. YOLO v5x with BTWD achieves 77.07% in average mAP, in comparison, only 70.78% without it on the Global Wheat Head Detection (GWHD) dataset.

Keywords: Object detection · Wheat head · Bag of Tricks for Wheat Detection (BTWD)

1 Introduction

Wheat is a cereal crop planted widely around the world and is one of the staple foods for humans. Farmers can adopt on-farm management measures corresponding to the growth of wheat, such as assessing the maturity and health of wheat based on the density and size of wheat heads in the images. In such cases, the use of visual object detection to measure wheat-related data becomes a strong alternative to manual processes.

In particular, the above three challenges are pronounced for accurate wheat head detection. The analysis is as follows:

(1) Various appearances. Each wheat plant may has different maturity, orientation, genotype, and head orientation.
(2) Cluttered arrangements. Wheat heads for detection are often densely arranged. To make matters worse, the wheat is disturbed by the wind, which makes the resulting images blurry.

© Springer Nature Switzerland AG 2020
A. Bartoli and A. Fusiello (Eds.): ECCV 2020 Workshops, LNCS 12540, pp. 450–460, 2020.
https://doi.org/10.1007/978-3-030-65414-6_31

(3) Different growing conditions. Since the wheat images in the Wheat Head Detection (GWHD) dataset [3] are from different countries and regions, there will be variability between the photos due to different varieties, planting densities, patterns, and field conditions.

In this paper we present a Bag of Tricks for Wheat Detection (BTWD), and our empirical evaluation shows that combining them in a logical way can further improve the performance of the model (see details in Table 4).

2   Bag of Tricks for Wheat Detection (BTWD)

To address the problems of blocking and weathering in the GWHD dataset, we add the MixUp to the image data augmentation, which is a weighted summation of the images to produce a overlapping effect on images. In addition, the Mosaic is also randomly used in our data augmentation. Before Non-Maximum Suppression (NMS), we need to remove the predicted boxes with scores less than a certain threshold through a filtering, in order to get a more reasonable threshold, so we try the Out-of-Fold (OOF). The post-processing method, we apply Weighted Box Fusion [10] (WBF) to avoid deleting the predictied boxes by mistake, which makes the result more reasonable by fusing the predicted boxes, instead of deleting the boxes with high overlap. Finally, in the testing phase, we also choose the Test-Time Augmentation (TTA) and Pseudo-Label to try to improve the predicted scores by enhancing the test images and adding pseudo-labels to the test images before adding the training set to train the network.

K-Means for Anchors. We adopt the idea of transfer learning and use the pre-training weights of the COCO dataset. However, the dimensions of the bounding boxes in the images of the GWHD dataset, which are smaller and the aspect ratio is around 1, are different from that in the COCO dataset. Therefore, K-Means is used to cluster the dimensions of the bounding boxes of the wheat training set and thereby set the anchor size again. At the beginning, we randomly select a bounding box as the centroid, and the distance from each remaining bounding box to the centroid is defined as:

$$d(\text{box}, \text{centroid}) = 1 - \text{IOU}(\text{box}, \text{centroid}) \tag{1}$$

where IOU is the IoU value when the center points of the two boxes of the box and the centroid overlap.

We choose 9 clustering centroids, and the scatter plot obtained from the K-Means clustering experiment shows that the pink dots represent the length and width positions of all bounding boxes, and the red triangles are the final clustering results, and the Table 1 shows the specific values. Figure 1 and presents the distribution scatter plot and the new anchors table.

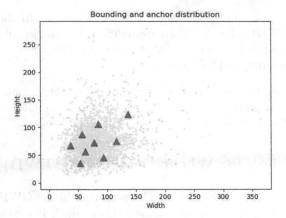

Fig. 1. The distribution of new anchors of GWHD dataset (Color figure online)

Table 1. Recommanded aspect ratios (width/height)

Width	Height	Height/Width
54.000	35.000	0.6
37.000	67.000	1.8
63.000	56.000	0.9
94.000	45.000	0.5
57.000	87.000	1.5
78.000	72.000	0.9
116.000	74.000	0.6
85.000	105.000	1.2
136.000	123.000	0.9

MixUp. Wheat heads overlap is a gordian knot in detection. However, overlapping images of wheat are limited. In order to solve this problem, we use the MixUp [13] augmentation algorithm. It makes use of existing images to generate a batch of similar overlapping data simply and efficiently through linear fusion. In the MixUp, each time we randomly sample two examples (x_i, y_i) and (x_j, y_j), where x is the image data, and y is the corresponding label (including the bounding boxes and the each object class of them). Then the MixUp augmentation image calculation formula is:

$$\hat{x} = \frac{1}{2}x_i + \frac{1}{2}x_j \tag{2}$$

$$\hat{y} = y_i \cup y_j \tag{3}$$

In training with MixUp, we only use the new example $(\hat{x}, \hat{y})$. The actual influence is as illustrated in Fig. 2.

CutOut. To enhance the robustness and overall performance of convolutional neural networks, we implement the CutOut [4] regularization technique of randomly masking out square regions of input during training, so that the convolutional neural network can make better use of global information. The influence of CutOut is shown in Fig. 2.

Mosaic. Since the wheat data comes from different places and different varieties, in order to eliminate the huge information differences between the data as much as possible, we use Mosaic data augmentation to form fusion data. Compared with CutMix [12] which only mixes two input images, Mosaic [1] mixes four different contexts. The sample images are shown in Fig. 2.

(a) MixUp   (b) CutOut   (c) Mosaic

Fig. 2. The influence of MixUp, CutOut and Mosaic

Pseudo-Label. In reality, there is little labeled data and a lot of unlabeled data. While in competitions, the training set is labeled and the test set is unlabeled. Compared to unsupervised learning, semi-supervised learning makes use of a portion of labeled data and a large amount of unlabeled data at the same time, which is also better suited to reality and competitions. Pseudo-Label [6] is the simpler way of training neural networks in a semi-supervised fashion to help models learn better from unlabeled information.

Figure 3 illustrates the method. Firstly, training the existing labeled data to get a model. Then the model trained is used to predict the unlabeled data. Finally, the Pseudo-Label data is added to the training set to train together.

We first pick up the class which has maximum predicted probability for each unlabeled sample.

$$y_i' = \begin{cases} 1 & \text{if } i = \text{argmax}_{i'} \, f_{i'}(x) \\ 0 & \text{otherwise} \end{cases} \tag{4}$$

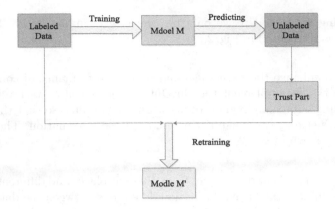

Fig. 3. The flow of Pseudo-Label

We add a pseudo-label to a sample when its maximum predictive probability is greater than the optimal threshold, otherwise it remains as an unlabeled sample. So the overall loss function is

$$L = \frac{1}{n} \sum_{m=1}^{n} \sum_{i=1}^{C} L\left(y_i^m, f_i^m\right) + \frac{1}{n'} \sum_{m=1}^{n'} \sum_{i=1}^{C} L\left(y_i'^m, f_i'^m\right) \tag{5}$$

where n is the number of mini-batch in labeled data for SGD, n' for unlabeled data, f_i^m is the output units of m's sample in labeled data, y_i^m is the label of that, $f_i'^m$ for unlabeled data, $y_i'^m$ is the pseudo-label of that for unlabeled data, and C is the number of labels.

The Loss of Bounding Box Regression. Mean Squared Error (MSE) Loss treats the four points as independent points, without considering the correlation between the four points, and the loss caused by the target detection bounding box of different scales is different. In order to make this issue processed better, some researchers recently proposed IoU Loss. Since IoU is not sensitive to scale and it can be directly used to determine the distance between the predicted box and ground truth bounding box, we try to use IoU Loss, but there are two problems with using IoU directly. (1) $IoU = 0$, it cannot reflect the distance between the two coincidence, there is no gradient back, and learning cannot be performed. (2) IoU cannot accurately reflect the degree of overlap between the predicted box and ground truth bounding box. As shown in the Fig. 4, the three cases' IoU are equal, but their coincidence degree is different. The regression effect of the left Fig. 4 is the best, and the right one is the worst.

Considering the IoU problem, the idea of GIoU [9] is to first calculate the minimum closure area which is the smallest box area that includes the two boxes, then calculate IoU, and then calculate the area of the closure area that does not

(a) Case1   (b) Case2   (c) Case3

Fig. 4. Different cases of overlap with the same IoU values

belong to the two boxes occupying the closure area Specific gravity, and finally subtract this specific gravity from IoU to get GIoU.

When the target bounding box completely wraps the predictied bounding box, GIoU degenerates to IoU, and its relative position relationship cannot be distinguished. Because DIoU [14] adds the normalized distance of the center point, it can better optimize such problems. DIoU Loss can directly optimize the distance between two boxes, which is faster than GIoU Loss. For the situation where the target bounding box wraps the prediction bounding box, DIoU Loss can converge very quickly, while GIoU Loss degenerates into IoU Loss with a slower convergence rate.

DIoU considers the overlap area and the center distance. CIoU [14] considers the aspect ratio on the basis of it. The three significant geometric factors of the target bounding box regression loss: the overlap area, the center distance and the aspect ratio are all considered, and its convergence accuracy is better.

$$CIoU = IoU - \frac{\rho^2\left(b, b^{gt}\right)}{c^2} - \alpha v \tag{6}$$

$$\alpha = \frac{v}{(1 - IoU) + v} \tag{7}$$

$$v = \frac{4}{\pi^2}\left(\arctan\frac{\omega^{gt}}{h^{gt}} - \arctan\frac{\omega}{h}\right)^2 \tag{8}$$

where α is trade-off parameter, b, b^{gt} respectively represent the center point of the predicted box and ground truth bounding box. $\rho(\cdot)$ represents the Euclidean distance. v measures the consistency of the aspect ratio. ω^{gt} and h^{gt} are defined as the length and width of the target box. ω and h represent the length and width of the predicted box.

Test-Time Augmentation (TTA). In order to enhance the detection accuracy of wheat during the test, we use the Test-Time Augmentation (TTA). When testing, the test set images are simply data augmentation mainly includes some common geometric transformations. Afterwards, these images are tested separately, and then the test results are merged to obtain the predicted results.

Out-of-Fold (OOF). In the predicted stage, the test image is augmented by TTA, and after inputting the model to make predictions, a lot of predicted bounding boxes and confidences are obtained. We need to initially filter the predicted bounding boxes due to the confidence score. The traditional way is to set a fixed confidence scoreThreshold (sThr), but this may miss a lot of valuable predicted bounding boxes and only using test data can not assess what sThr is appropriate. Therefore, we use a new Out-of-Fold (OOF) algorithm here. While testing, it will perform TTA on the validation set data and input it into the model, and filter the predicted bounding boxes obtained from the validation set according to the confidence score. At this time, the sThr is from 0 to 1, different thresholds filter out different predicted bounding boxes, and the final validation set mAP will also be different. According to the validation set mAP, the optimal sThr can be determined. Finally the sThr will be used in the test data set of this model (see the flow in Fig. 5).

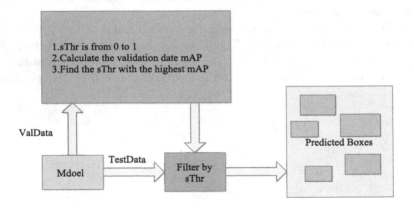

Fig. 5. The flow of OOF

Weighted Box Fusion (WBF). During the predicted of the N different models for the images, we use a Weighted Box Fusion [10] (WBF) ensembling algorithm to balance the different predicted boxes and enhance the quality of the last fused predicted box. Compared with NMS and SoftNMS [2], WBF uses the information of all predicted boxes instead of just keeping the largest IoU among many spatially overlapping predicted boxes. Figure 6 presents the performance of WBF.

(a) Before WBF   (b) After WBF

Fig. 6. The left and right images show the before and after effects of WBF. In the images, the red bounding box is the predicted bounding box and the green box is the ground truth bounding box (Color figure online)

3 Experiments

3.1 Dataset

In our experiments, we use the GWHD dataset, which contains 4,700 high-resolution RGB images and 190,000 labeled wheat heads collected from several countries around the world at different growth stages with a wide range of genotypes. 73% of the entire GWHD dataset images are divided into training sets, containing 3422 images from Europe and North America. The test dataset includes all images from Australia, Japan and China, representing 1276 images.

3.2 Experimental Setups

The hardware platform for this experiment is mainly an i7-76800K CPU with 16G of RAM and a built-in NVIDIA GeForce GTX1080Ti GPU. It uses NVIDIA's CUDA parallel computing architecture and CUDNN deep neural network-specific acceleration libraries to accelerate GPU-based deep learning training and inference processes. Run times are measured on an 1080Ti at 1024×1024. For YOLO v5x [5], we train the model for 100 epochs with batches of 2 in 18.3 h. When we run the model on the test set, the epochs of Pseudo-Label is 10, YOLO v5x runs in 6.5 h.

3.3 Influence of Different Features on BTWD

We mainly list the test results of different tricks on YOLO v5x and EfficientDet-D5 [11]. It can be seen from the table that a reasonable selection of different trick combinations can significantly improve the mAP. The Pseudo-Label trains the model by labeling on the images of the test set, and then adding these images to the training set, so that the test score of the model can be significantly improved. In comparison, some tricks can only play a minor role, such as OOF.

However, tricks cannot be combined simply. This may have counterproductive effects. For example, in the image data augmentation, when MixUp and Mosaic are used at the same time, the test score drops instead. Because the image data

Table 2. Influence BTWD on YOLO v5x

Mosaic	CutOut	MixUp	Pseudo-Label	TTA	OOF	WBF	Loss	mAP
							GIoU	0.7078
✓							GIoU	0.7180
	✓						GIoU	0.7065
		✓					GIoU	0.7102
			✓				GIoU	0.7230
				✓			GIoU	0.7178
					✓		GIoU	0.7079
						✓	GIoU	0.7183
✓			✓				GIoU	0.7053
✓							DIoU	0.7236
✓							CIoU	0.7328
✓	✓		✓				CIoU	0.7502
✓	✓		✓	✓			CIoU	0.7693
✓	✓		✓	✓	✓		CIoU	0.7697
✓	✓		✓	✓	✓	✓	CIoU	0.7707

Table 3. Influence BTWD on EfficientDet-D5

Mosaic	CutOut	MixUp	Pseudo-Label	TTA	OOF	WBF	Loss	mAP
							smoothL1 Loss	0.6876
✓							smoothL1 Loss	0.6953
	✓						smoothL1 Loss	0.6872
		✓					smoothL1 Loss	0.6923
			✓				smoothL1 Loss	0.7132
				✓			smoothL1 Loss	0.6992
					✓		smoothL1 Loss	0.6906
						✓	smoothL1 Loss	0.7024
✓			✓				smoothL1 Loss	0.7253
✓	✓		✓				smoothL1 Loss	0.7302
✓	✓		✓	✓			smoothL1 Loss	0.7368
✓	✓		✓	✓	✓		smoothL1 Loss	0.7401
✓	✓		✓	✓	✓	✓	smoothL1 Loss	0.7428

augmentation effect is too strong, the model cannot learn the characteristics of the image. Therefore, many attempts must be made in the trick selections. Finally, in terms of image data augmentation, we randomly select MixUp or Mosaic, and combine methods such as Pseudo-Label, TTA, OOF, CutOut and WBF to improve the performance of models. The various trained models are validated and analyzed on the test set, and the mAP is calculated as shown in Table 2 and Table 3.

After many attempts on the tricks, we finally choose the combination of Pseudo-Label, TTA, WBF, CutOut and OOF tricks, and the random selection of MixUp and Mosaic. Table 4 shows that the combination of tricks can upgrade the mAP effectively when applied to multiple detection network models, with YOLO v5x being the most effective, followed by EfficientDet-D5, Faster R-CNN [8] and YOLO v3 [7]. With BTWD, YOLO v5x can achieve 77.07% in average mAP, while it can only achieve 70.78% without BTWD.

Table 4. Influence of BTWD on different models

Method	Backbone	BTWD	mAP
YOLO v5x	BottleneckCSP	✓	0.7707
YOLO v5x	BottleneckCSP	×	0.7078
EfficientDet-D5	EfficientNet-B5	✓	0.7428
EfficientDet-D5	EfficientNet-B5	×	0.6876
YOLO v3	Darknet-53	✓	0.7352
YOLO v3	Darknet-53	×	0.6852
Faster R-CNN	ResNeSt-101	✓	0.7411
Faster R-CNN	ResNeSt-101	×	0.6723

4   Conclusion

We conduct a specific theoretical study and experimental validation of existing detection networks and applications of tricks in wheat detection, achieving 77.07% in mAP for the GWHD dataset. Furthermore, we provide a general bag of tricks for wheat detection that can be applied in different networks to enhance the performance. This facilitates the migration and learning of better detection networks for wheat detection in the future.

References

1. Bochkovskiy, A., Wang, C.Y., Liao, H.Y.M.: Yolov4: optimal speed and accuracy of object detection. arXiv preprint arXiv:2004.10934 (2020)
2. Bodla, N., Singh, B., Chellappa, R., Davis, L.S.: Improving object detection with one line of code. CoRR abs/1704.04503 (2017). http://arxiv.org/abs/1704.04503
3. David, E., et al.: Global wheat head detection (GWHD) dataset: a large and diverse dataset of high resolution RGB labelled images to develop and benchmark wheat head detection methods. arXiv preprint arXiv:2005.02162 (2020)
4. DeVries, T., Taylor, G.W.: Improved regularization of convolutional neural networks with cutout. arXiv preprint arXiv:1708.04552 (2017)
5. Jocher, G., et al.: ultralytics/yolov5: v3.0, August 2020. https://doi.org/10.5281/zenodo.3983579

6. Lee, D.H.: Pseudo-label: the simple and efficient semi-supervised learning method for deep neural networks. In: Workshop on Challenges in Representation Learning, ICML, vol. 3 (2013)
7. Redmon, J., Farhadi, A.: Yolov3: an incremental improvement. arXiv preprint arXiv:1804.02767 (2018)
8. Ren, S., He, K., Girshick, R., Sun, J.: Faster R-CNN: towards real-time object detection with region proposal networks. In: Advances in Neural Information Processing Systems, pp. 91–99 (2015)
9. Rezatofighi, H., Tsoi, N., Gwak, J., Sadeghian, A., Reid, I., Savarese, S.: Generalized intersection over union: a metric and a loss for bounding box regression. In: Proceedings of the IEEE Conference on Computer Vision and Pattern Recognition, pp. 658–666 (2019)
10. Solovyev, R., Wang, W.: Weighted boxes fusion: ensembling boxes for object detection models. arXiv preprint arXiv:1910.13302 (2019)
11. Tan, M., Pang, R., Le, Q.V.: EfficientDet: scalable and efficient object detection. In: Proceedings of the IEEE/CVF Conference on Computer Vision and Pattern Recognition, pp. 10781–10790 (2020)
12. Yun, S., Han, D., Oh, S.J., Chun, S., Choe, J., Yoo, Y.: CutMix: regularization strategy to train strong classifiers with localizable features. In: Proceedings of the IEEE International Conference on Computer Vision, pp. 6023–6032 (2019)
13. Zhang, H., Cisse, M., Dauphin, Y.N., Lopez-Paz, D.: Mixup: beyond empirical risk minimization. arXiv preprint arXiv:1710.09412 (2017)
14. Zheng, Z., Wang, P., Liu, W., Li, J., Ye, R., Ren, D.: Distance-IoU loss: faster and better learning for bounding box regression. In: AAAI, pp. 12993–13000 (2020)

W41 - Fair Face Recognition and Analysis

W41 - Fair Face Recognition and Analysis

The 2020 ChaLearn Workshop on Fair Face Recognition and Analysis (FairFaceRec 2020), held in conjunction with ECCV 2020, focused on bias analysis and mitigation methodologies, which will result in more fair face recognition and analysis systems. These advances will have a direct impact within society's equality of opportunity. We find it of crucial interest to centralize ideas, discuss them, and push the field to advance towards more fair systems for the good of society.

The Workshop accepted 5 papers, which contributed to push research on fair face recognition and/or analysis methods. The papers were reviewed by 3.25 reviewers on average. The Program Committee was formed by 44 experts, carefully selected due to their renewable contribution to the field of visual human analysis. Complementary to that, we also contributed pushing research in the field by releasing a large annotated dataset for fair face verification and running an associated challenge. A paper describing in detail the design of the challenge, the results, the description, and analysis of top winning solutions and their outcomes is included in the proceedings of the workshop. The challenge attracted a total of 151 registered participants, who made more than 1.8K submissions in total.

At the workshop, we invited eight distinguished researchers working on face recognition and/or analysis to give either a pre-recorded or live talk, in addition to a panel discussion following the ECCV 2020 online format. The keynotes and topics were: Kate Saenko ("Learning from Biased and Small Datasets"), Dimitris Metaxas ("Scalable Learning Methods to Generate, Analyze and Interpret Faces from Visual Data"), Matthew Turk ("Beyond Fairness"), Walter J. Scheirer ("Subjective Face Attributes: the Impossible and Possible"), Alice O'Toole ("Other race effects for face recognition algorithms: Strategies for measuring and minimizing bias"), Olga Russakovsky ("Fairness in Visual Recognition"), Rama Chellappa ("Gender Expressivity and Bias Reduction in Face Recognition") and Judy Hoffman ("Analyzing Bias in Computer Vision Systems"). Recorded sessions can be found in the associated (external) workshop page: http://chalearnlap.cvc.uab.es/workshop/37/description/

August 2020

Sergio Escalera
Rama Chellappa
Eduard Vazquez
Neil Robertson
Pau Buch-Cardona
Tomáš Sixta
Julio C. S. Jacques Junior

FairFace Challenge at ECCV 2020: Analyzing Bias in Face Recognition

Tomáš Sixta[1]([✉]), Julio C. S. Jacques Junior[2,3]([✉]), Pau Buch-Cardona[3,4], Eduard Vazquez[5], and Sergio Escalera[3,4]

[1] Czech Technical University in Prague, Prague, Czech Republic
tomas.sixta@gmail.com
[2] Universitat Oberta de Catalunya, Barcelona, Spain
jsilveira@uoc.edu
[3] Computer Vision Center, Barcelona, Spain
[4] Facultat de Matematiques i Informatica, Universitat de Barcelona, Barcelona, Spain
sergio@maia.ub.es
[5] Anyvision, London, UK
eduardov@anyvision.co

Abstract. This work summarizes the 2020 ChaLearn Looking at People Fair Face Recognition and Analysis Challenge and provides a description of the top-winning solutions and analysis of the results. The aim of the challenge was to evaluate accuracy and bias in gender and skin colour of submitted algorithms on the task of 1:1 face verification in the presence of other confounding attributes. Participants were evaluated using an in-the-wild dataset based on reannotated IJB-C, further enriched 12.5K new images and additional labels. The dataset is not balanced, which simulates a real world scenario where AI-based models supposed to present fair outcomes are trained and evaluated on imbalanced data. The challenge attracted 151 participants, who made more 1.8K submissions in total. The final phase of the challenge attracted 36 active teams out of which 10 exceeded 0.999 AUC-ROC while achieving very low scores in the proposed bias metrics. Common strategies by the participants were face pre-processing, homogenization of data distributions, the use of bias aware loss functions and ensemble models. The analysis of top-10 teams shows higher false positive rates (and lower false negative rates) for females with dark skin tone as well as the potential of eyeglasses and young age to increase the false positive rates too.

Keywords: Face verification · Face recognition · Fairness · Bias

---

These (T. Sixta and J. C. S. Jacques Junior) authors contributed equally to this work.

Electronic supplementary material The online version of this chapter (https://doi.org/10.1007/978-3-030-65414-6_32) contains supplementary material, which is available to authorized users.

A. Bartoli and A. Fusiello (Eds.): ECCV 2020 Workshops, LNCS 12540, pp. 463–481, 2020.
https://doi.org/10.1007/978-3-030-65414-6_32

1   Introduction

Automatic face recognition is a general topic that includes both face identification and verification [29]. Face identification is the process of identifying someone's identity given a face image. This process is generally known as 1-to-n matching and could be seen as asking to the system "who is this person?". Face verification, on the other hand, is concerned with validating a claimed identity based on the image of a face, and either accepting or rejecting the identity claim (1-to-1 matching). A simple example of face verification is when people unlock their smartphones using their faces (e.g., authentication), whereas searching for the identity of a given individual in a database of missing people, for instance, could be an example of face identification.

Fairness in face recognition recently started to receive increasing interest from different segments of scientific communities [19,36,38,44]. This is partially due to the huge impact new technologies have in our daily lives. Face recognition has been routinely utilized by both private and governmental organizations around the world [16,53]. Automatic face recognition can be used for legitimate and beneficial purposes (e.g. to improve security) but at the same time its power and ubiquity heightens a potential negative impact unfair methods can have for the society [46,52,54,55]. Recently, these concerns led several major companies to suspend distribution of their products to US police departments until a legislation regulating its deployment is passed by US Congress [56,57,59].

Although not sufficient, a necessary condition for a legitimate deployment of face recognition algorithms is equal accuracy for all demographic groups. A gold standard for testing commercial products is the Face Recognition Vendor Test (FRVT) performed by National Institute of Standards and Technology (NIST) [5,23]. However, this test is not designed for iterative and fast evaluation of new research directions. There is also a growing number of works that evaluate the algorithms on public data [3,13,50,64] and are therefore limited by what data is available, i.e., typically either small scale high quality datasets or large scale datasets with noisy annotations.

To motivate research on fair face recognition and provide a new challenging accurately annotated dataset, we designed and ran a computational face recognition challenge where participants were asked to provide solutions that maximize both accuracy and two fairness scores (minimize bias score). The submissions were evaluated on a reannotated version of IJB-C [37] database, enriched by newly collected 12,549 public domain images. The dataset contains large variations in head pose, face size and other attributes (detailed in Sect. 4.1). The dataset is not balanced with respect to different attributes, which imposes another challenge for the participants and is intended to stimulate usage of bias mitigation methods, also because the final ranking is defined by a weighted combination of accuracy and fairness (giving the bias scores a higher weight). To this end, we propose a new evaluation metric derived from a causal model by means of a causal effect of protected attributes to the accuracy of the algorithm, detailed in Sect. 4.2. The challenge attracted a total of 151 participants, who made more 1.8K submissions in total[1]. We expect the provided dataset and proposed fairness measure template to be a reference evaluation benchmark for face recognition systems, and that the outcomes of this challenge will help both to define priorities for future research as well as to help on the definition of technical requirements for real applications.

2 Ethics in Face Recognition

Face recognition methods have been researched for decades due to their wide number of scenarios for good[2]. They can be applied, e.g., in robotics, human-computer interaction, access and control, security, among others. Recently, face recognition research received additional attention due to the improved performance provided by deep learning architectures [24]. When it comes to public safety, past works raised the question about the efficacy of facial recognition systems for law enforcement following the apparent failure of the systems to identify suspects, reporting as possible reasons for failure problems like occlusions, angled facial shots, poor lighting or obscured facial features by hats or sunglasses [1]. However, recent studies show that automated methods for face analysis can also discriminate based on classes like gender and ethnicity [11], among others, which raised an additional focus of attention around such technologies. If face recognition methods are used to support decisions, erroneous but confident mis-identification can have serious consequences, and these possible and negative outcomes are making the society to rethink about what should be the limits of such technology, especially when it is applied at larger scales involving additional privacy concerns.

From a research point of view, a bottleneck to be solved is to develop methods that can work accurately for all target populations. While there is a need to promote good practices and reinforce regulations, we need to find the way to provide the required good (and fair) performance in practice, and if face recognition is to be applied, it should deal with the bias problem. Evidences show that the computer vision and machine learning research communities are starting to give visibility to different types of bias [10,22] and proposing different solutions to mitigate them (e.g., [6,11,21,58,65,68]). Nonetheless, additional efforts should be made to further reduce bias in future methods. This is precisely the main goal of the 2020 ChaLearn Looking at People Fair Face Recognition and Analysis Challenge, i.e., to stimulate and promote research on face recognition methods that produce fair outcomes.

---

[1] Data and winning solutions codes are available at http://chalearnlap.cvc.uab.es/challenge/38/description.

[2] For more information about ethics in AI you can visit the European guideline in the following link https://ec.europa.eu/digital-single-market/en/news/ethics-guidelines-trustworthy-ai.

3   Related Work

It is known that popular face recognition datasets like Labeled Faces in the Wild (LFW) [34], MegaFace [30], IJB-C [37], IMDB-WIKI [48,49], VGGFace2 [12] or MS-Celeb-1M [25] are imbalanced both in gender and skin colour [39]. To encourage research in fair face recognition there is growing number of datasets specifically designed with balance in mind and annotated for gender, ethnicity and potentially other attributes. Examples are Racial Faces in the Wild (RFW) [64] (40K images, 12K identities, subset of MS-Celeb-1M), Balanced Faces in the Wild (BFW) [47] (20K images, 0.8K identities, subjects sampled from VGGFace2) or DiveFace [40] (150K images, 24K identities, subset of Megaface). Even though these datasets are important step towards fairer face recognition, using labels for ethnicity does not in general allow for comparing models across datasets, because unlike for skin colour [9] there is no widely accepted definition of ethnicity groups and the labels instead rely on judgment of the annotators. Furthermore, balancing alone may not be enough to guarantee fair models [2], which motivates research of bias mitigation methods.

Nowadays, the gold standard for evaluating accuracy and bias of face recognition algorithms is the ongoing FRVT Test performed by NIST [5,23]. The submitted (mostly commercial) algorithms are evaluated on four datasets composed of photographs from various visa/benefits US governmental applications. In total, there are 18.27 million images of 8.49 million people. Besides FRVT, there are numerous small scale evaluations of bias in publicly and commercially available algorithms (e.g. [13,50,64]) as well as analysis of bias in models trained from scratch on publicly available datasets [3], that in most cases report better accuracy for men and people with light skin colour.

Traditional measures of fairness are based on calculating certain statistics related to the error rate of the algorithm. For example, Equalized Odds requires the true positive and false positive rates to be equal for all protected groups (see [19,60] for a comprehensive review). These measures are easy to calculate, but without having background in statistics it can be difficult to choose the "correct" one for the task at hand. This is a serious shortcoming, because certain traditional measures in general contradict each other [8,15,31] as it was dramatically demonstrated on the case of COMPAS (a system used in some US states to predict the risk of recidivism) [18,45]. Individual Fairness [20] tries to overcome these shortcomings by deriving a fairness measure from intuition, that *"similar individuals are treated similarly"*. However, it does not provide any general definition of similarity and only postpones the problem by proposing that it should be given by a regulatory body or a civil rights organization.

A growing number of state of the art measures is based on causal inference. They require a "model of the world" given as a causal diagram and the actual measure is then derived using this diagram, e.g. in terms of the causal effect of the protected attributes to the algorithm accuracy [41,69] or using counterfactuals [33], i.e., would the decision remain had the value of the protected attribute be different but everything else stayed the same. A crucial advantage of these approaches is that the underlying ethical views are encoded by the diagram in an easy to understand way, which exposes them to criticism and allows them to be changed if they prove to be inadequate. Furthermore, as these approaches are trying to identify the true causes of the unfairness, they can be used as a starting point for mitigating the bias in the real world.

Bias mitigation methods can be broadly divided based on what area of model deployment they target to pre-processing, in-processing and post-processing [7,43]. The most popular pre-processing technique is rebalancing the dataset [27,66], alternatively using synthetic data [32]. In-processing approaches include cost-sensitive training (higher weights for underrepresented groups) [27], adversarial learning for removing the sensitive information from the features [4,66], tuning parameters of a loss function for different protected groups [40,63] or attempts to learn bias free representations in unsupervised way [61]. Examples of post-processing techniques are renormalizing the similarity score of two feature vectors based on the demographic groups of the corresponding images [51] or attaching more fully connected layers to the feature extractor in order to remove the sensitive information from the representations [40]. The FairFace Recognition challenge, described in Sect. 4, did not impose any constrain to the participants to what model stage bias mitigation should be addressed. The best solutions rely on a combination of different strategies, detailed in Sect. 5.

4   Challenge Design

The participants were asked to develop their face verification methods aiming for a reduced bias in terms of the protected attributes (i.e., gender and skin color). Developed methods needed to output a list of confidence scores given test ID pairs to be verified (higher score means higher confidence, that the image pair contains the same person). The challenge[3] was managed using Codalab[4], an open source framework for running competitions that allows result or code submission.

The challenge ran from 4th April to 1st July 2020, and included two different phases: development and test. In the development phase, the participants were provided with public train data (with labels) and validation data (without labels, from which they should make predictions). At the test stage, the validation labels were released to all participants as well as the test data (without labels, considered for the final evaluation). The challenge attracted a total of 151 registered participants. During development phase we received 1330 submissions from 48 teams, and 476 submission from 36 teams at the test stage, resulting in more than 1800 submissions in total. Additional schedule details and participation statistics are provided in the supplementary material.

4.1   The Dataset

The dataset used in the challenge is a reannotated version of IJB-C [37], further enriched by newly collected 12,549 public domain images. In total, there are

---

[3] https://competitions.codalab.org/competitions/24184.
[4] https://competitions.codalab.org.

152,917 images from 6,139 identities. The images were annotated by Anyvision's internal annotation team for two protected attributes: gender (male, female) and skin colour (light corresponding to Fitzpatrick types I-III, dark corresponding to types IV-VI) and five legitimate attributes: age group (0–34, 35–64, 65+), head pose (frontal, other), image source (still image, video frame), wearing glasses and a bounding box size[5]. Detailed annotation instructions are in the supplementary material. Every attribute was annotated by at least 3 annotators (age and skin colour by 6, due to their subjectiveness, aiming to maximize the level of agreement). Labels for gender and skin colour were synchronized for each *identity* to the most prevalent ones, and labels of the other attributes were obtained by choosing for each *image* the most common label from the annotators.

For the purpose of the challenge, the dataset was split into training, validation and testing subsets containing 70%, 10% and 20% of identities. To facilitate evaluation of the submitted results we generated roughly half a million positive face image pairs (same identity) and half a million negative pairs for both validation and testing subsets. The pairs were selected such that the number of combinations of legitimate attributes is maximized. In the validation pairs there were 219 (positive) and 574 (negative) combinations, and test pairs contained 397 (positive) and 1162 (negative) combinations. Basic dataset statistics are summarized in Table 1. Few image samples and and statistics of the attributes are shown in Fig. 1 and Fig. 2, respectively.

Table 1. Dataset statistics.

	Train	Validation	Test	Total
Images	100,186	17,138	35,593	152,917
Unique identities	4,297	614	1,228	6,139
Positive pairs	–	448,119	500,176	–
Negative pairs	–	552,672	500,963	–

The images in the dataset have large variance in head pose, bounding box size and other attributes, which makes it challenging for face recognition. At the same time the distribution of these attributes is imbalanced, for example as seen in Fig. 2a, there is considerably more white males that dark females. Such imbalances are common in real world datasets and we intentionally have not reballanced the data to encourage research of bias mitigation methods.

---

[5] Attribute categories used in this work are imperfect for many reasons. For example, it is unclear how many skin colour and gender categories should be stipulated (or whether they should be treated as discrete categories at all). We base our definitions on widely accepted traditional categories and our methodology and findings are expected to be applied later to any re-defined and/or extended attribute category.

(a) Positive pairs   (b) Negative pairs

Fig. 1. Positive and negative samples of image pairs used in the challenge.

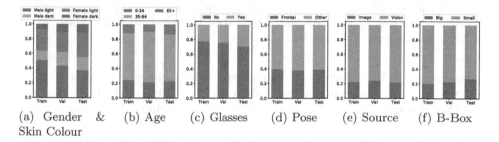

(a) Gender & (b) Age   (c) Glasses   (d) Pose   (e) Source   (f) B-Box
Skin Colour

Fig. 2. Distribution in percentage of attributes in training, validation and testing subsets of the dataset. Bounding box of a face was considered small if either its width or height was smaller than 224 px.

4.2 Evaluation Protocol

The challenge submissions were evaluated for bias in positive and negative pairs, and overall accuracy (given by AUC-ROC). The measure of bias/fairness was derived from a causal diagram shown in Fig. 3, in terms of a causal effect of protected attributes A (gender and skin colour) to the output $\hat{Y}$ of the algorithm. The diagram was chosen using the following principle: the accuracy of the algorithm might be influenced (caused) directly by gender and skin colour but in addition there might be other variables that influence the accuracy and depend on the protected attributes. Some of these additional variables are seen as legitimate causes for different accuracy, whereas the others are proxies for unfair discrimination. It should be emphasized that the structure of the diagram and designations of the additional variables are not learned from the data but selected to express ethical views on the real world. This does not allow to select an objectively best diagram but instead provides transparency needed for the public to review it and potentially change it based on democratic discussion.

Our definition of fairness is inspired by intuition, that an algorithm is fair if given fixed values of the *legitimate* attributes its outcome remained the same regardless of the values of the protected and proxy attributes. Distinguishing legitimate and proxy attributes is crucial for the definition of fairness. By denoting an attribute legitimate we chose to ignore, that it might have different prevalence in different protected groups (i.e., break the causal link from the protected attribute) and consequently that the algorithm has different accuracy for different groups. An example could be eyeglasses - as they can be easily removed, different accuracy caused by them is not seen as unfair even if they were worn

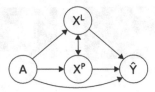

Fig. 3. Causal model used for our definition of fairness. A: protected attributes (gender and skin colour), X^L: legitimate attributes, X^P: proxy attributes, $\hat{Y}$: outcome of the algorithm. Note that in this challenge we deemed all additional attributes as legitimate, so X^P did not contain any variables.

more frequently by certain protected group. This is however not true for the proxy attributes - they are seen as mediators of potential unfair discrimination and therefore causal paths going over them must be included in the final objective.

Following the notation from [42], breaking causal links can be expressed by the $do()$ operator, which denotes an intervention on a variable. A prediction $\hat{Y}$ is fair with respect to protected attributes A and causal diagram M if for every pair of protected groups a, a' and value x^L of legitimate attributes

$$
\begin{aligned}
\sum_{X^P} P_M(\hat{Y}, X^P \mid do(A = a), do(X^L = x^L)) = \\
\sum_{X^P} P_M(\hat{Y}, X^P \mid do(A = a'), do(X^L = x^L)),
\end{aligned}
\tag{1}
$$

where X^P denotes the proxy variables. As described in [42], $do(X)$ in diagram M is equivalent to conditioning on plain X in mutilated diagram M^*, where links leading to X are removed. Furthermore, because in this challenge we deemed all additional attributes as legitimate, the criterion can be simplified to

$$
P_{M^*}(\hat{Y} \mid A = a, X^L = x^L) = P_{M^*}(\hat{Y} \mid A = a', X^L = x^L).
\tag{2}
$$

As probability of error depends on a recognition threshold, we replace it by AUC-ROC metrics and use $AUC(a; x^L)$ to denote accuracy for positive pairs from protected group a with legitimate attributes x^L (all negative pairs are used as the negative samples for the ROC curve; accuracy for negative pairs is obtained in the same way with the roles of positive and negative samples reversed). To obtain a single numerical measure of bias, we define the discrimination $d(a; x^L)$ for protected group a with legitimate attributes x^L as a difference in accuracy for this group and the best one:

$$
d(a; x^L) = \max_{a'} AUC(a'; x^L) - AUC(a; x^L).
\tag{3}
$$

The final measure of bias reported in the rankings is the difference between the average discriminations of the most and the least discriminated group:

$$
Bias = max_a \frac{1}{|X|} \sum_{x^L} d(a; x^L) - min_a \frac{1}{|X|} \sum_{x^L} d(a; x^L).
\tag{4}
$$

4.3   Ranking Strategy

Having the accuracy and the two bias scores (for positive and negative pairs), participants were ranked by the average rank position obtained on each of these 3 variables. This way, bias is receiving more weight than accuracy. However, to prevent a random number generator from winning the competition we require that the accuracy of the submissions must be higher than the accuracy of our baseline model (see Sect. 4.4). Similarly, the submission of constant values would return Bias score $= 0$, due to the *"max − min"* strategy defined in Sect. 4.2.

4.4   The Baseline

We provide a baseline in order to set a reference point. We implemented a well-known standard solution for the face verification task based on a Siamese network [14] over a ResNet50 [26] backbone architecture (pretrained on Faces [12] database). Standard bounding box regression network for face detection was applied to detect the face region in every single image. Training pairs were generated by considering a subset of the dataset with highest possible diversity in terms of legitimate attributes. These pairs were fed to the model in balanced batches of 16 samples. The system was optimized with respect to maximizing only face verification accuracy confidence. As training strategy, only the layers from the 4th convolutional block of ResNet50 have been fine-tuned, using Adam as optimizer, $lr = 0.0001$ and Binary Cross-Entropy Loss, for 300 epochs.

5   Challenge Results, Winning Methods and Bias Analysis

5.1   The Leaderboard

Results obtained by the top-10 winning solutions at the development phase[6] are reported in Table 2. As it can be seen, results are very good if only accuracy is considered. Thus, the Bias scores can be considered a relevant tiebreaker factor, as one of the goals of the challenge is to stimulate research and development of fair face recognition methods.

In Table 3, we present the results obtained by the top-10 participants at the test phase. Similarly as in the previous phase, results are still very good with even lower bias scores, at least for the top participants, suggesting that participants were able to further improve their methods after the end of development phase. Another important aspect that can be seen is that, compared to the development phase, participants made an overall smaller number of submission, which can be explained due to two main reasons: 1) they had around 1 week to make submissions to the test phase (to avoid cheating related issues, also verified at the code verification stage, as they would have access to the test data, i.e., without labels); 2) we fixed the maximum number of submissions per day to 5 to avoid participants to improve the results on the test set by try and error.

---

[6] The full leaderboards for both phases are shown in the supplementary material.

Table 2. Top-10 solutions on the development phase (and Baseline results). The number inside the parenthesis indicate the global rank position for that particular variable, used to compute the average ranking.

Participant	Average ranking	Entries	Bias (+ pairs)	Bias (− pairs)	Accuracy
ustc-nelslip	2.333333 (1)	30	0.000142 (1)	0.002956 (3)	0.999287 (3)
zheng.zhu	3.666667 (2)	133	0.000344 (3)	0.003781 (7)	0.999442 (1)
CdtQin	3.666667 (2)	72	0.000472 (5)	0.002334 (1)	0.998477 (5)
crisp	4.666667 (3)	14	0.000935 (8)	0.003193 (4)	0.999394 (2)
haoxl	4.666667 (3)	73	0.000348 (4)	0.003678 (6)	0.998699 (4)
cam_vision	5.000000 (4)	95	0.000731 (6)	0.002488 (2)	0.995621 (7)
Hyg	6.000000 (5)	33	0.000814 (7)	0.003305 (5)	0.998402 (6)
senlin11	9.333333 (6)	50	0.000165 (2)	0.010091 (16)	0.992093 (10)
hanamichi	10.666667 (7)	91	0.001631 (9)	0.006760 (10)	0.987382 (13)
paranoidai	12.000000 (8)	156	0.003779 (12)	0.007745 (13)	0.988359 (11)
Baseline	38.333333 (33)	1	0.057620 (40)	0.054311 (39)	0.889264 (36)

Table 3. Top-10 solutions on the test phase (and Baseline results). Top-3 winning solutions highlighted in bold. The number inside the parenthesis indicate the global rank position for that particular variable, used to compute the average ranking.

Participant	Average ranking	Entries	Bias (+ pairs)	Bias (− pairs)	Accuracy
paranoidai	1.333333 (1)	39	0.000059 (2)	0.000012 (1)	0.999966 (1)
ustc-nelslip	3.666667 (2)	12	0.000175 (4)	0.000172 (2)	0.999569 (5)
CdtQin	4.000000 (3)	25	0.000036 (1)	0.000405 (9)	0.999827 (2)
debias	4.666667 (4)	5	0.000036 (1)	0.000460 (10)	0.999825 (3)
zhaixingzi	5.000000 (5)	14	0.000116 (3)	0.000237 (8)	0.999698 (4)
bestone	5.333333 (6)	11	0.000175 (4)	0.000197 (5)	0.999565 (7)
haoxl	5.333333 (6)	31	0.000178 (6)	0.000195 (4)	0.999568 (6)
Early	5.333333 (6)	4	0.000175 (4)	0.000190 (3)	0.999547 (9)
lemoner20	7.000000 (7)	9	0.000176 (5)	0.000201 (6)	0.999507 (10)
ai	7.333333 (8)	14	0.000180 (7)	0.000217 (7)	0.999560 (8)
Baseline	34.666667 (28)	3	0.059694 (33)	0.058601 (36)	0.859175 (35)

5.2 Top Winning Approaches

This section briefly presents the top-winning approaches (shown in Table 3), specially those that agreed to share with the organizers the code (verified at the code verification stage) and fact sheets (containing detailed information about their methods), according to the rules of the challenge. Table 4 shows some general information about the top-3 winning approaches. The workflow diagrams of top-3 winning solutions are shown in the supplementary material.

Table 4. General information about the top-3 winning approaches.

Features/Team	1st: paranoidai	2nd: ustc-nelslip	3rd: CdtQin
Pre-trained models	–	√	√
External data	√	√	√
Regularization strategies	–	√	√
Handcrafted features	–	–	–
Face detection, alignment or segmentation strategy	√	√	√
Ensemble models	√	√	–
Different models for different protected groups	–	–	–
Explicitly classify the legitimate attributes	–	–	–
Explicitly classify other attributes (e.g., image quality)	–	–	–
Pre-processing bias mitigation (e.g. rebalancing training data)	–	√	√
In-processing bias mitigation (e.g. bias aware loss function)	√	–	√
Post-processing bias mitigation technique	√	–	√

1st place: *paranoidai* [70][7] team proposed an asymmetric-arc-loss training and multi-step fine-tuning. Their motivation was based on observation that even two different people have typically some similarity, and trying to minimise such similarity may make the model pay useless attention to easy negative samples. To address this problem, they alter the convergence target such that easy negative samples contribute less to the final gradient. They first train a general model (ResNet101 as backbone) and perform its multi-step fine-tuning. To improve the performance they also employ several tricks such as re-ranking, boundary cut and hard-sample model fusion. According to them, the hard-sample model fusion significantly helped to mitigate bias. For this, they assume that after getting a final model, there must be some data on the training set that the model cannot predict correctly. These are obvious hard samples. To address this problem, they propose a model fusion strategy, where a fine-tuned model is built for false-positive results, in addition to another model which performs better for those hard samples but worse in general cases. At the fusion step, they only take the result with extremely high confidence from the hard-sample model.

2nd place: *ustc-nelslip* [67][8] team addressed the problem focusing on data balancing and ensemble models. First, they tested different face detection algo-

---

[7] https://github.com/paranoidai/Fairface-Recognition-Solution.
[8] https://github.com/HaoSir/ECCV-2020-Fair-Face-Recognition-challenge_2nd_place_solution-ustc-nelslip-.

rithms to find an effective face cropped method [35]. Then, a data re-sampling method is used to balance the data distribution by under-sampling the majority class (based on gender and skin colour), combined with the use of external data. Next, different training data enhancement methods are used to increase the diversity of samples by means of image quality and light conditions, for instance, with the goal to improve performance. Finally, the prediction results of eight different models having different backbones (ResNet50 and ResNet152) and head loss (e.g., Arcface [17] and Cosface [62]) are linearly combined at test stage.

3rd place: *CdtQin*[9] team presented a multi-branch training approach, using a modified ResNet-101 as backbone, with similarity distribution constraints. The similarity distributions for these branches are estimated and constrained, with the goal of forcing the same kind of distribution among different groups to be closer and the distance between positive and negative distributions to be larger. To this end, hard positive pairs are defined offline, while top-k hard negative pairs are selected online for each branch. The cosine similarity of these pairs is computed, and the estimated distribution is obtained as in [28]. For the drawn distributions, three constrains are considered, specifically *kl_loss*, *order_loss* and *entropy_loss*. The first measures the KL Divergence of two different groups (e.g., females with dark *vs.* light skin colour). The *order_loss* measures the expected difference with respect to two distributions. Intuitively, it is desired a large margin between positive and negative distributions. So, this loss is applied on the positive and negative similarity distributions for each branch. Finally, *entropy_loss* measures the negative entropy of a single distribution, designed to allows the similarity distribution near the threshold to have lower variance, promoting a better separation. The final loss is defined by a linear combination of these losses in addition to the ArcFace Loss [17].

5.3   Bias Analysis

In this section we analyze biases in the results of top-10 teams and discuss their possible causes. To conduct the analysis, we removed from the test set two error images found after the test stage was closed (one non-face, one wrong identity), which reduced the number of positive matches by 56 but did not affect the number of combinations of legitimate attributes. The changes to the calculated values of bias and accuracy were therefore very small and did not affect the findings nor changed the ranking of the top-3 teams. Detailed analyses are provided in the supplementary material. Main findings are summarized next.

Breakdown of Average Discrimination: A discrimination d, as defined by Eq. 3, quantifies the difference in accuracy between a given protected group and the best achieved one. High average discrimination of certain protected group therefore indicates that the accuracy of the algorithm is lower than for other

---

[9] https://github.com/CdtQin/FairFace.

protected groups. The character of bias we found in the algorithms of the top teams was not that they would have higher accuracy in all circumstances for certain groups and lower for others, but instead that they consider people from certain groups more similar to each other that individuals from other groups. Specifically, even though the differences were small, the algorithms consistently had difficulties distinguishing females with dark skin colour. This resulted in the lowest values of discrimination in the positive samples and the highest in the negative ones. Considering the averages over the top-10 teams, in positive samples the group with the highest discrimination were males with dark skin colour: $d = 4.748e{-}04$ (males with white skin colour were very close with $d = 4.690e{-}04$) and females with dark skin colour were the least discriminated: $d = 2.349e{-}04$. Conversely, in the negative samples females with dark skin colour were the most discriminated: $d = 1.783e{-}04$ and males with light skin colour least with $d = 0.475e{-}04$. Note however, that there were some exceptions from this trend. For example, for team *paranoidai* the least discriminated group in positive samples were not females with dark skin colour, but males with dark skin colour.

In addition to the absolute values of discrimination we also calculated for each protected group the frequency how often it was the most discriminated one (over all combinations of legitimate attributes). Even if a group is the most discriminated in 100% of the cases, the actual differences from the other groups might still be negligible. Nevertheless, it is convenient for showing trends as it allows to filter out outliers For the top-10 teams, in positive samples males with light skin colour were the most often discriminated group (42.2% cases) whereas females with dark skin colour the least often (11.2%). This was almost perfectly reversed in negative samples: females with dark skin colour were the most frequently the group with the highest discrimination value (45.5%) whereas males with light skin colour were the least often group (12.6%). The exception was *paranoidai*, with the lowest frequency for females with dark skin colour in both positive and negative samples.

Impact of Legitimate Attributes on Average Discrimination: To analyze the effect of legitimate attributes we split their combinations into as many subgroups as there are possible values of the chosen attribute. For example, for glasses there are three subsets, the first one contains all samples where none of the images contain glasses, second group consists of samples where both images contain glasses and the third group are the remaining ones. We found that for some teams, wearing glasses makes individuals in both positive and negative samples look more similar in the sense that the differences in accuracy in positive samples tend to be the smallest if both images contain glasses and in negative samples the largest (note that in positive samples, teams *ustc-nelslip*, *bestone*, *haoxl*, *ai* are exceptions from this observation but in negative samples it holds for all top-10 teams). This is to a large extent an expected result: glasses cover part of the face which is one of the most important for recognition and therefore make people look more similar to each other.

Age was another attribute that clearly influenced the magnitude of bias: all top-10 teams exhibited higher values of discrimination in positive samples where both individuals were younger than 35 years and for the majority of the teams this was reversed in the negative samples, where the largest differences were obtained for the oldest subset (both individuals older than 65 years; exceptions are teams *paranoidai*, *CdtQin* and *debias*). This corresponds to findings of [50] and those from FRVT test [23] (which however emphasizes frequent exceptions). By analyzing the results further we found that in the competition dataset young individuals are less likely to wear glasses than the older ones. When considering only combinations of legitimate attributes where both individuals are younger than 35 years, only in 16% of them both individuals wear glasses but this ratio increases to 27.3% and 53.23% for the middle age and old subsets. Given the findings we made for the glasses attribute it is conceivable, that these two attributes act as magnifiers for each other.

Furthermore, we analyzed the effect of the remaining three legitimate attributes, i.e., head pose, image source and bounding box size. We did not find any clear trends shared by majority of the top-10 teams.

Hardest Samples: Hardest samples for top-3 teams are shown in Fig. 4. Even though the samples are different for different teams, they share common characteristics. The hardest positive samples are often composed from one "normal" image and one with extreme head pose or appearance variation, which makes them look differently. In the hardest negative samples on the other hand both images have often extreme head pose or glasses, which obscure parts of the faces important for the recognition and makes them look similar to each other.

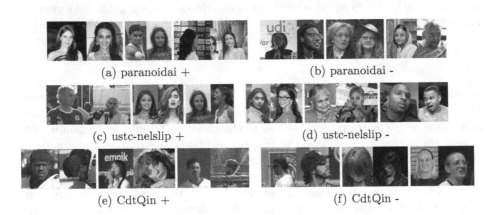

(a) paranoidai +   (b) paranoidai -

(c) ustc-nelslip +   (d) ustc-nelslip -

(e) CdtQin +   (f) CdtQin -

Fig. 4. Most difficult samples for the top-3 teams: positive samples with lowest score (+), negative samples with highest score (−).

6   Conclusions

This work presented the design and results of the FairFace Recognition Challenge at ECCV'2020. The submissions were evaluated on a reannotated version of IJB-C [37] database enriched by newly collected 12,549 public domain images. The participants were ranked using a novel evaluation protocol where both accuracy and bias scores were considered. The challenge attracted 151 participants. Top winning solutions obtained high performance in terms of accuracy ($\geq$0.999 AUC-ROC) and bias scores. The post challenge analysis showed that top winning solutions applied a combination of different strategies to mitigate bias, such as face pre-processing, homogenization of data distributions, the use of bias aware loss functions and ensemble models, among others, suggesting there is not a general approach that works better for all the cases. Despite the high accuracy none of the methods was free of bias. By analysing the results of top-10 teams we found that their algorithms tend to have higher false positive rates for females with dark skin tone and for samples where both individuals wear glasses. In contrast there were higher false negative rates for males with light skin tone and for samples where both individuals are younger than 35 years. We also found that in the dataset individuals younger than 35 years wear glasses less often than older individuals, resulting in a combination of effects of these attributes.

Acknowledgment. This work has been partially supported by the Spanish projects RTI2018-095232-B-C22 and PID2019-105093GB-I00 (MINECO/FEDER, UE), ICREA under the ICREA Academia programme, and CERCA Programme/Generalitat de Catalunya. We gratefully acknowledge the support of NVIDIA Corporation with the donation of the GPU used for this research.

References

1. Facial recognition tech under spotlight after Boston bombings. Biometric Technology Today **2013**(5), 1 (2013)
2. Albiero, V., Bowyer, K.W., Vangara, K., King, M.C.: Does face recognition accuracy get better with age? Deep face matchers say no. In: Winter Conference on Applications of Computer Vision (WACV), pp. 250–258 (2020)
3. Albiero, V., Krishnapriya, K.S., Vangara, K., Zhang, K., King, M.C., Bowyer, K.W.: Analysis of gender inequality in face recognition accuracy. CoRR abs/2002.00065 (2020)
4. Alvi, M.S., Zisserman, A., Nellåker, C.: Turning a blind eye: explicit removal of biases and variation from deep neural network embeddings. CoRR abs/1809.02169 (2018)
5. Snow, J.: Amazon's Face Recognition Falsely Matched 28 Members of Congress With Mugshots. American Civil Liberties Union, July 2018. https://www.aclu.org/blog/privacy-technology/surveillance-technologies/amazons-face-recognition-falsely-matched-28. Accessed 5 Aug 2020
6. Anne Hendricks, L., Burns, K., Saenko, K., Darrell, T., Rohrbach, A.: Women also snowboard: Overcoming bias in captioning models. In: European Conference on Computer Vision (ECCV), pp. 793–811 (2018)

7. Bellamy, R.K.E., et al.: AI fairness 360: an extensible toolkit for detecting, understanding, and mitigating unwanted algorithmic bias. CoRR abs/1810.01943 (2018)
8. Berk, R., Heidari, H., Jabbari, S., Kearns, M., Roth, A.: Fairness in criminal justice risk assessments: the state of the art. Sociological Methods & Research (2018)
9. Bino, S., Bernerd, F.: Variations in skin colour and the biological consequences of ultraviolet radiation exposure. Br. J. Dermatol. **169**(s3), 33–40 (2013)
10. Bird, S., Hutchinson, B., Kenthapadi, K., Kıcıman, E., Mitchell, M.: Fairness-aware machine learning: practical challenges and lessons learned. In: Companion Proceedings of the 2019 World Wide Web Conference, pp. 1297–1298 (2019)
11. Buolamwini, J., Gebru, T.: Gender shades: intersectional accuracy disparities in commercial gender classification. In: Proceedings of the 1st Conference on Fairness, Accountability and Transparency. Proceedings of Machine Learning Research, vol. 81, pp. 77–91. PMLR (2018)
12. Cao, Q., Shen, L., Xie, W., Parkhi, O.M., Zisserman, A.: VGGFace2: a dataset for recognising faces across pose and age. In: International Conference on Automatic Face Gesture Recognition (FG), pp. 67–74 (2018)
13. Cavazos, J.G., Phillips, P.J., Castillo, C.D., O'Toole, A.J.: Accuracy comparison across face recognition algorithms: where are we on measuring race bias? CoRR abs/1912.07398 (2019)
14. Chopra, S., Hadsell, R., LeCun, Y.: Learning a similarity metric discriminatively, with application to face verification. In: Conference on Computer Vision and Pattern Recognition (CVPR), vol. 1, pp. 539–546 (2005)
15. Chouldechova, A.: Fair prediction with disparate impact: a study of bias in recidivism prediction instruments. Big Data **5**(2), 153–163 (2017)
16. Davies, B., Innes, M., Dawson, A.: An Evaluation of South Wales Police's Use of Automated Facial Recognition, September 2018. https://static1.squarespace.com/static/51b06364e4b02de2f57fd72e/t/5bfd4fbc21c67c2cdd692fa8/1543327693640/AFR+Report+%5BDigital%5D.pdf. Accessed 5 Aug 2020
17. Deng, J., Guo, J., Xue, N., Zafeiriou, S.: ArcFace: additive angular margin loss for deep face recognition. In: Conference on Computer Vision and Pattern Recognition (CVPR) (2019)
18. Dieterich, W., Mendoza, C., Brennan, T.: Compas risk scales: demonstrating accuracy equity and predictive parity performance of the compas risk scales in broward county, July 2016. https://go.volarisgroup.com/rs/430-MBX-989/images/ProPublica_Commentary_Final_070616.pdf. Accessed 5 Aug 2020
19. Drozdowski, P., Rathgeb, C., Dantcheva, A., Damer, N., Busch, C.: Demographic bias in biometrics: a survey on an emerging challenge. CoRR abs/2003.02488 (2020)
20. Dwork, C., Hardt, M., Pitassi, T., Reingold, O., Zemel, R.S.: Fairness through awareness. CoRR abs/1104.3913 (2011)
21. Escalante, H.J., et al.: Modeling, recognizing, and explaining apparent personality from videos. IEEE Trans. Affect. Comput. (2020)
22. Friedler, S.A., Scheidegger, C., Venkatasubramanian, S., Choudhary, S., Hamilton, E.P., Roth, D.: A comparative study of fairness-enhancing interventions in machine learning. In: Proceedings of the Conference on Fairness, Accountability, and Transparency, pp. 329–338. ACM (2019)
23. Grother, P., Ngan, M., Hanaoka, K.: Face Recognition Vendor Test (FRVT) Part 3: Demographic Effects. Technical report, National Institute of Standards and Technology (NIST) Interagency/Internal Report (NISTIR) - 8280 (2019)
24. Guo, G., Zhang, N.: A survey on deep learning based face recognition. Comput. Vis. Image Underst. **189**, 102805 (2019)

25. Guo, Y., Zhang, L., Hu, Y., He, X., Gao, J.: MS-Celeb-1M: a dataset and bench-mark for large-scale face recognition. CoRR abs/1607.08221 (2016)
26. He, K., Zhang, X., Ren, S., Sun, J.: Deep residual learning for image recognition. In: Conference on Computer Vision and Pattern Recognition (CVPR), pp. 770–778 (2016)
27. Huang, C., Li, Y., Loy, C.C., Tang, X.: Deep imbalanced learning for face recognition and attribute prediction. CoRR abs/1806.00194 (2018)
28. Huang, Y., et al.: Improving face recognition from hard samples via distribution distillation loss. CoRR abs/2002.03662 (2020)
29. Jayaraman, U., Gupta, P., Gupta, S., Arora, G., Tiwari, K.: Recent development in face recognition. Neurocomputing **408**, 231–245 (2020)
30. Kemelmacher-Shlizerman, I., Seitz, S.M., Miller, D., Brossard, E.: The megaface benchmark: 1 million faces for recognition at scale. In: Conference on Computer Vision and Pattern Recognition (CVPR), pp. 4873–4882 (2016)
31. Kleinberg, J., Mullainathan, S., Raghavan, M.: Inherent trade-offs in the fair deter-mination of risk scores. CoRR abs/1609.05807 (2016)
32. Kortylewski, A., Egger, B., Schneider, A., Gerig, T., Morel-Forster, A., Vetter, T.: Analyzing and reducing the damage of dataset bias to face recognition with synthetic data. In: Conference on Computer Vision and Pattern Recognition Work-shops (CVPRW), pp. 1–8 (2019)
33. Kusner, M.J., Loftus, J., Russell, C., Silva, R.: Counterfactual fairness. In: Guyon, I., et al. (eds.) Advances in Neural Information Processing Systems 30, pp. 4066–4076. Curran Associates, Inc. (2017)
34. Learned-Miller, E., Huang, G.B., RoyChowdhury, A., Li, H., Hua, G.: Labeled faces in the wild: a survey. In: Kawulok, M., Celebi, M.E., Smolka, B. (eds.) Advances in Face Detection and Facial Image Analysis, pp. 189–248. Springer, Cham (2016). https://doi.org/10.1007/978-3-319-25958-1_8
35. Li, J., et al.: DSFD: dual shot face detector. In: Conference on Computer Vision and Pattern Recognition (CVPR) (2019)
36. Lo Piano, S.: Ethical principles in machine learning and artificial intelligence: cases from the field and possible ways forward. Humanit. Soc. Sci. Commun. **7**(9), 1–7 (2020)
37. Maze, B., et al.: IARPA Janus benchmark - C: face dataset and protocol. In: International Conference on Biometrics (ICB), pp. 158–165 (2018)
38. Mehrabi, N., Morstatter, F., Saxena, N., Lerman, K., Galstyan, A.: A survey on bias and fairness in machine learning. CoRR abs/1908.09635 (2019)
39. Merler, M., Ratha, N.K., Feris, R.S., Smith, J.R.: Diversity in faces. CoRR abs/1901.10436 (2019)
40. Morales, A., Fiérrez, J., Vera-Rodríguez, R.: Sensitivenets: learning agnostic rep-resentations with application to face recognition. CoRR abs/1902.00334 (2019)
41. Nabi, R., Shpitser, I.: Fair inference on outcomes. CoRR abs/1705.10378 (2017)
42. Pearl, J.: Causal inference in statistics: an overview. Stat. Surv. **3**, 96–146 (2009)
43. Pessach, D., Shmueli, E.: Algorithmic fairness. CoRR abs/2001.09784 (2020)
44. Pierce, J., Wong, R.Y., Merrill, N.: Sensor illumination: exploring design qualities and ethical implications of smart cameras and image/video analytics. In: Confer-ence on Human Factors in Computing Systems, pp. 1–19 (2020)
45. Angwin, J., Larson, J., Mattu, S., Kirchner, L.: Machine bias: there's soft-ware used across the country to predict future criminals and it's biased against blacks. ProPublica, May 2016. https://www.propublica.org/article/machine-bias-risk-assessments-in-criminal-sentencing. Accessed 5 Aug 2020

46. Raji, I.D., Gebru, T., Mitchell, M., Buolamwini, J., Lee, J., Denton, E.: Saving face: investigating the ethical concerns of facial recognition auditing. In: Proceedings of the AAAI/ACM Conference on AI, Ethics, and Society, pp. 145–151 (2020)
47. Robinson, J.P., Livitz, G., Henon, Y., Qin, C., Fu, Y., Timoner, S.: Face recognition: too bias, or not too bias? In: Conference on Computer Vision and Pattern Recognition Workshops (CVPRW), pp. 1–10 (2020)
48. Rothe, R., Timofte, R., Gool, L.V.: DEX: deep expectation of apparent age from a single image. In: International Conference on Computer Vision Workshops (ICCVW), pp. 252–257 (2015)
49. Rothe, R., Timofte, R., Gool, L.V.: Deep expectation of real and apparent age from a single image without facial landmarks. Int. J. Comput. Vision **126**(2–4), 144–157 (2018)
50. Srinivas, N., Ricanek, K., Michalski, D., Bolme, D.S., King, M.: Face recognition algorithm bias: performance differences on images of children and adults. In: Conference on Computer Vision and Pattern Recognition Workshops (CVPRW), pp. 2269–2277 (2019)
51. Terhörst, P., Kolf, J.N., Damer, N., Kirchbuchner, F., Kuijper, A.: Post-comparison mitigation of demographic bias in face recognition using fair score normalization. CoRR abs/2002.03592 (2020)
52. Wang, X.: China testing facial-recognition surveillance system in Xinjiang - report. The Guardian, January 2018. https://www.theguardian.com/world/2018/jan/18/china-testing-facial-recognition-surveillance-system-in-xinjiang-report. Accessed 5 Aug 2020
53. Valentino-DeVries, J.: How the Police Use Facial Recognition, and Where It Falls Short. The New York Times, January 2020. https://www.nytimes.com/2020/01/12/technology/facial-recognition-police.html. Accessed 5 Aug 2020
54. Mozur, P.: Inside China's Dystopian Dreams: A.I., Shame and Lots of Cameras. The New York Times, July 2018. https://www.nytimes.com/2018/07/08/business/china-surveillance-technology.html. Accessed 5 Aug 2020
55. Mozur, P.: One Month, 500,000 Face Scans: How China Is Using A.I. to Profile a Minority. The New York Times, April 2019. https://www.nytimes.com/2019/04/14/technology/china-surveillance-artificial-intelligence-racial-profiling.html. Accessed 5 Aug 2020
56. Greene, J.: Microsoft won't sell police its facial-recognition technology, following similar moves by Amazon and IBM. The Washington Post, June 2020. https://www.washingtonpost.com/technology/2020/06/11/microsoft-facial-recognition. Accessed 5 Aug 2020
57. Krishna, A.: IBM CEO's Letter to Congress on Racial Justice Reform. THINKPolicy Blog, June 2020. https://www.ibm.com/blogs/policy/facial-recognition-sunset-racial-justice-reforms. Accessed 5 Aug 2020
58. Torralba, A., Efros, A.A.: Unbiased look at dataset bias. In: Conference on Computer Vision and Pattern Recognition (CVPR), pp. 1521–1528 (2011)
59. US Day One Blog: We are implementing a one-year moratorium on police use of rekognition, June 2020. https://blog.aboutamazon.com/policy/we-are-implementing-a-one-year-moratorium-on-police-use-of-rekognition. Accessed 5 Aug 2020
60. Verma, S., Rubin, J.: Fairness definitions explained. In: Proceedings of the International Workshop on Software Fairness, pp. 1–7 (2018)
61. Vowels, M.J., Camgoz, N.C., Bowden, R.: NestedVAE: isolating common factors via weak supervision. In: Conference on Computer Vision and Pattern Recognition (CVPR), pp. 9202–9212 (2020)

62. Wang, H., et al.: CosFace: large margin cosine loss for deep face recognition. In: Conference on Computer Vision and Pattern Recognition (CVPR) (2018)
63. Wang, M., Deng, W.: Mitigating bias in face recognition using skewness-aware reinforcement learning. In: Conference on Computer Vision and Pattern Recognition (CVPR), pp. 9322–9331 (2020)
64. Wang, M., Deng, W., Hu, J., Tao, X., Huang, Y.: Racial faces in-the-wild: reducing racial bias by information maximization adaptation network. CoRR abs/1812.00194 (2018)
65. Wang, T., Zhao, J., Yatskar, M., Chang, K.W., Ordonez, V.: Balanced datasets are not enough: estimating and mitigating gender bias in deep image representations. In: International Conference on Computer Vision (ICCV), pp. 5310–5319 (2019)
66. Wang, Z., et al.: Towards fairness in visual recognition: effective strategies for bias mitigation. In: Conference on Computer Vision and Pattern Recognition (CVPR), pp. 8919–8928 (2020)
67. Yu, J., Hao, X., Xie, H., Yu, Y.: Fair face recognition using data balancing, enhancement and fusion. In: Proceedings of the European Conference on Computer Vision (ECCV) Workshops (ECCVW) (2020, in press)
68. Yucer, S., Akcay, S., Al-Moubayed, N., Breckon, T.P.: Exploring racial bias within face recognition via per-subject adversarially-enabled data augmentation. In: Conference on Computer Vision and Pattern Recognition Workshops (CVPRW) (2020)
69. Zhang, L., Wu, Y., Wu, X.: A causal framework for discovering and removing direct and indirect discrimination. CoRR abs/1611.07509 (2016)
70. Zhou, S.: AsArcFace: asymmetric additive angular margin loss for fairface recognition. In: Proceedings of the European Conference on Computer Vision Workshops (ECCVW) (2020, in press)

AsArcFace: Asymmetric Additive Angular Margin Loss for Fairface Recognition

Shengyao Zhou[✉], Junfan Luo[✉], Junkun Zhou[✉], and Xiang Ji[✉]

Ruiyan Technology, Chengdu, China
{zhoushengyao,luojunfan,zhoujunkun,jixiang}@ruiyanai.com

Abstract. Fairface recognition aims to the mitigate the bias between different attributes in face recognition task while maintaining the state-of-art accurancy. It is a challenging task due to high variances between different attributes and unbalancement of data. In this work, we provide an approach to make a fairface recognition by using asymmetric-arc-loss training and multi-step finetuning. First, we train a general model with an asymmetric-arc-loss, and then, we make a mutli-step finetuning to get higher auc and lower bias. Besides, we propose another viewpoint on reducing the bias and use bag of tricks such as reranking, boundary cut and hard-sample model ensembling to improve the performance. Our approach achieved the first place at ECCV 2020 ChaLearn Looking at People Fair Face Recognition Challenge.

Keywords: Asymmetric-arc-loss · AsArcFace · Face recognition · Metric learning

1 Introduction

Face recognition has been widely used and researched and a lot of class-level losses such as Softmax, SphereFace [10], CosineFace [15] and ArcFace [2] are used to improve the performance. All of these losses are trying to minimize between-class similarity s_n and maximize within-class similarity s_p. However, we don't always need to minimize between-class similarity extremely since there are also similar faces between class, in this situation, try to minimize the between-class similarity extremely may lead to noise in gradient and potentially lead to worse convergence. Besides, the previous face-recognition approach focus more on the auc on the whole test-set and less on bias between attributes. Based on these ideas, we proposed a fairface recognition approach aiming at higher accuracy and lower bias. Our major contribution can be summarized into four aspects:

- **First, an asymmetric-arc-loss.** From the previous class-level loss analysis, we propose an asymmetric-arc-loss which is a combination of arc-face loss and circle-loss. Besides, we modify the loss contribution from negative pairs similarities and make it asymmetric to positive similarities, which means we

© Springer Nature Switzerland AG 2020
A. Bartoli and A. Fusiello (Eds.): ECCV 2020 Workshops, LNCS 12540, pp. 482–491, 2020.
https://doi.org/10.1007/978-3-030-65414-6_33

don't minimize the negative similarity extremely, which finally decrease the gradient contribution from easy negative samples.

- **Second, a multi-step finetuning.** We propose a multi-step finetuing method to minimize the bias between different protected attributes and this is controllable and stable in improving the model's performance on most discriminated protected-attribute data.
- **Third, bag of tricks.** We use bag of tricks such as reranking, boundary cut and hard-sample model fusion to get higher accuracy and lower bias. And the hard-sample model fusion are quite significant for bias mitigation.
- **Finally, another viewpoint on bias mitigation.** We give another viewpoint on bias mitigation. And it's easy to implement and can decrease the bias even as whatever you want.

2   Related Work

2.1   Deep Face Recognition

Deep Face Recognition benefits from the development of Convolutional Neural Networks compared to previous handcrafted features. There are a lot of methods to use CNN models in face recognition such as Siamese network [1], but usually, we treat face recognition as a kind of metric learning, we train a cnn model as a feature extractor and then use the Euclidean distance or cosine distance to get the similarity between two faces. There are four main factors that have influence on the performance of face recognition models:

- **Data.** There are various distributed dataset [6,8], which are quite significant for the face recognition task and help in both training and testing. But it is still difficult to get a balanced data with a bulk of subject IDs, and making fusion on different source data is still a challenging task since face data from different source may have same subject ID.
- **Backbone.** Various backbone networks were designed for the purpose of imporving performance in image classification such as Alexnet [9], VGG [11], Inception [13], Resnet [7], Resnest [16], they also imporve the performance in face recognition. Recently, NAS(Neural Architecture Search) was also introduced in backbone network design such as EfficientNet [14] but these network may suffer from the domain adaptation problem, so hand crafted backbone networks are more widely used.
- **Loss.** Loss is very important in face recognition and in this paper we'll focus on it. Losses in face recognition can be summarized as two types: class-level loss and pair-wise loss. For class-level loss, most of them came from the original softmax-loss, which is most widely used. And these modified class-level loss try to add margin between positive and negative samples to get better decision, Let's start with the arc loss [2]:

$$L_{arc} = -\frac{1}{m} \sum_{i=1}^{m} \log \frac{e^{s(\cos(\theta_{y_i}+m))}}{e^{s(\cos(\theta_{y_i}+m))} + \sum_{j=1, j \neq y_i}^{n} e^{s \cos \theta_j}}$$

subject to

$$W_j = \frac{W_j}{\|W_j\|}, x_i = \frac{x_i}{\|x_i\|}, \cos\theta_j = W_j^T x_i$$

We assume θ_{y_i} as θ_p and others as θ_n. We can see that arc loss give additive angular margin between θ_p and θ_n and it's easy to analyze that the loss is monotonically increasing to the θ_p while $\theta_p + m < \pi$ and monotonically decreasing to θ_n, so, as shown in Fig. 1, it's convergence target is to maximize θ_n and to minimize θ_p.

Then we take a look at Circle loss [12], which is:

$$L_{cir} = \log[1 + \sum_{j=1}^{L} \exp(\gamma\alpha_n^j(s_n^j - \Delta_n)) \sum_{i=1}^{K} \exp(-\gamma\alpha_p^i(s_p^i - \Delta_p))]$$

where s_n means negative similarity and s_p means positive. And in the class-level style, there is only one s_p so the loss can be shown as:

$$L_{cir} = -\frac{1}{m} \sum_{i=1}^{m} \log \frac{e^{\gamma\alpha_p^{y_i}(s_p^{y_i} - \Delta_p)}}{e^{\gamma\alpha_p^{y_i}(s_p^{y_i} - \Delta_p)} + \sum_{j=1,j\neq y_i}^{n} e^{\gamma\alpha_n^j(s_n^j - \Delta_n)}}$$

subject to

$$\begin{cases} \alpha_p^i = \left|O_p - s_p^i\right|_+ \\ \alpha_n^j = \left|s_n^j - O_n\right|_+ \end{cases}$$

$$O_p = 1 + m, O_n = -m, \Delta_p = 1 - m, \Delta_n = m$$

$$W_j = \frac{W_j}{\|W_j\|}, x_i = \frac{x_i}{\|x_i\|}, s^j = W_j^T x_i$$

Circle-loss provide the self-weighted for s_n and s_p. We can also analyze that the loss is monotonically increasing to the s_n and monotonically decreasing to s_p, while both s_p and s_n are in (0,1). From the angel view, as shown in Fig. 1, it's convergence target is to maximize θ_n to $\pi/2$ and to minimize θ_p to 0.

- **PostProcess.** Reranking [17] is well known in person-reid task and also make difference to the face recognition task.

3   Method

3.1   Asymmetric Arc Loss

Based on the previous analysis, we know that arc loss give additive angular margin between θ_p and θ_n with a convergence target is to maximize θ_n and to minimize θ_p, while circle-loss provide the self-weighted for s_n and s_p with a convergence target is to maximize θ_n to $\pi/2$ and to minimize θ_p to 0. So we can get two insights on improving the loss function.

- **Combination of advantages.** As mentioned before, we can make a combination for these two loss to use both of their advantages and propose a kind of loss function with additive angular margin and self-weighted coefficient.
- **Convergence target shift.** From the previous anylasis, the convergence target of circle-loss is to maximize θ_n to $\pi/2$ and the convergence target of arc-loss is even maximize θ_n to π. But in fact, we don't always need to maximize θ_n to $\pi/2$ or π. Since in face recognition situation, we can't make sure that people in different sub ids are not similar at all, it's usual that two different people have some similarity, like 0.3 or 0.2, and try to minimise this similarity may make model pay useless attention on easy negative samples. To solve this problem, we give a shift on the convergence target for negative and make easy negative samples contribute less to the final grad.

So we proposed an asymmetric-arc-loss. The asymmetric-arc-loss can be shown like this:

$$L_{asymmetric-arc} = -\frac{1}{m}\sum_{i=1}^{m}\log \frac{e^{\gamma \alpha_p^{y_i} \cos(\theta_p^{y_i}+\Delta_p)}}{e^{\gamma \alpha_p^{y_i} \cos(\theta_p^{y_i}+\Delta_p)} + \sum_{j=1,j\neq y_i}^{n} e^{\gamma \alpha_n^{j} \cos(\theta_n^{j}+\Delta_n)}}$$

subject to

$$\begin{cases} \alpha_p^i = \left| O_p + \theta_p^i \right|_+ \\ \alpha_n^j = \left| O_n - \theta_n^j \right|_+ \end{cases}$$

$$O_p = \pi - tm, O_n = tm, \Delta_p = tm, \Delta_n = \pi - tm$$

$$W_j = \frac{W_j}{\|W_j\|}, x_i = \frac{x_i}{\|x_i\|}, \cos\theta^j = W_j^T x_i$$

Where γ and tm are hyperparameters and $\theta_p^{y_i} + \Delta_p$, $\theta_n^j + \Delta_n$ are clip to $(0, \pi)$.

Let's make a analysis on this loss. First, just like circle-loss, θ_n and θ_p get self-weighted based on their own value via α. Since O_n and O_p are fixed, the higher value of θ_p, which is more difficult get higher weights and lower value of θ_n, also difficult, get more weights. And turn to the easy samples, for positive, the weights are still kept, and for negative samples, if $\theta_n^j > O_n$, their weights will become 0. Then we can see that this loss give a margin on θ instead of similarity, just like the arc-loss, to get an additive cosine margin. The decision boundary is achieved at

$$\gamma(\alpha_p \cos(\theta_p + \Delta_p) - \alpha_n \cos(\theta_n + \Delta_n)) = 0$$

What's more, seen from the grad, we take a look at item about θ_n, we assume that $v_n = \alpha_n \cos(\theta_n + \Delta_n) = (tm - \theta_n) \cos(\theta_n + \pi - tm)$ and $\frac{\partial v_n}{\partial \theta_n} = \cos(\theta_n - tm) - (\theta_n - tm) \sin(\theta_n - tm)$ so the loss get min value for $\cos(\theta_n - tm) - (\theta_n - tm) \sin(\theta_n - tm) = 0$, in our hyperparameter setting where tm $= 0.65\pi$, the θ_n is at about 0.38π, and this target can shift base on the value of tm.so this loss can focus less on easy negative samples since their grad are smaller.

Figure 1 shows the different convergence target between these loss, and we can find that our loss's negative convergence target shift to left, compared to arc-loss and circle-loss, which means easy nagetive samples contribute less to the grad.

3.2   Multi-step and Multi-model Finetuning

Since we get a asymmetric-arc-loss trained general model, we still need to finetune it in the target domain. At the beginning of the finetuing, we freeze all layers but the last fc-layer because the grad generated at the beginning are noisy to other layers. And then train all layers to get a better adapted model. After we get a well finetuned model, we use this model to make a prediction on training set and get the data with most-discriminated protected-attributes, and continue to make a slight finetuning on those data. This step is quite important for the bias mitigation and easy to understand. Training a model on one attribute data can directly improve its performance on this attribute data and therefore dismiss bias. But this step needs careful tuning because the model will overfit on this attributes and lead to decreasing in acc and increasing in bias.

After we get a final finetuned model, there must be some data on the fairface training set that the model can't truely predict. These are obvious hard samples. According to the existing conclusions, pay much attention to most-hard smaples will lead to bad performance, for example, in triplet loss traning, we select semihard samples. But the hard-samples problem also needs to be solved, so we proposed a model fusion strategy. We get the false-predicted ids from the model, which means the predicted argmax id is not equal to the annotation id, and then finetune a model from step1 general model with just picked ids. We then get a model which performs better for those hard-sample but worse in general cases, so at the fusion step, we only take the result with extremely high confidence from the hard-sample model.

3.3   Postprocess

- **Reranking.** It's widely used in person-reid task [17] and we just use a very simple edition at this work. For template id A and B, we can get their original similarity score produced by the model, and by traversal the predictions file,

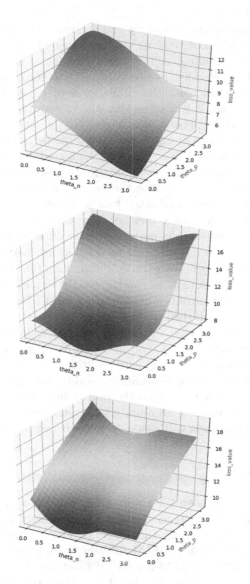

Fig. 1. Loss value with θ_n and θ_p for arc-face-loss, circle-loss and asymmetric-arc-loss.

we can get a set consists of k template ids which have highest similarity score to A and B, then we compare the two sets, if some template id C are both in A's top k set and B's top k set, we add a similarity score to another similarity score between A and B called top-k similarity. And the final similarity of A and B is a weighted sum of original similarity and top-k similarity.

- **BoundaryCut.** For some template ids in the predictions, there is a obvious boundary between the positive samples and negative samples, so, we increase the similarity score up the boundary and decrease the similarity score under the boundary. For a template id A, we first traversal the predictions file, and find all pairs that contain A as a itemlist, and found its boundary. We sort the itemlist and get itemlist[1] - itemlist[i+1] as grad, and find the lowest boundary. And then increase the similarity score if it is greater than the boundary and decrease it if it s less then the boundary.

- **Hard-Sample Fusion.** For a pair A and B, we get two similarity scores from step3 finetuned model and hard-sample model, and we only take the hard-sample model when it has extremely high confidence.

- **Random Noise.** We use a protected-attribute based random noise at development phase and it can decrease the bias as whatever you want. In test phase, this method is not used since bias is low enough. But from research view, this is what we thought as another viewpoint to bias and acc. For example, if we know that this model performs better on attribute A, but worse on B, just for fair purpose, making this model perform better on B is equal to making this model perform worse on A, but making this model perform worse on A is quite easy. Random noise can be thought as a method, it can be done by follow steps: first we train a classification model for two attributes, and a face-recognition model, then we evaluate the face-recognition model on the real-used domain test dataset, and we can get the bias between two attributes. Assume we perfom better on A, and in actual using we can make a attribute classification and choose relatively sure result, for example, confidence-score is greater than 0.95 and set a probability to modify the original face-recognition result opposite and dynamic adjust the probability based on the performance in actul using. We can get a absolutely fair system by this way. So, if we get a model with high accuracy, it's easy to make it fair to different groups.

4   Experiments

We make a data preprosses on the training set, using a open-source retina-face model [3–5] to make a face-detection and get landmarks, then we align the face using standard 5-point landmarks and get a 112*112 size aligned face. And then we train a general model using MS1M-ArcFace (85K ids/5.8M images) [6] and our self-owned face(10K ids/0.5M images) dataset for this stage. We train the general model using 4T-v100 gpus at batch-size 1600, the starting learning rate is 0.1 and then decreasing to 0.01 after 100000 steps then decreasing to 0.001 at 160000 steps and finally decreasing to 0.0001 at 200000 steps. We use fp16 data-fomat in training to speed-up the training and maximize the batch-size. The hyperparameter setting is 0.65π for tm and 64 for γ. The And we made an ablation experiment on backbone and loss, the results are as shown in Table 1 and Table 2, all training use the same settings.

Table 1. Loss and results using resnet101 on validation

Loss	PosiBias	NegBias	Auc
Arcface	0.021367	0.017595	0.976372
Circle-loss	0.008559	0.009488	0.991284
Asymmetric-arc-loss	0.005070	0.006807	0.992260

Table 2. Backbone and results using asymmetric-arc-loss on validation.

Backbone	PosiBias	NegBias	Auc
Mobilefacenet	0.012916	0.017778	0.982363
Resnet50	0.009215	0.010823	0.988492
Resnet101	0.005070	0.006807	0.992260
ResNeSt101	0.005939	0.007511	0.990803

Then we use asymmetric-arc-loss to make a multi-step and multi-model fine-tuning on the model. We set same hyperparameters for asymmetric-arc-loss. At the beginning of the finetuing, we freeze all layers but the last fc-layer for three epochs because the grad generated at the beginning are noisy to other layers. And after three epochs, we train all layers. The learning rate is set to 0.002 for finetuing and decreas to 0.0002 after 3000 steps since we start to train all layers. And then we finetune the model on most discriminated attributes data. As mentioned before, we also finetune another model on weak performanced subject IDs and make some post process. The ablation experiment results are shown in Table 3 and Table 4:

Table 3. Finetuning step and postprocess on validation

Step	PosiBias	NegBias	Auc
Asymmetric-arc-loss training	0.005070	0.006807	0.992260
Asymmetric-arc-loss finetuing	0.004988	0.005699	0.994518
Select attribute finetuing	0.005414	0.001250	0.995319
Select attribute finetuing with reranking and BoundaryCut	0.002707	0.000697	0.996075

Table 4. Finetuning step and postprocess on test

Step	PosiBias	NegBias	Auc
Asymmetric-arc-loss finetuing with reranking and BoundaryCut	0.000299	0.000115	0.999899
Select attribute finetuing with reranking and BoundaryCut	0.000273	0.000079	0.999910
Select attribute finetuing with reranking and BoundaryCut and hard-sample fusion	0.000012	0.000059	0.999966

5   Conclusions

In this paper, we made contribution to faceface recognition from loss function, training methods and post processes. We've (1) proposed an asymmetric-arc-loss and made an anylasis on the relationship between positive samples and negative samples and how convergence target of negative samples affects on the final performance in face recognition task. (2) given a training, multi-step, multi-model finetuning pipeline and made ablation experiments to improve their significance. (3) given some meaningful post process methods and achieved state of art performance in fair face recognition dataset.

References

1. Chopra, S., Hadsell, R., LeCun, Y.: Learning a similarity metric discriminatively, with application to face verification. In: 2005 IEEE Computer Society Conference on Computer Vision and Pattern Recognition (CVPR 2005), vol. 1, pp. 539–546 (2005)
2. Deng, J., Guo, J., Niannan, X., Zafeiriou, S.: Arcface: additive angular margin loss for deep face recognition. In: CVPR (2019)
3. Deng, J., Guo, J., Yuxiang, Z., Yu, J., Kotsia, I., Zafeiriou, S.: Retinaface: single-stage dense face localisation in the wild. In: arxiv (2019)
4. Deng, J., et al.: The menpo benchmark for multi-pose 2D and 3D facial landmark localisation and tracking. IJCV **127**(6–7), 599–624 (2018)
5. Guo, J., Deng, J., Xue, N., Zafeiriou, S.: Stacked dense u-nets with dual transformers for robust face alignment. In: BMVC (2018)
6. Guo, Y., Zhang, L., Hu, Y., He, X., Gao, J.: MS-CELEB-1M: a dataset and benchmark for large-scale face recognition (2016)
7. He, K., Zhang, X., Ren, S., Sun, J.: Deep residual learning for image recognition. In: The IEEE Conference on Computer Vision and Pattern Recognition (CVPR), June 2016
8. Huang, G.B., Ramesh, M., Berg, T., Learned-Miller, E.: Labeled faces in the wild: a database for studying face recognition in unconstrained environments. Technical report, 07-49, University of Massachusetts, Amherst, October 2007

9. Krizhevsky, A., Sutskever, I., Hinton, G.E.: Imagenet classification with deep convolutional neural networks. In: Pereira, F., Burges, C.J.C., Bottou, L., Weinberger, K.Q. (eds.) Advances in Neural Information Processing Systems 25, pp. 1097–1105. Curran Associates, Inc. (2012). http://papers.nips.cc/paper/4824-imagenet-classification-with-deep-convolutional-neural-networks.pdf

10. Liu, W., Wen, Y., Yu, Z., Li, M., Raj, B., Song, L.: Sphereface: deep hypersphere embedding for face recognition. In: The IEEE Conference on Computer Vision and Pattern Recognition (CVPR), July 2017

11. Simonyan, K., Zisserman, A.: Very deep convolutional networks for large-scale image recognition (2014)

12. Sun, Y., et al.: Circle loss: a unified perspective of pair similarity optimization. In: The IEEE/CVF Conference on Computer Vision and Pattern Recognition (CVPR), June 2020

13. Szegedy, C., Vanhoucke, V., Ioffe, S., Shlens, J., Wojna, Z.: Rethinking the inception architecture for computer vision. In: Proceedings of the IEEE Conference on Computer Vision and Pattern Recognition (CVPR), June 2016

14. Tan, M., Le, Q.V.: Efficientnet: rethinking model scaling for convolutional neural networks (2019)

15. Wang, H., et al.: Cosface: large margin cosine loss for deep face recognition. In: The IEEE Conference on Computer Vision and Pattern Recognition (CVPR), June 2018

16. Zhang, H., et al.: Resnest: split-attention networks (2020)

17. Zhong, Z., Zheng, L., Cao, D., Li, S.: Re-ranking person re-identification with k-reciprocal encoding. In: The IEEE Conference on Computer Vision and Pattern Recognition (CVPR), July 2017

Fair Face Recognition Using Data Balancing, Enhancement and Fusion

Jun Yu[1], Xinlong Hao[1(✉)], Haonian Xie[1], and Ye Yu[2]

[1] University of Science and Technology of China, Hefei, China
harryjun@ustc.edu.cn, {haoxl,xie233}@mail.ustc.edu.cn
[2] School of Computer and Information, Hefei University of Technology, Hefei, China
yuye@hfut.edu.cn

Abstract. Racial bias is an important issue in biometrics, while has not been thoroughly studied in deep face recognition. By reducing the influence of gender and skin colour, this paper proposes a fair face recognition system with low bias. First, multiple preprocessing methods are added to improve the dual shot face detector for obtaining target face from a given test image. Then, a data re-sampling approach is employed to balance the data distribution and reduce the bias based on the analysis of training data. Moreover, multiple data enhancement methods are used to increase the accuracy performance. Finally, a linear-combination strategy is adopted to benefit from mutil-model fusion. ChaLearn Looking at People Fair Face Recognition challenge is supported by ECCV 2020. Our team (ustc-nelslip) ranked 1st in the development stage and 2nd in the test stage of this challenge. The code is available at https://github.com/HaoSir/ECCV-2020-Fair-Face-Recognition-challenge_2nd_place_solution-ustc-nelslip-.

Keywords: Racial bias · Face detector · Data re-sampling · Data enhancement · Linear-combination strategy

1 Introduction

Face recognition achieved a high level of performance in recent years [6,9,14,30], along with the development of convolutional neural network [15,22,26]. However, as its wider and wider application, its potential for unfairness is raising alarm [2, 4,21,24]. For instance, according to [11], a year-long research investigation across 100 police departments revealed that African-American individuals are more likely to be stopped by law enforcement. Obviously, it is particularly important to obtain a fair face recognition system.

And as point out by some works [2,3,32], the main cause of the model bias between well-represented groups and under-represented groups is the distribution of training dataset. As shown in [32], we can easily observe that the commonly used face recognition datasets [5,13,16] are dominated by Caucasian identities, since the dataset is mainly formed by Caucasian subjects, face recognition

© Springer Nature Switzerland AG 2020
A. Bartoli and A. Fusiello (Eds.): ECCV 2020 Workshops, LNCS 12540, pp. 492–505, 2020.
https://doi.org/10.1007/978-3-030-65414-6_34

models performance on Caucasian outperforms that on other groups of people, such as African, Asian, and Indian. Similarly, gender is another aspect of face recognition datasets' imbalance, dataset mainly consists of male faces. To solve these problems, many efforts on face recognition aim to tackle the class imbalance problem on training data. For example, in prior-DNN era, Zhang et al. [38] proposed a cost-sensitive learning framework to reduce misclassification rate of face identification. To correct the skew of separating hyperplanes of SVM on imbalanced data, Liu et al. [20] proposed Margin-Based Adaptive Fuzzy SVM that obtains a lower generalization error bound. In the DNN era, face recognition models are trained on large-scale face datasets with highly-imbalanced class distribution. Range Loss [37] learns a robust face representation that makes the most use of every training sample. To mitigate the impact of insufficient class samples, center-based feature transfer learning [35] and large margin feature augmentation [33] are proposed to augment features of minority identities and equalize class distribution. Besides, the FRVT 2019 [12] shows the demographic bias of over 100 face recognition algorithms. To uncover deep learning bias, Alexander et al. [3] developed an algorithm to mitigate the hidden biases within training data. Wang et al. [32] proposed a domain adaptation network to reduce racial bias in face recognition. They recently extended their work using reinforcement learning to find optimal margins of additive angular margin based loss functions for different races [31].

In this paper, we present a face recognition method to achieve fair face recognition. Giving an image with a loosely cropped face roughly in the center, the first thing we need to do is obtaining the face from the image. As a fundamental step for various facial applications, like face alignment [28], parsing [7], recognition [34], and verification [9], face detection achieves great progress inspired by deep convolutional neural network (CNN). Previous state-of-the-art face detectors can be roughly divided into two categories. The first one is mainly based on the Region Proposal Network (RPN) adopted in Faster RCNN [25] and employs two stage detection schemes [29]. RPN is trained end-to-end and generates high quality region proposals which are further refined by Fast R-CNN detector. The other one is Single Shot Detector (SSD) [19] based one-stage methods, which get rid of RPN, and directly predict the bounding boxes and confidence [8]. Recently, one-stage face detection framework has attracted more attention due to its higher inference efficiency and straightforward system deployment. We test multiple face detectors, such as MTCNN [36], Retinaface [10], DSFD [17], and propose an improved method based on DSFD.

After obtaining the face from image, as diversity between different attributes is very large, a series of face preprocessing methods are used to reduce bias and improve accuracy at the same time. For instance, we use a data re-sampling method to balance the data distribution by under-sampling the majority class. Train data enhancement and test time augmentation are used for obtaining improved accuracy. Then train data would be used to train face recognition models, trained models are used to extract features of test data. Next, by calculating the cosine similarity between two feature vectors, confidence scores of test

Table 1. Data split, positive/negative pairs.

	Train	Validation	Test	Total
Images	100,186	17,138	35,593	152,917
Unique identities	4297	614	1,228	6,139
Positive pairs	-	448,119	500,176	-
Negative pairs	-	552,672	500,963	-

data would be generated, which indicate the degree two faces belong to the same person. Finally, after obtaining several different prediction scores based on different models, the final prediction score would be generated by linear combination of them. The main contributions of this paper are summarized as follows.

(1) A new face detection method is proposed to obtain faces from the given image by employing multiple preprocessing methods, instead of using the detected faces directly.
(2) To acquire low bias and high accuracy, not only multiple train data enhancement methods and multi-model fusion strategy are used, but also a data re-sampling approach is adopted to balance the data distribution.
(3) Our team (ustc-nelslip) has made remarkable achievements in the ECCV 2020 ChaLearn Looking at People Fair Face Recognition challenge [27], where we achieve 1st place in the development stage, and 2nd place in the test stage.

2   Dataset Description

The given dataset is a new collected dataset with 13k images from 3k new subjects along with a reannotated version of IJB-C [23] (140k images from 3.5k subjects), totalling 153k facial images from 6.1k unique identities. Both databases have been accurately annotated for gender and skin colour (protected attributes) as well as for age group, eyeglasses, head pose, image source and face size. Its motivation is to develop fair face verification methods aiming for a reduced bias in terms of gender and skin color (protected attributes).

The dataset is splited to train data, validation data and test data, they have different number of images and unique identities, moreover, positive and negative pairs are defined for validation and test sets, used in development and test phases. The basic information about the dataset is shown in Table 1. And there is a huge class imbalance problem in given data, for instance, as for gender&skin color, the number of male persons with bright skin is five times bigger than the number of female persons with dark skin, as for age, middle age people is more than old people. The detail infromation is shown in Fig. 1.

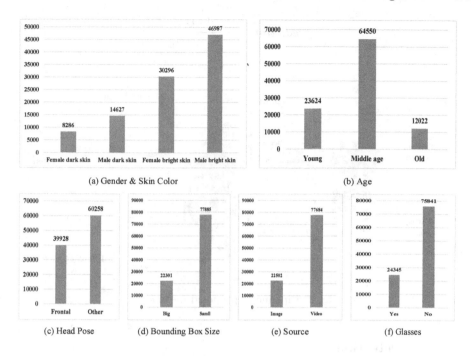

(a) Gender & Skin Color   (b) Age

(c) Head Pose   (d) Bounding Box Size   (e) Source   (f) Glasses

Fig. 1. Distribution of train data.

3   Proposed Method

Automated face recognition has achieved remarkable success with the rapid developments of deep learning algorithms. Despite the improvement in the accuracy of face recognition, there was a big problem of fairness. It has been observed that many face recognition systems have lower performance for certain demographic groups than others, therefore, it is necessary to reduce the bias of face recognition. In this work, we propose a deep face recognition method with low bias.

3.1   Pipeline

The proposed approach consists of four parts, they are face detection, face preprocess, train module and inference module, as shown in Fig. 2. In this work, we use a method based on DSFD [17] as face detector, multiple enhanced methods are used in face preprocess stage. Besides, IR_50 [15] or IR_152 [15] are used as backbone for feature extraction, which are pre-trained on the MS-Celeb-1M [13] and DeepGlint dataset [1]. In train module, we use two heads: Arcface [9] and Cosface [30]. Then in inference module, test data feature vectors are generated based on trained backbone. Next, the final prediction is obtained by calculating the cosine similarity between two feature vectors. The prediction is defined as follows, A represents a feature vector, B represents another feature vector.

$$prediction = \frac{A \cdot B}{\|A\| \times \|B\|} \tag{1}$$

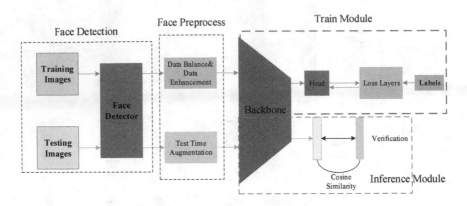

Fig. 2. Pipeline of proposed method.

3.2 Face Detection

Giving an image with a loosely cropped face roughly in the center, the first thing we need to do is obtaining face from the image. There are many great face detection methods in recent years, such as MTCNN [36], Retinaface [10], DSFD [17], we use their pretrained models in this paper and it is found detection rates of different face detectors are diverse under the same parameter setting: confidence_threshold = 0.7, nms_iou_threshold = 0.3. For train data, MTCNN met- hod gets 4% missing rate, Retinaface gets 2% missing rate, while DSFD method gets lower missing rate for 1%, so we choose DSFD as our basic face detector.

Our network input size should be 112×112, multiple methods are used to improve the effect of face detector, as shown in Fig. 3, and our method would get better crop effect considering vision aspect. As for too small image to detect face, proposed method would crop face in the center of the image with a rate of 7/12 directly. As for general image, detected bounding box was enlarged by factor 1.2, by this way, the obtained face would contain more information for recognition.

3.3 Face Preprocess

In train stage, it is found that the distribution of train data is imbalanced, so the most direct idea is balancing the data firstly. Two methods are proposed to solve this problem, one is a data re-sampling method which balances the data distribution by under-sampling the majority class, the other is using extra data to ease this problem.

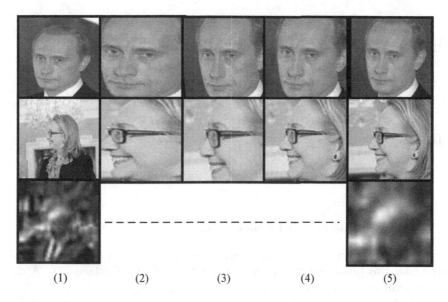

(1)   (2)   (3)   (4)   (5)

Fig. 3. Comparison with different face detection methods. (1) Original images. (2) Face images which are resized to 112*112 directly. (3) Aligned faces which are expended to 112*112. (4) Cropped faces by the square extended box based on detection bounding box. (5) The outputs of the proposed method. As for too small image to detect face, proposed method would crop face in the center of the image with a rate of 7/12 directly. As for general image, detected bounding box was enlarged by factor 1.2.

As shown in Fig. 1, the distribution of protected attribute A (gender and skin colour - male bright, male dark, female bright, female dark) is imbalanced, for instance, the number of men with bright skin is five times bigger than the number of women with dark skin, so we mainly reduce the weight of male bright skin class and female bright skin class considering the amount of their faces. One choice of our weight rate is female dark skin:male dark skin:female bright skin:man bright skin = 1:0.7:0.5:0.3. Besides, it is found that using extra data to balance data is also effective, we train a classifier of protected attribute A (gender and skin colour), which is used to collect extra data from other datasets such like MS-Celeb-1M [13] and DeepGlint dataset [1]. Then some of those face images are added to the given train data for increasing numbers of women with dark skin and men with dark skin. Both methods are effective, but the re-sample method could help us get better score. The ablation study is shown in Table 2.

In addition to data balance, multiple attempts have been made in the data enhancement phase. For train data enhancement, multiple enhanced methods, such as RandomHorizontalFlip (RHF), ColorJitter (CJ) and RandomBlur(RB), are used to improve data diversity. The ablation study of data enhancement methods is shown in Table 3.

In inference stage, test time augmentation is used, we put original face and the face which is filpped horizontally into backbone respectively, then both of the output feature vectors would be added together to produce the final feature vector.

Table 2. Comparison of data balance methods.

Backbone	Data Balance	Verification Score		
		Bias(positive pairs)	Bias(negative pairs)	Accuracy
IR_50	No use	0.023193	0.002516	0.981941
	Add data	0.018124	**0.001872**	0.983367
	Re-sample	**0.013624**	0.002072	**0.988642**
IR_152	No use	0.019356	0.002272	0.985961
	Add data	0.015411	0.001679	0.987578
	Re-sample	**0.010013**	**0.001360**	**0.991659**

3.4   Train Module and Inference Module

In train module, two different backbones are used, they are IR_50 and IR_152, meanwhile, two heads are tried out, i.e. Arcface [9] and Cosface [30]. Arcface uses an additive angular margin m:

$$L_{arc} = -\frac{1}{N} \sum_{j=1}^{N} log \frac{e^{s_c}(cos(\theta_y^{(j)}{}_j + m))}{e^{s_c}(cos(\theta_y^{(j)}{}_j + m)) + \sum_{i=1,i\neq y^{(j)}}^{n} e^{s_c cos\theta_{ij}}} \tag{2}$$

and Cosface uses an additive cosine margin m:

$$L_{cos} = -\frac{1}{N} \sum_{j=1}^{N} log \frac{e^{s_c}(cos(\theta_y^{(j)}{}_j) - m)}{e^{s_c}(cos(\theta_y^{(j)}{}_j) - m) + \sum_{i=1,i\neq y^{(j)}}^{n} e^{s_c cos\theta_{ij}}} \tag{3}$$

where θ_{ij} is the angle between the weight W_i and the feature x_j, $x_j \in \mathbb{R}^d$ denotes the deep feature of the $j-th$ sample, belonging to the $y^{(j)}-th$ class, and $W_i \in \mathbb{R}^d$ denotes the $i-th$ column of the weight $W \in \mathbb{R}^{d \times n}$. N is the batch size, n is the number of classes, and s_c is the scale factor.

Besides, the focal loss [18] is used to alleviate the significant imbalance of the proportion of positive and negative samples, which can be formulated as:

$$\mathcal{L}_{focal} = -a_t(1 - p_t)^\gamma log(p_t) \tag{4}$$

$$a_t = \begin{cases} a, \; if \; y = 1 \\ 1 - a, \; otherwise \end{cases} \tag{5}$$

$$p_t = \begin{cases} p, \; if \; y = 1 \\ 1 - p, \; otherwise \end{cases} \tag{6}$$

where γ is a focusing parameter, a is a weight that controls the contribution of positive and negative samples to the total loss, and p is the output of the sigmoid activation function. When $a = 0.5$, the focal loss is equivalent to BCE loss.

In inference module, after obtaining features, Eq. 1 is used to calculate cosine similarity scores, and eight great predictions would be generated, which would be divided into two groups based on their score. Finally, a hyper-parameter λ would be used to balance two groups and output better final prediction results.

4   Experiments

A workstation with Intel i7-7700K 4.2G CPU, 192 G memory and 8*NVIDIA TITAN V 12 G GPUs is used for experiments. We evaluate our method on the given dataset, and essential ablation studies are elaborately designed, as well as quantitative evaluations with other contestants.

For implementation details, all face images are resized to 112×112 pixels as the inputs. The number of image pairs or number of image for the training set, validation set, and test set are shown in Tabel 1. For optimization, we choose the SGD optimizer, where the initial learning rate is set to 0.01 and weight_decay is set to $5e-4$, momentum parameter is set to 0.9. The model is trained with a batch size of 256 and over 60 epochs, which is based on pytorch. In addition, we evaluate our method on the validation set of given dataset and compare it to other contestants on the validation set and test set of given dataset.

In this challenge, as for evaluation metric, for given values of the legitimate attributes $X = x$ (e.g. age, wearing glasses, head pose etc.), organizers define a discrimination $d_a(X = x)$ of a protected group $A = a$ (gender and skin colour) as a difference in accuracy for this group and a group with the best accuracy:

$$d_a(X = x) = \max_{a'} \text{Accuracy}_{a'}(X = x) - \text{Accuracy}_a(X = x) \tag{7}$$

The final measure of unfairness is then the difference between the average discrimination of the most discriminated group and the least discriminated one: where $|X|$ denotes the number of combinations of values of legitimate attributes. Organizers use AUC-ROC to measure the accuracy.

$$\text{Bias} = \max_a \frac{1}{|X|} \sum_x d_a(X = x) - \min_a \frac{1}{|X|} \sum_x d_a(X = x) \tag{8}$$

4.1   Ablation Studies

In this work, we implement the ablation studies of our face recognition method based on 2 backbones (i.e., IR_50 and IR_152), 2 heads (i.e., Arcface and Cosface) and 4 types of train data augmentation. (i.e., RandomHorizontalFlip (RHF), ColorJitter (CJ), RandomHorizontalFlip (RHF) plus ColorJitter (CJ), RandomHorizontalFlip (RHF) plus ColorJitter (CJ) plus RandomBlur (RB)). As shown in Table 3, a total of 20 methods are evaluated on the validation set of given dataset, where the best four results are marked red and four second-best results are marked blue. When the backbone is IR_152, the head is Arcface, and the data enhancement operation is RHF+CJ+RB, we obtain the best verification score, which is lower in bias and higher in accuracy. In addition, experimental results show that Arcface performs better than Cosface in most cases, and IR_152 is superior to IR_50 as a backbone in our face recognition network.

Table 3. Ablation studies of our face recognition method (**RHF** represents RandomHorizontalFlip, **CJ** represents ColorJitter, **RB** is the short for RandomBlur).

Backbone	Head	Data Enhancement	Verification Score		
			Bias(positive pairs)	Bias(negative pairs)	Accuracy
IR_50	CosFace	Baseline	0.015785	0.002898	0.985795
		+RHF	0.010751	0.001598	0.988343
		+CJ	0.011134	0.000980	0.989941
		+RHF+CJ	0.007641	0.000780	0.993353
		+RHF+CJ+RB	0.005064	0.000547	0.994754
	ArcFace	Baseline	0.013624	0.002072	0.988642
		+RHF	0.009579	0.001359	0.992217
		+CJ	0.007471	0.000876	0.993394
		+RHF+CJ	0.003982	0.000442	0.997401
		+RHF+CJ+RB	0.004006	0.000368	0.997853
IR_152	CosFace	Baseline	0.011624	0.001650	0.989982
		+RHF	0.007682	0.001175	0.992034
		+CJ	0.007213	0.001046	0.993359
		+RHF+CJ	0.004425	0.000647	0.997468
		+RHF+CJ+RB	0.004865	0.000580	0.997780
	ArcFace	Baseline	0.010013	0.001360	0.991659
		+RHF	0.007249	0.000854	0.993254
		+CJ	0.005442	0.000543	0.995270
		+RHF+CJ	0.003687	0.000321	0.998054
		+RHF+CJ+RB	0.003945	0.000269	0.998653

4.2   The Effect of Hyper-parameter λ

Eight great models are obtained by using ablation studies in Table 3, they are divided into two groups, the red ones has better performance while the blue ones are also useful for impoving the final score. For each group, the group score is the average of four different model scores. In our face recognition network, the hyper-parameter λ are set to balance two groups and output better final prediction results. The final confidence score is defined as:

$$Score = S_{group1} + \lambda S_{group2} \qquad (9)$$

As shown in Table 4, changing λ has an effect on positive pairs bias, negative pairs bias and accuracy. With the increase of λ, the accuracy first increases and then decreases. One possible explanation is that the group2 has better performance in some class than group1, so it would be helpful at start, but it would also bring its disadvantage as the increase of λ, which would lead to overall performance degradation. When $\lambda = 0.4$, the accuracy and negative pairs bias are best, while positive pairs bias is best when $\lambda = 0.6$.

Table 4. The effect of hyper-parameter λ.

λ	Verification Score		
	Bias(positive pairs)	Bias(negative pairs)	Accuracy
0.0	0.003125	0.000201	0.999059
0.2	0.003025	0.000176	0.999132
0.4	0.002956	**0.000142**	**0.999287**
0.6	**0.002847**	0.000169	0.999264
0.8	0.003102	0.000252	0.999110
1.0	0.003425	0.000353	0.999012

Table 5. Leaderboard of Fair Face Recognition Challenge at development stage (Top 10 contestants and baseline).

User	Ranking	Verification Score		
		Bias(positive pairs)	Bias(negative pairs)	Accuracy
Ustc-nelslip	3.333333(1)	0.002956(4)	0.000142(3)	0.999287(3)
Zheng.zhu	4.666667(2)	0.003781(8)	0.000344(5)	0.999442(1)
CdtQin	4.666667(2)	0.002334(2)	0.000472(7)	0.998477(5)
Haoxl	5.666667(3)	0.003678(7)	0.000348(6)	0.998699(4)
Crisp	6.000000(4)	0.003193(5)	0.000935(11)	0.999394(2)
Cam vision	7.333333(5)	0.003305(6)	0.000814(10)	0.998402(6)
Hyg	7.333333(5)	0.003305(6)	0.000814(10)	0.998402(6)
Senlin11	11.000000(6)	0.010091(19)	0.000165(4)	0.992093(10)
Hanamichi	12.333333(7)	0.006760(12)	0.001631(12)	0.987382(13)
Paranoidai	14.333333(8)	0.007745(15)	0.003779(17)	0.988359(11)
Baseline	41.333333(-)	0.057620(-)	0.054311(-)	0.889264(-)

4.3 Quantitative Evaluations

The leaderboard of ECCV 2020 ChaLearn Looking at People Fair Face Recognition challenge is illustrated in Table 5 and Table 6, where we show the top 10 contestants and the baseline. For our method, our team (ustc-nelslip) ranks the 1st place in the development stage leaderboard of Looking at People Fair Face Recognition challenge, and ranks the 2nd place in the test stage leaderboard. The line chart of verification score is drawn in Fig. 4, which obviously shows that the accuracy scores of top contestants are all great, negative pairs bias is always lower than positive pairs bias for both development stage and test stage, besides, it is found that comparing to development stage, the bias is much lower and accuracy is higher in test stage, indicated the test data may be easier than validation data.

Table 6. Leaderboard of Fair Face Recognition Challenge at test stage (Top 10 contestants and baseline).

User	Ranking	Verification Score		
		Bias(positive pairs)	Bias(negative pairs)	Accuracy
Paranoidai	1.333333(1)	0.000012(1)	0.000059(2)	0.999966(1)
Ustc-nelslip	3.666667(2)	0.000172(2)	0.000175(4)	0.999569(5)
CdtQin	4.000000(3)	0.000405(9)	0.000036(1)	0.999827(2)
Debias	4.666667(4)	0.000460(10)	0.000036(1)	0.999825(3)
Zhaixingzi	5.000000(5)	0.000237(8)	0.000116(3)	0.999698(4)
Bestone	5.333333(6)	0.000197(5)	0.000175(4)	0.999565(7)
Haoxl	5.333333(6)	0.000195(4)	0.000178(6)	0.999568(6)
Early	5.333333(6)	0.000190(3)	0.000175(4)	0.999547(9)
Lemoner20	7.000000(7)	0.000201(6)	0.000176(5)	0.999507(10)
Ai	7.333333(8)	0.000217(7)	0.000180(7)	0.999560(8)
Baseline	34.666667(-)	0.058601(-)	0.059694(-)	0.859175(-)

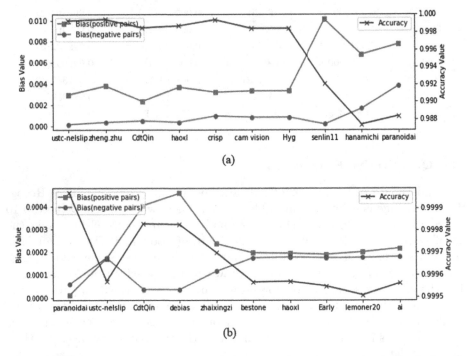

(a)

(b)

Fig. 4. Verification score line chart, including top 10 contestants in the leaderboard. (a) Development stage. (b) Test stage.

5   Conclusions

This paper proposes a fair face recognition method with low bias considering gender and skin colour. By the operations of face detection and face preprocess, feature extraction and similarity quantification, the output score can quantify the similarity of two faces, then a linear-combination strategy is used to output the final fairer prediction. Our team (ustc-nelslip) ranked 1st in development stage and ranked 2nd in test stage of ECCV 2020 ChaLearn Looking at People Fair Face Recognition challenge.

References

1. http://trillionpairs.deepglint.com/overview
2. Alvi, M., Zisserman, A., Nellåker, C.: Turning a blind eye: explicit removal of biases and variation from deep neural network embeddings. In: Proceedings of the European Conference on Computer Vision (ECCV) (2018)
3. Amini, A., Soleimany, A.P., Schwarting, W., Bhatia, S.N., Rus, D.: Uncovering and mitigating algorithmic bias through learned latent structure. In: Proceedings of the 2019 AAAI/ACM Conference on AI, Ethics, and Society, pp. 289–295 (2019)
4. Buolamwini, J., Gebru, T.: Gender shades: intersectional accuracy disparities in commercial gender classification. In: Conference on Fairness, Accountability and Transparency, pp. 77–91 (2018)
5. Cao, Q., Shen, L., Xie, W., Parkhi, O.M. and Zisserman, A.: Vggface2: a dataset for recognising faces across pose and age. In: 2018 13th IEEE International Conference on Automatic Face & Gesture Recognition (FG 2018), pp. 67–74. IEEE (2018)
6. Chen, K., Wu, Y., Qin, H., Liang, D., Liu, X. and Yan, J.: R3 adversarial network for cross model face recognition. In: Proceedings of the IEEE Conference on Computer Vision and Pattern Recognition, pp. 9868–9876 (2019)
7. Chen, Y., Tai, Y., Liu, X., Shen, C., Yang, J.: Fsrnet: end-to-end learning face super-resolution with facial priors. In:Proceedings of the IEEE Conference on Computer Vision and Pattern Recognition, pp. 2492–2501 (2018)
8. Chi, C., et al.: Selective refinement network for high performance face detection. In: Proceedings of the AAAI Conference on Artificial Intelligence, vol. 33, pp. 8231–8238 (2019)
9. Deng, J., Guo, J., Xue, N., Zafeiriou, S.: Arcface: additive angular margin loss for deep face recognition. In Proceedings of the IEEE Conference on Computer Vision and Pattern Recognition, pp. 4690–4699 (2019)
10. Deng, J., Guo, J., Xue, N. and Zafeiriou, S.: Retinaface: single-stage dense face localisation in the wild. arXiv preprint arXiv:1905.00641 (2019)
11. Garvie, C.: The perpetual line-up: Unregulated police face recognition in America. Center on Privacy & Technology, Georgetown Law (2016)
12. Grother, P.J., Ngan, M.L., Hanaoka, K.K.: Face recognition vendor test part 3: Demographic effects (2019)
13. Guo, Y., Zhang, L., Hu, Y., He, X., Gao, J.: MS-Celeb-1M: a dataset and benchmark for large-scale face recognition. In: Leibe, B., Matas, J., Sebe, N., Welling, M. (eds.) ECCV 2016. LNCS, vol. 9907, pp. 87–102. Springer, Cham (2016). https://doi.org/10.1007/978-3-319-46487-9_6

14. Han, C., Shan, S., Kan, M., Wu, S., Chen, X.: Face recognition with contrastive convolution. In Proceedings of the European Conference on Computer Vision (ECCV), pp. 118–134 (2018)
15. He, K., Zhang, X., Ren, S. and Sun, J.: Deep residual learning for image recognition. In: Proceedings of the IEEE Conference on Computer Vision and Pattern Recognition, pp. 770–778 (2016)
16. Huang, G.B., Mattar, M., Berg, T., Learned-Miller, E.: Labeled faces in the wild: A database forstudying face recognition in unconstrained environments (2008)
17. Li, J., et al.: DSFD: dual shot face detector. In: Proceedings of the IEEE Conference on Computer Vision and Pattern Recognition, pp. 5060–5069 (2019)
18. Lin, T.Y., Goyal, P., Girshick, R., He, K., Dollár, P.: Focal loss for dense object detection. In: Proceedings of the IEEE International Conference on Computer Vision, pp. 2980–2988 (2017)
19. Liu, W., Anguelov, D., Erhan, D., Szegedy, C., Reed, S., Fu, C.-Y., Berg, A.C.: SSD: Single shot MultiBox detector. In: Leibe, B., Matas, J., Sebe, N., Welling, M. (eds.) ECCV 2016. LNCS, vol. 9905, pp. 21–37. Springer, Cham (2016). https://doi.org/10.1007/978-3-319-46448-0_2
20. Liu, Y.-H., Chen, Y.-T.: Face recognition using total margin-based adaptive fuzzy support vector machines. IEEE Trans. Neural Networks 18(1), 178–192 (2007)
21. Steve, L.: Facial recognition is accurate, if you're a white guy. New York Times 9, 8 (2018)
22. Ma, N., Zhang, X., Zheng, H.T., Sun, J.: Shufflenet v2: practical guidelines for efficient CNN architecture design. In: Proceedings of the European Conference on Computer Vision (ECCV), pp. 116–131 (2018)
23. Maze, B., et al. Iarpa janus benchmark-c: face dataset and protocol. In: 2018 International Conference on Biometrics (ICB), pp. 158–165. IEEE (2018)
24. Orcutt, M.: Are face recognition systems accurate? depends on your race. MIT Technology Review 2016 (2016)
25. Ren, S., He, K., Girshick, R. and Sun, J.: Faster r-cnn: Towards real-time object detection with region proposal networks. In: Advances in Neural Information Processing Systems, pp. 91–99 (2015)
26. Sandler, M., Howard, A., Zhu, M., Zhmoginov, A., Chen, L.C.: Mobilenetv 2: Inverted residuals and linear bottlenecks. In: Proceedings of the IEEE Conference on Computer Vision and Pattern Recognition, pp. 4510–4520 (2018)
27. Sixta, T., Junior, J., Jacques, C.S., Buch-Cardona, P., Vazquez, E., Escalera, S.: Fairface challenge at ECCV 2020: analyzing bias in face recognition. In: Proceedings of the European Conference on Computer Vision (ECCV) Workshops (2020)
28. Tai, Y., et al.: Towards highly accurate and stable face alignment for high-resolution videos. In: Proceedings of the AAAI Conference on Artificial Intelligence, vol. 33, pp. 8893–8900 (2019)
29. Wang, H., Li, Z., Ji, X., Wang, Y.: Face r-cnn. arXiv preprint arXiv:1706.01061 (2017)
30. Wang, H., et al.: Cosface: large margin cosine loss for deep face recognition. In: Proceedings of the IEEE Conference on Computer Vision and Pattern Recognition, pp. 5265–5274 (2018)
31. Wang, M., Deng, W.: Mitigating bias in face recognition using skewness-aware reinforcement learning. In: Proceedings of the IEEE/CVF Conference on Computer Vision and Pattern Recognition, pp. 9322–9331 (2020)
32. Wang, M., Deng, W., Hu, J., Tao, X., Huang, Y.: Racial faces in the wild: reducing racial bias by information maximization adaptation network. In: Proceedings of the IEEE International Conference on Computer Vision, pp. 692–702 (2019)

33. Wang, P., Fei, S., Zhao, Z., Guo, Y., Zhao, Y., Zhuang, B.: Deep class-skewed learning for face recognition. Neurocomputing **363**, 35–45 (2019)
34. Xie, J., Yang, J., Qian, J.J., Tai, Y., Zhang, H.M.: Robust nuclear norm-based matrix regression with applications to robust face recognition. IEEE Trans. Image Process. **26**(5), 2286–2295 (2017)
35. Yin, X., Yu, X., Sohn, K., Liu, X., Chandraker, M.: Feature transfer learning for face recognition with under-represented data. In: Proceedings of the IEEE Conference on Computer Vision and Pattern Recognition, pp. 5704–5713 (2019)
36. Zhang, K., Zhang, Z., Li, Z., Qiao, Y.: Joint face detection and alignment using multitask cascaded convolutional networks. IEEE Signal Process. Lett. **23**(10), 1499–1503 (2016)
37. Zhang, X., Fang, Z., Wen, Y., Li, Z., Qiao, Y.: Range loss for deep face recognition with long-tailed training data. In: Proceedings of the IEEE International Conference on Computer Vision, pp. 5409–5418 (2017)
38. Zhang, Y., Zhou, Z.-H.: Cost-sensitive face recognition. IEEE Trans. Pattern Anal. Mach. Intell. **32**(10), 1758–1769 (2009)

Investigating Bias and Fairness in Facial Expression Recognition

Tian Xu[1]($\boxtimes$), Jennifer White[1], Sinan Kalkan[2], and Hatice Gunes[1]

[1] Department of Computer Science and Technology, University of Cambridge, Cambridge, UK
{Tian.Xu,Hatice.Gunes}@cl.cam.ac.uk, jw2088@cam.ac.uk
[2] Department of Computer Engineering, Middle East Technical University, Ankara, Turkey
skalkan@metu.edu.tr

Abstract. Recognition of expressions of emotions and affect from facial images is a well-studied research problem in the fields of affective computing and computer vision with a large number of datasets available containing facial images and corresponding expression labels. However, virtually none of these datasets have been acquired with consideration of fair distribution across the human population. Therefore, in this work, we undertake a systematic investigation of bias and fairness in facial expression recognition by comparing three different approaches, namely a baseline, an attribute-aware and a disentangled approach, on two well-known datasets, RAF-DB and CelebA. Our results indicate that: (i) data augmentation improves the accuracy of the baseline model, but this alone is unable to mitigate the bias effect; (ii) both the attribute-aware and the disentangled approaches equipped with data augmentation perform better than the baseline approach in terms of accuracy and fairness; (iii) the disentangled approach is the best for mitigating demographic bias; and (iv) the bias mitigation strategies are more suitable in the existence of uneven attribute distribution or imbalanced number of subgroup data.

Keywords: Fairness · Bias mitigation · Facial expression recognition

1 Introduction

Automatically recognising expressions and affect from facial images has been widely studied in the literature [27,33,42]. Thanks to the unprecedented advances in machine learning field, many techniques for tackling this task now use deep learning approaches [27] which require large datasets of facial images labelled with the expression or affect displayed.

An important limitation of such a data-driven approach to affect recognition is being prone to *bias*es in the datasets against certain demographic groups [5,10,14,25,34,38]. The datasets used for training do not necessarily contain an even distribution of subjects in terms of demographic attributes such as

© Springer Nature Switzerland AG 2020
A. Bartoli and A. Fusiello (Eds.): ECCV 2020 Workshops, LNCS 12540, pp. 506–523, 2020.
https://doi.org/10.1007/978-3-030-65414-6_35

race, gender and age. Moreover, majority of the existing datasets that are made publicly available for research purposes do not contain information regarding these attributes, making it difficult to assess bias, let alone mitigate it. Machine learning models, unless explicitly modified, are severely impacted by such bias since they are given more opportunities (training samples) for optimizing their objectives towards the majority demographic group in the dataset. This leads to lower performances for the minority groups, i.e., subjects represented with less number of samples [5,10,12,14,25,34,38,41]. To address these issues, many solutions have been proposed in the machine learning community to tackle the problem at the data level with data generation or sampling approaches [2,9,20, 21,32,36,39,49], at the feature level using adversarial learning [1,35,49,51] or at the task level using multi-domain/task learning [11,49].

Bias in facial analysis have attracted increasing attention from both the general public and the research communities. For example, many studies have investigated bias and mitigation strategies for face recognition [5,12–14,35,38,41], gender recognition [8,12,43,50], age estimation [6,8,12,16,43], kinship verification [12] and face image quality estimation [44]. However, bias in facial expression recognition has not been investigated, except for [9,49], that only focused on the task of smiling/non-smiling using the CelebA dataset.

In this paper, we undertake a systematic investigation of bias and fairness in facial expression recognition. To this end, we consider three different approaches, namely a baseline deep network, an attribute-aware network and a representation-disentangling network (following [1,35,49,51]) under the two conditions of with and without data augmentation. As a proof of concept, we conduct our experiments on RAF-DB and CelebA datasets that contain labels in terms of gender, age and/or race. To the best of our knowledge, this is the first work (i) to perform an extensive analysis of bias and fairness for facial expression recognition, beyond the binary classes of smiling/non-smiling [9,49], (ii) to use the sensitive attribute labels as input to the learning model to address bias, and (iii) to extend the work of [31] to the area of facial expression recognition in order to learn fairer representations as a bias mitigation strategy.

2 Related Work

Bias in the field of Human-Computer Interaction (HCI), and in particular issues arising from the intersection of gender and HCI have been discussed at length in [4]. However, studies specifically analysing, evaluating and mitigating race, gender and age biases in affect recognition have been scarce. We therefore provide a summary of related works in other forms of facial analysis including face and gender recognition, and age estimation.

2.1 Bias and Mitigation in Machine Learning

Attention to bias and fairness in machine learning (ML) has been rapidly increasing with the employment of ML applications in everyday life. It is now well accepted that ML models are extremely prone to biases in data [5,22], which has raised substantial concern in public such that regulatory actions are being as preventive measure; e.g. European Commission [7] requires training data for such applications to be "sufficiently broad," and to reflect "all relevant dimensions of gender, ethnicity and other possible grounds of prohibited discrimination".

Bias mitigation strategies in ML generally take inspiration from data or class balancing approaches in ML, a very related problem which directly pertains to imbalance in the task labels. Bias can be addressed in an ML model in different ways [3,10]: For example, we can balance the dataset in terms of the demographic groups, using under-sampling or over-sampling [20,49], sample weighting [2,21, 39], data generation [9,36], data augmentation [32] or directly using a balanced dataset [46,47]. However, it has been shown that balancing samples does not guarantee fairness among demographic groups [48].

Another strategy to mitigate bias is to remove the sensitive information (i.e. gender, ethnicity, age) from the input at the data level (a.k.a. "fairness through unawareness") [1,49,51]. However, it has been shown that the remaining information might be implicitly correlated with the removed sensitive attributes and therefore, the residuals of the sensitive attributes may still hinder fairness in the predictions [11,17]. Alternatively, we can make the ML model more aware of the sensitive attributes by making predictions independently for each sensitive group (a.k.a. "fairness through awareness") [11]. Formulated as a multi-task learning problem, such approaches allow an ML model to separate decision functions for different sensitive groups and therefore prohibit the learning of a dominant demographic group to negatively impact the learning of another one.Needless to say this comes at a cost - it dramatically increases the number of parameters to be learned, as a separate network or a branch needs to be learned for each sensitive group.

2.2 Bias in Facial Affect Recognition

It has been long known that humans' judgements of facial expressions of emotion are impeded by the ethnicity of the faces judged [23]. In the field of automatic affect recognition, systematic analysis of bias and the investigation of mitigation strategies are still in their infancy. A pioneering study by Howard et al. [19] investigated how using a cloud-based emotion recognition algorithm applied to images associated with a minority class (children's facial expressions) can be skewed when performing facial expression recognition on the data of that minority class. To remedy this, they proposed a hierarchical approach combining outputs from the cloud-based emotion recognition algorithm with a specialized learner. They reported that this methodology can increase the overall recognition results by 17.3%. Rhue [40], using a dataset of 400 NBA player photos, found systematic racial biases in Face++ and Microsoft's Face API. Both systems assigned to African American players more negative emotional scores on average, regardless of how much they smiled.

When creating a new dataset is not straightforward, and/or augmentation is insufficient to balance an existing dataset, Generative Adversarial Networks

(GAN) have been employed for targeted data augmentation. Denton et al. [9] present a simple framework for identifying biases in a smiling attribute classifier. They utilise GANs for a controlled manipulation of specific facial characteristics and investigate the effect of this manipulation on the output of a trained classifier. As a result, they identify which dimensions of variation affect the predictions of a smiling classifier trained on the CelebA dataset. For instance, the smiling classifier is shown to be sensitive to the Young dimensions, and the classification of 7% of the images change from a smiling to not smiling classification as a result of manipulating the images in this direction. Ngxande et al. [36] introduce an approach to improve driver drowsiness detection for under-represented ethnicity groups by using GAN for targeted data augmentation based on a population bias visualisation strategy that groups faces with similar facial attributes and highlights where the model is failing. A sampling method then selects faces where the model is not performing well, which are used to fine-tune the CNN. This is shown to improve driver drowsiness detection for the under-represented ethnicity groups. A representative example for non-GAN approaches is by Wang et al. [49] who studied the mitigation strategies of data balancing, fairness through blindness, and fairness through awareness, and demonstrated that fairness through awareness provided the best results for smiling classification on the CelebA dataset.

3   Methodology

To investigate whether bias is a problem for the facial expression recognition task, we conduct a comparative study using three different approaches. The first one acts as the baseline approach and we employ two other approaches, namely the Attribute-aware Approach and the Disentangled Approach, to investigate different strategies for mitigating bias. See Fig. 1 for details.

3.1   Problem Definition and Notation

We are provided with a facial image $\mathbf{x}_i$ with a target label y_i. Moreover, each $\mathbf{x}_i$ is associated with sensitive labels $\mathbf{s}_i = <s_1, ..., s_m>$, where each label s_j is a member of an attribute group, $s_j \in \mathcal{S}_j$. For our problem, we consider $m = 3$ attribute groups (race, gender, age), and $\mathcal{S}_j$ is illustratively defined as {Caucasian, African-American, Asian} for j = race. The goal then is to model $p(y_i | \mathbf{x}_i)$ without being affected by $\mathbf{s}_i$.

3.2   The Baseline Approach

Our baseline is a Residual Network (ResNet) [18], a widely used architecture which achieved high performance for many automatic recognition tasks. We utilise a 18-layer version (ResNet-18) for our analyses. We train this baseline

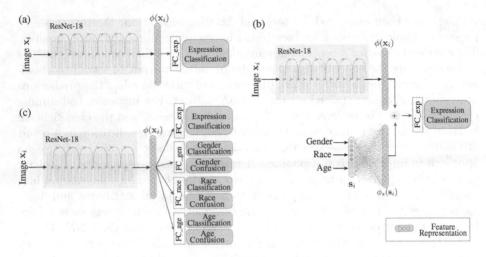

Fig. 1. An illustration of the three approaches. (a) The Baseline Approach. (b) The Attribute-aware Approach. (c) The Disentangled Approach.

network with a Cross Entropy loss to predict a single expression label y_i for each input $\mathbf{x}_i$:

$$\mathcal{L}_{exp}(\mathbf{x}_i) = -\sum_{k=1}^{K} \mathbb{1}[y_i = k] \log p_k, \tag{1}$$

where p_k is the predicted probability for $\mathbf{x}_i$ being assigned to class $k \in K$; and $\mathbb{1}[\cdot]$ is the indicator function.

3.3   The Attribute-Aware Approach

Inspired by the work described in [11,16], we propose an alternative "fairness through awareness" approach. In [11,16], separate networks or branches are trained for each sensitive attribute, which is computationally more expensive. In our attribute-aware solution, we provide a representation of the attributes as another input to the classification layer (Fig. 1(b)). Note that this approach allows us to investigate how explicitly providing the attribute information can affect the expression recognition performance and whether it can mitigate bias.

To be comparable with the baseline approach, ResNet-18 is used as the backbone network for extracting a feature vector $\phi(\mathbf{x}_i)$ from image $\mathbf{x}_i$. In order to match the size of $\phi(\mathbf{x}_i)$, which is 512 in the case of ResNet-18, the attribute vector $\mathbf{s}_i$ is upsampled through a fully-connected layer: $\phi_s(\mathbf{s}_i) = \mathbf{W}_s\mathbf{s}_i + \mathbf{b}_s$. Then, the addition $\phi_s(\mathbf{s}_i) + \phi(\mathbf{x}_i)$ is provided as input to the classification layer (Fig. 1). The network is trained using the Cross Entropy loss in Eq. 1.

3.4   The Disentangled Approach

The main idea for this approach is to make sure the learned representation $\phi(\mathbf{x}_i)$ does not contain any information about the sensitive attributes $\mathbf{s}_i$; in other words, we cannot predict $\mathbf{s}_i$ from $\mathbf{x}_i$ while being able to predict the target label y_i. To do that, we utilise the disentangling approach described in [1, 29].

For this purpose, we first extract $\phi(\mathbf{x}_i)$ using ResNet-18 to be consistent with the first two approaches. Then we split the network into two sets of branches: One primary branch for the primary classification task (expression recognition) with the objective outlined in Eq. 1, and parallel branches designed to ensure that $\phi(\mathbf{x}_i)$ cannot predict $\mathbf{s}_i$.

The parallel branches use a so-called *confusion* loss to make sure that sensitive attributes cannot be predicted from $\phi(\mathbf{x}_i)$:

$$\mathcal{L}_{conf} = -\sum_{\mathcal{S}_j \in \mathcal{S}} \sum_{s \in \mathcal{S}_j} \frac{1}{|\mathcal{S}_j|} \log p_s, \tag{2}$$

which essentially tries to estimate equal probability $(1/|\mathcal{S}_j|)$ for each sensitive attribute. If this objective is satisfied, then we can ensure that $\phi(\mathbf{x}_i)$ cannot predict $\mathbf{s}_i$. However, the network can easily learn a trivial solution to not map $\phi(\mathbf{x}_i)$ to $\mathcal{S}$ even when $\phi(\mathbf{x}_i)$ contains sensitive information. To avoid this trivial solution, an attribute predictive Cross Entropy loss is also used [29]:

$$\mathcal{L}_s = -\sum_{\mathcal{S}_j \in \mathcal{S}} \sum_{s \in \mathcal{S}_j} \mathbb{1}[y_s = s] \log p_s, \tag{3}$$

which functions as an adversary to $\mathcal{L}_{conf}$ in Eq. 2.

These tasks share all the layers till the final fully-connected layer $\phi()$. At the final fully-connected layer parallel branches are created for specific tasks (Fig. 1(c)). The difference between the training of the primary expression classification task and the sensitive attribute classification tasks is that the gradients of $\mathcal{L}_s$ are only back-propagated to $\phi()$, but do not update the preceding layers similar to [29].

The overall loss is then defined as:

$$\mathcal{L} = \mathcal{L}_{exp} + \mathcal{L}_s + \alpha \mathcal{L}_{conf}, \tag{4}$$

where α controls the contribution of the confusion loss.

By jointly minimizing $\mathcal{L}_{exp}$, $\mathcal{L}_s$ and $\mathcal{L}_{conf}$, the final shared feature representation ϕ is forced to distill the facial expression information and dispel the sensitive attribute information. Alvi et al. [1] claim that this approach can improve the classification performance when faced with an extreme bias.

4   Implementation Details

In this section, we provide details about how we evaluate and compare the three methods in terms of their performance for expression recognition and fairness.

4.1   Datasets

Majority of the affect/expression datasets do not contain gender, age and ethnicity labels. Therefore, as a proof of concept we conduct our investigations on two well-known datasets, RAF-DB [28] and CelebA [30], that meet the following criteria: (1) providing labels for expressions of emotions and/or affect; (2) providing labels for gender, age and/or ethnicity attributes for each sample; and (3) being large enough to enable the training and evaluation of the state-of-the-art deep learning models.

RAF-DB [28] is a real-world dataset, with diverse facial images collected from the Internet. The images are manually annotated with expression labels and attribute labels (i.e. race, age and gender of the subjects). For our experiments, we chose a subset of the dataset with basic emotion labels - 14,388 images, with 11,512 samples used for training and 2,876 samples used for testing. The task we focus on using this dataset is to recognize the seven categories of facial expressions of emotion (i.e. Anger, Disgust, Fear, Happiness, Sadness, Surprise and Neutral).

CelebA [30] is a large-scale and diverse face attribute dataset, which contains 202,599 images of 10,177 identities. There are 40 types of attribute annotations in the dataset. Three attributes are chosen in our experiments, including "Smiling" , "Male" and "Young", corresponding to "Facial Expression", "Gender" and "Age" information. Although CelebA does not contain full range of expression annotations, it provides "Gender" and "Age" information that can be utilized to investigate bias and fairness. To the best of our knowledge, there is no other large-scale real-world dataset that can be used for this purpose. Thus, we conduct additional experiments on CelebA as a supplement. Officially, CelebA is partitioned into three parts (training set, validation set and testing set). Our models are trained using the training set and evaluated on the testing set. The task we focus on using this dataset is to recognize the expression of "Smiling" .

4.2   Image Pre-processing, Augmentation and Implementation

For both RAF-DB and CelebA datasets, images are first aligned and cropped, so that all faces appear in approximately similar positions, and normalized to a size of 100×100 pixels. These images are fed into the networks as input.

Deep networks require large amounts of training data to ensure generalization for a given classification task. However, most facial expression datasets available for research purposes do not contain sufficient number of samples to appropriately train of a model, and this may result in overfitting. Therefore, we use data augmentation in this study, which is a commonly employed strategy in many recognition studies when training a network. During the training step, two strategies of image augmentation are applied to the input. For strategy one, the input samples are randomly cropped to a slightly smaller size (i.e. 96×96); randomly rotated with a small angle (i.e. range from $-15°$ to $15°$); and horizontally mirrored in a randomized manner. For strategy two, histogram equalization is applied to increase the global contrast of the images. Following [26], we adopt a weighted summation approach to take advantage of both strategy one and two.

All three methods are implemented using PyTorch [37] and trained with the Adam optimizer [24], with a mini-batch size of 64, and an initial learning rate of 0.001. The learning rate decays linearly by a factor of 0.1 every 40 epochs for RAF-DB and every 2 epochs for CelebA. The maximum training epochs is 200, but early stopping is applied if the accuracy does not increase after 30 epochs for RAF-DB and 5 epochs for CelebA.

4.3 Evaluation Metrics

When performing a classification task, the most commonly used metrics are accuracy, precision and recall, with most studies reporting comparative results using the accuracy metric. However, these metrics are not sufficient in exposing differences in performance (bias) in terms of gender, age and ethnicity attributes. Therefore, we propose to evaluate the three algorithms introduced using two evaluation metrics: Accuracy and Fairness.

Accuracy is simply the fraction of the predictions that the model correctly predicts. **Fairness** indicates whether a classifier is fair to the sensitive attributes. There are various definitions of fairness [45]. For this study, we use the Fairness of "equal opportunity" as described in [17]. The main idea is that the classifier should ideally provide similar results across different demographic groups.

Let $\mathbf{x}$, y be the variables denoting the input and the ground truth label targeted by the classifier; let $\hat{y}$ be the predicted variable; and let s be a sensitive attribute (i.e. $s \in \{s_0, s_1, ..., s_n\}$ with $n > 1$). Suppose there are C classes: $c = 1, ..., C$, and let $p(\hat{y} = c|\mathbf{x})$ denote the probability that the predicted class is c. Then the fairness can be defined in terms of the largest accuracy gap among all demographic groups. Let d denote the dominant sensitive group (i.e. the group which has the highest overall per-class accuracy). This is calculated by summing up the class-wise accuracy. Then the fairness measure $\mathcal{F}$ is defined as:

$$\mathcal{F} = \min \left(\frac{\sum_{c=1}^{C} p(\hat{y} = c|y = c, s = s_0, \mathbf{x})}{\sum_{c=1}^{C} p(\hat{y} = c|y = c, s = d, \mathbf{x})}, ..., \frac{\sum_{c=1}^{C} p(\hat{y} = c|y = c, s = s_n, \mathbf{x})}{\sum_{c=1}^{C} p(\hat{y} = c|y = c, s = d, \mathbf{x})} \right). \quad (5)$$

5 Experiments and Results

5.1 Experiments on RAF-DB

We first present a bias analysis on RAF-DB and perform experiments to investigate whether it is possible to mitigate this bias through the three approaches we proposed in the Methodology section.

Table 1. RAF-DB data distribution of the test set (Cau: Caucasian, AA: African-American, pct.: percentage).

	Gender		Race			Age					pct.
	Male	Female	Cau	AA	Asian	0–3	4–19	20–39	40–69	70+	
Surprise	138	159	260	16	21	36	39	180	36	6	10.3%
Fear	43	36	61	5	13	3	7	50	16	3	2.7%
Disgust	69	89	125	6	27	3	13	106	28	8	5.5%
Happy	429	712	855	98	188	43	216	581	264	37	39.7%
Sad	147	239	291	30	65	51	97	164	61	13	13.4%
Anger	119	45	144	10	10	2	16	115	28	3	5.7%
Neutral	312	339	489	39	123	20	83	458	68	22	22.6%
pct	43.7%	56.3%	77.4%	7.1%	15.5%	5.5%	16.4%	57.5%	17.4%	3.2%	

Table 2. Class-wise accuracy of the models by expression labels on RAF-DB.

	Without Augmentation			With Augmentation		
	Baseline	Attri-aware	Disentangle	Baseline	Attri-aware	Disentangle
Mean	65.3%	66.9%	62.2%	73.8%	74.6%	**74.8%**
Surprise	75.8%	79.7%	77.0%	**82.8%**	82.5%	81.8%
Fear	40.5%	47.4%	40.5%	54.4%	**55.7%**	53.8%
Disgust	41.1%	41.1%	33.1%	51.6%	53.8%	**54.1%**
Happy	91.6%	92.7%	92.4%	**93.6%**	92.7%	93.3%
Sad	63.2%	64.2%	55.4%	73.1%	**80.6%**	77.7%
Anger	66.5%	62.8%	53.7%	73.8%	74.4%	**81.0%**
Neutral	78.7%	80.4%	83.5%	**87.6%**	82.2%	82.1%

RAF-DB Bias Analysis. RAF-DB contains labels in terms of facial expressions of emotions (Surprise, Fear, Disgust, Happy, Sad, Anger and Neutral) and demographic attribute labels along gender, race and age. The provided attribute labels are as follows - *Gender*: Male, Female, Unsure; *Race*: Caucasian, African-American, Asian; and *Age*: 0–3 years, 4–19 years, 20–39 years, 40–69 years, 70+ years. For simplicity, we exclude images labelled as Unsure for Gender. We performed an assessment of how images in the dataset represent the different race, age and gender categories. Table 1 shows a detailed breakdown for the testing data. Note that the distribution of the training data are kept similar to the testing data. Looking at Table 1, we observe that with 77.4%, the vast majority of the subjects in the dataset are Caucasian, 15.5% are Asian, and only 7.1% are African-American. 56.3% of the subjects are female, while 43.7% are male. Most subjects are in the 20–39 years age category, with the 70+ years category containing the fewest images. This confirms what we have mentioned earlier, that majority of the affect/expression datasets have been acquired without a consideration for containing images that are evenly distributed across the attributes of

Table 3. Mean class-wise accuracy of the models, broken down by attribute labels on RAF-DB (Cau: Caucasian, AA: African-American, M: Male, F: Female).

	Without Augmentation			With Augmentation		
	Baseline	Attri-aware	Disentangle	Baseline	Attri-aware	Disentangle
Male	65.3%	67.4%	62.5%	72.3%	73.7%	**74.2%**
Female	63.5%	64.9%	61.0%	74.1%	74.1%	**74.4%**
Cau	65.9%	68.3%	63.4%	74.7%	74.9%	**75.6%**
AA	68.1%	62.8%	58.4%	76.3%	76.3%	**76.6%**
Asian	60.0%	59.8%	54.4%	67.8%	69.9%	**70.4%**
0–3	63.6%	59.9%	56.7%	**80.2%**	71.9%	65.0%
4–19	59.5%	58.8%	57.0%	61.1%	63.7%	**69.9%**
20–39	65.9%	68.2%	62.9%	74.9%	75.8%	**76.4%**
40–69	65.0%	63.4%	60.1%	73.8%	**74.4%**	72.1%
70+	51.3%	53.6%	51.6%	60.8%	54.3%	**62.2%**
M-Cau	65.3%	69.3%	63.6%	73.3%	73.9%	**74.5%**
M-AA	77.0%	70.4%	63.2%	66.4%	**80.2%**	78.7%
M-Asian	61.2%	58.6%	56.2%	67.8%	68.4%	**70.2%**
F-Cau	64.1%	66.2%	62.2%	74.7%	74.9%	**75.5%**
F-AA	61.6%	57.9%	62.8%	**87.6%**	75.8%	74.6%
F-Asian	59.1%	59.5%	52.4%	65.6%	68.4%	**69.0%**

gender, age and ethnicity. Therefore the goal of our experiment is to investigate whether it is possible to mitigate this bias through the three approaches we proposed in the Methodology section: the Baseline Approach, the Attribute-aware Approach, and the Disentangled Approach.

Expression Recognition. For each method, we trained two versions, with and without data augmentation. The performance of all six models are presented in Table 2. We observe that data augmentation increases the accuracy and this applies to almost all the expression categories. The baseline model with data augmentation provides the best performance, but the difference compared to the attribute-aware approach and the disentangled approach with data augmentation are minimal. When comparing the performance across all expression categories, we observe that the accuracy varies and this variation is closely associated with the number of data points available for each expression category (See Table 1). The expression category of "Happiness" is classified with the highest accuracy, while the categories of "Fear" and "Disgust" are classified with the lowest accuracy.

The accuracy breakdown provided in Table 2 by expression labels cannot shed light on the performance variation of the classifiers across different demographic groups. Thus, in Table 3 we provide a detailed comparison of the accuracies broken down by each demographic group. To further shed light on the inter-play between the gender and race attributes, in Table 3 we also include results for

Table 4. Fairness measure of the models, broken down by attribute labels on RAF-DB (G-R: Joint Gender-Race groups).

	Without Augmentation			With Augmentation		
	Baseline	Attri-aware	Disentangle	Baseline	Attri-aware	Disentangle
Gender	97.3%	96.3%	97.5%	97.6%	<u>99.5%</u>	**99.7%**
Race	88.1%	87.5%	85.8%	88.8%	<u>91.6%</u>	**91.9%**
Age	77.7%	78.6%	**82.1%**	75.8%	71.6%	<u>81.4%</u>
G-R	76.7%	82.2%	83.0%	74.8%	<u>85.3%</u>	**87.7%**

joint Gender-Race groups. Note that the accuracy presented in Table 3 is class-wise accuracy, which refers to the accuracy for each expression category. In this way, we ensure that the weights of all categories are the same. This enables a more accurate analysis of fairness, not affected by the over-represented classes in the dataset. Note that there are only a few data points for certain subgroups (i.e. Age 0–3 years, Age 70+, African-American), so the results obtained for these groups are likely to be unstable. From Table 3, we can see that for class-wise accuracy, the disentangled approach with data augmentation provides the best accuracy, with the attribute-aware approach being the runner-up. This suggests that both the attribute-aware and the disentangled approaches can improve the class-wise accuracy when equipped with data augmentation.

Assessment of Fairness. To provide a quantitative evaluation of fairness for the sensitive attributes of age, gender and race, we report the estimated fairness measures (obtained using Eq. 5) for the three models in Table 4. We observe that, compared to the baseline model, both the attribute-aware and the disentangled approaches demonstrate a great potential for mitigating bias for the unevenly distributed subgroups such as Age and Joint Gender-Race. We note that the observed effect is not as pronounced when the distribution across (sub)groups is more or less even (e.g., results for gender, which has less gap in comparison – Table 1). For the baseline model, applying data augmentation improves the accuracy by approximately 7% (Table 2), but this alone can not mitigate the bias effect. Instead, both the attribute-aware and the disentangled approaches when equipped with data augmentation achieve further improvements in terms of fairness. We conclude that, among the three approaches compared, the disentangled approach is the best one for mitigating demographic bias.

Discussion. We provided a comprehensive analysis on the performance of the three approaches on RAF-DB in Tables 2, 3, 4. We observed that although the accuracies achieved by the three models are relatively similar, their abilities in mitigating bias are notably different. The attribute-aware approach utilizes the attribute information to allow the model to classify the expressions according to the subjects sharing similar attributes, rather than drawing information from the whole dataset, which may be misleading. The drawback of this approach is that it requires explicit attribute information apriori (e.g., the age, gender and race of the subject) which may not be easy to acquire in real-world applications. Instead, the disentangled approach mitigates bias by enforcing the model to learn a representation that is indistinguishable for different subgroups and does not require attribute labels at test time. This approach is therefore more appropriate and efficient when considering usage in real-world settings. In [19], facial expressions of Fear and Disgust have been reported to be less well-recognized than those of other basic emotions and Fear has been reported to have a significantly lower recognition rate than the other classes. Despite the improvements brought along with augmentation and the attribute-aware and disentangled approaches, we observe in Table 2 similar results for RAF-DB.

To further investigate the performance of all six models, we present a number of challenging cases in Table 5. We observe that many of these have ambiguities in terms of the expression category they have been labelled with. Essentially, learning the more complex or ambiguous facial expressions that may need to be analysed along/with multiple class-labels, remains an issue for all six models. This also relates to the fact that despite its wide usage, the theory of six-seven basic emotions is known to be problematic in its inability to explain the full range of facial expressions displayed in real-world settings [15].

Table 5. Results of a number of challenging cases from RAF-DB (Cau: Caucasian, AA: African-American, M: Male, F: Female, D.A.: Data Augmentation). Correct estimations are highlighted in bold.

	Gender	Female	Male	Female	Male	Female	Female	Male	Female
	Race	Asian	Cau	Cau	Cau	AA	Asian	Cau	Cau
	Age	20-39	40-69	70+	20-39	4-19	20-39	40-69	0-3
Actual Label		Fear	Anger	Disgust	Neutral	Fear	Anger	Surprise	Sad
w/o D.A.	Baseline	Sad	**Anger**	Anger	**Neutral**	Neutral	Sad	**Surprise**	**Sad**
	Attri.	**Fear**	**Anger**	Sad	**Neutral**	Happy	Happy	**Surprise**	**Sad**
	Disen.	Happy	Disgust	**Disgust**	**Neutral**	Surprise	Happy	**Surprise**	**Sad**
w D.A.	Baseline	**Fear**	**Anger**	**Disgust**	**Neutral**	Disgust	Happy	Fear	Neutral
	Attri.	**Fear**	**Anger**	**Disgust**	Sad	**Fear**	Happy	Fear	**Sad**
	Disen.	Sad	Disgust	Anger	Sad	**Fear**	**Anger**	**Surprise**	**Sad**

5.2 Experiments on CelebA Dataset

We follow a similar structure, and first present a bias analysis on CelebA Dataset and subsequently perform experiments using the three approaches in Sect. 3.

CelebA Bias Analysis. CelebA dataset contains 40 types of attribute annotations, but for consistency between the two studies, here we focus only on the attributes "Smiling" , "Male" and "Young". Table 6 shows the test data breakdown along these three attribute labels. Compared to RAF-DB, CelebA is a much larger dataset. The target attribute "Smiling" is well balanced, while the sensitive attributes Gender and Age are not evenly distributed.

Table 6. CelebA data distribution of the test set.

	Gender		Age		Percentage
	Female	Male	Old	Young	
Not Smiling	5354	4620	2358	7616	50.0%
Smiling	6892	3094	2489	7497	50.0%
Percentage	61.4%	38.6%	24.3%	75.7%	

Expression (Smiling) Recognition. The task on CelebA is to train a binary classifier to distinguish between "Smiling" and "Non-Smiling". The Baseline Approach, the Attribute-aware Approach, and the Disentangled Approach (Sect. 3) are trained and tested on this task. Again, evaluation is performed with and without data augmentation and performance is reported in Table 7. As this is a relatively simple task with sufficient samples, the accuracies of all six models do not show significant differences. In Table 8, all of them provide comparable results for class-wise accuracy broken down by attribute labels. The fairness measures reported in Table 9 are also very close to one other.

Table 7. Accuracy of the models, broken down by smiling labels on CelebA dataset (NoSmi: Not Smiling)

	Without Augmentation			With Augmentation		
	Baseline	Attri-aware	Disentangle	Baseline	Attri-aware	Disentangle
Mean	93.0%	**93.1%**	92.8%	92.9%	93.0%	92.9%
NoSmi	93.9%	**94.1%**	**94.1%**	93.9%	93.7%	93.7%
Smiling	92.1%	92.1%	91.6%	91.9%	**92.2%**	**92.2%**

Table 8. Mean class-wise accuracy of the models, broken down by attribute labels on CelebA dataset (M: Male, F: Female)

	Without Augmentation			With Augmentation		
	Baseline	Attri-aware	Disentangle	Baseline	Attri-aware	Disentangle
Female	93.5%	93.5%	93.4%	**93.6%**	**93.6%**	**93.6%**
Male	91.8%	**91.9%**	91.6%	91.2%	91.6%	91.5%
Old	**91.6%**	**91.6%**	91.4%	91.0%	91.4%	91.5%
Young	93.4%	**93.6%**	93.3%	93.5%	93.5%	93.4%
F-Old	92.1%	92.0%	92.0%	92.3%	**92.7%**	92.5%
F-Young	93.7%	**93.8%**	93.6%	**93.8%**	93.7%	93.7%
M-old	**90.7%**	90.6%	90.4%	89.6%	90.0%	90.2%
M-young	92.5%	**92.8%**	92.3%	92.3%	92.5%	92.3%

Discussion. The results obtained for CelebA-DB could potentially be due to several reasons. Firstly, it is more than ten times larger than RAF-DB, thus the trained models do not benefit as much from data augmentation. Secondly, the bias mitigation approaches are more suitable in the context of an uneven attribute distribution or imbalanced number of data points for certain subgroups, which is not the case for CelebA-DB. Thirdly, the task is a simple binary classification task, and therefore both accuracy and fairness results are already high with little to no potential for improvement. In light of these, presenting visual examples of failure cases does not prove meaningful and does not provide further insights. We did however manually inspect some of the results and observed that the binary labelling strategy may have introduced ambiguities. In general, when labelling affective constructs, using a continuous dimensional approach (e.g., labelling Smiling using a Likert scale or continuous values in the range of $[-1,+1]$) is known to be more appropriate in capturing the full range of expressions [15].

Table 9. Fairness measure of the models, broken down by attribute labels on CelebA dataset (G-A: Joint Gender-Age groups)

	Without Augmentation			With Augmentation		
	Baseline	Attri-aware	Disentangle	Baseline	Attri-aware	Disentangle
Gender	98.2%	**98.3%**	98.1%	97.4%	97.8%	97.8%
Age	**98.1%**	97.8%	97.9%	97.4%	97.8%	98.0%
G-A	**96.9%**	96.6%	96.6%	95.5%	96.0%	96.3%

6   Conclusion

To date, there exist a large variety and number of datasets for facial expression recognition tasks [27,33]. However, virtually none of these datasets have been acquired with consideration of even distribution across the human population in terms of sensitive attributes such as gender, age and ethnicity. Therefore, in this paper, we first focused on quantifying how these attributes are distributed across facial expression datasets, and what effect this may have on the performance of the resulting classifiers trained on these datasets. Furthermore, in order to investigate whether bias is a problem for facial expression recognition, we conducted a comparative study using three different approaches, namely a baseline, an attribute-aware and a disentangled approach, under two conditions w/ and w/o data augmentation. As a proof of concept we conducted extensive experiments on two well-known datasets, RAF-DB and CelebA, that contain labels for the sensitive attributes of gender, age and/or race.

The bias analysis undertaken for RAF-DB showed that the vast majority of the subjects are Caucasian and most are in the 20–39 years age category. The experimental results suggested that data augmentation improves the accuracy of the baseline model, but this alone is unable to mitigate the bias effect. Both the attribute-aware and the disentangled approach equipped with data augmentation perform better than the baseline approach in terms of accuracy and fairness, and the disentangled approach is the best for mitigating demographic bias. The experiments with the CelebA-DB show that the models employed do not show significant difference in terms of neither accuracy nor fairness. Data augmentation does not contribute much as this is already a large dataset. We therefore conclude that bias mitigation strategies might be more suitable in the existence of uneven attribute distribution or imbalanced number of subgroup data, and in the context of more complex recognition tasks.

We note that the results obtained and the conclusions reached in all facial bias studies are both data and model-driven. Therefore, our study should be expanded by employing other relevant facial expression and affect datasets and machine learning models to fully determine the veracity of the findings. Utilising inter-relations between other attributes and gender, age and race, as has been done by [49], or employing generative counterfactual face attribute augmentation and investigating its impact on the classifier output, as undertaken in [9], might be also able to expose other more implicit types of bias encoded in a dataset. However, this requires the research community to invest effort in creating facial expression datasets with explicit labels regarding these attributes.

Acknowledgments. The work of T. Xu and H. Gunes is funded by the European Union's Horizon 2020 research and innovation programme, under grant agreement No. 826232. S. Kalkan is supported by Scientific and Technological Research Council of Turkey (TÜBİTAK) through BIDEB 2219 Program.

References

1. Alvi, M., Zisserman, A., Nellåker, C.: Turning a blind eye: explicit removal of biases and variation from deep neural network embeddings. In: Proceedings of the European Conference on Computer Vision (ECCV) (2018)
2. Amini, A., Soleimany, A.P., Schwarting, W., Bhatia, S.N., Rus, D.: Uncovering and mitigating algorithmic bias through learned latent structure. In: AAAI/ACM Conference on AI, Ethics, and Society, AIES (2019)
3. Bellamy, R.K., et al.: Ai fairness 360: an extensible toolkit for detecting, understanding, and mitigating unwanted algorithmic bias. arXiv preprint arXiv:1810.01943 (2018)
4. Breslin, S., Wadhwa, B.: Gender and Human-Computer Interaction, chap. 4, pp. 71–87. Wiley (2017). https://onlinelibrary.wiley.com/doi/abs/10.1002/9781118976005.ch4
5. Buolamwini, J., Gebru, T.: Gender shades: intersectional accuracy disparities in commercial gender classification. In: Conference on fairness, accountability and transparency, pp. 77–91 (2018)
6. Clapes, A., Bilici, O., Temirova, D., Avots, E., Anbarjafari, G., Escalera, S.: From apparent to real age: gender, age, ethnic, makeup, and expression bias analysis in real age estimation. In: Proceedings of the IEEE Conference on Computer Vision and Pattern Recognition Workshops, pp. 2373–2382 (2018)

7. Commission, E.: White paper on artificial intelligence-a european approach to excellence and trust (2020)
8. Das, A., Dantcheva, A., Bremond, F.: Mitigating bias in gender, age and ethnicity classification: a multi-task convolution neural network approach. In: Proceedings of the European Conference on Computer Vision (ECCV) (2018)
9. Denton, E.L., Hutchinson, B., Mitchell, M., Gebru, T.: Detecting bias with generative counterfactual face attribute augmentation. ArXiv abs/1906.06439 (2019)
10. Drozdowski, P., Rathgeb, C., Dantcheva, A., Damer, N., Busch, C.: Demographic bias in biometrics: A survey on an emerging challenge. IEEE Transactions on Technology and Society (2020)
11. Dwork, C., Hardt, M., Pitassi, T., Reingold, O., Zemel, R.: Fairness through awareness. In: Proceedings of the 3rd Innovations in Theoretical Computer Science Conference, pp. 214–226 (2012)
12. Georgopoulos, M., Panagakis, Y., Pantic, M.: Investigating bias in deep face analysis: the kanface dataset and empirical study. Image and Vision Computing, in press (2020)
13. Gong, S., Liu, X., Jain, A.K.: Mitigating face recognition bias via group adaptive classifier. arXiv preprint arXiv:2006.07576 (2020)
14. Grother, P., Ngan, M., Hanaoka, K.: Ongoing face recognition vendor test (frvt) part 3: demographic effects. National Institute of Standards and Technology, Tech. Rep. NISTIR 8280 (2019)
15. Gunes, H., Schuller, B.: Categorical and dimensional affect analysis in continuous input: current trends and future directions. Image Vis. Comput. **31**, 120–136 (2013)
16. Guo, G., Mu, G.: Human age estimation: what is the influence across race and gender? In: 2010 IEEE Computer Society Conference on Computer Vision and Pattern Recognition-Workshops, pp. 71–78. IEEE (2010)
17. Hardt, M., Price, E., Srebro, N.: Equality of opportunity in supervised learning. In: Advances in Neural Information Processing Systems, pp. 3315–3323 (2016)
18. He, K., Zhang, X., Ren, S., Sun, J.: Deep residual learning for image recognition. In: Proceedings of the IEEE Conference on Computer Vision and Pattern Recognition, pp. 770–778 (2016)
19. Howard, A., Zhang, C., Horvitz, E.: Addressing bias in machine learning algorithms: a pilot study on emotion recognition for intelligent systems. In: 2017 IEEE Workshop on Advanced Robotics and its Social Impacts (ARSO), pp. 1–7 (2017)
20. Iosifidis, V., Ntoutsi, E.: Dealing with bias via data augmentation in supervised learning scenarios. Jo Bates Paul D. Clough Robert Jäschke 24 (2018)
21. Kamiran, F., Calders, T.: Data preprocessing techniques for classification without discrimination. Knowl. Inf. Syst. **33**(1), 1–33 (2012)
22. Khiyari, H., Wechsler, H.: Face verification subject to varying (age, ethnicity, and gender) demographics using deep learning. Journal of Biometrics & Biostatistics 07 (2016). https://doi.org/10.4172/2155-6180.1000323
23. Kilbride, J.E., Yarczower, M.: Ethnic bias in the recognition of facial expressions. J. Nonverbal Behav. **8**(1), 27–41 (1983)
24. Kingma, D.P., Ba, J.: Adam: A method for stochastic optimization. arXiv preprint arXiv:1412.6980 (2014)
25. Koenecke, A., et al.: Racial disparities in automated speech recognition. Proc. National Acad. Sci. **117**(14), 7684–7689 (2020)
26. Kuo, C.M., Lai, S.H., Sarkis, M.: A compact deep learning model for robust facial expression recognition. In: Proceedings of the IEEE Conference on Computer Vision and Pattern Recognition Workshops, pp. 2121–2129 (2018)

27. Li, S., Deng, W.: Deep facial expression recognition: a survey. In: IEEE Transactions on Affective Computing, pp. 1–1 (2020)
28. Li, S., Deng, W., Du, J.: Reliable crowdsourcing and deep locality-preserving learning for expression recognition in the wild. In: Proceedings of the IEEE Conference on Computer Vision and Pattern Recognition, pp. 2852–2861 (2017)
29. Liu, Y., Wei, F., Shao, J., Sheng, L., Yan, J., Wang, X.: Exploring disentangled feature representation beyond face identification. In: Proceedings of the IEEE Conference on Computer Vision and Pattern Recognition, pp. 2080–2089 (2018)
30. Liu, Z., Luo, P., Wang, X., Tang, X.: Deep learning face attributes in the wild. In: Proceedings of International Conference on Computer Vision (ICCV), December 2015
31. Locatello, F., Abbati, G., Rainforth, T., Bauer, S., Schölkopf, B., Bachem, O.: On the fairness of disentangled representations. In: Advances in Neural Information Processing Systems, pp. 14611–14624 (2019)
32. Lu, K., Mardziel, P., Wu, F., Amancharla, P., Datta, A.: Gender bias in neural natural language processing. ArXiv abs/1807.11714 (2018)
33. Martinez, B., Valstar, M.F., et al.: Automatic analysis of facial actions: a survey. IEEE Trans. Affective Comput. 10(3), 325–347 (2019)
34. Mayson, S.G.: Bias in, bias out. YAle lJ 128, 2218 (2018)
35. Morales, A., Fierrez, J., Vera-Rodriguez, R.: Sensitivenets: Learning agnostic representations with application to face recognition. arXiv preprint arXiv:1902.00334 (2019)
36. Ngxande, M., Tapamo, J., Burke, M.: Bias remediation in driver drowsiness detection systems using generative adversarial networks. IEEE Access 8, 55592–55601 (2020)
37. Paszke, A., et al.: Automatic differentiation in pytorch (2017)
38. Phillips, P.J., Grother, P., Micheals, R., Blackburn, D.M., Tabassi, E., Bone, M.: Face recognition vendor test 2002. In: 2003 IEEE International SOI Conference. Proceedings (Cat. No. 03CH37443), p. 44. IEEE (2003)
39. du Pin Calmon, F., Wei, D., Vinzamuri, B., Ramamurthy, K.N., Varshney, K.R.: Optimized pre-processing for discrimination prevention. In: Advances in Neural Information Processing Systems (NIPS) (2017)
40. Rhue, L.: Racial influence on automated perceptions of emotions. Available at SSRN. https://doi.org/10.2139/ssrn.3281765 (2018)
41. Robinson, J.P., Livitz, G., Henon, Y., Qin, C., Fu, Y., Timoner, S.: Face recognition: too bias, or not too bias? In: Proceedings of the IEEE/CVF Conference on Computer Vision and Pattern Recognition Workshops, pp. 0–1 (2020)
42. Sariyanidi, E., Gunes, H., Cavallaro, A.: Automatic analysis of facial affect: a survey of registration, representation, and recognition. IEEE Trans. Pattern Anal. Mach. Intell. 37(6), 1113–1133 (2015). https://doi.org/10.1109/TPAMI.2014.2366127
43. Shin, M., Seo, J.H., Kwon, D.S.: Face image-based age and gender estimation with consideration of ethnic difference. In: 2017 26th IEEE International Symposium on Robot and Human Interactive Communication (RO-MAN), pp. 567–572. IEEE (2017)
44. Terhörst, P., Kolf, J.N., Damer, N., Kirchbuchner, F., Kuijper, A.: Face quality estimation and its correlation to demographic and non-demographic bias in face recognition. arXiv preprint arXiv:2004.01019 (2020)
45. Verma, S., Rubin, J.: Fairness definitions explained. In: 2018 IEEE/ACM International Workshop on Software Fairness (FairWare), pp. 1–7. IEEE (2018)

46. Wang, M., Deng, W.: Mitigate bias in face recognition using skewness-aware reinforcement learning. CoRR abs/1911.10692 (2019), http://arxiv.org/abs/1911.10692

47. Wang, M., Deng, W., Hu, J., Tao, X., Huang, Y.: Racial faces in the wild: reducing racial bias by information maximization adaptation network. In: 2019 IEEE/CVF International Conference on Computer Vision, ICCV 2019, Seoul, Korea (South), October 27 - November 2, 2019. pp. 692–702. IEEE (2019). https://doi.org/10.1109/ICCV.2019.00078

48. Wang, T., Zhao, J., Yatskar, M., Chang, K., Ordonez, V.: Balanced datasets are not enough: Estimating and mitigating gender bias in deep image representations. In: 2019 IEEE/CVF International Conference on Computer Vision (ICCV), pp. 5309–5318 (2019)

49. Wang, Z., et al.: Towards fairness in visual recognition: effective strategies for bias mitigation. In: Proceedings of the IEEE/CVF Conference on Computer Vision and Pattern Recognition, pp. 8919–8928 (2020)

50. Wu, W., Michalatos, P., Protopapaps, P., Yang, Z.: Gender classification and bias mitigation in facial images. In: 12th ACM Conference on Web Science, pp. 106–114 (2020)

51. Zhang, B.H., Lemoine, B., Mitchell, M.: Mitigating unwanted biases with adversarial learning. In: Proceedings of the 2018 AAAI/ACM Conference on AI, Ethics, and Society, pp. 335–340 (2018)

Disguised Face Verification Using Inverse Disguise Quality

Amlaan Kar[1], Maneet Singh[1], Mayank Vatsa[2], and Richa Singh[2(✉)]

[1] IIIT-Delhi, New Delhi, India
[2] IIT Jodhpur, Jodhpur, India
richa@iitj.ac.in

Abstract. Research in face recognition has evolved over the past few decades. With initial research focusing heavily on constrained images, recent research has focused more on unconstrained images captured in-the-wild settings. Faces captured in unconstrained settings with disguise accessories persist to be a challenge for automated face verification. To this effect, this research proposes a novel deep learning framework for disguised face verification. A novel Inverse Disguise Quality metric is proposed for evaluating amount of disguise in the input image, which is utilized in likelihood ratio as a quality score for enhanced verification performance. The proposed framework is model-agnostic and can be applied in conjunction with existing state-of-the-art face verification models for obtaining improved performance. Experiments have been performed on the Disguised Faces in Wild (DFW) 2018 and DFW 2019 datasets, with three state-of-the-art deep learning models, where it demonstrates substantial improvement compared to the base model.

Keywords: Disguised face verification · Biometric fusion

1   Introduction

Face recognition has witnessed substantial research interest with large number of applications, especially in social media, biometric authentication, social security, and enhanced user experience in the product domain. Several successful state-of-the-art face recognition models, including VGGFace [3], Residual Networks (ResNet) [4] and ArcFace [5], have been proposed in the literature. Research with these models has focused on covariates such as pose, illumination, expression, ageing, and heterogeneity, however, disguise variations have received limited research attention. Disguises can be considered as external "noise" which challenge the robustness of the face verification systems. Earlier research on disguised face recognition focused mostly on datasets prepared under constrained settings, thus failing to capture the real world scenario. Recently, the focus has shifted towards adapting face recognition to in-the-wild datasets like Disguised Faces in the Wild (DFW) 2018 [1] and 2019 [2], which incorporate both intentional and unintentional disguises [17,18].

© Springer Nature Switzerland AG 2020
A. Bartoli and A. Fusiello (Eds.): ECCV 2020 Workshops, LNCS 12540, pp. 524–540, 2020.
https://doi.org/10.1007/978-3-030-65414-6_36

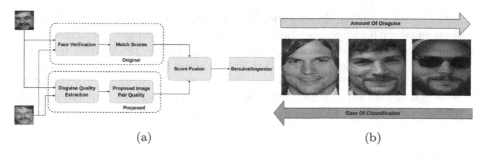

(a)   (b)

Fig. 1. (a) Proposed face verification algorithm where extracted image pair Inverse Disguise Quality is fused with match scores generated. (b) Inverse Disguise Quality denotes the biometric quality of a sample, i.e. on how easy it is to classify based on the "amount of disguise" present in the image.

While deep neural networks such as VGGFace [3] and ResNet trained on face images have produced superlative performance on popularly used face recognition databases, their performance is subpar on disguise databases. Analysis of the VGGFace model pre-trained on the VGGFace dataset, ResNet-50 model trained on the MS-Celeb-1M [34] and VGGFace2 dataset [35] (referred to as VGGFace2), and the ResNet-100 model trained on MS1M-ArcFace dataset demonstrates verification accuracies of 33.76%, 66.97% and 65.51%, respectively, on the DFW 2018 dataset [1]. The poor performance of state-of-the-art face recognition models thus suggests a requirement for robust models invariant to disguise variations. Automated disguise face verification suffers from the challenge of both intentional and unintentional disguises. For example, concealing the identity, impersonating someone using glasses, moustache, beard, different hairstyles, scarfs or caps and makeup or even unintentionally changing the appearance as seen with hair growth or removal. These variations often result in reduced inter-class distance between subjects and increased intra-class variations, thus rendering the problem of disguised face recognition further challenging.

This research proposes a novel face verification framework for authenticating face images under disguise variations. The proposed framework utilizes the "amount of disguise" in an image to improve the performance of the face recognition algorithm in an attempt to make it more robust, secure, and usable in real-life scenarios. Figure 1 shows how this "amount of disguise" is used as an image quality for this purpose. The contributions of this research are as follows:

– A novel face verification framework is proposed which utilizes an *Inverse Disguise Quality* metric for quantifying the biometric quality of a face image. The proposed framework is model-agnostic and can be applied in conjunction with existing state-of-art deep learning based face verification models for obtaining enhanced performance.
– Inverse Disguise Quality is derived by performing disguise detection on the face images. Semantic segmentation has been applied to recognize disguise and non-disguise patches, followed by a combination of the confidence scores

for generating the quality metric. Further, the Inverse Disguise Quality has been fused with face verification match scores for improved performance.

– Experiments performed using state-of-the-art models pre-trained on large-scale face datasets demonstrate that the proposed framework yields improved performance as compared to the baseline models. For example, on the DFW 2018 dataset at 0.1% FAR, VGGFace [3], VGGFace2 [35] and ArcFace [5] models yield an overall increase of 63.41%, 38.11% and 35.07%, respectively.

2   Related Work

The proposed framework presents a novel technique for disguised face recognition. Disguise detection is performed on face images using semantic segmentation, followed by the estimation of Inverse Disguise Quality metric. The quality metric is fused with the match scores obtained via a face recognition model. This section presents the related work for the concepts of disguised face recognition, semantic segmentation, and likelihood ratio-based biometric score fusion.

2.1   Disguised Face Recognition

Initial research on disguised face recognition utilized datasets captured in constrained settings [11,19–22] and [23]. Chellappa et al. [11] studied the facial similarity for several variations including disguise by forming two eigenspaces from two halves of the face, one using the left half and other using the right half. From the test image, the optimally illuminated half face is chosen and projected into the eigenspace. Silva et al. [12] used the concept of eigeneyes to ensure that any change in facial features other than eyes does not affect the recognition performance. Singh et al. [13] used 2D log polar Gabor transform to extract phase features from faces, which were then divided into frames and matched using Hamming distance. Dhamecha et al. [14] classified the local facial regions of both visible and thermal face images into biometric and non-biometric classes. Yoon et al. [15] detected partially occluded faces using a SVM to characterize suspicious ATM users. From 2018, with the release of the Disguised Faces in the Wild (DFW) datasets [1] and [2], research started focusing more on unconstrained disguised face recognition. Smirnov et al. [16] proposed several ways to create auxiliary embeddings and used them to increase the number of potentially hard positive and negative examples. Zhang et al. [17] extracted features for generic faces using two networks. PCA based on the DFW 2018 dataset was applied to attempt a form of transfer learning. Deng et al. [18] applied the ArcFace [5] loss on the DFW 2018 [1] and DFW 2019 [2] datasets. They improved their generic implementation [5] by combining hard sample mining with the intra-loss and inter-loss. While the domain of disguised face recognition has attracted research focus in recent times, not-so-high performance of state-of-the-art algorithms as compared to non-disguised datasets demonstrate a need for further research.

2.2   Semantic Segmentation

Given an input image, semantic segmentation is the process of assigning a class label to each pixel for the purpose of object detection. Ciresan et al. [6] used a sliding widow to train a network, which would predict the class label of each pixel. This was done by providing a patch around that pixel as input. More recent approaches [7] utilized multiple resolutions of images to allow the use of localization and neighbourhood context at the same time. Long et al. [8] used fully convolutional networks to define a novel architecture that combines semantic and appearance information from different layers for segmentation. Ronneberger et al. [9] modified and extended this architecture and used an encoder-decoder model with skip connections to combine high resolution features with upsampled ones. Girshick et al. [36] combined several object detection predictions in images for semantic segmentation. He et al. [37] extended this by training a multi-branch network for bounding box recognition and object mask prediction. Semantic segmentation of faces has mostly focused on identifying different facial regions of the face image. Khan et al. [10] performed multi-class semantic segmentation on faces to separate various parts of the face using random forests. Jackson et al. [41] performed landmark localisation and then used it to guide semantic part segmentation. Kalayeh et al. [42] performed facial part parsing to transfer localisation properties to improve face attribute detection. Zhou et al. [43] combine fully-convolutional network, super-pixel information and CRF model to perform semantic segmentation. Lin et al. [44] used Mask RCNN [37] and FCN [8] branches for semantic labeling of the inner face and hair regions.

2.3   Likelihood Ratio and Quality in Biometric Fusion

At a conceptual level, the quality of any input (e.g. image or text) is a measure of the suitability of the input for automated analysis. For our task of face verification, a high quality sample is one which gets classified as genuine or imposter with relative ease as compared to a poor quality sample. Bharadwaj et al. [24] discussed several definitions and interpretations of biometric quality, and Singh et al. [32] performed a comprehensive survey of biometric fusion techniques. Bigun et al. [25] performed multimodal biometric fusion using a statistical framework that combined multiple verification scores along with corresponding quality metrics defined by human users. Fierrez-Aguilar et al. [26] used the quality of biometric samples as sample weights for training a SVM. Poh et al. [27] defined the quality of a biometric sample as how close the corresponding biometric sample is to the decision boundary that satisfies the equal error rate criterion. Nandakumar et al. [28] determined the quality of local regions in fingerprint and iris images to derive an overall quality of the match between each pair of template and query images. This quality was used to estimate a joint density between biometric match scores and their corresponding quality. Vatsa et al. [30] proposed applying Redundant Discrete Wavelet Transform (RDWT) to quantify distinguishing information present in an image. As far as the image

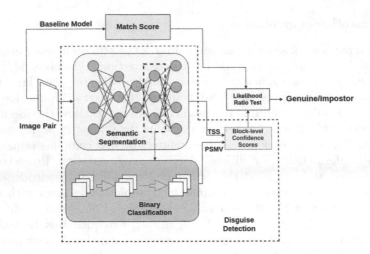

Fig. 2. The complete pipeline consisting of Disguise Detection and the usage of Inverse Disguise Quality in Face Verification.

quality in face images is concerned, Subasic et al. [38] used 17 tests to determine face quality. These included image properties like resolution, brightness, aspect ratio, and sharpness, and facial properties like position and tilt of eyes and head. Gao et al. [39] used facial symmetry, lighting and pose to define face quality metrics. Zhang and Wang [40] extracted scale-invariant SIFT features from faces to define face asymmetry-based quality.

The proposed disguised face verification framework utilizes semantic segmentation for identifying disguised patches in a given input image. Based on the "amount of disguise", a novel Inverse Disguise Quality metric is calculated, followed by biometric fusion with the match scores obtained via the base network. The proposed framework is model-agnostic and can be applied in conjunction with existing deep learning based face verification models.

3   Proposed Disguised Face Verification Framework

The proposed framework (Fig. 2) consists of three components: (i) disguise detection on the input face image using thresholding on semantic segmentation to obtain block-wise semantic labelling, followed by the classification of block-wise learned features as disguise or non-disguise, (ii) computation of a quality metric based on the detected disguise which gives a quantitative measure of the amount of disguise in a face, and (iii) combination of the image quality into biometric pair quality for fusion with face verification scores obtained from a base network. The following subsections elaborate upon each component of the framework.

Fig. 3. Sample pairs of detected face and the corresponding semantic segmentation along with block-level annotation. We annotate each block in an image as disguise/non-disguise. Images are taken from the DFW 2018 [1] dataset.

3.1 Disguise Detection

In this research, the task of disguise detection involves predicting the regions of disguise or non-disguise in an image. The proposed pipeline for disguise detection broadly consists of three steps: (i) The input image is divided into 8×8 patches and semantic segmentation is performed on the image in order to classify each patch as disguise or non-disguise, (ii) A binary classifier is utilized for classifying the learned feature as disguise or non-disguise, and (iii) Weighted majority voting is performed by combining the 8×8 patches into 4×4 blocks for incorporating the neighbourhood information during disguise detection.

Step-1: Semantic Segmentation of Faces. Semantic segmentation refers to the process of classifying each pixel in an image for the purpose of object detection. The task to be performed here is disguise detection in face images. In this research, semantic segmentation is applied to label blocks of the input image as *disguised* or *non-disguised* in a manner similar to [14]. The image is divided into 8×8 patches or blocks, and block-level labeling is performed. Thus, for an input image of dimension 224×224, 64 blocks are obtained. The U-net architecture [9,31] is used for the said task. U-net is an encoder-decoder framework which performs semantic segmentation by downsampling an image in the encoder using convolution and up-sampling it using transpose convolutions to obtain pixel-level classification. It consists of skip connections between the encoder and decoder at equal levels of feature size to get precise locations by preventing information loss. Since U-net provides pixel-level labels, post training, thresholding is applied to transfer predicted pixel labels to their corresponding 28×28 block, to obtain the block-wise labels for disguise/non-disguise regions. In order to further strengthen the predictions at block-level, features are extracted from the trained model, followed by block-level classification into disguise/non-disguise.

Step-2: Binary Block Classification. As demonstrated in Fig. 3, the semantic segmentation model described in the previous section provides a good representation of disguised regions in an image. In order to further enhance the performance, block-level features are extracted from the trained U-net model for each image. As we separate the blocks of the image for classification, they become individual entities and retain no properties of being part of a larger image. This causes the neighbourhood of a block to no longer be explored, thus losing the structure of the image. On the other hand, semantic segmentation utilizes the whole image and not just different blocks. Thus, the features of semantic segmentation for a block also encode the neighbourhood of the block, therefore resulting in more descriptive and informative features.

The features for each block are provided as input to a binary block classifier, which classifies it as disguised or non-disguised. This module consists of six convolutional layers and two fully-connected layers. Further, since the number of non-disguised blocks are much more in number as compared to the disguised blocks, a weighted binary cross-entropy loss is applied to this model. We calculated the loss for weighted samples, i.e., individual weights are introduced for disguised samples, as shown in Eq. 1.

$$C = -\frac{1}{n} \sum_x [w_x(y \ln a + (1 - y) \ln(1 - a))] \tag{1}$$

where a, y are the predicted and ground-truth labels, respectively. x denotes the input and w_x is the weight corresponding to a given input. The above Equation allows us to increase the representation of disguised blocks in our training, as we are manually instructing the classification module to focus more on reducing the loss incurred due to the mis-classification of disguised blocks.

Step-3: Parent-Sibling Majority Voting. In order to further incorporate the structure of the facial image, a parent-sibling majority voting is performed for obtaining the final disguise detection predictions. It is important to note that this component of the framework does not involve any training, and is applied directly on the predicted outputs of the binary block classifier. After getting predictions of all the 8×8 blocks, we classify each block on the basis of the predictions of blocks in its neighbourhood. Parent and siblings are the two types of neighbours considered here. Each block is seen as a part of a 4×4 block and each 4×4 block consists of four 8×8 blocks. For each block, we consider the corresponding 4×4 block as its parent. The other three 8×8 blocks corresponding to the 4×4 block are defined as the siblings of the block under consideration.

We follow majority voting among the four 8×8 blocks that constitute a 4×4 block to determine its label. Random selection of labels acts as a tie-breaker. A binary classifier, similar to the one described in the previous step, is used to classify 4×4 blocks as disguise/non-disguise. Once we get predictions of the 8×8 and 4×4 blocks, we find all the non-disguise confidence scores for 8×8 blocks. For all the blocks classified with confidence scores below a threshold, we follow a weighted majority voting between the disguise and non-disguise confidence scores of the siblings and parent of that block. The 4×4 block is given a weight of 0.7, while the siblings are given weights of 0.1 each. This weight assignment has been done to take into account the fact that the 4×4 block best represents the neighbourhood of a given 8×8 block. The proposed voting mechanism helps low confidence samples to be represented via a more confident label obtained from its neighbourhood. Mathematically, this is denoted as:

$$Class = \arg\max_{C} \left\{ \left(w * D^{4 \times 4} + \left(\frac{1-w}{3} \right) * \sum_{n} [D_n^{8 \times 8}] \right), \right.$$
$$\left. \left(w * N^{4 \times 4} + \left(\frac{1-w}{3} \right) * \sum_{n} [N_n^{8 \times 8}] \right) \right\}$$

where C, w, n, D and N denote the set of classes, hierarchy level weight, 8×8 neighbours, disguise and non-disguise confidence score, respectively.

3.2 Inverse Disguise Quality for Face Verification

The result obtained from the disguise detection module are utilized to compute a novel *Inverse Disguise Quality* metric for quantifying the quality of the input image. The inverse disguise quality score is further fused with the match scores obtained via a base face verification model for obtaining improved performance.

Inverse Disguise Quality. The quality of biometric samples has a significant impact on the accuracy of a matcher. Poor quality biometric samples often lead to incorrect matching results since the features extracted from them are not reliable. Therefore, assigning weights to the predicted output of a face verification model based on the quality of the input sample can improve the overall recognition performance. In this research, a novel Inverse Disguise Quality metric is used to quantify the "amount of disguise" in an image. As disguise proves to be an obstruction in face recognition or verification, the amount of disguise in an image is an ideal metric to quantify whether the input is of good biometric quality or not. Hence, we define the Inverse Disguise Quality of an image as the sum of the non-disguised confidence scores of each block of the image.

$$\mathcal{D} = \sum_{i=1}^{64} [(Non - DisguiseConfidence)_i] \tag{2}$$

where, $\mathcal{D}$ denotes the inverse disguise quality. If the amount of non-disguised blocks are higher, the inverse disguise quality metric will also be higher, thus suggesting more confidence in the feature extraction by the base model.

Inverse Disguise Quality Based Likelihood Ratio Test. The Inverse Disguise Quality is fused with the likelihood ratio for enhanced disguised face verification. The likelihood ratio test utilizes the match score densities obtained for the genuine and impostor classes. In this research, we have used a Gaussian distribution [29] to estimate the match score densities. Let s, $f_{gen}(s_{gen})$ and $f_{imp}(s_{imp})$ denote the vector of match scores and the conditional densities of the genuine and impostor match scores, respectively. Traditionally, let Ψ be a statistical test for testing. For an input I_s: H_0: I_s corresponds to an impostor and H_1: I_s corresponds to a genuine user. As per the Neyman-Pearson theorem [33], for testing H_0 against H_1, there exists a test Ψ and a constant η such that:

$$P(\Psi(s) = 1|H_0) = \alpha \tag{3}$$

$$\Psi(I_s) = \begin{cases} 1, & \frac{f_{gen}(I_s)}{f_{imp}(I_s)} \geq \eta \\ 0, & \frac{f_{gen}(I_s)}{f_{imp}(I_s)} \leq \eta \end{cases} \tag{4}$$

If a test satisfies these two equations for some η, then it is the most powerful test for testing H_0 against H_1 at level α. If the false accept rate (FAR) α is fixed, the optimal test for deciding which class I_s belongs to is the likelihood ratio test $\Psi(I_s)$ if $f_{gen}(I_s)$ and $f_{imp}(I_s)$ are estimated properly. In other words, any data point with likelihood ratio $\geq \eta$ will be classified as belonging to the genuine class. All other data samples will be classified as impostor.

In the proposed framework, the Inverse Disguise Quality metric is incorporated into the likelihood ratio test for face verification. For this purpose, we have used two quality metrics for image pairs: (i) Average Image Pair Quality (AIPQ), where we take the average of the inverse disguise image qualities of the image pair; (ii) Normalized Image Pair Quality (NIPQ), where the image qualities are normalized in the range of [0, 1]. The quality metric is incorporated into the likelihood ratio test by multiplying each element of the log-likelihood ratio vector with the corresponding image pair quality, taking from Poh and Bengio [27], where the independent match scores are combined with the corresponding sample quality in the authentication phase. Mathematically, it is expressed as:

$$\Psi(I_s) = \begin{cases} 1, & Q(I) * \ln \frac{f_{gen}(I_s)}{f_{imp}(I_s)} \geq \eta_1 \\ 0, & Q(I) * \ln \frac{f_{gen}(I_s)}{f_{imp}(I_s)} < \eta_1 \end{cases} \tag{5}$$

where $Q(I)$ denotes AIPQ or NIPQ of the input image pair I. Multiplying the log-likelihood ratios helps in enhanced face verification performance since the inherent concept of quality of an image is that a poor quality sample will be difficult to classify as genuine or impostor. By multiplying log-likelihood ratios with the corresponding image pair quality, we explicitly weigh the resulting log-likelihood ratios with their corresponding image pair qualities. For example, if face verification is performed using NIPQ, we observe that multiplication of the log-likelihood ratios and NIPQ leads to downscaling of the LRs as NIPQ lies between 0 and 1. Multiplication helps us achieve the purpose of trusting good quality samples more and poor quality samples less, as pairs with low quality are downscaled to an extent greater than that of good quality pairs.

4   Dataset and Implementation Details

The proposed framework has been evaluated for disguised face verification on two recent challenging datasets: DFW 2018 and DFW 2019 datasets. Details regarding the datasets, protocols, and implementation details are as follows.

4.1   Datasets

DFW 2018 Dataset was released as part of the DFW Workshop in CVPR 2018 [1]. The dataset contains 11,157 face images belonging to 1000 subjects having unconstrained disguise variations. Out of these 1000 subjects, 400 subjects belong to the training set, and 600 belong to the testing set. Each subject contains at least 5 and at most 26 face images of normal, validation, disguise and impersonator types. The dataset has been released with three protocols for evaluation. For all protocols, face verification is to be performed between a gallery and a probe image for classification as genuine or imposter.

DFW 2019 Dataset: It was released as part of the DFW Workshop in ICCV 2019 [2]. The dataset was released as a test set only, while encouraging researchers to utilize the DFW 2018 dataset as the training and validation set. The DFW 2019 dataset contains images pertaining to 600 subjects. Further, the dataset contains variations due to bridal make-up, and out of the 600 subjects, 250 subjects demonstrate variations due to plastic surgery. As compared to DFW 2018, an additional protocol related to plastic surgery has also been added. The DFW 2019 dataset has been released with four protocols for evaluation. For all four protocols, face verification is to be performed between a gallery and a probe image for classification as genuine or imposter. Detailed description of each protocol [1,2] is given as:

– **Protocol-1 (Impersonation)** is defined for the purpose of differentiating between genuine users and impersonators. A genuine pair of gallery and probe images consists of same subject images while an imposter pair consists of pairs of impersonator images with the other types of images for a subject. This protocol is present in both DFW 2018 and DFW 2019 datasets.
– **Protocol-2 (Obfuscation)** evaluates the ability of a face verification framework to perform accurately even with obfuscated face images. Genuine pairs include disguised images of a subject paired with normal, validation and other disguised images of the same subject. The imposter set is created by combining cross-subject normal, validation, and disguised images. This protocol is present in both DFW 2018 and DFW 2019 datasets.
– **Protocol-3 (Plastic Surgery)** evaluates the ability of a face verification framework to identify faces despite changes in them due to plastic surgery. The genuine pairs are made of same subject pre-surgery and post-surgery images while the imposter set contains cross-subject pre-surgery and post-surgery images. This protocol is present only in the DFW 2019 dataset.

- **Protocol-4 (Overall)** is used to evaluate the performance of any face recognition algorithm on the entire DFW dataset. The genuine and imposter sets created in the above mentioned protocols are combined to generate the data for this protocol. The overall protocol is Protocol 3 in DFW 2018 and is formed by combining Protocols 1 and 2. In DFW 2019, it is formed by combining all three protocols described above.

4.2   Implementation Details

For both the datasets, face detection is performed using the face coordinates provided by the authors. Since there does not exist any dataset with annotated disguise patches, manual annotation is performed on the training images for obtaining the ground-truth labels for the disguise detection module. The face images are resized to 224×224, and divided into blocks of 8×8. Each block is annotated as *disguised* or *non-disguised*, where a disguised block is defined as a block with at least 25% coverage of the disguise accessory. The annotation procedure is performed in an absolute manner for external objects obstructing the face like caps, hats, scarves, glasses, etc. Any such external accessory is marked as disguise. Factors like facial hair and hair colour are considered relative to the normal image. Every block in a normal image which is not affected by an external accessory is considered as non-disguised. For validation and disguised images, blocks containing facial hair and hair colour are annotated with respect to the normal image. Impersonator image blocks containing facial hair are marked as disguise if facial hair is present in the corresponding region in the normal image as well. The annotated images are passed through a U-Net for semantic segmentation, as described in Sect. 3.1. The model is trained using the Adam optimizer and the dice coefficient loss function with a learning rate of $1e-3$. Features of the last convolutional layer are separated into 8×8 blocks for binary classification, which is done using a classification network consisting of six convolutional layers and two fc-layers. As described in Sect. 3.1, training is done using weighted binary cross entropy loss. The weights of the disguised blocks for the separate 8×8 and 4×4 networks are selected as 7 and 21 respectively for the best average between block-level accuracies of disguised and non-disguised blocks. Training was done using the Adam optimizer at a learning rate of $1e-5$.

5   Experimental Results and Analysis

The experiments for disguise detection are performed on the DFW 2018 dataset, and face verification experiments on the DFW 2018 and DFW 2019 datasets. For face verification, experiments are performed on all the protocols of both the datasets. The efficacy of the proposed framework is demonstrated by applying it on existing pre-trained models. Three base models are used: (i) VGGFace model pre-trained on the VGGFace dataset [3], (ii) ResNet-50 model trained on the MS-Celeb-1M and VGGFace2 dataset [35] (referred to as VGGFace2), and (iii) ArcFace model (ResNet-100) trained on the MS1M dataset [5].

Table 1. Block-level disguise detection accuracy (%) on the DFW 2018 dataset. Results have been computed for the thresholding on semantic segmentation (TSS), binary classification (BC) and parent-sibling majority voting (PSMV).

Technique	TSS	BC (8×8)	BC (4×4)	PSMV
Disguised blocks (%)	56.43%	68.66%	76.45%	69.32%
Non-disguised blocks (%)	89.66%	77.47%	66.54%	77.67%

Table 2. Verification accuracy (%) for the impersonation (P-1), obfuscation (P-2), and overall (P-3) protocols of the DFW 2018 dataset. The proposed technique with NIPQ scores demonstrates improved performance as compared to the baseline model across different False Acceptance Rates (FAR).

Prtcl.	Algo.	GAR@1% FAR			GAR@0.1% FAR		
		VGGFace	VGGFace2	Arcface	VGGFace	VGGFace2	Arcface
P-1	Base	52.77%	80.17%	87.22%	27.05%	48.23%	55.79%
	AIPQ	80.00%	92.60%	91.93%	71.43%	86.05%	77.64%
	NIPQ	95.24%	97.98%	**98.46%**	93.10%	**96.97%**	95.12%
P-2	Base	31.52%	66.32%	64.10%	15.72%	43.79%	45.48%
	AIPQ	54.65%	77.81%	75.56%	37.92%	66.75%	61.58%
	NIPQ	89.26%	89.85%	**90.52%**	79.82%	**80.89%**	80.23%
P-3	Base	33.76%	66.97%	65.51%	17.74%	44.05%	47.51%
	AIPQ	56.39%	78.51%	76.03%	40.28%	67.27%	62.81%
	NIPQ	91.47%	91.92%	**92.43%**	81.14%	82.16%	**82.58%**

5.1 Disguise Detection Results

Table 1 summarizes the block classification accuracies on the DFW 2018 dataset obtained by thresholding on semantic segmentation (TSS), binary classification (BC) and parent-sibling majority voting (PSMV). BC improves the disguised block accuracy by 12.23%, as compared to the traditional semantic segmentation. While the non-disguised block accuracy comes down by 12.19%, this is because BC removes the bias present due to TSS. PSMV improves disguised and non-disguised block accuracies by 0.66% and 0.20%, respectively and achieves the best performance for disguise detection.

5.2 Face Verification Performance

The proposed framework has been evaluated on the DFW 2018 and DFW 2019 test datasets. Figure 4 presents the Receiver-Operator Characteristics (ROC) curves for the respective best model on the protocols of the DFW 2018 and the DFW 2019 datasets. Table 2 shows the verification accuracies of the baseline models and the proposed frameworks for the protocols of the DFW 2018 dataset. Table 3 shows the same for the protocols of the DFW 2019 dataset.

Table 3. Verification accuracy (%) for the impersonation (P-1), obfuscation (P-2), plastic surgery (P-3), and overall (P-4) protocols of the DFW 2019 dataset. The proposed technique with NIPQ scores demonstrates improved performance as compared to the baseline model across different FARs.

Prtcl.	Algo.	GAR@0.01% FAR			GAR@0.1% FAR		
		VGGFace	VGGFace2	Arcface	VGGFace	VGGFace2	Arcface
P-1	Base	14.80%	27.20%	7.20%	26.00%	57.60%	47.20%
	AIPQ	48.00%	63.60%	9.20%	86.80%	72.80%	65.60%
	NIPQ	78.40%	**92.80%**	24.80%	86.80%	**98.00%**	88.80%
P-2	Base	3.71%	35.55%	12.33%	10.05%	55.37%	25.34%
	AIPQ	22.92%	60.85%	26.55%	26.88%	75.61%	42.75%
	NIPQ	73.68%	**90.22%**	55.47%	73.68%	**93.57%**	74.97%
P-3	Base	7.20%	35.60%	34.40%	14.00%	60.40%	54.40%
	AIPQ	14.80%	40.40%	38.00%	18.80%	68.00%	59.60%
	NIPQ	50.80%	**72.80%**	59.60%	50.80%	**82.80%**	72.00%
P-4	Base	3.11%	34.12%	14.12%	9.52%	54.70%	27.16%
	AIPQ	25.48%	62.21%	28.48%	27.41%	73.03%	44.59%
	NIPQ	74.81%	**90.59%**	57.03%	74.81%	**93.80%**	75.94%

Analysis with Different Base Models: On the DFW 2018 dataset, the ArcFace model gives the best baseline verification accuracies. On application of NIPQ, it is observed from Table 2 that the performance of VGGFace and VGGFace2 is comparable to that of ArcFace. At 0.1% FAR, the performance of the three models improve by 63.41%, 38.11% and 35.07% respectively. VGGFace2 easily outperforms the other two models on DFW 2019 dataset for both the baseline and NIPQ. At 0.1% FAR, the performance of the three models improve by 65.29%, 39.10% and 48.78% respectively. The improved performance after applying the proposed framework motivates its usage with different models.

Protocol-Specific Performance: All three base models show substantial improvements on all the protocols in both datasets. On DFW 2018 dataset, VGGFace2 is the best performing model for Protocols 1 and 2 with NIPQ showing improvements of 48.74% and 37.10% respectively over the baseline at 0.1% FAR. ArcFace is the best performing model on the overall protocol with a corresponding improvement of 35.07%. On the DFW 2019 dataset, Table 3 shows that VGGFace2 is the best performing model on all four protocols, with NIPQ improving over the baseline by 40.40%, 38.20%, 22.40% and 39.10% respectively at 0.1% FAR.

Effect of Quality Metric: As described in Sect. 3.2, we have shown results for AIPQ and NIPQ, where AIPQ is the average of the Inverse Disguise Qualities of an image pair while NIPQ is obtained by normalizing AIPQ to lie between 0 and 1. ArcFace, the best performing model on DFW 2018 dataset, shows an improvement of 10.92%, 14.14% and 14.89% when NIPQ is used instead of

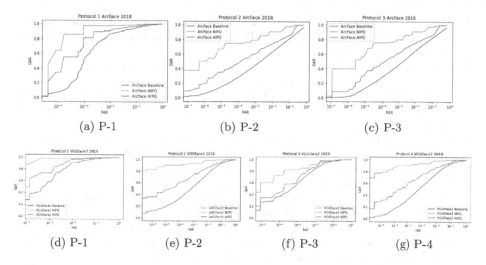

Fig. 4. ROC curves for the best models on all protocols of the DFW 2018 dataset ((a)–(c)) [1] and the DFW 2019 dataset ((d)–(g)) [2].

AIPQ on Protocols 1, 2 and 3 of DFW 2018 dataset at 0.1% FAR respectively. The corresponding results for VGGFace2 on the DFW 2019 dataset are 25.20%, 17.96%, 14.80% and 10.77% respectively. The results motivate the inclusion of NIPQ in the proposed framework.

6   Conclusion

This research proposes a novel framework for disguised face verification incorporating the proposed Inverse Disguise Quality metric. The framework is model-agnostic and can be applied in conjunction with existing deep learning based face verification models. Disguise detection is performed on the input face image using a combination of semantic segmentation and binary classification models. Based on the predictions, the Inverse Disguise Quality Metric is computed which provides an estimate of the image's quality. The proposed metric has been incorporated into the likelihood-ratio based verification process for obtaining enhanced performance. Experiments have been performed on the recently released Disguised Faces in the Wild (DFW) 2018 and DFW 2019 datasets. Analysis is drawn using three state-of-the-art models: (i) VGGFace, (ii) VGGFace2, and (iii) ArcFace, where, substantial improvement is observed in the verification performance upon applying the proposed framework.

References

1. Singh, M., Singh, R., Vatsa, M., Ratha, N.K., Chellappa, R.: Recognizing disguised faces in the wild. IEEE Trans. Biom. Behav. Identity Sci. 1(2), 97–108 (2019)

2. Singh, M., Chawla, M., Singh, R., Vatsa, M., Chellappa, R.: Disguised faces in the wild 2019. In: Proceedings of the IEEE International Conference on Computer Vision Workshops (2019)
3. Parkhi, O.M., Vedaldi, A., Zisserman, A., et al.: Deep face recognition. In: British Machine Vision Conference, vol. 1, p. 6 (2015)
4. He, K., Zhang, X., Ren, S., Sun, J.: Deep residual learning for image recognition. In: Proceedings of the IEEE Conference on Computer Vision and Pattern Recognition, pp. 770–778 (2016)
5. Deng, J., Guo, J., Xue, N., Zafeiriou, S.: ArcFace: additive angular margin loss for deep face recognition. In: Proceedings of the IEEE Conference on Computer Vision and Pattern Recognition, pp. 4690–4699 (2019)
6. Ciresan, D.C., Gambardella, L.M., Giusti, A., Schmidhuber, J.: Deep neural networks segment neuronal membranes in electron microscopy images. In: Neural Information Processing Systems, pp. 2852–2860 (2012)
7. Seyedhosseini, M., Sajjadi, M., Tasdizen, T.: Image segmentation with cascaded hierarchical models and logistic disjunctive normal networks. In: IEEE International Conference on Computer Vision (ICCV), pp. 2168–2175 (2013)
8. Long, J., Shelhamer, E., Darrell, T.: Fully convolutional networks for semantic segmentation. In: Proceedings of the IEEE Conference on Computer Vision and Pattern Recognition, pp. 3431–3440 (2015)
9. Ronneberger, O., Fischer, P., Brox, T.: U-Net: convolutional networks for biomedical image segmentation. In: Navab, N., Hornegger, J., Wells, W.M., Frangi, A.F. (eds.) MICCAI 2015. LNCS, vol. 9351, pp. 234–241. Springer, Cham (2015). https://doi.org/10.1007/978-3-319-24574-4_28
10. Khan, K., Mauro, M., Leonardi, R.: Multi-class semantic segmentation of faces. In: IEEE International Conference on Image Processing (ICIP), pp. 827–831 (2015)
11. Ramanathan, N., Chowdhury, A.R., Chellappa, R.: Facial similarity across age, disguise, illumination and pose. In: Proceedings of International Conference on Image Processing, vol. 3, pp. 1999–2002 (2004)
12. Silva, P.Q., Santa Rosa, A.N.C.: Face recognition based on eigeneyes. Pattern Recogn. Image Anal. **13**(2), 335–338 (2003)
13. Singh, R., Vatsa, M., Noore, A.: Face recognition with disguise and single gallery images. Image Vis. Comput. **27**(3), 245–257 (2009)
14. Dhamecha, T.I., Nigam, A., Singh, R., Vatsa, M.: Disguise detection and face recognition in visible and thermal spectrums. In: The International Conference on Biometrics, pp. 1–8 (2013)
15. Kim, J., Sung, Y., Yoon, S.M., Park, B.G.: A new video surveillance system employing occluded face detection. In: Ali, M., Esposito, F. (eds.) IEA/AIE 2005. LNCS (LNAI), vol. 3533, pp. 65–68. Springer, Heidelberg (2005). https://doi.org/10.1007/11504894_10
16. Smirnov, E., Melnikov, A., Oleinik, A., Ivanova, E., Kalinovskiy, I., Luckyanets, E.: Hard example mining with auxiliary embeddings. In: Proceedings of the IEEE Conference on Computer Vision and Pattern Recognition Workshops, pp. 37–46 (2018)
17. Zhang, K., Chang, Y.-L., Hsu, W.: Deep disguised faces recognition. In: Proceedings of the IEEE Conference on Computer Vision and Pattern Recognition Workshops, pp. 32–36 (2018)
18. Deng, J., Zafeririou, S.: ArcFace for disguised face recognition. In: IEEE International Conference on Computer Vision Workshop on Disguised Faces in the Wild (2019)

19. Martinez, A.M.: The AR face database. CVC Technical Report 24 (1998)
20. Li, B.Y., Mian, A.S., Liu, W., Krishna, A.: Using Kinect for face recognition under varying poses, expressions, illumination and disguise. In: IEEE Workshop on Applications of Computer Vision, pp. 186–192 (2013)
21. Wang, T.Y., Kumar, A.: Recognizing human faces under disguise and makeup. In: IEEE International Conference on Identity, Security and Behavior Analysis, pp. 1–7 (2016)
22. Raghavendra, R., Vetrekar, N., Raja, K.B., Gad, R., Busch, C.: Detecting disguise attacks on multi-spectral face recognition through spectral signatures. In: International Conference on Pattern Recognition, pp. 3371–3377 (2018)
23. Singh, A., Patil, D., Reddy, M., Omkar, S.: Disguised face identification (DFI) with facial keypoints using spatial fusion convolutional network. In: Proceedings of the IEEE International Conference on Computer Vision, pp. 1648–1655 (2017)
24. Bharadwaj, S., Vatsa, M., Singh, R.: Biometric quality: a review of fingerprint, iris, and face. EURASIP J. Image Video Process. **1**, 34 (2014)
25. Bigun, J., Fierrez-Aguilar, J., Ortega-Garcia, J., Gonzalez-Rodriguez, J.: Multimodal biometric authentication using quality signals in mobile communications. In: 12th International Conference on Image Analysis and Processing, pp. 2–11 (2003)
26. Fierrez-Aguilar, J., Ortega-Garcia, J., Gonzalez-Rodriguez, J., Bigun, J.: Discriminative multimodal biometric authentication based on quality measures. Pattern Recogn. **38**(5), 777–779 (2005)
27. Poh, N., Bengio, S.: Improving fusion with margin-derived confidence in biometric authentication tasks. In: Kanade, T., Jain, A., Ratha, N.K. (eds.) AVBPA 2005. LNCS, vol. 3546, pp. 474–483. Springer, Heidelberg (2005). https://doi.org/10.1007/11527923_49
28. Nandakumar, K., Chen, Y., Jain, A.K., Dass, S.C.: Quality-based score level fusion in multibiometric systems. In: International Conference on Pattern Recognition, pp. 473–476 (2006)
29. Nandakumar, K., Chen, Y., Dass, S.C., Jain, A.: Likelihood ratio-based biometric score fusion. IEEE Trans. Pattern Anal. Mach. Intell. **30**(2), 342–347 (2008)
30. Vatsa, M., Singh, R., Noore, A.: Integrating image quality in 2ν-SVM biometric match score fusion. Int. J. Neural Syst. **17**(05), 343–351 (2007)
31. https://github.com/ZFTurbo/ZF_UNET_224_Pretrained_Model
32. Singh, M., Singh, R., Ross, A.: A comprehensive overview of biometric fusion. Inf. Fusion **52**, 187–205 (2019)
33. Lehmann, E.L., Romano, J.P.: Testing Statistical Hypotheses. STS. Springer, New York (2005). https://doi.org/10.1007/0-387-27605-X_15
34. Guo, Y., Zhang, L., Hu, Y., He, X., Gao, J.: MS-Celeb-1M: a dataset and benchmark for large-scale face recognition. In: Leibe, B., Matas, J., Sebe, N., Welling, M. (eds.) ECCV 2016. LNCS, vol. 9907, pp. 87–102. Springer, Cham (2016). https://doi.org/10.1007/978-3-319-46487-9_6
35. Cao, Q., Shen, L., Xie, W., Parkhi, O.M., Zisserman, A.: VGGFace2: a dataset for recognising faces across pose and age. In: IEEE International Conference on Automatic Face and Gesture Recognition, pp. 67–74 (2018)
36. Girshick, R., Donahue, J., Darrell, T., Malik, J.: Rich feature hierarchies for accurate object detection and semantic segmentation. In: Computer Vision and Pattern Recognition, pp. 580–587 (2014)
37. He, K., Gkioxari, G., Dollar, P., Girshick, R.: Mask R-CNN. In: International Conference on Computer Vision (ICCV), pp. 2961–2969 (2017)

38. Subasic, M., Loncaric, S., Petkovic, T., Bogunovic, H., Krivec, V.: Face image validation system. In: International Symposium on Image and Signal Processing and Analysis (ISPA), pp. 30–33 (2005)
39. Gao, X., Li, S., Liu, R., Zhang, P.: Standardization of face image sample quality. In: Proceedings of Advances in Biometrics, pp. 242–251 (2007)
40. Zhang, G., Wang, Y.: Asymmetry-based quality assessment of face images. In: Bebis, G., et al. (eds.) ISVC 2009. LNCS, vol. 5876, pp. 499–508. Springer, Heidelberg (2009). https://doi.org/10.1007/978-3-642-10520-3_47
41. Jackson, A.S., Valstar, M., Tzimiropoulos, G.: A CNN cascade for landmark guided semantic part segmentation. In: Hua, G., Jégou, H. (eds.) ECCV 2016. LNCS, vol. 9915, pp. 143–155. Springer, Cham (2016). https://doi.org/10.1007/978-3-319-49409-8_14
42. Kalayeh, M.M., Gong, B., Shah, M.: Improving facial attribute prediction using semantic segmentation. In: Proceedings of the IEEE Conference on Computer Vision and Pattern Recognition, pp. 6942–6950 (2017)
43. Zhou, L., Liu, Z., He, X.: Face Parsing via a Fully-Convolutional Continuous CRF Neural Network. arXiv-1708 (2017)
44. Lin, J., Yang, H., Chen, D., Zeng, M., Wen, F., Yuan, L.: Face parsing with RoI Tanh-warping. In: Proceedings of the IEEE Conference on Computer Vision and Pattern Recognition, pp. 5654–5663 (2019)

W44 - Perception Through Structured Generative Models

W44 - Perception Through Structured Generative Models

Perception Through Structured Generative Models (PTSGM) is a workshop aimed at exploring how generative models can facilitate perception, and in particular, how to design and use structured generative models (of images, video, and 3D data) for computer vision inference applications.

On the workshop day, we had six invited talks:

- Max Welling (University of Amsterdam, The Netherlands): "Combining Generative and Discriminative Models"
- J. Kevin O'Regan (University of Paris, France): "Thinking about vision in a different way: the world as an outside memory"
- Peter Battaglia (DeepMind, UK): "Structured understanding and interaction with the world"
- Sanja Fidler (University of Toronto, Canada): "AI for 3D Content Generation"
- Ruslan Salakhutdinov (Carnegie Mellon University, USA): "Geometric Capsule Autoencoders for 3D Point Clouds"
- Carl Vondrick (Columbia University, USA): "Data and Task Generalization"

Each talk was 40 minutes long. The talks were recorded for posterity, and are now available to watch for free, through the workshop website.

We solicited original contributions relating to the themes of the workshop. We asked for 4-page-long submissions, to encourage concise papers and works in progress. We used OpenReview to organize the submission and reviewing process. We received four qualifying submissions. Each submission was reviewed by two of the workshop organizers in single-blind fashion, using the ECCV review format. Papers with an average positive score were accepted into the workshop. Three papers met this criteria. Each accepted paper was included in poster presentation sessions, on the workshop day. We also encouraged the authors to submit 5-minute presentations of their work.

We invite the reader to visit the workshop website, watch the recorded talks, and see the paper presentations: http://generativeperception.com/

August 2020

Adam W. Harley
Shubham Tulsiani
Katerina Fragkiadaki

Toward Continuous-Time Representations
of Human Motion

Weiyu Du[(✉)], Oleh Rybkin, Lingzhi Zhang, and Jianbo Shi

University of Pennsylvania, Philadelphia, USA
{weiyudu,oleh,zlz,jshi}@seas.upenn.edu

Abstract. For human motion understanding and generation, it is common to represent the motion sequence via a hidden state of a recurrent neural network, learned in an end-to-end fashion. While powerful, this representation is inflexible as these recurrent models are trained with a specific frame rate, and the hidden state is further hard to interpret. In this paper, we show that we can instead represent the *continuous* motion via latent parametric curves, leveraging techniques from computer graphics and signal processing. Our parametric representation is powerful enough to faithfully represent continuous motion with few parameters, easy to obtain, and is effective when used for downstream tasks. We validate the proposed method on AMASS and Human3.6M datasets through reconstruction and on a downstream task of point-to-point prediction, and show that our method is able to generate realistic motion. See our demo at www.github.com/WeiyuDu/motion-encode.

Keywords: Motion generation · 3D Motion · Sequence representation

1 Introduction

Human motion understanding and generation techniques commonly employ recurrent neural networks [2,6] that aggregate information across the temporal domain by processing each frame step-by-step [1,4,5,11,14]. While this allows using powerful black-box neural networks for the task at hand, such as action classification or future prediction, the representation is also inflexible in several key ways. It requires operating at a fixed frame rate, while real-world data often have different or even variable frame rates, which makes current systems cumbersome in practice. Similarly, when used for motion generation, the generated motion is restricted to the training frame rate, and different networks need to be trained for different temporal resolution.

Instead of taking this per-frame perspective, we argue that human motion should be represented holistically in a *continuous* manner, using latent parametric curves. We introduce two motion encoding schemes based on classical techniques from computer graphics and signal processing, namely Bezier and Sine Motion Encoding that represent the motion with Bezier and Sine curves respectively. Crucially, we apply this parametrization in a latent space instead of

© Springer Nature Switzerland AG 2020
A. Bartoli and A. Fusiello (Eds.): ECCV 2020 Workshops, LNCS 12540, pp. 543–548, 2020.
https://doi.org/10.1007/978-3-030-65414-6_37

the original joint space. This combination of a powerful per-step latent encoding with a simple and interpretable parametric temporal encoding makes our approach both powerful and flexible. Our approach can be used with input sequences of any framerate, or even variable framerate. When used for generation, it further enables us to generate frames at any desired rates and timestamps. Moreover, as these curves only require a few parameters, they provide a compact representation compared to the original full-size embedding.

We also study the task of controllable human motion generation with endpoints as an application of our proposed method. In animation production, in order to generate an animated motion, artists usually define key frames for the character's pose and design a trajectory using spline interpolation. Motion generation with end points can greatly expedite this process as it can automatically fill in the blanks between two poses with relatively long, realistic motion and reduces the number of key frames needed from the artists.

Experiments on AMASS [10] and Human3.6M [7] data show that our model with Bezier and Sine Motion Encoding beats latent linear interpolation baseline by a large margin both visually and with joint angle mean squared error. Our model generates realistic and smooth motion on the AMASS and Human3.6 datasets. Please see video results on the demo website.

2 Method

We want to encode and represent a motion sequence $x_{1:M}$ and time stamps $t_{1:M}$, where $x_m \in R^N$ is an individual pose at time t_m. First, we trained a Variational Auto-Encoder (VAE, [8,13]) that encodes individual frames $x_{1:M}$ into per-frame latent codes $z_{1:M}$. Given this per-frame encoding, we want to find a representation of the continuous sequence F such that $F(t)$ approximates the latent sequence $z_{1:M}$.

2.1 Bezier Motion Encoding

We first evaluate Bezier Motion Encoding, a technique inspired by classic computer graphics techniques [3]. The encoding is defined as the set of control points P for Bezier curves in the latent space of pose VAE. To ensure maximal expressiveness, we model a time channel in addition to the latent dimensions. To generate pose at different timestamps, we discretize the curve by taking 1000 samples in time channel and take the latent code with closest matching time. Formally, Bezier Motion Encoding is defined as $F_{bezier}(t) = B(s)$ s.t. $t = T(s), B(s) = \sum_{i=0}^{n} \binom{n}{i}(1-s)^{n-i}s^i P_i$ where t is time, $T(s)$ is the matching process in time domain, $0 \le s \le 1$, P is the set of control points and n is its size. We do not use s to represent time because s is not evenly distributed along the curve. Taking s as time flattens the curve, which limits the expressiveness of the encoding.

Bezier Motion Encoding has several advantages: 1) The curve begins at P_0 and ends at P_n, which is desirable in the controllable motion generation application. 2) Displacement of control point in a direction corresponds to a smooth drag of the curve. However, Bezier curve has global control points, which makes it hard to adjust the curve locally in detail. This can result in overly smooth latent trajectories, which have trouble modeling highly subsampled motion sequences.

2.2   Sine Motion Encoding

Motivated by the shortcomings of the Bezier encoding, we further evaluate Sine Motion Encoding that represents the curve via the most salient frequences in the frequency domain, as common in signal processing techniques [12]. The Sine encoding is defined as a linear combination of Sine curves $F_{sin}(t) = \sum_{i=0}^{n} A \sin(\omega t + \phi)$, where n is the number of Sine curves, A is amplitude, ω is angular frequency and ϕ is phase. Sine curves are periodic and smooth and a linear combination of them can model complex signals with few parameters. We can also increase the level of complexity and details in encoding by using more Sine curves. This is hard to achieve by Bezier encoding.

2.3   Optimization

Given motion sequence $x_{0:T}$, we obtain Bezier or Sine Motion Encoding from the following optimization: $\min_F \sum_m ||F(t_m) - z_m||^2$ where F is either F_{bezier} or F_{sin} defined in the above sections.

2.4   Controllable Human Motion Generation

To evaluate the proposed encodings on a downstream task, we study the task of controllable human motion generation with endpoints. Given a pair of poses (x_1, x_M), the task is to fill in the motion sequence $x_{2:M-1}$ in between.

We first embed the input pose pair with the pre-trained pose VAE to latent codes (z_0, z_M). Then we use a Multilayer Perceptron (MLP) that takes this as input and outputs a Gaussian distribution for control points P in the case of Bezier Motion Encoding and A, ω, ϕ in the case of Sine Motion Encoding. We use the ground truth latent trajectory and motion sequence as supervision. The loss is formulated as follows

$$\mathcal{L}_{gen} = ||F_{0:M} - z_{0:M}||^2 + ||\text{Dec}(F_{1:M}) - x_{1:M}||^2 - \text{D}_{\text{KL}}(\mathcal{N}(\hat{\mu}, \hat{\sigma})||\mathcal{N}(0,1)) \quad (1)$$

where Dec is the pose VAE decoder, $\hat{\mu}, \hat{\sigma}$ is the output of MLP. The first two terms are reconstruction loss and the last term helps as regularization.

3   Experiments

We experiment on Human3.6M [7] and AMASS [10] datasets in SMPL [9] format. We use linear interpolation in the latent space of pre-trained pose VAE as baseline. We split the AMASS data into 30-frame sequences with interval of 10 frames. On Human3.6M, we subsample the data 5 times to evaluate our method

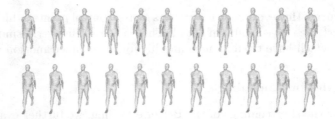

Fig. 1. Visualization on AMASS test set. Our Sine Motion Encoding (top) encodes walking motion while baseline (bottom) fails. On the website, we show further generations, including generating frames at a higher frequency than training data.

on long-term motion, then split it using the same scheme. We use 4 free control points for Bezier Motion Encoding and 3 curves for Sine Motion Encoding. To model more local and detailed movements, we experiment with adding 3 more curves to Sine Motion Encoding, where they only contribute to local portions of the sequence. We use MSE on joint angles as evaluation metric.

3.1   Motion Representation

We first evaluate the ability of our method to represent continuous motion. Our results are shown in Table 1, first row, as well as on the demo website. We see that our method is able to faithfully reproduce the encoded motion, demonstrating potential for sequence representation learning for many downstream tasks.

Table 1. Row 1: motion reconstruction error from Bezier and Sine encoding on AMASS test set. Row 2–3: Motion generation error in joint angle MSE per sequence. Our motion encodings generate motions with better visual quality and smaller error.

	LERP	Bezier	Sine (3 curves)	Sine (6 curves)
Reconstruction	–	1.89	2.87	1.32
AMASS	194.13	15.91	14.90	**14.43**
Human3.6M	307.66	30.25	29.28	**27.19**

3.2   Controllable Motion Generation

We show that our representation is suitable for point-to-point motion generation in Table 1. In Fig. 1 and on the demo website, we see that latent linear interpolation baseline (LERP) can only generate smooth transitions, while parametric latent curves capture the diversity in the data, and can even model cyclic motion where the initial and ending poses are similar. Also qualitatively, we observe that only the Sine (6 curves) representation models the Human3.6 data faithfully.

This method is more powerful, and better captures high-frequency details (such as in fast walking) that contribute little to quantitative metrics but are crucial for visual quality.

3.3   Variable Frequency Motion Generation

An advantage of our approach for motion generation is that it is possible to sample the generated curve at a frequency different than training data. To demonstrate this, we show the same predicted sequence that is sampled with 12 frames per second, 30-frame-long (training data) and 24 frames per second, 60-frame-long from our predicted representation. Our method produces good quality motion even on higher temporal resolution. This improves the visual quality of the motion by making it smoother. The video result of this experiment is shown on the demo website.

References

1. Aksan, E., Kaufmann, M., Hilliges, O.: Structured prediction helps 3D human motion modelling. In: Proceedings of the IEEE International Conference on Computer Vision, pp. 7144–7153 (2019)
2. Cho, K., et al.: Learning phrase representations using RNN encoder-decoder for statistical machine translation. In: Proceedings of the 2014 Conference on Empirical Methods in Natural Language Processing (EMNLP), Doha, Qatar, pp. 1724–1734. Association for Computational Linguistics, October 2014. https://doi.org/10.3115/v1/D14-1179. https://www.aclweb.org/anthology/D14-1179
3. Foley, J.D., et al.: Computer Graphics: Principles and Practice, vol. 12110. Addison-Wesley Professional, Boston (1996)
4. Fragkiadaki, K., Levine, S., Felsen, P., Malik, J.: Recurrent network models for human dynamics. In: Proceedings of the IEEE International Conference on Computer Vision, pp. 4346–4354 (2015)
5. Ghosh, P., Song, J., Aksan, E., Hilliges, O.: Learning human motion models for long-term predictions. In: 2017 International Conference on 3D Vision (3DV), pp. 458–466. IEEE (2017)
6. Hochreiter, S., Schmidhuber, J.: Long short-term memory. Neural Comput. **9**, 1735–1780 (1997). https://doi.org/10.1162/neco.1997.9.8.1735
7. Ionescu, C., Papava, D., Olaru, V., Sminchisescu, C.: Human3.6M: large scale datasets and predictive methods for 3D human sensing in natural environments. IEEE Trans. Pattern Anal. Mach. Intell. **36**(7), 1325–1339 (2014)
8. Kingma, D.P., Welling, M.: Auto-encoding variational Bayes. CoRR abs/1312.6114 (2014)
9. Loper, M., Mahmood, N., Romero, J., Pons-Moll, G., Black, M.J.: SMPL: a skinned multi-person linear model. ACM Trans. Graphics (Proc. SIGGRAPH Asia) **34**(6), 248:1–248:16 (2015)
10. Mahmood, N., Ghorbani, N., Troje, N.F., Pons-Moll, G., Black, M.J.: AMASS: archive of motion capture as surface shapes. In: International Conference on Computer Vision, pp. 5442–5451, October 2019
11. Martinez, J., Black, M.J., Romero, J.: On human motion prediction using recurrent neural networks. In: Proceedings of the IEEE Conference on Computer Vision and Pattern Recognition, pp. 2891–2900 (2017)

12. Proakis, J.G.: Digital Signal Processing: Principles Algorithms and Applications. Pearson Education India (2001)
13. Rezende, D.J., Mohamed, S., Wierstra, D.: Stochastic backpropagation and variational inference in deep latent Gaussian models. In: International Conference on Machine Learning, vol. 2 (2014)
14. Wang, T.H., Cheng, Y.C., Lin, C.H., Chen, H.T., Sun, M.: Point-to-point video generation. In: Proceedings of the IEEE International Conference on Computer Vision, pp. 10491–10500 (2019)

DisCont: Self-Supervised Visual Attribute Disentanglement Using Context Vectors

Sarthak Bhagat$^{(\boxtimes)}$, Vishaal Udandarao, Shagun Uppal, and Saket Anand

IIIT Delhi, Delhi, India
{sarthak16189,vishaal16119,shagun16088,anands}@iiitd.ac.in

Abstract. Disentangling underlying feature attributes within an image with no prior supervision is a challenging task. Models that can disentangle attributes well, provide greater interpretability and control. In this paper, we propose a self-supervised framework: *DisCont* to disentangle multiple attributes by exploiting structural inductive biases within images. Motivated by a recent surge in contrastive learning frameworks, our model bridges the gap between self-supervised contrastive learning algorithms and unsupervised disentanglement. We evaluate the efficacy of our approach, qualitatively and quantitatively, on four benchmark datasets. The code is available at https://github.com/sarthak268/DisCont.

Keywords: Self-supervision · Contrastive learning · Disentanglement

1 Introduction

In this work, we present an intuitive self-supervised framework *DisCont* to disentangle multiple feature attributes from images using meaningful data augmentations. We hypothesize that applying stochastic transformations to an image can be used to recover underlying feature attributes. Consider an example of data possessing two attributes, *i.e*, color and position. To this image, if we apply color transformation (*eg.* color jittering, gray-scale), only the underlying color attribute should change but position should be preserved. Similarly, on applying a translation and/or rotation to the image, the position attribute should vary keeping the color attribute intact.

It is known that there are several intrinsic variations present within different independent attributes [5], for *eg.* facial images containing discrete hair colors and continuously varying poses. To aptly capture these variations, we introduce 'Attribute Context Vectors' (refer Sect. 2). We posit that by constructing

---

S. Bhagat, V. Udandarao and S. Uppal—Equal Contribution. Ordered Randomly.

Electronic supplementary material The online version of this chapter (https://doi.org/10.1007/978-3-030-65414-6_38) contains supplementary material, which is available to authorized users.

© Springer Nature Switzerland AG 2020
A. Bartoli and A. Fusiello (Eds.): ECCV 2020 Workshops, LNCS 12540, pp. 549–553, 2020.
https://doi.org/10.1007/978-3-030-65414-6_38

attribute-specific context vectors that capture the entire variability within that attribute, we can learn richer representations.

Our major contributions in this work can be summarised as follows:

- We propose a self-supervised method *DisCont* to disentangle multiple visual attributes by using inductive biases in images via data augmentations. We leverage composite stochastic transformations for robust disentanglement.
- We present the idea of 'Attribute Context Vectors' to capture and utilize intra-attribute variations in an extensive manner.
- We impose attribute clustering constraint that is commonly used in distance metric learning, and show that it promotes attribute disentanglement.

2   Methodology

Assume we have a dataset $X = \{x_1, x_2, ..., x_N\}$ containing N images, where $x_i \in \mathbb{R}^{C \times H \times W}$, consisting of k factors of variations $y = \{y_1, y_2...y_K\}$.

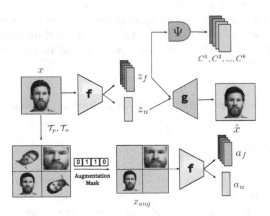

Fig. 1. Overview of our architecture *DisCont*. Given a batch of images x, we generate an augmented batch x_{aug} by sampling a set of stochastic transformations. We then encode x and x_{aug} to extract z_f, z_u and a_f, a_u respectively. z_f is then used to construct 'Attribute Context Vectors' $C^1, C^2, ..., C^k$ corresponding to each feature attribute. The C^i's, $z_f \cup z_u$'s and $a_f \cup a_u$'s are then used to optimize our disentanglement objective.

Model Description. To learn disentangled attributes, we propose to view the underlying latent space of an auto-encoder network as two disjoint subspaces.

- $Z_f \subset \mathbb{R}^{d \times k}$: denotes the feature attribute space containing the disentangled and interpretable attributes, where d and k denote the dimensionality of the space and the number of feature attributes respectively.
- $Z_u \subset \mathbb{R}^d$: denotes the unspecified attribute space containing background and miscellaneous attributes, where d is the dimensionality of the space. We enforce a $\mathcal{N}(0, I)$ prior over this space.

Assume that we have an encoding function f parameterized by θ, then each image x can be encoded in the following way: $z_f, z_u = f_\theta(x), z_f \in Z_f, z_u \in Z_u$, where we can index z_f to recover the independent feature attributes *i.e.* $z_f = [z_{f,1}, z_{f,2}, ..., z_{f,k}]$. To project the latent encodings back to image space, we make use of a decoding function g parameterized by ϕ. Therefore, we can get image reconstructions $\hat{x}$ using the separate latent encodings, $\hat{x} = g_\phi(z_f, z_u)$.

Composite Data Augmentations. Following our hypothesis of recovering latent attributes using stochastic transformations, we formulate a mask-based augmentation approach leveraging positive and negative transformations. Assume that we have two sets of stochastic transformations $\mathcal{T}_p = \{p_1, p_2, ..., p_k\}$ and $\mathcal{T}_n = \{n_1, n_2, ..., n_k\}$ that can augment an image into a correlated form *i.e.* $x_{aug} = t(x), t : \mathbb{R}^{C \times H \times W} \rightarrow \mathbb{R}^{C \times H \times W}, t \in \mathcal{T}_p \cup \mathcal{T}_n$. $\mathcal{T}_p$ denotes the set of positive transformations that do not change any underlying attributes of the image, whereas $\mathcal{T}_n$ denotes the set of negative transformations changing a single underlying attribute, *i.e.*, when n_i is applied to an image x, it should lead to a change only in the $z_{f,i}$ attribute and all other attributes should be preserved.

For every batch of images $x = \{x_1, x_2 ... x_B\}$, we sample a subset of transformations to apply compositionally to x and retrieve an augmented batch x_{aug} and a mask vector $m \in \{0, 1\}^k$. This is further detailed in the appendix.

Attribute Context Vectors. Inspired by [3], we propose attribute context vectors $C^i \; \forall i \in \{1, 2, ..., k\}$. A context vector C^i is formed from each individual feature attribute z_f^i through a non-linear projection. This is to encapsulate variability of the ith attribute in C^i. Hence, each individual context vector should capture independent disentangled feature space of the factors of variation. Assume a non-linear mapping function $\Psi : \mathbb{R}^{d \times B} \rightarrow \mathbb{R}^c$, where c denotes the dimensionality of each context vector C^i and B denotes the size of a sampled mini-batch. We construct context vectors by aggregating all the ith feature attributes locally within the sampled minibatch, *i.e.*, $C^i = \Psi\left([z_{f,1}^i, z_{f,2}^i, ..., z_{f,B}^i]\right) \forall i \in \{1, 2, ..., k\}$.

Loss Functions. We describe the losses that we use to enforce disentanglement:

- **ELBO Objective.** We use the standard reconstruction loss ($\mathcal{L}_R$) with a KL objective ($\mathcal{L}_{KL}$) on the z_u attribute.
- **Center Loss ($\mathcal{L}_{cen}$).** We enforce clustering using a center loss [4] in the Z_f space by differentiating inter-attribute features. This promotes accumulation of feature attributes into distantly-placed clusters by providing self-supervision through pseudo-labels obtained from the context vectors.
- **Augmentation Consistency Loss ($\mathcal{L}_A$).** To ensure that a specific negative augmentation n_i enforces change in only i^{th} feature attribute z_f^i, we encourage all other specified factors apart from $a_f^{i\,1}$ and z_f^i to be close in latent space. This ensures that feature attribute is invariant to specific augmentations. Since background attributes of images should be preserved irrespective of augmentations, we also encourage proximity of z_u and a_u [1].

---

[1] We obtain a_f, a_u by encoding x_{aug}, refer Fig. 1 and appendix.

The overall loss of our model is a weighted sum of these constituent losses[2].

$$\mathcal{L}_{total} = \mathcal{L}_R + \lambda_{KL}\mathcal{L}_{KL} + \lambda_{cen}\mathcal{L}_{cen} + \mathcal{L}_A \qquad (1)$$

3   Experiments

3.1   Quantitative Results

To ensure a robust evaluation of our disentanglement models using unsupervised metrics, we compare informativeness scores [1] of our model's latent feature attributes in Fig. 2 with the state-of-the-art unsupervised disentanglement model presented in [2] (referred to as MIX). Further details are in the appendix.

3.2   Qualitative Results

We present attribute transfer visualizations to construe disentanglement qualitatively. The images in first two rows in each grid are randomly sampled from the test set. The bottom row images are formed by swapping one specified attribute from top row image with corresponding attribute chunk in second row image, keeping all other attributes fixed. Additional results are shown in the appendix.

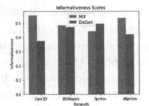 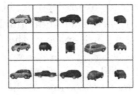 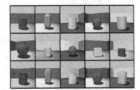

Fig. 2. (Left) Informativeness scores (lower is better) for DisCont and MIX, **(Center)**: Attribute transfer results for Cars3D; Specified Attribute: Color, **(Right)**: Attribute transfer results for 3DShapes; Specified Attribute: Orientation

4   Conclusion

In this paper, we propose a self-supervised attribute disentanglement framework *DisCont* which leverages specific data augmentations to exploit spatial inductive biases in images. Our results show that such a framework can be readily used to recover semantically meaningful attributes independently.

---

[2] For more technical details of loss functions, refer to the appendix.

References

1. Do, K., Tran, T.: Theory and evaluation metrics for learning disentangled representations. ArXiv abs/1908.09961 (2020)
2. Hu, Q., Szabó, A., Portenier, T., Zwicker, M., Favaro, P.: Disentangling factors of variation by mixing them. In: 2018 IEEE/CVF Conference on Computer Vision and Pattern Recognition, pp. 3399–3407 (2018)
3. Oord, A., Li, Y., Vinyals, O.: Representation learning with contrastive predictive coding. ArXiv abs/1807.03748 (2018)
4. Wen, Y., Zhang, K., Li, Z., Qiao, Y.: A discriminative feature learning approach for deep face recognition. In: Leibe, B., Matas, J., Sebe, N., Welling, M. (eds.) ECCV 2016. LNCS, vol. 9911, pp. 499–515. Springer, Cham (2016). https://doi.org/10.1007/978-3-319-46478-7_31
5. Zhang, J., Huang, Y., Li, Y., Zhao, W., Zhang, L.: Multi-attribute transfer via disentangled representation. In: Proceedings of the AAAI Conference on Artificial Intelligence, vol. 33, pp. 9195–9202 (2019)

3D Noise and Adversarial Supervision Is All You Need for Multi-modal Semantic Image Synthesis

Vadim Sushko[1], Edgar Schönfeld[1(✉)], Dan Zhang[1], Jürgen Gall[2], Bernt Schiele[3], and Anna Khoreva[1]

[1] Bosch Center for Artificial Intelligence, Tübingen, Germany
edgarschonfeld@live.de
[2] University of Bonn, Bonn, Germany
[3] Max Planck Institute for Informatics, Saarbrücken, Germany

Abstract. Semantic image synthesis models suffer from training instabilities and poor image quality when trained with adversarial supervision alone. Historically, this was alleviated via an additional VGG-based perceptual loss. Hence, we propose a new simplified GAN model, which needs only adversarial supervision to achieve high-quality results. In doing so, we also show that the VGG supervision decreases image diversity and can hurt image quality. We achieve the improvement by re-designing the discriminator as a semantic segmentation network. The resulting stronger supervision makes the VGG loss obsolete. Moreover, in contrast to previous work, we enable high-quality multi-modal image synthesis through a novel noise sampling scheme. Compared to the state of the art, we achieve an average improvement of 6 FID and 7 mIoU.

1 Introduction

Semantic image synthesis is the task of generating realistic images from semantic label maps [5,8]. State-of-the-art models [4,5,8] still suffer from training instabilities and poor image quality when trained only with adversarial supervision. This is commonly overcome by using an additional perceptual loss [8], that matches intermediate generator features of the synthetic and real images, estimated via an external VGG perception network [7], pre-trained on ImageNet. However, the perceptual loss causes an additional computational overhead and a potential bias towards ImageNet, which decreases image diversity and impacts quality, as we show in our experiments. Hence, in this work we propose a novel, simplified GAN model that achieves state-of-the-art results without requiring a perceptual loss. We argue that the previously used encoder-shaped discriminators are not capable of providing the rich supervision necessary for the task. Therefore, we re-design the discriminator as an encoder-decoder semantic segmentation network [6], which exploits the given semantic label maps as ground truth via an

---

V. Sushko and E. Schönfeld—Equal Contribution.

© Springer Nature Switzerland AG 2020
A. Bartoli and A. Fusiello (Eds.): ECCV 2020 Workshops, LNCS 12540, pp. 554–558, 2020.
https://doi.org/10.1007/978-3-030-65414-6_39

N+1-class cross-entropy loss (N real semantic classes and 1 fake class). Since the discriminator has to look at each patch in detail to assign it to the right semantic class, it learns semantic-aware fine-grained representations. In contrast, the previous encoder-shaped discriminators require concatenating the label map to the input image, allowing them to ignore parts of the label map, as they are not penalized for predicting wrong semantic classes. Second, we introduce a LabelMix regularization, which helps the discriminator to focus on the semantic and structural differences of real and fake images. Our changes lead to a much stronger discriminator which makes the perceptual loss supervision superfluous. Third, we enable multi-modal image synthesis via a new noise sampling scheme. Previously, direct input noise to the generator was ignored or led to poor image quality [5,8]. In contrast, our model synthesizes diverse multi-modal outputs by simply re-sampling an input 3D noise tensor. As our noise tensor is spatially sensitive, we can re-sample it both globally (by spatially replicating a 1D noise vector) and locally (pixel-wise, allowing to change the appearance of the whole scene or specific image regions (see Fig. 2). We call our model OASIS, as it needs only adversarial supervision for semantic image synthesis.

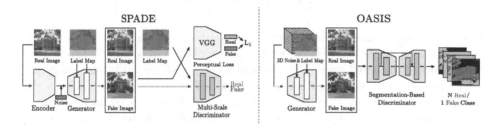

Fig. 1. SPADE vs. OASIS. OASIS outperforms SPADE, while being simpler and lighter: it uses only adversarial loss supervision and a single segmentation-based discriminator.

2   OASIS Model

Our model builds on SPADE [5], with modifications to the discriminator. The discriminator architecture follows U-Net [6] for semantic segmentation. Real images use the given semantic label maps with N classes as the ground truth, while all pixels of the fake images are categorized as one extra class. Overall, we have $N + 1$ classes in the semantic segmentation problem, and thus propose to use a weighted $(N+1)$-class cross-entropy loss for training:

$$\mathcal{L}_D = -\mathbb{E}_{(x,t)}\left[\sum_{c=1}^{N} \alpha_c \sum_{i,j}^{H \times W} t_{i,j,c} \log D(x)_{i,j,c}\right] - \mathbb{E}_{(z,t)}\left[\sum_{i,j}^{H \times W} \log D(G(z,t))_{i,j,c=N+1}\right],$$

(1)

where x denotes the real image; (z, t) is the noise-label map pair; and $D(x)$ is per-pixel $(N+1)$-class prediction. The ground truth label map t has three dimensions: the spatial position $(i, j) \in H \times W$, and one-hot vector encoding the class $c \in \{1, .., N+1\}$. The class balancing weight α_c is the inverse of its per-pixel class frequency.

To encourage our discriminator to focus on differences in structure between fake and real classes, we propose a LabelMix regularization. Based on the semantic layout, we generate a binary mask M to mix a pair $(x, \hat{x})$ of real and fake images conditioned on the same label map: $\text{LabelMix}(x, \hat{x}, M) = M \odot x + (1 - M) \odot \hat{x}$. We train the discriminator to be equivariant under the LabelMix operation, via $\mathcal{L}_{cons} = \left\| D_{\text{logits}}\left(\text{LabelMix}(x, \hat{x}, M)\right) - \text{LabelMix}\left(D_{\text{logits}}(x), D_{\text{logits}}(\hat{x}), M\right) \right\|^2$. The mask M is generated according to the label map and therefore encourages the generator to respect semantic class boundaries. The OASIS generator loss is

$$\mathcal{L}_G = -\mathbb{E}_{(z,t)} \left[\sum_{c=1}^{N} \alpha_c \sum_{i,j}^{H \times W} t_{i,j,c} \log D(G(z,t))_{i,j,c} \right]. \tag{2}$$

Further, we enable multi-modal synthesis through noise sampling. For this, we construct a 64-dimensional noise vector and replicate it to match the spatial dimensions of the label map. The channel-wise concatenation of the noise and label map forms a 3D tensor used as input and also for conditioning at the SPADE-norm layers. As the 3D noise is channel-wise and pixel-wise sensitive, at test time, one can sample the noise tensor globally, per-channel, and locally, per-pixel, allowing controlled synthesis of the whole scene but also of specific semantic classes (Fig. 2). Lastly, we remove the first generator block, decreasing the parameter count by 24M without a noticeable performance loss.

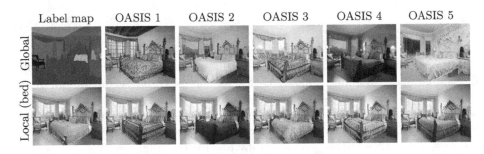

Fig. 2. OASIS multi-modal synthesis results. The 3D noise can be sampled globally, or locally. For the latter, we only re-sample noise in the bed segment area (in red). (Color figure online)

Table 1. Comparison with other methods.

Method	# param	VGG	ADE20K		ADE-outd.		Cityscapes		COCO-stuff	
			FID↓	mIoU↑	FID↓	mIoU↑	FID↓	mIoU↑	FID↓	mIoU↑
Pix2pixHD [8]	183M	✓	81.8	20.3	97.8	17.4	95.0	58.3	111.5	14.6
CC-FPSE [4]	131M	✓	31.7	43.7	n/a	n/a	54.3	65.5	19.2	41.6
SPADE [5]	102M	✓	33.9	38.5	63.3	30.8	71.8	62.3	22.6	37.4
SPADE+	102M	✓	32.9	42.5	51.1	32.1	47.8	64.0	21.7	38.8
		✗	60.7	21.0	65.4	22.7	61.4	47.6	99.1	16.1
OASIS	94M	✗	**28.4**	**50.6**	**48.4**	**41.5**	**47.7**	**69.3**	**16.7**	**45.5**

3 Experiments

We conduct experiments on ADE20K, ADE-Outdoors [12], COCO-stuff [1] and Cityscapes [2]. We evaluate quantitatively using the Fréchet Inception Distance (FID) [3], mean Intersection-over-Union (mIoU) and MS-SSIM [9] with LPIPS [11] for multi-modal synthesis. We follow the experimental setting of [5]. We did not apply the GAN feature matching loss and the VGG perceptual loss. All our models use an exponential moving average (EMA) with 0.9999 decay [10]. We use SPADE [5] as our baseline, trained without the feature matching loss and using EMA [10] (further referred to as SPADE+).

OASIS outperforms the current state of the art on all datasets with an average improvement of 6 FID and 7 mIoU points (Table 1). Importantly, OASIS achieves the improvement via adversarial supervision alone. On the contrary, SPADE+ does not produce images of high visual quality without the perceptual loss.

In contrast to previous work, OASIS produces diverse images by re-sampling input 3D noise. The image can be re-sampled both globally and locally (Fig. 2). We observed that OASIS improves over SPADE+ with an image encoder, both in diversity (MS-SSIM 0.65 and LPIPS 0.35 vs 0.85 and 0.16) and quality (FID 28.3 and mIoU 48.8 vs 33.4 and 40.2). Using 3D noise increases diversity both for SPADE+ and OASIS. We noticed a quality-diversity tradeoff for SPADE+: 3D noise improves diversity at the cost of quality, and the VGG loss acts vice versa. Such tradeoff is not observed in OASIS.

4 Conclusion

We propose OASIS, a semantic image synthesis model relying only on adversarial supervision to achieve high fidelity image synthesis. This is achieved via detailed spatial and semantic-aware supervision from our novel segmentation-based discriminator. OASIS can easily generate diverse multi-modal outputs by re-sampling the 3D noise, both globally and locally, allowing to change the appearance of the whole scene and of individual objects. OASIS significantly improves over the state of the art in terms of image quality and diversity, while being simpler and more lightweight than previous methods.

References

1. Caesar, H., Uijlings, J., Ferrari, V.: COCO-stuff: thing and stuff classes in context. In: Conference on Computer Vision and Pattern Recognition (CVPR) (2018)
2. Cordts, M., et al.: The cityscapes dataset for semantic urban scene understanding. In: Conference on Computer Vision and Pattern Recognition (CVPR) (2016)
3. Heusel, M., Ramsauer, H., Unterthiner, T., Nessler, B., Hochreiter, S.: GANs trained by a two time-scale update rule converge to a local Nash equilibrium. In: Advances in Neural Information Processing Systems (NeurIPS) (2017)
4. Liu, X., Yin, G., Shao, J., Wang, X., et al.: Learning to predict layout-to-image conditional convolutions for semantic image synthesis. In: Advances in Neural Information Processing Systems (NeurIPS) (2019)
5. Park, T., Liu, M.Y., Wang, T.C., Zhu, J.Y.: Semantic image synthesis with spatially-adaptive normalization. In: Conference on Computer Vision and Pattern Recognition (CVPR) (2019)
6. Ronneberger, O., Fischer, P., Brox, T.: U-net: convolutional networks for biomedical image segmentation. In: MICCAI (2015)
7. Simonyan, K., Zisserman, A.: Very deep convolutional networks for large-scale image recognition. In: International Conference on Learning Representations (ICLR) (2015)
8. Wang, T.C., Liu, M.Y., Zhu, J.Y., Tao, A., Kautz, J., Catanzaro, B.: High-resolution image synthesis and semantic manipulation with conditional GANs. In: Conference on Computer Vision and Pattern Recognition (CVPR) (2018)
9. Wang, Z., Simoncelli, E.P., Bovik, A.C.: Multiscale structural similarity for image quality assessment. In: Asilomar Conference on Signals, Systems and Computers (2003)
10. Yazici, Y., Foo, C.S., Winkler, S., Yap, K.H., Piliouras, G., Chandrasekhar, V.: The unusual effectiveness of averaging in GAN training. arXiv preprint arXiv:1806.04498 (2018)
11. Zhang, R., Isola, P., Efros, A.A., Shechtman, E., Wang, O.: The unreasonable effectiveness of deep features as a perceptual metric. In: Conference on Computer Vision and Pattern Recognition (CVPR) (2018)
12. Zhou, B., Zhao, H., Puig, X., Fidler, S., Barriuso, A., Torralba, A.: Scene parsing through ADE20K dataset. In: Conference on Computer Vision and Pattern Recognition (CVPR) (2017)

Author Index

Printed in the United States
By Bookmasters